LABORATORY . . . .

SAFETY OFFICER . . . .

DEPUTY SAFETY OFFICER OR PERSON RESPONSIBLE
FOR MAINTAINING FIRST-AID BOX . . . .

NEAREST HOSPITAL FOR CASUALTIES . . . .

. . . . Telephone . . . .

NEAREST AMBULANCE SERVICE . . . .

. . . . Telephone . . . .

NEAREST FIRE SERVICE . . . .

. . . . Telephone . . . .

OR DIAL 999 FOR AMBULANCE/FIRE SERVICE;
THEN GIVE EXACT LOCATION OF CASUALTY/FIRE

NEAREST DOCTORS . . . .

. . . . Telephone . . . .

. . . .

. . . . Telephone . . . .

PERSONS ON PREMISES TRAINED IN FIRST AID . . . .

. . . .

. . . .

PERSON TO WHOM ACCIDENTS MUST BE REPORTED

. . . .

OTHER INFORMATION . . . .

. . . .

. . . .

For addresses of poison information centres see p 139.

HAZARDS
IN THE
CHEMICAL LABORATORY

HAZARDS
IN THE
CHEMICAL LABORATORY

edited by

L. BRETHERICK, BSc, CChem, FRSC

*Consultant; formerly Senior Project Leader,
BP Research Centre, Sunbury*

LONDON
THE ROYAL SOCIETY OF CHEMISTRY

The information contained in this book has been compiled by recognised authorities from sources believed to be reliable and to represent the best opinion on the subject as of 1985. However, no warranty, guarantee, or representation is made by The Royal Society of Chemistry as to the correctness or sufficiency of any information herein, and the Publisher assumes no responsibility in connection therewith; nor can it be assumed that all necessary warnings and precautionary measures are contained in this publication, or that other or additional information or measures may not be required or desirable because of particular or exceptional conditions or circumstances, or because of new or changed legislation.

British Library Cataloguing in Publication Data

Hazards in the chemical laboratory—4th ed.
    1. Chemical laboratories—Safety measures
    I. Bretherick, L. II. Royal Society of
    Chemistry
    363.1'79   QD 51
    ISBN 0-85186-489-9

First published September 1971; second edition, March 1977; third edition, April 1981; fourth edition, September 1986

Published by The Royal Society of Chemistry, Burlington House, London W1V 0BN

Printed in Great Britain at the Alden Press, Oxford.

CONTENTS

J. M. L. GILKS, MA, BM, FFOM, DIH, *Chief Medical Adviser, Kuwait Petroleum Company; formerly Chief Adviser, Occupational Health and Hygiene Division, Shell International Petroleum Company; Regional Employment Medical Adviser, NE Region, UK Health and Safety Executive*

L. BRETHERICK, BSc, CChem, FRSC, *Safety Consultant*
G. D. MUIR, BSc, PhD, CChem, FRSC, *Consultant; formerly Technical Development Manager, BDH Chemicals Ltd, Poole, Dorset*

S. B. OSBORN, BSc, PhD, FInstP, *Radiological Protection Consultant, SE Thames Regional Health Authority; formerly Director, Department of Medical Physics, King's College Hospital and Medical School, London and Consultant to the World Health Organisation.*

Contents

Preface to the Fourth Edition

The continuing developments in many technical and legislative matters which have taken place in the past 5 years since publication of the third edition, and which impinge upon safety aspects of chemical laboratory practice, call for further revision of many of the significant points of detail herein. The good reception accorded to the previous edition, coupled with the nature of these developments, led the existing team of authors and myself, after consideration with appropriate RSC Management, to the view that its format and general content are still sound and capable of adequate updating together with some extension and this has now been done.

Mr Luxon has revised both his 'Introduction' in Chapter 1 and his Chapter 2 on the 'Health and Safety at Work Act 1974' to reflect the steady developments in the effects of this key legislation upon laboratory life. Chapter 3 now draws attention to the likely effects of the forthcoming regulations on the Control of Substances Hazardous to Health, and to the need to make effective provision for laboratory waste disposal. Several of the detailed references and regulations related to 'Fire prevention' have been revised, and these have all been included by Messrs Warwicker and Sheldon in their revision of Chapter 4. Chapter 5 has been extended to include a brief account of instrumental and computational methods of assessing and predicting chemical reaction hazards, and the examples given are now keyed to the page numbers of the 3rd edition of *Handbook of Reactive Chemical Hazards* published at the end of 1985.

Dr Magos has recast and partially rewritten his Chapter 6 to emphasize the key role of exposure factors and their consequences in the whole area of 'Chemical hazards and toxicology'. Chapter 7 has been considerably revised by Dr Gilks to take account of some changes of detail and emphasis in several areas of his speciality topic of 'Health care and first aid'. The many changes, largely of minor detail necessary to update fully the major Chapter 8 on 'Hazardous chemicals', ('Yellow pages'), have arisen from three separate sources. Translation of the EEC Directive on Classification and Labelling into UK legislation as the *The Classification, Packaging and Labelling of Dangerous Substances Regulations 1984* now permits the officially accepted risk classification phrases to be included in most of the nearly 500 monographs forming the heart of the Chapter. The significant change in the occupational exposure limit values in Guidance Note EH 40/85 away from the TLV to the dual concepts of *Control limits* and *Recommended limits* has led to many changes in the monographs, though not necessarily of the numerical values quoted. A selection of new reactive hazards drawn from the third edition of the *Handbook* mentioned above has been added, and all the page numbers referring to that new text have been changed. CAS Registry numbers have now been added to the title lines for each chemical entry, and where possible each monograph ends with a reference to the appropriate Laboratory Hazard Data Sheet originating from the RSC Information Services at Nottingham.

Dr Osborn has completely recast his Chapter 9, 'Precautions against radiations' and associated bibliography to match the requirements of the long awaited and complicated *Ionising Radiations Regulations 1985* and its *Approved Code of Practice 1985*. Finally we welcome to the team a new author and old friend from the USA, Howard H. Fawcett, who outlines authoritatively for us in Chapter 10 (An American view) some of the principal distinctions between US and UK legislation and practice in chemical laboratories.

It is again a pleasure to record the great encouragement and assistance extended by the Editorial and Information Services staff of the Royal Society of Chemistry, in particular by Dr. Ivor A. Williams, Dr Philip G. Gardam and Mr Peter W. Shallis who were jointly responsible for converting promptly our combined offerings into this new volume.

L. BRETHERICK
January 1986

Preface to the First Edition

The present volume is a successor to the Royal Institute of Chemistry's *Laboratory Handbook of toxic agents*, first published in 1960, and issued in a revised edition in 1966. Before the second edition went out of print, the future of the publication was considered by both the Institute's Publications Committee, and by the previous editor, Professor C. H. Gray, and myself. It was generally felt that, rather than merely revise the existing material, it would be preferable to alter the underlying philosophy of the book by changing its scope from toxic hazards to a consideration of all hazards likely to be encountered in the chemical laboratory.

The general format remains the same, with the major part of the book (printed on tinted paper) being an alphabetical guide to hazardous chemicals and measures to be taken in the event of accidents in their use. However, whereas previous editions have included details of measures to be taken against the toxic hazards of such chemicals, the present edition also includes methods for spillage disposal and extinguishing of fires where appropriate. Once more, an alphabetical listing is adopted to enable the use of the book as a speedy reference in the case of emergency.

Much assistance was required in preparing this chapter, now extended to over 430 hazardous chemicals, and the preceding one on first aid. I must record my special thanks to my colleague Mr W. G. Moss for his collaboration when we prepared these chapters for the first edition of the earlier book, and BDH Chemicals Ltd for permission to use their extensive records on the hazards, handling and disposal of chemicals; also my colleague Dr P. Mostyn Williams who has added considerably to the earlier medical advice of Dr W. B. Rhodes.

Dr D. P. Duffield and Dr K. P. Whitehead of Imperial Chemical Industries Ltd have also provided important medical advice on up-to-date first aid practice, particularly on the treatment of cyanide and phenol poisoning. The chemical world must always be grateful for the pioneer efforts of ICI in encouraging chemical factory and laboratory safety and we would record again our thanks to Dr A. J. Amor and Dr A. Lloyd Potter for their interest in the first book and to pay special tribute to the work of Dr L. J. Burrage who has contributed so much to promote laboratory safety in this country.

Many other firms have given us the benefit of special knowledge of certain of their products and we are also grateful to James North & Sons Ltd for permission to reproduce their chart advising on the types of glove to be used when handling different classes of chemicals.

Despite extensive practical experience of chemical hazards, the writers of a book such as this lean heavily on the authors of major works on industrial toxicology. Not many may have had the privilege of knowing the charm and intelligence of that great lady, Dr Ethel Browning, who guided the Institute and some of the authors when the first book was conceived and drafted, and wrote two renowned works upon which we draft extensively. *Toxicity and metabolism of industrial solvents* and *Toxicity of industrial metals*, as well as editing the important series of monographs on toxic agents in which they appeared. Her death last year ended a long life of devoted service to industrial safety.

The valuable publications of the Chemical Industries Association—*Marking containers of hazardous chemicals* and *Exposure to gases and vapours*—have been referred to frequently and we would like to thank the Association for the privilege of perusing the text of the latter at the proofing stage. The *Laboratory waste disposal manual* published by the Association's counterpart in the US, the Manufacturing Chemists' Association, was also consulted extensively in preparing chapter 6, as were the following works.

Dangerous properties of industrial materials by N. Irving Sax
Extra pharmacopoeia (Martindale) edited by R. G. Todd
Industrial hygiene and toxicology edited by Frank A. Patty
Industrial toxicology by L. S. Fairhall
Poisoning by drugs and chemicals by P. Cooper
Poisons by Brookes and Jacobs
Toxicology of drugs and chemicals by W. B. Deichmann and H. W. Gerarde.
Other acknowledgements appear in chapter 5.

Of the remainder of the book, new chapters have been provided by Mr

Ackroyd, Dr Taylor and Mr Sheldon on fire protection and by Mr Neill and Dr Russell Doggart on the particular hazards facing chemical workers in hospital biochemistry laboratories. In addition, Mr Luxon, of the Department of Employment, has contributed an entirely new introduction to replace the one by the late Sir Roy Cameron which appeared in the earlier editions. Mr Beard and Dr Osborn have thoroughly revised chapters 2 and 7 respectively. To all these authors, I am extremely grateful for the time they have spent on and interest shown in this project.

Dr Farago and his staff in the Editorial Office of the Institute have my sincerest thanks for their painstaking help and I am particularly indebted to Dr Martin Sherwood for his close collaboration in co-ordinating our efforts, sharing in our proof reading, and carrying out all the necessary negotiations with the printers. With my co-authors he has made the task of editing not only light, but both stimulating and enjoyable.

Finally, I should like to pay tribute to the immense amount of work which Professor Gray put into the planning and production of both editions of the *Laboratory handbook of toxic agents*. Although pressure of work prevented Professor Gray from taking a full part in the editing of this volume, I have had the benefit of his wise advice throughout its preparation. Without this and the substantial contribution he made to the earlier book, it is unlikely that this volume would have been possible.

G. D. MUIR
April 1971

Chapter 1

Introduction

Since the publication of the third edition a number of major chemical disasters have occurred which have involved not only those workers immediately concerned with the processes but also the general population in the immediate vicinity of the plant. This has brought renewed public pressure to bear on the relevant regulatory authorities for a more formal assessment of the hazards associated with the storage and use of chemicals of all kinds. Fortunately, the general requirements of the Health & Safety at Work Act of 1974 provide the necessary framework by which some additional more comprehensive controls can be effected. The recent spate of proposals for new regulations to be made under the enabling powers contained in this Act may be considered as but one response to that increased public awareness of the dangers that chemicals pose to the environment.

The Act has now been on the Statute Book for some ten years and as indicated in Chapter 2, its general purpose is to provide for one comprehensive and integrated system of law dealing with the health and safety of all persons engaged in work activities and the health and safety of others who may be affected by such activities. Reference has been made to one of the principal features of recent legislation, *i.e.*, the provisions relating to the making of detailed Regulations and Codes of Practice. Such requirements have now been laid down in respect of a number of hazardous substances, for example, Lead

Compounds and Asbestos and others are under active consideration. It is expected that over the next decade this will continue, many of the earlier regulatory requirements being updated, simplified and replaced. In the meantime the advice given in the previous edition is still valid, *i.e.*, guidance as to what may be required in laboratories may be drawn from existing parallel legislation in other areas such as factories, and from codes of practice such as that on the use of ionising radiation. More general guidance and the reasoning underlying such guidance is contained in this manual.

Particular attention should be paid to systems of work and to the clear delegation of specific responsibilities to those organising units of laboratory activity. A further important aspect of the legislation lies in the requirements in respect of consultation and this places a particular responsibility on chemists who are best able to advise the layman on the hazardous nature and properties of chemical substances, and the precautions necessary to ensure their safe use. It is, therefore, important that everyone concerned makes an evaluation of all aspects of health and safety and puts in train any necessary steps to ensure that his house is in order.

Regrettably, perhaps, in the past chemists, who habitually handle dangerous substances, have been inclined to disregard the hazards associated with their use particularly if such hazards are of a long-term nature. Every human being, and a chemist is no exception, tends towards the view that although an accident may happen to another, it will never happen to him personally because he is too wise and knowledgeable. Experience shows that nothing can be further from the truth. During work in the laboratory, many persons have suffered injury to their health which because of the insidious symptoms, may never have been associated with their work activity. It is only when permanent injury has occurred that many persons come to realise that the observance of even elementary precautions could have prevented such injury or, in extreme cases, premature death.

It is against this background that one should look at this edition of the handbook. The contributors have attempted to indicate and discuss the dangers likely to arise in the laboratory and have offered practical advice on their avoidance. The work will, I believe, also prove most useful in devising precautionary techniques in respect of the many reagents and substances which, for reasons of space, have not been included in this work.

At the same time the manual has become a much more useful and complete work not only for chemists in the laboratory but also for all those who handle hazardous substances on a small scale, *e.g.*, in industry. Additionally, the work will be useful in schools and higher eduction establishments where training in the correct use and handling of the substances should be considered an integral part of the curriculum of students in science subjects, and where the Health and Safety at Work *etc* Act has extended specific obligations in respect of safety and health matters.

The control of hazards of the laboratory is well known: the enforcement of safe work systems; the need for mechanical safety involving the guarding of dangerous parts of machines, even if driven by only fractional horse power motors, so that injury from contact with moving or trapping parts is prevented; the need to provide safe means of access to every place where anyone is at work even if the work is only undertaken on rare occasions; the need for good housekeeping to minimise the possibility of accidents occurring through persons striking or being struck by objects, the need for care in handling glassware; the need to protect electrical conductors and to provide or use low voltage supplies or adequate earthing; and, of course, the matters with which this handbook is intimately concerned—the prevention of injury from fire, explosion or from exposure to hazardous substances.

Identification

Perhaps the most important single step we can take in securing the safe handling of chemicals is to ensure that a proper system of labelling is used that will identify the substance, indicate the hazards involved, and set out the simple precautionary measures to be followed. There has, of recent years, been considerable international discussion concerning labelling systems and the original work carried out by the Council of Europe has now been taken up and extended by the European Economic Commission. Directives have now been published dealing with the commoner pure substances and with some formulations such as solvents, paints, varnishes and printing inks. The Treaty of Rome setting up the EEC requires that such instruments be given legal force in the UK. Chapter 8 details these labelling requirements where they are directly applicable and in respect of other chemicals applies similar parameters. Failure to give some such simple warning of hazards is inexcusable, particularly in the laboratory where many chemicals may at some time be handled by inexperienced and unskilled persons who are not members of the lab staff, *e.g.*, cleaners and maintenance workers. Accordingly all chemicals should be labelled following the general guidelines used in these systems as set out eleswhere in this work.

Management's task

Safety and health is the responsibility of management and must be set out in a policy statement. Not only must the manager and all members of the staff know the hazards involved, but they must all be clearly seen to be directly interested and involved in the promotion of a safe and healthy environment. Strict procedures should be written into analytical and other methods. Where such methods do not exist, the work should be immediately supervised by a responsible person who is aware of the dangers and precautions to be followed

3

both during normal working and in any emergency that may arise. In larger laboratories a safety officer and hygienist should be appointed to provide advice and general supervision and, not least, to look critically at the procedures involved from outside the group undertaking the project. Experience shows that such a view dissociated from the actual scientific work is invaluable in bringing to light relatively simple hazards which may have been overlooked.

General management and supervision must be tight to ensure that work is conducted in a predetermined and orderly manner, that unauthorised actions are checked and that proper care and attention is given to the minute-by-minute operation of processes or experiments. In particular, at each meal or tea break and at the end of the day a thorough check must be made to see that everything can be left safely. If there is any doubt, arrangements should be made for the continued supervision of the operations still in progress.

The overall aim should be to design out hazards so that the whole system can operate in such a way that any possible human error is eliminated as far as is practicable. Chapter 3 deals with this general aspect of the problem.

Fire and explosion hazards

The dangers of fire are well known, but again we must remember the maxim 'familiarity breeds contempt'. The very large number of fires in laboratories proves the seriousness of this problem. Chapter 4 gives detailed advice on such risks and the text under each substance in Chapter 8 indicates properties on which an assessment of the fire hazard can be based and makes suggestions as to the selection of fire fighting equipment.

It should be clearly understood that when a liquid is used having a flash point below the highest normal ambient temperature it can, in suitable circumstances, liberate a sufficient quantity of vapour to give rise to a flammable mixture with air. This can accumulate in the workroom to such an extent as to give rise to the possibility of a serious explosion by ignition of the vapour/air mixture from an ignition source already present some distance away, causing a flash-back to the original source. There is then the consequent possibility of disastrous fire.

A flammable gas or vapour must be present in a concentration of the order of 1 per cent or more by volume if its mixture with air is to be flammable, so it is a relatively simple matter to check whether or not a dangerous concentration is likely to be present in closed plants such as ovens *etc*. During normal working it is desirable to ensure that one-quarter of the lower flammable limit is never exceeded. The amount of flammable vapour or gas in the workroom air should, of course, never approach this concentration during normal working procedures. Account must, however, be taken of possible leakages and spillages so that although it is, perhaps, unnecessary to provide special precautions such as flameproof electrical equipment in normal circumstances, these may be very desirable to provide against contingencies arising from the unexpected loss of flammable solvents having a flash point of less than 32 °C and particularly those having a flash point of less than 21 °C. Our aim, therefore, should be to restrict the use of such liquids to situations in which they are absolutely necessary and even then to reduce the quantity involved as far as possible. This is particularly important with solvents used for routine operations where in every case we should carefully consider whether or not it is possible to substitute an alternative having a flashpoint above the highest possible ambient temperature (32 °C).

The quantity of all flammable materials and of solvents in particular should be kept to the absolute minimum. There is often a tendency to disregard this and allow large quantities of solvents that are used only occasionally to accumulate in the laboratory. When flammable substances are not in use there must be adequate supervision to ensure that they are kept in a properly constructed fireproof store.

Suitable fire fighting equipment should be readily available and adequate means of escape provided (*see* Chapter 4). All personnel should be trained and familiar with the use of the equipment so that a small fire can be quickly localised and prevented from spreading, while in the event of the fire getting out of hand everyone must know how to escape safely.

Reactive chemical hazards

Particular care must be exercised when using highly reactive or unstable substances that may be liable to cause an explosion. The quantities used should be kept to a minimum and, if necessary, several reactions carried out on a smaller scale. Consideration should be given to limiting the effect of the explosion—should one occur—by the provision of suitable reliefs venting to a safe place. In all cases protective screens should be provided or the experiment operated by remote control so as to ensure that the operator will not be injured. It should be remembered that such substances may be produced as a result of side reactions or in residues standing over a period of time.

Exothermic reactions should be carefully controlled and monitored to ensure that there is no failure of the cooling or stirring systems. Again quantities should be kept to a minimum and suitable screening provided. No operation of this kind should be entrusted to anyone other than a highly skilled and competent chemist knowledgeable in the dangers involved and the precautions to be taken. All these matters are discussed in detail in Chapter 5.

Toxic hazards

Toxic substances can act in three ways causing poisoning by ingestion,

5

percutaneous absorption and inhalation. Our first thought should always be: can a harmless or less hazardous substance be used instead of the substance under consideration? Such a step removes or reduces the danger in an infallible manner and this possibility should therefore never be overlooked, or dismissed without very careful consideration.

The dangers of ingestion by contamination of the hands and food can be virtually eliminated if there is proper attention to personal cleanliness. Washing accommodation of a high standard should be provided together with a means of drying that is always available. This precaution is, of course, equally applicable to other health risks and promotion of personal cleanliness should never be neglected. Another common, but inexcusable, danger in this category is the use of mouth pipettes. Such methods should never be used for pipetting hazardous liquids and we should train school pupils in the use of rubber bulbs.

The contact of corrosive substances with the skin is generally obvious and so this is somewhat less dangerous than contact with percutaneous poisons. Nevertheless gloves (*see* Chapter 3, p. 26) and, where necessary, protective clothing should always be worn where this hazard is present. If contact with the skin occurs, the affected parts should be washed immediately with soap and water. Special attention is necessary to protect the eyes, and where corrosive substances are regularly used, eye protection should be worn as a routine precaution. It must always be remembered that it is often a bystander and not the person actually carrying out the process who suffers injury.

Substances that can be imperceptibly absorbed through the skin present a more insidious hazard, particularly if the contact is repeated or prolonged. Great vigilance is necessary to ensure that the dangers are appreciated by everyone concerned. The aim should be to prevent contact. The use of protective gloves has been found in many cases to be of doubtful value as, unless strict working procedures are observed, contamination can occur on the inner surfaces through pinholes or by careless removal and replacement. Any splashes on the unprotected skin should be washed off immediately with soap and water. Where the more hazardous substances in this group are used regularly it is advisable to make arrangements for the periodic medical examination of the persons involved.

Inhalation of harmful vapours, gases, dusts and liquid aerosols is another insidious and widespread danger in the laboratory. Many persons have at one time or another been exposed to excessive quantities of vapours such as mercury, benzene and carbon tetrachloride or to dusts such as lead and beryllium. This is the more serious because it is deceptive, there often being no sensible perception of danger. Additionally, since it is all-pervading, once it has entered the air of the workroom, the contaminant must be breathed. Everyone is inclined to judge the danger by the short-term effects whereas it is the long-term effects that are the more serious and may give rise to permanent and irreversible injury. Unfortunately, such effects may not be directly attributable to exposure to toxic chemicals because the affected person may have changed his employment or be no longer working with the hazardous materials. Therefore, there is little statistical evidence as to the incidence of ill health brought about by such exposures in the laboratory and in consequence the very real and serious dangers tend to be disregarded.

It is the duty of every responsible person to see that substances having irreversible effects are only used when it is absolutely necessary. If such a hazardous substance must be used, then adequate instructions must be given, proper supervision assured and the whole process carried out in such a way as to ensure that the material is contained and does not enter the air of the workroom where it may be breathed. A properly-designed fume cupboard should always be used and all steps in the process or experiment carried out therein. Where the hazardous substance giving rise to long-term effects is used regularly, the possibility of monitoring the atmosphere at regular intervals should be given serious consideration. If the sample can be taken in the breathing zone of the worker by means of a suitable personal sampler then confirmation can be obtained as to the efficiency of the precautionary measures and the worker reassured as to the absence of any long-term health risk. The assessment of toxicity is a complex matter, but general parameters are set out in Chapter 6 which, as indicated earlier, lock into the labelling system recommended in Chapter 8.

While the hazards from a toxic chemical may be difficult enough to assess it must always be remembered that two relatively harmless liquids or substances may, when in contact, liberate an unexpected poisonous gas. Perhaps the commonest example is bleaching powder and acid lavatory cleaners which liberate chlorine gas. Another common example is alkaline cyanides which, in contact with acid, liberate hydrocyanic acid gas. The phenomenon often occurs in sink traps and other parts of the drainage system or during the use of containers that have not been properly washed out from a previous operation. It is particularly important when any such substances are used by unqualified persons, *e.g.*, cleaners, to consider in advance whether or not chance contact between such types of chemicals could produce a more serious hazard. If this is possible steps should be taken to avoid any possible contact. Working instructions should include the safe disposal of waste and require the routine cleaning of containers for chemicals.

The subject of carcinogenicity is a difficult one. Many substances if implanted repeatedly in animals will produce a carcinogenic reaction in the fullness of time, but they may not be hazardous when used in the normal way. On the other hand there are substances such as 2-naphthylamine that will almost certainly produce carcinogenic effects if ingested into the human body. Between these two extremes lie a large number of potentially hazardous materials. Where a known carcinogenic risk exists it is indicated in the text.

Substitution of another reagent should always be considered, but, if any such substance must be used, all practical precautions should be taken to reduce all exposures to as low a level as is practicable. The Carcinogenic Substances Regulations 1967 prohibit the use industrially, and the importation of, certain carcinogenic compounds, namely 2-naphthylamine, benzidine, 4-aminodiphenyl, 4-nitrodiphenyl and their salts and substances containing any of these compounds other than in very small concentrations.

General environment

As with all chemical hazards good housekeeping, *i.e.*, the cleaning up and removal of spillages and lost material, is a cornerstone to safe working. In this context all spilt material which may become a hazard must be completely removed or made chemically inert. Liquids may be removed with water if they are soluble. If insoluble, detergents or solvents may be used. In the last resort it may be necessary to react the material chemically to an inert form as a method of decontamination.

Toxic dusts are particularly hazardous in this respect since if lost material is not removed it will become repeatedly airborne whenever it is disturbed either by movement of persons or materials or by strong air currents. Such dust, if present, must be removed at frequent intervals by an industrial vacuum cleaner fitted with a high efficiency filter to prevent recirculation of the fine particles in a breathing zone.

Past experience clearly indicates that many cases of poisoning in laboratories are due to background contamination brought about by a neglect of these general principles.

Coupled with good housekeeping is the quality of the general working environment—in particular, lighting, ventilation and heating. Good lighting is important because hazards become immediately apparent; there is less need when carrying out intricate manipulations for the operative to approach close to the danger area to obtain a better view of the operation and malfunction of the equipment or instrumentation is immediately apparent, permitting remedial measures to be taken before danger occurs.

Heating and ventilation go hand in hand. Good general ventilation is essential in laboratories where toxic materials are handled. A general standard of at least five air changes per hour should be the aim. Adequate heating must be provided during the winter months or such ventilation will not be used.

We may sum up all these general matters by saying that where hazardous materials are handled we should provide a pleasant, clean working environment properly planned with good access and means of escape, and a high standard of lighting, ventilation and heating with regular cleaning to ensure removal of any spilt or lost material.

After a hazardous substance has been used there remains the problem of disposal. The first method to be considered is its return to stock *via* reprocessing. If this is not possible small quantities of combustible material may be burnt in a controlled manner so that a hazard will not arise from fire or toxic decomposition products. Small quantities of less hazardous chemicals can be flushed away to the drain with copious quantities of water, but if larger quantities are involved the local authority should be consulted. Highly hazardous chemicals should be reacted in a fume cupboard to break them down into less hazardous compounds which are more readily disposable. It must be clearly understood that the responsibility for safe disposal rests with the user and under no circumstances should substances be disposed of in such a way as to constitute a hazard either inside or outside laboratories. In the last resort specialist companies are available who will undertake the task of disposal for a fee.

Where laboratories are situated in a built-up area or where the effluent from exhaust systems must be discharged at a low level, steps must be taken to ensure that means of trapping the dangerous substances are provided so that the exhausted air does not contain a dangerous quantity of harmful substance. Care should be taken to discharge the effluent at a safe height where it will not recirculate into the laboratory or cause a nuisance or danger to other persons in the vicinity.

Radiation hazards

Ionising radiation
There is increasing interest in the subject of radiological protection and the effect of low doses in particular. While this is in part generated by the nuclear power programme there is no doubt that any incident in which radioactive substances are involved will evoke a demand for a critical investigation.

As a result of initiatives by the European Community new Regulations together with Approved Codes of Practice have been drafted and laid before Parliament. These regulations and codes will apply to all work activities (*see* Chapter 2) and the documents give detailed practical guidance on the steps to be taken when using radioactive substances or machines generating ionising radiation. Their general requirements are reviewed in Chapter 9.

The essential points to be noted are that notification to the Regulatory Authority is required unless only trivial amounts (as quantified in the appropriate regulatory documents) are used.

Non-ionising radiation
Since the publication of the previous edition much progress has been made in evaluating the hazards of non-ionising radiation. Standards for exposure to microwaves, ultraviolet and lasers have been published by responsible bodies

in the United States and in particular by the American Conference of Governmental Industrial Hygienists. Discussions have also taken place within the framework of the EEC but no firm European initiatives have yet emerged.

Conclusions

With the passage of time knowledge of the toxicological effects of exposure to chemical substances is rapidly advancing and each year there is an improved understanding of the long-term hazards associated with their use. Massive documentation now exists and for the great majority of substances some general assessment of the hazard can be made. Nevertheless, the general policy should be to reduce exposure as far as is practicable in every case. This has become particularly important as the safe levels of substances are constantly being reduced and, in addition, new and unexpected toxic effects become apparent.

It must be the aim of everyone involved to ensure that all chemicals are handled safely without either immediate or long-term dangers. A clean, healthy general working environment must be provided and individuals encouraged to become safety conscious. The subsequent parts of this book provide detailed information on particular types of hazard.

In Chapter 8, details are given of the properties and dangers of each substance, together with the RL or CL (recommended limit or control limit) and flashpoint where applicable. This will enable the reader to decide for himself what should be done. At the same time it must be emphasised that, given a wise, commonsense approach to the problem, the general elementary precautions indicated above will enable any normal operation to be carried out safely. One last piece of advice cannot be overemphasised. It is essential before carrying out any new procedure to stop, stand back, and consider what hazards may arise, what precautions should be taken and what emergency procedures may be necessary.

Chapter 2

Health and Safety at Work Act 1974

General

The Health and Safety at Work Act was founded very largely on the principal recommendations of the Robens Committee on Safety and Health at Work whose report was published in 1972.

The main purpose of the Act is to provide one comprehensive and integrated system of law dealing with the health, safety and welfare of workers, and the health and safety of the public as affected by work activities. The Act can be described as the most significant statutory advance in the field of health and safety at work since the Shaftesbury Factory Acts of 1833. It aims to change radically not only the scope of the provisions, but also the way in which those provisions are enforced and administered. It thus extends obligations and protection to five million or more people who have never before come within the scope of this kind of legislation including workers in health, education and research establishments. It also covers for the first time the self-employed.

The Act, of course, does not seek to cover every eventuality nor does it try to spell out rules for each and every work situation. It is an enabling instrument, whose foundation is the concept of a general duty of care in respect of people engaged in or associated with work activities. It adopts a flexible general approach and thus provides legislation capable of being expanded and adapted to deal with the risks and problems associated with the current technological changes in industry at any particular time.

The general umbrella of the Act provides for an interaction of responsibility between individuals and organisations associated with work or touched by its immediate consequences. The employer now has a duty to his employees with regard to their safety and health, and those employees in turn have a duty to one another. So, too, the self-employed person has a duty to other people around him. There are two significant advances which go much further even than the spirit of previous legislation. The general public is now entitled to a duty of care (in terms of safety and health) from people carrying out work activities, so that an employer, for example, not only has to ensure that his workers are safe, but also that members of the public who might be affected by any hazard from his work activities are not at risk. The legislation also includes an innovation requiring the incorporation of safeguards at an early stage—earlier than had been possible before—placing obligations on suppliers, importers *etc* of machinery, plants, substances *etc* to make sure that they will be safe when properly used.

Scope

The Act applies to all persons at work: employers, the self-employed and employees, the only exception being domestic servants in private employment. Many of these people were, however, covered by earlier safety and health legislation and in particular the Factories Acts, which applied to process laboratories. The 1974 Act extends this cover to all types of laboratory. New regulations and approved codes of practice will eventually largely replace most of the existing legislation on safety and health, including the Explosives Act, the Alkali *etc* Works Regulation Act, the Public Health (Smoke Abatement) Act, the Petroleum (Consolidation) Act, the Agricultural (Poisonous Substances) Act, the Mines Act, the Factories Act, the Offices, Shops and Railway Premises Act and a number of other minor Acts. With such a massive area of legislation to replace, it is obvious that the whole field cannot be magically changed overnight and indeed this is not necessary. Some of the legislation is relatively recent and much more is still very relevant. Many of the detailed requirements of these Acts therefore continue in force by virtue of the new Act until they are eventually replaced by new legislation.

General obligations

The obligations set out in the new Act, important to each and everyone working in the laboratory, are in addition to, and not in diminution of, the existing obligations under health and safety legislation, *i.e.*, the Factory Acts will remain and will be enforced where they are applicable. In addition much earlier legislation may be taken as providing guidelines as to the interpretation of the general requirements of the 1974 Act in respect of other premises where similar hazards arise. Indeed the general obligations set out in Section 2 of the Act may be taken as giving effect to such requirements.

It will be seen that to obtain flexibility many of the duties imposed by the Act and related legislation are qualified by the words 'so far as is reasonably practicable'. Other duties require people concerned to use the 'best practicable means'. The latter expression is used in the context of requirements for controlling sources of environmental pollution, such as emissions into the atmosphere.

If someone were to be prosecuted for failing to comply with a duty 'so far as is reasonably practicable', it would be the responsibility of the accused to show the court that it was *not* reasonably practicable for him to do *more* than he had in fact done to comply with the duty. Similarly, in a case involving the expression 'best practicable means', it is for the accused to show the court that there was no *better* practicable means than the one he used to comply with the relevant duty.

Although neither of the expressions is defined in the Act, both have acquired quite clear meaning through long established interpretations by the courts. Someone who is required to do something 'so far as is reasonably practicable' must assess, on the one hand, the risks of a particular work activity or environment and, on the other hand, the physical difficulties, time, trouble and expense which would be involved in taking steps to avoid the risks. If, for example, the risks to health and safety of a particular work process are very low, and the cost of technical difficulties in taking certain steps to avoid those risks are very high, it might *not* be reasonably practicable on balance to take those steps. However, if the risks are very high, then less weight can be given to the cost of measures needed to avoid those risks. The comparison does not take into account the current financial standing of the employer. A precaution which is 'reasonably practicable' for a prosperous employer is equally 'reasonably practicable' for the less well off. The expression 'best practicable means' involves similar considerations: the person on whom the duty is imposed must use the most effective means to comply with the duty, taking into account local conditions and circumstances, the current state of technical knowledge, and the financial implications.

Duties of employers to their employees

The general duty imposed on every employer is to ensure, so far as is reasonably practicable, the health, safety and welfare at work of all his employees. It applies to all employers, both those who already have certain duties under earlier legislation, such as the Factories Act, and those who have never before been covered by health and safety legislation, *e.g.*, in research laboratories.

Employers' general duties

The employer must, so far as is reasonably practicable, provide machinery, equipment and other plant that is safe and without risks to health, and must maintain them in that condition. He must also ensure that, so far as is reasonably practicable, the systems of work are safe and without risks to health. 'Systems of work' means the way in which the work is organised and includes, for example, the layout of the workplace, the order in which work is carried out, and any special precautions that may have to be taken before carrying out hazardous tasks. This duty, therefore, means that, for example, a machine itself and the way it is operated must both be safe.

Articles and substances for use at work

The use of particular articles and substances at work, machinery and chemicals for example, may give rise to risks to employees' health and safety. There is a chain of responsibility involving various people with duties to ensure that those risks are reduced as far as possible. In brief, the manufacturer must ensure that, so far as is reasonably practicable, the materials are safe and without risks to health when properly used, and users must be given sufficient information about proper use and about any hazards. Once the materials have reached the place of work there is also a duty on the employer to ensure that, so far as is reasonably practicable, employees' health and safety are not put at risk by contact with the materials. In particular, he is required to ensure that the ways in which the materials are used, handled, stored and transported are safe and without risks to health. He should ensure that attention is paid to any information given by the manufacturer or supplier about safe handling and storage. His duty extends to the end products as well as to the materials used during the work process. He must ensure that the ways in which end products are transported and stored before leaving the workplace are safe and without risks to the health of his employees. He may of course also have duties as a supplier in turn.

Training

The employer should provide for all his employees the information, training and supervision necessary to ensure, so far as is reasonably practicable, their health and safety at work. The information to be supplied must include information about hazards at the workplace and methods of avoiding them. In particular, the employer should make sure that employees are given information made available by manufacturers and suppliers of materials used at work about risks attached to the materials and about safe ways of handling them. Health and safety training might include such things as instruction in safety and emergency procedures such as routine checking of equipment, fire drills and first-aid; special training for work involving a high degree of risk; and retraining when the work changes or new safety methods are introduced. It is the employer's duty to ensure that all his employees are competent to carry out their jobs in a safe manner, *i.e.*, with the minimum of risk to themselves or others. Employers must ensure that managers understand their responsibility and have the necessary knowledge and skills to carry them out. Training is necessary not only for operatives, but also for supervisors and managers at all levels. Good supervision is vital in health and safety terms for spotting potential hazards and ensuring that safety rules are complied with.

The local emergency services should also be informed of any potential hazards which might affect their members so that they can provide suitable training and instruction.

Employers' safety policies

The Act requires every employer to prepare a written statement of the safety policy in his undertaking, except where there are less than five employees. However, it should be noted that 'undertaking' does not have the same meaning as 'establishment'. An employer may operate a number of small establishments, each employing less than five employees. If all the establishments form part of one undertaking and if the total number of employees is five or more the employer must prepare a policy statement.

The purpose of the safety policy requirement is to ensure that the employer carefully considers the nature of the hazards at the workplace and what should be done to reduce those hazards and to make the workplace safe and healthy for his employees. The statement should set out the employer's aims and objectives for improving health and safety at work. It should also set out the organisation and arrangements currently in force for achieving those objectives. 'Organisation' can be taken to mean people and their responsibilities, and 'arrangements' systems and procedures.

Another purpose of the statement is to increase employees' awareness of the employer's policy and arrangements for safety. For this reason, the employer is required to bring the statement to the notice of all his employees. In some cases the best and easiest way of effecting this may be to give a copy to every employee and ensure that new employees are given copies during induction training. Alternatively, copies may be posted on notice boards where they can be easily seen and read.

In most organisations, working conditions are continually changing. New hazards arise, control measures alter. The safety policy should therefore be kept up-to-date. All revisions in the statement must also be brought to the notice of employees.

Safety representatives and safety committees

The 1978 Regulations give recognised trade unions the right to appoint safety representatives to represent the employees in consultations with the employer about health and safety matters. The Regulations also provide for the possibility of employers being required by safety representatives to set up safety committees that would have the job of keeping under review measures to ensure health and safety at the workplace. Full details are contained in the booklet Safety Representatives and Safety Committees published by HMSO which contains the Regulations, Approved Codes of Practice and guidance notes. Approved Codes of Practice give practical guidance on the functions of safety representatives, the information to be provided to them by employers and the time off with pay to be allowed for training approved by the TUC or by individual unions.

Duties to people who are not employees

One of the major innovations of the Act is that an employer has duties not only to his own workers but also to outside contractors, workers employed by them and to members of the public whether within or outside the workplace who may be affected by work activities. Both employers and the self-employed are required to carry out their undertakings in such a way as to ensure, so far as is reasonably practicable, that they do not expose people who are not their employees to risks to their health and safety. The duty extends to, for example, risks to the public outside the workplace from fire or explosion, or from the release of harmful substances into the atmosphere. The duty of employers and the self-employed also applies to people who may be inside the workplace, such as visitors, outside contractors and their employees working on the premises on a permanent basis, for example maintenance men, or another employer's workers temporarily visiting the premises. It should also be noted that outside contractors, whether employed or self-employed, will themselves have responsibilities under this duty for the health and safety of workers and others on the premises they enter to carry out the contract work.

In general, the standard of protection required for visitors and others within a workplace will be similar to those an employer should give his employees. For example, if machinery and substances are used in such a way as to ensure the safety of employees, that will usually also be sufficient to ensure the safety of non-employees. There may be, however, a need to apply different criteria to achieve these standards in view of the fact that certain people, such as the very young or disabled, may be more vulnerable than others and that people visiting a work-place may have less knowledge of the potential hazards and how to avoid them.

Every employee must take reasonable care for the health and safety of himself and of other persons who may be affected by what he does or fails to do at work. This duty implies not only avoiding obviously silly or reckless behaviour, but also taking positive steps to understand the hazards in the workplace, to comply with safety rules and procedures, and to ensure that nothing he does or fails to do puts himself or others at risk.

Regulations and approved codes of practice

The relationship between the HSW Act and earlier Acts dealing with health and safety, has been outlined above. Most of the provisions still in force are those laying down specific standards of health, safety and welfare for particular circumstances. One of the most important aspects of the Act is that it contains powers to modify, replace and repeal the earlier legislation. The eventual aim is to bring all health and safety requirements into a single system of regulations and approved codes of practice under the Act. The regulations specify requirements to supplement the general duties imposed by the Act, while approved codes of practice give practical guidance about how compliance with the general duties or regulations might be achieved.

Approved codes of practice do not themselves lay down legal requirements. No-one can be prosecuted for failing to follow the guidance contained therein. However, every approved code has a special legal status similar to the status of the Highway Code under road traffic laws. If someone is being prosecuted for breach of any requirement of the Act or related legislation, any approved code which appears to the court to be relevant to the case is admissible in evidence. If the guidance of the approved code has not been followed, it is up to the defendant to show that he has satisfactorily complied with the requirements in some other way.

Anyone who chooses not to follow the guidance in a particular approved code of practice must therefore realise that if legal proceedings are taken against him for breach of the requirements that are the subject of the approved code he must be able to prove to the court that he has nevertheless satisfactorily fulfilled those requirements by some other means. Such proof will often be very difficult. For example, the Safety Representatives Regulations require [Reg 7(2)] the employer to make available to safety representatives the information within his knowledge that is necessary to enable them to fulfill their functions. The Approved Code of Practice says that such information should include, for example, information about the plans and performance of the undertaking and any changes proposed insofar as they affect the health and safety at work of the employees. If an employer is prosecuted for failing to provide information in accordance with the

Regulations, and if it were shown that he had information about plans and performance of the undertaking and had not disclosed it, he would probably find it difficult to show that he had compiled with the Regulations in another way.

The making of Regulations and Approved Codes of Practice under the Act has been somewhat slow due to necessary consultation proceedures. Recently, however, three sets of Regulations have been published relevant to Laboratory Health and Safety, namely those dealing with Electricity, Pressure Vessels, and Ionising Radiation. While all these Documents may be said to formalise current standards of good working practice, the current proposals for the control of Substances Hazardous to Health are likely to have the greatest impact in view of their universal application to laboratories. They propose that a written assessment be made by a competent person of any risks to health arising from the use of chemical substances. The means for controlling any such risks so identified must be determined and put into effect. The supporting General Approved Code of Practice gives detailed practical guidance as to good occupational hygiene practice and forms a valuable document in its own right.

Chapter 3

Safety Planning and Management

Effective and safe operation of a chemical laboratory involves three essential and complementary components. These are: well-designed premises and equipment suitable for the intended purpose; a well-designed and administered operating system suitable both for the intended work and the personnel executing it; and a clearly stated progressive and positively-implemented safety policy that will motivate all those involved with the supervision and execution of laboratory work to contribute on a long-term basis to the effective and safe use of the available facilities. This latter is undoubtedly the most important factor.

Shortcomings in the premises and equipment may, to some extent, be offset by extra attention to the operating system, but an inadequate or poorly administered safety policy surely will vitiate the effects of the first two components mentioned above.

The particular selection of detailed aspects necessary as constituent parts of these three major components will differ for individual laboratories and it is not feasible to attempt to cover here all the possibilities of variation in all three components. Those topics which are dealt with are considered to be of sufficiently wide validity to be relevant to a majority of laboratory installations. It will almost invariably be necessary for individual laboratories to provide (and revise), as part of their operating system, written details covering points of considerable local significance which are not dealt with here.

Safety policy and implementation

To comply with the requirements of the Health and Safety at Work *etc* Act 1974, employers must provide their employees with clearly written and unambiguous statements on their current health and safety policy, and on how it is intended to be implemented in practice by their management and employees. Because it must apply to all the employees (some or many of whom may not work in laboratories), the policy document will necessarily be concerned with general health and safety objectives, such as overall organisation, training, areas of responsibility and matters of consultation. In places of work where several distinct and specialised types of work are carried on it may well be necessary to provide the more detailed implementation documents specific to each type of work, and a laboratory is an obvious case in point.

Depending on the width of variety of work undertaken in a laboratory and on the number of personnel employed, the implementation document will need to cover in more or less detail the topics discussed below.

Responsibilities
Responsibilities must clearly be specified at all levels within a particular organisation, with an indication of permissible degrees of delegation of authority. Responsibility may be for people, operations, equipment, or a work area, or some combination of these. Particular care is needed in defining and assigning responsibility for shared facilities or working areas. An organigram may well assist in their clear specification of responsibility, but it must be both readily available to staff and up-to-date to be useful.

One of the main responsibilities that line management has to employees is to ensure that adequate training and instruction in safe methods of working is available, and is given, to employees of lower levels of experience. Such training is normally given 'on the job' by more experienced employees and it is part of the laboratory manager's duty to ensure that this is being done effectively. It will probably also be necessary to ensure that suitable written outline material is available for this purpose.

Consultation
Consultation with employees on matters of safety is an important aspect—and duty in the larger establishment—of laboratory management. It is essential that consultation should be effected in a way that will encourage employees to accept and exercise their responsibilities in a constructive way.

The usual vehicle for consultation is, of course, the safety committee and its composition should attempt to represent all different grades of workers and sections in a laboratory. It would normally be chaired by the senior manager and include the safety officer as his adviser. The composition of the safety committee in larger and unionised laboratories is now governed by legislation.

Safety audits
Periodic safety audits are an essential part of the continuing process of assessing systematically the effectiveness and current relevance of all aspects of the overall safety organisation. Personnel working in a particular laboratory or section of a laboratory will necessarily be involved in the audit, as will the safety officer and a safety committee member. It is, however, desirable to include an outsider in the audit team to ensure an unbiased view during the discussion and tour of inspection. Suitable detailed procedures for laboratory safety audits are available elsewhere.[1]

Information
The need to provide adequate information on working materials and methods to employees is stressed early in the 1974 Act and much effort has gone into this book to provide such information. However, no single publication can hope to cover all possible safety aspects relevant to the many differenty types of laboratory, some highly specialised, which exist. It may well be necessary for additional written material to be produced for a specific laboratory to supplement published information. Normally, a selection of published information is available to laboratory personnel, either directly, or *via* the safety officer. Some sources of additional general information are given in the bibliography to this chapter, and sources of specialised information are referred to in other chapters.

Laboratory safety officer
An excellent and extensive definition of the qualities and overall role expected of a laboratory safety officer is given in the *Code of practice for chemical laboratories.*[2] The main attributes and duties may be identified thus:
Tact and persuasiveness coupled with some formal training in safety and good practice
Wide technical knowledge and professional experience and the ability to communicate these
Advising all levels of responsibility on major issues including training
Co-operating with various specialists and emergency services organisations
Acting as a focus for relevant information from many sources (including accident reports)
Organising safety audits.

It is recommended that wherever possible the appointment of safety officer should be a fulltime one.

Other matters

Other matters with strong safety connotations which must be considered and organised integrally by laboratory management are routine maintenance of equipment, installations and services, and arrangements for supervision or monitoring if experiments or equipment are to be run outside normal working hours. Although chemicals in use for laboratory experiments are excluded from the specifically restrictive provisions of the poisons legislation, it is nevertheless essential to maintain adequate control over the storage, issue and use of scheduled poisons by a nominated responsible person. If carcinogenic materials are used it is mandatory for close control to be exercised by a responsible person, with a locked storage cupboard and an appropriate register of issues and distribution.

The precise implications of the forthcoming Code of Practice for the Control of Substances Hazardous to Health are not yet clear, particularly in educational laboratories. However, it is likely that more effort will be required to assess possible risks, to advise workers and students of these, and in keeping records of significant exposures.

Laboratory premises, services and equipment

When the construction of a new laboratory is contemplated, or when existing premises are to be adapted to meet new or changed laboratory requirements, considerable planning effort is necessary. This phase must involve the maximum of consultation between architect, consulting engineers and staff who are to use the laboratories, including the safety officer, to ensure a full appreciation by the former as to what is really needed.

Due consideration must be given to those factors (ordering of materials, storage of samples, chemicals and equipment, laboratory cleaning, disposal of laboratory wastes, *etc*) ancillary to the main work of the laboratory.

Space requirements

The working area required for comfortable working conditions for a given number of laboratory personnel will vary widely with the type of work that is to be undertaken. The current tendency towards equipment-intensive work, represented by small-scale instrumental analysis (especially if automated), will tend to decrease the free floor area in relation to that required for the more labour-intensive type of laboratory work, typified by traditional wet analytical methods or preparative chemistry. However, adequate space for free circulation by people as well as for installed equipment, permanent facilities, storage of portable equipment, samples and consumable materials must be allowed to permit safe and effective working. Although there are as yet no official guidelines on free floor area requirements, a minimum total area of 4.5 m² per person has been recommended.[3] A minimum free floor space of 1.2 m appears desirable in front of each benching run or item of installed equipment (*i.e.*, 2.4 m between island installations), unless particular accessibility requirements dictate more than this minimum. A recent review covers all these design points in some detail.[4]

Surface finishes
Lack of adequate attention at the planning stage to the question of finish of working surfaces, and especially floors, may incur subsequently very considerable expenditure and inconvenience in upgrading surface finishes to meet the demands of actual working conditions in a particular laboratory.

The selection of laboratory flooring will often involve compromise between comfort, durability and resistance to water, solvents or acids, and ease of decontamination after a spill. The multiple jointing of wood or composition blocks or vinyl tiles make thorough decontamination difficult or impossible, *e.g.*, after a mercury spill, and clearly excludes them from most laboratory situations. Cork linoleum, well-maintained with a non-skid polish, will have rather limited applicability in light-service situations. Impervious PVC sheeting with welded joints and coved edges is finding increasing use as medium-service laboratory floor surface not subject to heavy solvent contamination or wheeled traffic. For heavier duty, resin-screeds or ceramic tiles set in resin cement may be necessary for long-term durability, with suitable grades to floor drains where regular washing down or decontamination may be involved.

It is now generally accepted that the traditional (and very expensive) teak or other solid hardwood bench surface is far from ideal for laboratories where harzardous chemicals are used. Unless exceptionally well-maintained, the fissures and surface damage that will arise in use make thorough cleaning or decontamination very difficult, even on a polyurethane-sealed surface.[5] Press-bonded melamine-faced laminated boards possess most of the attributes necessary for a general laboratory working surface and are coming into widespread use.[6] For severe corrosive service, such as in fume cupboards, epoxy resin surfaced boards have found application. Though less durable than acid-resisting ceramic tiles set in resin cement, these lead to a noticeably high breakage rate for glassware. A chemical lead sheet surface, though resistant to most aggressive reagents, may be easily and extensively perforated by small spills of mercury or its soluble salts. Many plastic sheet materials possess severe temperature limitations (*see Table 3.1*) and will need local protection from heat sources. Stainless steel is particularly vulnerable to rapid corrosion by hydrochloric acid, and inorganic chlorides or bromides.

Table 3.1 Resistance of plastic laboratory ware

	Polythene	Polythene h.d.	Polypropylene	PVC	Polystyrene	Polyacrylic	Polycarbonate	Polymethyl-pentene ('TPX')	Polytetra-fluoroethylene
Upper limit °C									
Short periods	90	105	140	75	70	90	130	200	300
Continuous	80	95	130	65	60	75	110	180	260
Resistance rating									
Dilute acids	G	E	E	E	G	E	G	E	E
Strong acids	F	F	G	G	F	F	F	G	E
Dilute alkalis	E	E	E	G	G	E	F	E	E
Strong alkalis	G	E	E	G	F	E	X	E	E
Organic solvents	F	F	G	L	X	L	X	G	E
Sunlight during two years	G	G	G	G	E	G	E	G	E

Resistance rating code: E—Excellent; F—Useable; G—Resistant; L—Limited periods; X—Attacked (at 20 °C)
The information above is by courtesy of Azlon Products Ltd., and more detailed information on the resistance to 153 chemicals of 15 plastic materials used for chemical apparatus is tabulated in a brochure[7] available from the same source.

Ventilation and fume cupboards

Effective laboratory ventilation and fume extraction is one of the most important factors likely to have long-term benefits for the health of laboratory workers, and much attention is now rightly concentrated on the best practical means to achieve this end. To maintain a reasonable level of comfort in a well-ventilated laboratory (10–12 air changes per hour) containing one or more fume cupboards it is necessary to match the extraction rate with the heat input. Provision of a ducted supply of unheated make-up air to the fume cupboard may be necessary to minimise loss of heated air from the laboratory where high extraction rates are involved for hazardous materials. Several aerodynamically efficient designs of fume cupboards are now commercially available that will maintain the recommended minimum face velocity of 0.5 m s^{-1} irrespective of the width of opening of the sash.[8]

Fume cupboards must be sited away from doors—to prevent face-turbulence arising from cross draughts—and in a position where a minimum run of fire- and corrosion-resistant ducting to the extract fan will be required. Preferably, each fume cupboard should have its own extract fan located vertically above to eliminate the possibility of cross-contamination which may exist where several cupboards are manifolded into one extract fan. However, if such an arrangement is unavoidable, provision of devices to indicate a satisfactory extraction rate from each fume cupboard is desirable. Where significant runs of horizontal ducting inside a laboratory building are involved, fire dampers may be necessary (p. 47).

The materials of construction of the inner surfaces of both fume cupboards and ductwork will, of course, depend upon what is likely to be used in the fume cupboard and prolonged use of some specially hazardous materials, *e.g.*, perchloric acid or hydrofluoric acid, may necessitate the provision of wash-down or air-scrubbing arrangements.

External control of the service outlets inside the fume cupboard enclosure is essential for safe use of the cupboard and electricity outlet sockets are additionally located externally to minimise corrosion of the contact surfaces. Such sockets are best located towards the end of the fascia to avoid leads dangling across the doors of the (extracted) storage cupboards below the working area.

The provision of small horizontally sliding glass panels in the front sash offers the facility of gaining occasional local access for manual adjustment of equipment with minimal reduction of overall containment and protection afforded by the closed sash. Toughened glass is the preferred glazing material for fume cupboards, wired glass is not suitable where the possibility of explosion exists.

Much more detail of this rather specialised aspect of basic laboratory design and equipment is to be found in three chapters of a particularly useful book,[9] and there is a draft standard for safety requirements for fume cupboards, performance testing and recommendations on installation and use.[10] An experimental study of performance has been published.[11]

Safety equipment

The provision and siting of fire extinguishers and other protective equipment in a laboratory is sufficiently important to be included in the early stages of the overall laboratory design. This will ensure that appropriate extinguishers, fire blanket, face mask, respiratory protection, gloves or other relevant safety equipment can be installed (preferably in a custom-built safety unit located near to the door), so that all the emergency equipment relevant to the laboratory operations is obviously and immediately to hand and ready for use. The laboratory safety officer should expect to be involved to a considerable degree to ensure adequate consultation with the local fire authority in matters both of principle and detail. Multi-storey laboratory buildings require special consideration, *e.g.*, provision of dry rising mains for foam injection, alarm, evacuation and smoke-detection systems, and escape routes. The selection of fire extinguishers is dealt with in Chapter 4.

There is now a wide choice of well-designed equipment available for personal protection against all toxic or corrosive materials. Comfort in wearing personal protective equipment is perhaps the most important factor in determining whether it will be used in practice.

The prime importance of protection of the eyes is reflected in the Protection of Eyes Regulations 1974, which requires the use of adequate eye protection in areas of risk. Many laboratories where corrosive or hazardous chemicals are used, or where glassware is used under vacuum or pressure, will be areas of risk, and must be clearly labelled at the entrances as eye protection areas. The eye protection deemed to be appropriate will depend upon the degree of hazard associated with both the type and quantity of material and may be selected from protective spectacles with side pieces, ventilated goggles, or full face mask. Protective spectacles (if necessary with toughened prescription lenses) or protective goggles (some of which may be worn over ordinary spectacles) should conform to the British Standard specification.[12] Protection of the eyes against non-ionizing radiation is also an important consideration (Chapter 9).

In laboratories where eye protection must be adopted against substantial amounts of corrosive or toxic chemicals, it is likely that the hands may also need protecting with gloves, of which there are many types available. Selection of the type or types of gloves to be made available in a laboratory will involve consideration not only of the material to be used but also the nature of the operations in which it is to be used. Use of gloves thicker than surgical or disposable polythene types may considerably reduce the manual dexterity and sense of touch, so that even these gloves may pose problems if very fine operations with high-hazard materials are contemplated. Where relatively large amounts of aggressive materials are to be used, stout gloves (with apron and goggles) with appropriate resistance will be necessary and the consequent reduction in manual dexterity should be allowed for in the equipment and operations which are to be used while the gloves are worn. *Table 3.2* will assist in the choice of gloves for particular applications, but note that some solvents tend to penetrate all gloves.

Relatively loose fitting and stiff gloves appear to be more acceptable in practice because of their ease in donning and removal and degree of ventilation, as compared with closer-fitting light surgical or household gloves (which should be internally dusted with talc before use). It is particularly important to encourage wearers to wash and dry the outside of the latter types after use and before removal. This will prevent contamination of the inside of the left wrist when the right thumb is inserted (by a right-handed person) to remove the left-hand glove.

Routine respiratory protection of laboratory workers is best effected by conducting operations with volatile or dusty hazardous materials within the confines of the fume cupboard or ventilated glove box. Second best is the use of partial enclosure or local ('spot') ventilation to capture contaminants at the source[13] and some of the more resistant types of flexible plastic air ducting are useful for this purpose. If neither of these methods can be applied, personal respiratory protection appropriate to the hazard must be considered. For particulate contaminants one of the many approved dust-masks will be suitable,[14] but remember that some dusts can also irritate exposed skin.

Where relatively low concentrations of gases or vapours may be present in a laboratory atmosphere, arising from small leaks and/or use of small quantities of toxic materials, a canister respirator specific to the expected contaminant may give the limited protection necessary to deal with a minor emergency.[15] However, if relatively large quantities of toxic gases or vapours are in use, the protection afforded by a canister respirator will be quite inadequate and an air-breathing set must be available for emergency use. The small (satchel) type of breathing set will provide only a few minutes' use, and is intended to allow protected short-term access, *e.g.*, to close a cylinder valve or to rescue a colleague. For more extended use, or fire-fighting purposes, a back-pack air-breathing set (which requires a trained operator) will be needed. In all such cases it is essential for a second person to know that respiratory protection is being used, so that appropriate supervision can be given. Only self-contained air-sets or air-line breathing equipment are of use in an oxygen-deficient atmosphere.[15]

Laboratory furniture and equipment
The choice of equipment and furniture for a laboratory from the very wide range available will, of course, be closely related to the intended work. There are, however, some general safety considerations which apply to all equipment and furniture, as well as those specific to individual items.

Laboratory furniture units are usually modules made to fit under the continuous runs of impervious benching now preferred for freedom from contamination. Metal under-bench units are not usually considered suitable except for particularly clean laboratories (pharmaceutical formulation, scintillation counting, *etc*) where the possibility of corrosion is remote. Wooden furniture is of proven durability in most laboratory atmospheres and, if well made, will not disort in the rather wide variations in humidity that may be caused by the use of steam or water baths or drying ovens. Wide drawer units (> 1 m) invariably cause jamming problems unless the drawers are a really good fit in their runners and may lead to breakage of glassware. Cupboard door catches should have a positive, but smooth, action, again to prevent glassware breakage, and magnetic catches are advantageous in this respect. Top surfaces of benches or support units for tall laboratory instrumentation or equipment may need to be set below the usual bench height (0.92 m) to provide ready access to the equipment for safe manipula-

Table 3.2 Glove resistance ratings (by courtesy of James North & Sons Ltd)
Resistance rating code: E—Excellent; F–Fair; G—Good; NR—Not recommended.

Chemical	Natural rubber	Neoprene	Nitrile	Normal PVC	High-grade PVC
Organic acids					
Acetic acid	E	E	E	E	E
Citric acid	E	E	E	E	E
Formic acid	E	E	E	E	E
Lactic acid	E	E	E	E	E
Lauric acid	E	E	E	E	E
Maleic acid	E	E	E	E	E
Oleic acid	E	E	E	E	E
Oxalic acid	E	E	E	E	E
Palmitic acid	E	E	E	E	E
Phenol	E	E	G	E	E
Propionic acid	E	E	E	E	E
Stearic acid	E	E	E	E	E
Tannic acid	E	E	E	E	E
Inorganic acids					
Arsenic acid	G	G	G	E	E
Carbonic acid	G	G	G	E	E
Chromic acid (up to 50%)	G	F	F	E*	G
Fluorosilicic acid	G	G	G	E	G
Hydrochloric acid (up to 40%)	G	G	G	E	G
Hydrofluoric acid	G	G	G	E*	G
Hydrogen sulphide (acid)	F	F	G	E	E
Hydrogen peroxide	G	G	G	E	E
Nitric acid (up to 50%)	NR	NR	NR	G*	F*
Perchloric acid	F	G	F	E*	G
Phosphoric acid	G	G	G	E	G
Sulphuric acid (up to 50%)	G	G	F	E*	G
Sulphurous acid	G	G	G	E	E
Saturated salt solutions					
Ammonium acetate	E	E	E	E	E
Ammonium carbonate	E	E	E	E	E
Ammonium lactate	E	E	E	E	E

* Resistance not absolute, but the best available.

Chemical	Natural rubber	Neoprene	Nitrile	Normal PVC	High-grade PVC
Ammonium nitrate	E	E	E	E	E
Ammonium nitrite	E	E	E	E	E
Ammonium phosphate	E	E	E	E	E
Calcium hypochlorite	NR	G	G	E	E
Ferric chloride	E	E	E	E	E
Magnesium chloride	E	E	E	E	E
Mercuric chloride	G	G	G	E	E
Potassium chromate	E	E	E	E	E
Potassium cyanide	E	E	E	E	E
Potassium dichromate	E	E	E	E	E
Potassium halides	E	E	E	E	E
Potassium permanganate	E	E	E	E	E
Sodium carbonate	E	E	E	E	E
Sodium chloride	E	E	E	E	E
Sodium hypochlorite	NR	F	F	E	E
Sodium nitrate	E	E	E	E	E
Solutions of copper salts	G	G	G	E	E
Stannous chloride	E	E	E	E	E
Zinc chloride	E	E	E	E	E
Alkalis					
Ammonium hydroxide	E	E	E	E	E
Calcium hydroxide	E	E	E	E	E
Potassium hydroxide	E	G	G	E	E
Sodium hydroxide	E	G	G	E	E
Aliphatic hydrocarbons					
Hydraulic oil	F	G	F	G	E
Paraffins	F	G	E	G	E
Petroleum ether	F	G	E	F	G
Pine oil	G	G	E	G	E

Table 3.2 *continued*
Table 3.2 Glove resistance ratings (by courtesy of James North & Sons Ltd)
Resistance rating code: E—Excellent; F–Fair; G—Good; NR—Not recommended.

Chemical	Natural rubber	Neoprene	Nitrile	Normal PVC	High-grade PVC	Chemical	Natural rubber	Neoprene	Nitrile	Normal PVC	High-grade PVC
Aromatic hydrocarbons†						**Alcohols**					
Benzene	NR	F	G	F	G	Amyl alcohol	E	E	E	E	E
Naphtha	NR	F	F	F	G	Butyl alcohol	E	E	E	E	E
Naphthalene	G	G	E	G	E	Ethyl alcohol	E	E	E	E	E
Toluene	NR	F	G	F	G	Ethylene glycol	G	G	E	E	E
Turpentine	F	G	E	F	G	Glycerol	G	G	E	E	E
Xylene	NR	F	G	F	G*	Isopropyl alcohol	E	E	E	E	E
						Methyl alcohol	E	E	E	E	E
Halogenated hydrocarbons†						**Amines**					
Benzyl chloride	F	F	G	F	G	Aniline	F	G	E	E	E
Carbon tetrachloride	F	F	G	F	G	Butylamine	G	G	E	E	E
Chloroform	F	F	G	F	G	Ethylamine	G	G	E	E	E
Ethylene dichloride	F	F	G	F	G	Ethylaniline	F	G	E	E	E
Methylene chloride	F	F	G	F	G	Methylamine	G	G	E	E	E
Perchloroethylene	F	F	G	F	G	Methylaniline	F	G	E	E	E
Trichloroethylene	F	F	G	F	G	Triethanolamine	G	E	E	E	E
Esters						**Miscellaneous**					
Amyl acetate	F	G	G	F	G	Animal fats	F	G	G	G	E
Butyl acetate	F	G	G	F	G	Bleaches	NR	G	G	G	E
Ethyl acetate	F	G	G	F	G	Carbon disulphide	NR	F	G	F	G
Ethyl butyrate	F	G	G	F	G	Degreasing solution	F	F	G	F	G
Methyl butyrate	F	G	G	F	G	Diesel fuel	NR	F	G	F	E
						Hydraulic fluids	F	G	G	G	E
Ethers						Mineral oils	F	G	E	G	E
Diethyl ether	F	G	E	F	G	Ozone resistance	F	E	G	E	E
						Paint and varnish removers	F	G	G	F	G
Aldehydes						Petrol	NR	G	G	F	G
Acetaldehyde	G	E	E	E	E	Photographic solutions	G	E	E	G	E
Benzaldehyde	F	F	E	G	E	Plasticizers	F	G	E	G	E
Formaldehyde	F	F	E	G	E	Printing inks	G	G	E	G	E
						Refrigerant solutions	G	G	E	F	G
Ketones						Resin oil	F	G	G	G	E
Acetone	G	G	G	F	G	Vegetable oils	F	G	G	G	E
Diethyl ketone	G	G	G	F	G	Weed killers	G	E	E	G	E
Methyl ethyl ketone	G	G	G	F	G	White spirit	F	G	G	F	G
						Wood preservatives	NR	G	G	F	G

† Aromatic and halogenated hydrocarbons will attack all types of natural and synthetic gloves. Should swelling occur, switch to another pair, allowing swollen gloves to dry and return to normal.

tion of samples *etc.* There is a British Standard on other aspects of laboratory furniture.[16]

Modern laboratory equipment and instrumentation has become increasingly complex in the past decade or so, and many laboratories may now contain a selection of such equipment, each item of which may involve several service connections and more than one potential hazard. Most equipment is electrically powered and will require regular examination of lead, plug and fuse for mechanical and electrical integrity (see below). Additionally, flammable gases and oxidants (hydrogen, acetylene, nitrous oxide or oxygen) may be involved in analytical instruments, such as a gas chromatograph or an atomic absorption spectrometer. Secure and permanent gas connection lines and fittings of appropriate materials are essential for safe operation and leak-checking should be undertaken on a routine basis. The flexible tubing on peristaltic pumps will need regular replacement with the correct grade of material to withstand possible aggressive reagents. Flexible leads for supply of water (cold or hot), vacuum, or compressed air or nitrogen to laboratory equipment must all be of adequate durability for the particular service,[17] and must, of course, be inspected regularly and replaced when necessary.

Bench services and lighting
Benches or working areas in many laboratories may be supplied with a considerable number of piped services to cater for the needs of sophisticated modern equipment and several of the commonly supplied services may lead to potential hazards if not properly installed and used.

The most widely used service is mains electricity, usually 240 volt single phase, but occasionally 400 volt 3-phase for high consumption equipment. As well as the obvious need for regular inspection of the flexible leads and checking of fuse ratings, insulation and earth resistance of plugged appliances, the earth resistance of socket outlets should also be checked if atmospheric corrosion of the contacts appears likely. Switches that have become worn or faulty may lead to excessive sparking or arcing when operated and increase the possibility of ignition of flammable materials. If substantial amounts of highly flammable materials are normally present, a flameproof installation may be required. The use of a temporary electrical wiring installation, a serious potential hazard, must be discouraged as far as possible, and rigorously controlled and inspected regularly if its use is unavoidable.

Potential hazards associated with laboratory water and drainage services arise most frequently from the use of perished or insecurely attached flexible tubing used to feed condensers, water baths, *etc.* These are most likely to split or blow off water taps when the mains water pressure rises fairly sharply shortly after normal working hours. It follows that such incidents are likely to go undetected for some hours and flood damage to electrical equipment or the building fabric may be extensive, particularly in multi-storey buildings. For the same reason, water-reactive materials must be kept in covered storage above floor level. Regular inspection and replacement of flexible water leads will minimise such incidents.

If the drainage system has been well designed and installed, with corrosion resistant materials (glass, polythene or polypropylene, properly supported) and adequate cleanouts and precautions against blockage included (grids for sink wastes, gulleys or tundishes), no problems should be encountered. In multi-storey buildings fume cupboard drainage should be directly connected into the drainage downcomer, with any side bench lines trapped to prevent back-diffusion of volatile toxic vapours on lower floors.

Although the use of natural gas for laboratory heating is much less prevalent now than previously, gas burners are still found in many chemical laboratories for various essential purposes. The obvious hazard of ignition of highly flammable materials may be greatly reduced by prominent display of appropriate notices to ban either flames or flammables in a particular area.

Of the other commonly available piped services—general vacuum, compressed air and steam—the latter two have obvious potential for hazard. Compressed air is often available at pressures up to 7 bar with simple on–off valves or Schrader connectors and a flow restrictor (a short section of capillary tubing) may be needed downstream of the control valve to prevent excessive air flow to small laboratory equipment.

Steam burns (scalds) are particularly serious because of the very high latent heat released when steam condenses. Asbestos lagging of steam-pipes may constitute an insidious hazard if the exterior surface of the lagging becomes damaged, because the thoroughly dry asbestos compound can generate airborne dust if disturbed.

As well as natural (or manufactured) fuel gas, many other 'chemical' gases may now be available as piped services in laboratories. When a glassblowing torch is fed with oxygen, a non-return valve must be incorporated in the fuel gas line to prevent back-flow of oxygen and risk of explosion.

Preferably, these service gases will be piped in from cylinders located outside the laboratory to minimise hazards from leaking fittings or regulators on cylinders of flammable or toxic gases, and to eliminate a major hazard in case of fire in the laboratory.

Lighting is an important laboratory service in that the level must be adequate for clear vision in all parts of the laboratory, particularly where hazardous operations may be in progress. Fume cupboards must, therefore, be particularly well-lit and it is advantageous for their lights to be separately controlled, so that the fume cupboard light alone may be left on when overnight operations are in progress.

Vacuum and pressure equipment
The differential pressure created when laboratory apparatus is operated at

31

pressures above or below that of the atmosphere may give rise to several different potential hazards.

In vacuum systems, particularly when glassware is being thermally stressed by local heating or cooling, screens should be used to protect glassware from accidental impact and to contain any glass fragments caused by mechanical failure of the glassware. Leakage (or accidental admission) of air into a vacuum distillation system containing hot organic material may lead to rapid oxidation or ignition.

The use of internal pressure in glassware is inevitably attended by some risk, because glass is naturally weak in tension. If metal equipment cannot be used, glassware that will be stressed must be checked for strain and well screened before cautious application of the minimum pressure possible. Work at substantially higher pressures may require an isolation cell (p. 67).

Laboratory instrumentation
The great increase in availability and variety of laboratory instrumentation in recent years has meant a corresponding increase in the number of potential hazards, some of them quite unusual. All modern instruments are electrically powered, and will thus need proper installation with effective isolation switches and earthing, as well as subsequent routine inspection and maintenance to avoid the obvious electrical hazards (p. 30).

Gas chromatographs with flame ionisation detection are connected to a hydrogen supply (check for leaks, especially on the high pressure side of the regulator) and the flame is likely to serve as an exposed ignition source if flammable vapours are present in the laboratory. If other detector systems are used, any highly toxic components in analysed samples will be swept out unchanged with the carrier gas into the laboratory atmosphere, so adequate ventilaton is important. Hypodermic syringes used for injecting samples (and other laboratory purposes) must be handled carefully to minimise personal risk.

Atomic absorption spectrometers may use high intensity flames or electric heating (furnace or plasma) to vaporise the sample. The former type may use hydrogen, propane or acetylene as fuels and oxygen or nitrous oxide as oxidants; all are potentially hazardous. Some of the old, acetylene-fed burners may cause an explosion if air is drawn up the liquid sample drainage line. Both types of source are usually of sufficient intensity for eye protection to be used (p. 578).

High pressure liquid chromatographs may be used with flammable or toxic solvents, and even a small leak may produce a very finely dispersed spray of liquid or vapour, or a liquid jet which may penetrate the skin with very serious consequences. Adequate containment and ventilation and isolation from ignition sources (or automatic fire protection) are essential in such circumstances.

Instruction manuals for such instruments will now identify all the known hazards and will obviously be worth close study.

Glass- and plastic-ware

The most common material to be seen in chemical laboratories has always been glass and failure to appreciate the shortcomings in some of its mechanical properties has been a major cause of laboratory injuries and accidents. While strong in compression, glass is weak in tension and most failures occur because of this fact. Spherical flasks will readily withstand complete evacuation because the external stress (100 kN m^{-2}, 1 bar) is even and compressive, but will fail at a modest internal pressure, because the walls are then in tension. Surface scratches concentrate stress and will lead to earlier failure. Proper cleaning techniques (p. 77) will prevent scratching.

The flat base of a conical flask will be in tension and so tend to collapse whether under vacuum or pressure. Buchner suction flasks and Carius (pressure) tubes have thick and even walls capable of withstanding the stresses normally imposed in service. When inserting glass tubes into bungs, the hands *must* be protected, because it is impossible to ensure that the manually applied pressure will be sufficiently even to keep the glass tube under a purely compressive stress, even when the ends are fire polished and a lubricant is used.

Because of its brittle nature, care is necessary when clamping glass, particularly for large assemblies. Use only clamps with intact cork or rubber facings and arrange for limited flexibility to cater for expansion or vibration. Large assemblies may need cross-braced retort stands or a scaffolding frame for proper support.

To eliminate breakage of glassware, many laboratory items are now available made from various plastics. Apart from the upper and lower temperature limitations of all plastics in comparison with borosilicate glass, many of the more powerful solvents or aggressive reagents will soften or dissolve some plastics or lead to stress cracking, so considerable care must be exercised to ensure compatibility of container and contents. *Table 3.1* will serve as a useful guide here. However, plastic T- or Y-piece connectors should always be used in preference to their glass counterparts to extend flexible leads for water, vacuum, compressed air or nitrogen to eliminate the risk of cut hands when the tubes are pushed over the connectors. The connectors must be secured, like the ends of the leads, with wire or tube clips.

Laboratory materials: handling, storage and disposal

Most hazards in a chemical laboratory derive from the properties of the various chemical materials that are used therein. It follows that most laboratory hazards may be avoided by matching adequate knowledge of the various properties and the implications of those properties, with knowledge of protective equipment and preventive measures. In practical terms, relevant information on chemical properties (Chapter 8) coupled with that on reactive hazards (Chapter 5) must be related to fire protection measures (Chapter 4) and the use of protective equipment (this Chapter). The examples which follow illustrate this need to interpret and combine data when some specially hazardous materials are stored and handled.

Specially hazardous chemicals
Carbon disulphide (p. 228) being very volatile and extremely flammable poses a severe fire risk and must only be used in a fume cupboard. It has a very low autoignition temperature—below 100 °C on corroded (catalytic) metal surfaces—and unusually wide flammability limits, so the vapour can readily ignite on a hotplate or steam bath. Vapour containment and good ventilation are important factors in its use and a fire extinguisher should be available at the fume cupboard.

Hydrogen sulphide (p. 363) is sometimes underrated as a hazardous material, but it is in fact as toxic as hydrogen cyanide by inhalation. More importantly, it rapidly desensitises the sense of smell, so that one may be misled into believing that the concentration is diminishing. If the source of a sizeable and continuing hydrogen sulphide leak is a cylinder, don a respirator before attempting to stop the leak.

Hydrofluoric acid (p. 358) is exceptional among the corrosive acids in that the lack of immediate pain from a skin burn can lead to deep penetration and a serious wound. For this reason, it is essential to test protective gloves for pinhole leaks by inflation under water, to prevent the possibility of internal contamination by the acid.

Mercury (p. 390) is a particularly insidious toxic material,[18] and determined efforts should be made to prevent long term contamination of laboratory premises by careful design and (properly documented) working techniques. These will invariably include a sheet of siliconised release paper on the bench under each site where mercury is to be used,[19] or less conveniently, a deep containment tray. Mercury manometers used for indicating pressures above atmospheric should be fitted with an outlet trap to prevent expulsion of mercury droplets.[20]

Nitric acid (p. 419) is undoubtedly the material most widely involved in laboratory accidents and without exception the hazards arise from its oxidizing power which is still considerable at ambient temperature and at reasonable dilutions. If organic materials are present considerable volumes of gas may be evolved including toxic oxides of nitrogen (p. 424) and carbon dioxide. Mixtures of nitric acid with other acids and/or organic solvents have been used as cleaning or etching agents, but they are not stable in storage, particularly after use, and may decompose with evolution of gases that will burst a closed container. Care is necessary to ensure that the inert liner remains in place in the plastic screw-cap of a winchester of fuming nitric acid to prevent attack on the polythene and development of gas pressure.

Gas cylinders

These may be regarded as a special class of materials where the mechanical hazard represented by the large store of kinetic energy can be augmented by the hazards that may arise from the toxic or reactive properties of the particular gas. Careful handling and firm support are essential at all times to prevent damage to the control valves. If a cylinder falls over and shears off its valve, the sudden release of the stored energy will lead to the violent and possible fatal results to be expected of an unguided heavy missile.

The pressure reducing assembly normally required to reduce the cylinder pressure and control the gas flow to valves suitable to experimental use must be made of materials appropriate to the gas in use. The main cylinder valve and the regulator should both be opened slowly to prevent pressure surges and the regulator should be inspected often for signs of corrosion or mechanical damage. Regulators and valves must be kept clean and free from dirt, oil or grease—this is an absolute requirement for oxygen service.

The contents of gas cylinders are identified both by a stencilled name near the neck and by colour coding of the cylinder. A small version of the BS colour coding chart is available gratis.[21]

Storage

The conditions under which reactive chemicals are stored may have a considerable effect upon their subsequent utility in reactions, and in some cases potential hazards may arise from inappropriate storage conditions. Reputable suppliers of reactive chemicals go to considerable lengths to pack air- and moisture-reactive materials in suitable containers with appropriate caps or seals, and some care should be exercised to ensure that the caps are properly replaced once the bottle or jar has been opened to maintain the seals in good condition. Sealed ampoules of moisture-sensitive materials may often by resealed if just the tip is cracked off and the sample is withdrawn with a syringe to avoid contaminating the glass before resealing.

Bottles of particularly moisture-sensitive materials (anhydrous aluminium chloride, phosphorus tribromide and the like, see p. 73) may best be kept in a desiccated container at atmospheric pressure to minimise ingress of moisture through the bottle seals and development of internal gas pressure. A snap-seal plastic container containing silica gel makes a fairly effective desiccator, especially if the lid is also sealed round with PVC (not cellulose) tape.

Handling of winchester quantities of any chemicals must take into account the possible fragility and weight if the contents are dense (5 kg for sulphuric acid), when the neck may break off if it is used as a handle. Winchesters should be carried in carriers or baskets that will give proper protection and support. Some suppliers now use plastic-encapsulated winchesters for corrosive chemicals, which in the event of dropping and breakage will retain the contents.

Air-sensitive (peroxidisable) liquids are best stored with inhibitor(s) in sealed brown glass containers in a cool, dark location to minimise autoxidation and the development of hazardous peroxides (see p. 72).

Materials with relatively high vapour pressures at ambient temperature need cool, or in some cases refrigerated, storage. Highly volatile materials supplied in sealed glass ampoules should be cooled in ice before opening, then transferred to screw-cap bottle for refrigerated storage. Attention is drawn to the high risk involved in attempting to store flammable liquids in domestic refrigerators, as detailed on p. 48.

A common material that often leads to vapour pressure problems in warm weather is concentrated (0.880, 35 per cent w/v) ammonia solution (see p. 174). Such problems can be avoided by use of the rather less concentrated solution (0.910, 25 per cent w/v) where this is technically feasible.

Gases in cylinders may also be considered as materials of high vapour pressure, and storage of these in a cool place is similarly desirable. Adequate ventilation is also necessary to prevent accumulation of any leaking gases in a cylinder store.

The need to segregate materials in storage is indicated later (p. 76), and gas cylinders should be stored separately from flammable materials. These in turn need segregation from oxidizing agents (p. 74), which should be located as far as possible from reducing agents. Fire protection aspects of storage of chemicals are specifically dealt with on pp. 43–46. Storage of radioactive materials is specifically covered by legislation, which is detailed elsewhere (p. 566). The whole subject of safe storage of laboratory chemicals has been reviewed in detail.[22]

Waste disposal
The disposal of some chemical waste comes within the provisions of the Deposit of Poisonous Waste Act 1972, which prohibits the deposition of poisonous, noxious or polluting wastes whose presence on land would be liable to lead to an environmental hazard. Although laboratory management is most unlikely to be directly concerned with ultimate land disposal operations, it will be involved with aspects closer to the point where waste is first produced and handled in the laboratory, and usually with the public sewerage system. The need to minimise and control the generation of laboratory waste and its effective disposal as an integral part of laboratory management has been stressed in a recent and comprehensive book.[23]

Different categories of expected waste should be established for each laboratory or group, and arrangements made for their segregation and disposal, using a contractor if large quantities are involved. Appropriate documentation is necessary for the latter case.

Highly-flammable and water-immiscible solvents and oils form one fairly convenient category for segregation, and safety cans in each laboratory are the usual method of collection. The capacity of the cans should be limited to prevent accumulation of large amounts, and advice on disposal is given on p. 55. Multi-storey groups of laboratories using relatively large amounts of flammable solvents or oils may have a metal piped and adequately trapped collection system leading to external storage tanks.

It is usual to exclude chlorinated solvents from either can or piped collection systems, because of their relatively high volatility and toxicity, the reactivity of chloroform and to minimise corrosion on incineration of the collected flammable solvents. So a separate container will be necessary.

Relatively small volumes of aqueous or water-soluble liquid wastes of low acidity or alkalinity may be flushed down the sink with plenty of water to give the necessary degree of dilution. Strong acids or alkalis should be diluted cautiously (eye protection!) and, if possible, largely neutralised before flushing away. The degree of dilution necessary to comply with the requirements of the local water authority will depend upon the average flow of effluent which is being discharged to the sewer from the laboratory premises. If microbiologically toxic heavy metals (silver, cadmium, chromium, mercury and lead) are likely to be used, acceptable dilution levels should be established and appropriate instructions issued to staff.

Many of the methods recommended in Chapter 8 for dealing with accidental spillage of chemicals may be adapted for use in normal disposal procedures of the same chemicals. Detailed disposal procedures for many more individual chemicals have been published elsewhere,[23,24] and the current situation on disposal has been reviewed.[22] Methods for the disposal of small amounts of carcinogenic materials at the point of use have been developed and are to be preferred, but if this is not possible such materials must be adequately packed and labelled for disposal. The specific requirements for radio-active disposals are outlined on p. 573.

Laboratory techniques and procedures

While management can provide premises and equipment that are conducive to safe working in a laboratory, it is the various techniques and procedures that are actually used by the workers in that laboratory that will determine whether work is conducted safely. This section attempts to give examples of good laboratory practice designed to achieve that end, arranged on the basis of some common laboratory operations.

Measuring liquids

As a general rule, the mouth should never be used to suck liquids into pipettes, to avoid the danger of contamination from toxic solutions or vapours. Either a loose-fitting vacuum line or rubber bulb may be used to fill the pipette, and the palm-sized bulbs with moulded-in valves are particularly convenient. Fuming corrosive liquids are best measured in a hypodermic syringe or glass piston pipette, used in the fume cupboard.

When pouring corrosive liquids from a bottle, keep the label uppermost so that any drips formed will not run onto and damage the label. If liquid is to be poured from a large storage bottle into a small measuring-cylinder, clamp the latter and use a funnel.

Heating

Heating of reaction mixtures is probably the most common laboratory operation and where flammable materials are being heated there is some risk of fire. The risk may be minimised by using sound glass vessels (check for star cracks), indirect heating (steam bath, oil bath or electric heating mantle with a controller, rather than a direct gas flame), with a safety tray of adequate volume underneath if 1 l or more of flammable liquid is being heated.

Electric hotplates are widely used to boil off water and some of the less volatile solvents. The relatively high surface temperature of the hotplate may, however, lead to ignition of some vapours with exceptionally low autoignition temperatures, such as carbon disulphide (95 °C on a corroded surface), hydrazine (132 °C on iron, 23 °C on rust), acetaldehyde (140 °C), diethyl ether or dioxane (180 °C). Further information on ignition sources is given on p. 47.

When solvents are boiled under a reflux condenser, the water leads must be sound and the ends securely attached, preferably with brass tube clips. Bare copper wire will cause the ends of rubber tubes to perish rapidly (copper–sulphur interaction) unless sleeving or electrical tape is used to prevent direct contact. If reflux is to proceed overnight, fit a short length of metal tube in the outlet lead to prevent it being whipped out of the drain by the passage of slugs of air which occasionally occur in waterpipes.

In the absence of stirring, anti-bumping granules or glass beads should be added at an early stage of the heating sequence, so that smooth and regular boiling will begin and prevent superheating of the liquid. Under no circumstances should granules (or charcoal) be added to solutions close to their boiling points, as violent or explosive boiling may set in.

Stirring

Stirring of mixtures is common in preparative chemistry to mix heterogeneous reaction systems and an electric motor most frequently serves as the power source. Many of the small AC/DC motors used for this purpose have an extensive spark source at the commutator. These are therefore unsuitable for direct stirring of reactions involving highly flammable liquids under reflux or flammable effluent gases, unless a connection from the top of the condenser or vent is taken well away from the motor. The spark-free AC-only motors or compressed air motors are safe for such applications.

When stirring viscous liquids or those containing much suspended solid, it is essential to use a geared electric motor with high torque. High-speed laboratory stirrer motors will burn out under these conditions and may lead to a fire.

Long hair must be kept away even from small laboratory stirrer motors, as an exposed coupling will easily tear out any hair which may become entangled.

Cooling

Solid carbon dioxide and liquid nitrogen are widely used as cryogenic coolants and both will burn the skin on close contact, so eye protection and impervious gloves must be worn when they are used. Eye protection is particularly important when transferring liquid nitrogen into or from glass vacuum flasks, because on breaking glass is scattered over a wide area by the rush of evaporating gas.

Solid carbon dioxide ('Cardice' or 'Drikold', $-78\ ^\circ$C) is used with a solvent to ensure good thermal contact in cooling baths. The considerable evolution of gas inevitably leads to evaporation of solvent from the bath, so relatively toxic chlorinated solvents should not be used. Propan-2-ol has an appreciably higher flash point ($12\ ^\circ$C) than the widely used, but highly flammable, acetone ($-18\ ^\circ$C). Both nitrogen and carbon dioxide coolants tend to displace air so must only be used in well ventilated areas. For this reason, neither coolant must be transported in passenger lifts in multi-storey buildings.

A hazard specific to the use of liquid nitrogen is its ability to condense oxygen as liquid from atmospheric air. When liquid nitrogen is used to cool vacuum traps, air must not be admitted after using the traps until the coolant has been lowered away. This avoids the dangerous condensation of liquid oxygen on to organic trap residues, which may lead to violent explosion.

Vacuum distillation

This procedure has an unusual combination of potential hazards. These are the mechanical stress on the glassware due to the differential pressure (p. 33), possibly in combination with the thermal stresses of high temperature distillation and of low temperature cryogenic high vacuum traps (above); the possibility of ingress of air; and contact with hot, air-reactive distillation residues. The risk of failure of glassware can be minimised by careful inspection for cracks or scratches, using a strain viewer if the apparatus has previously been strongly heated. Use of an oil bath as heat source and high vacuum may be necessary to prevent overheating and decomposition of labile

39

materials, and a nitrogen bleed will minimise decomposition of air-sensitive materials or distillation residues. Wide bore connections are essential in high vacuum distillation to minimise pressure drop between the pump and the surface of the boiling liquid. For particularly high boiling or unstable materials, a short-path ('molecular') still may be necessary. Eye protection is, of course, essential in vacuum distillation and an additional safety screen is desirable to contain fragments if a distillation set-up fails in use.

References

1 *Safety audits—A guide for the chemical industry*, London: Chemical Industries Association, 1976.
2 *Code of practice for chemical laboratories*, London: RIC, 1976 (RSC update in press)
3 K. Everett and D. Hughes, *A guide to laboratory design*, p. 6, London: Butterworths, 1975.
4 W.J.H. Gray, *Safety in the design of laboratories*, J. Inst. Water Eng. Sci., 1981, **35**(6), 483.
5 Comparative figures on ease of decontamination of various surface materials are given on p. 16 of reference 3.
6 *Decorated laminated plastics sheet*, BS 3794:1973.
7 *Properties of plastics commonly used in the manufacture of scientific products*, London, Azlon Products, 1983.
8 *A guide to the design and installation of laboratory fume cupboards*, Ann. Occ. Hyg., 1975, **18**, 273–291.
9 Reference 3 chapters 6–8.
10 DD80:1982 (3 parts)
11 W.G. Mikell and L.R. Hobbs, *Laboratory hood studies*, J. Chem. Educ., 1981, **58**(5), A155.
12 *Industrial eye protectors*, BS 2092:1967 with later amendments.
13 The limitations are fully discussed in pp. 53–55 of reference 3.
14 *Respirators for protection against harmful dusts, gases and scheduled agricultural chemicals*, BS 2091:1969.
15 *Recommendations on selection, use and maintenance of respiratory protection devices*, BS 4275:1974.
16 *Recommendations on laboratory furniture and fittings*, BS 3202:1959.
17 *Rubber water hose for low pressure*, BS 5119:1975; *General purpose rubber water hose for medium pressure*, BS 3716:1964.
18 D.J.Moore and A.E.Timbs, *Mercury – a health risk*, Chem. Br., 1984, 622.
19 D.J.Ward, *Handling mercury on siliconised release paper*, Chem. Ind., 1983, 211.
20 L.Bretherick, *An outlet trap for mercury manometers*, Lab. Practice, 1973, 533.
21 *Identification of contents of industrial gas cylinders*, BS 349:1973 (as miniature chart 12765000), New Malden: Air Products, 1976.
22 D.A. Pipitone (ed.), *Safe storage of laboratory chemicals*, Chichester, Wiley, 1984.
23 *Laboratory chemical disposal manual*, revised 2nd edn. Washington, DC: Manufacturing Chemists' Association, 1972. (now withdrawn)
24 *The Aldrich catalogue handbook of fine chemicals* 1985–86, Gillingham: Aldrich Chemical Co. Ltd, 1985.
25 M.J. Pitt and E. Pitt, *Handbook of laboratory waste disposal*, Chichester, Ellis Horwood, 1985.

Bibliography

F. Grover and P. Wallace, *Laboratory organisation and management*, London: Butterworths, 1979.

K. Guy, *Laboratory organisation and administration*, 2nd edn. London: Butterworths, 1973.

A.J.D. Cooke, *A guide to laboratory law*, London: Butterworths, 1976.

Handbook of laboratory safety, 2nd edn (N.V. Steere, ed.), Cleveland: The Chemical Rubber Co., 1971.

Manufacturing Chemists' Association General Committee, *Guide for safety in the chemical laboratory*, New York: Van Nostrand, 2nd edn, 1972.

Safety representatives and safety committees, London: Health and Safety Executive, 1977. This contains the Regulations on Safety Representative and Safety Committees, effective 1978, and the Code of Practice on Safety Representatives approved by the HSE, together with guidance notes on these.

Handling chemicals safely 1980, Amsterdam, Dutch Association of Safety Experts *et al.*, 2nd edn, 1980 (550 industrial chemicals).

Compendium of safety data sheets for research and industrial chemicals, L.H. Keith and D.B. Walters (eds.), Weinheim, VCH, 3 vols., 1985 (850 chemicals).

Chapter 4

Fire Protection

Introduction

Chemical laboratories vary greatly in size and operation, but most are production, research and development (including pilot plant), analytical or teaching laboratories, and each of these may present a variety of fire hazards. Therefore, each laboratory is unique and to implement fire protection an intimate knowledge of the working of the laboratory and the elements of fire prevention are required. Many research workers and laboratory staff, although highly trained in their particular fields are often unacquainted with the basis of fire prevention.

Managers must accept responsibility for providing a safe working environment, but those working in laboratories should appraise the fire hazards of their own particular work. If there is a safety committee, part of its task should be to establish what the fire hazards are. In a multi-occupancy building, a liaison committee should be set up for all the occupants to coordinate their fire precautions, because the spread of fire and smoke from one part of a building to another is the concern of all the occupants and if fire is to be contained or its spread delayed, concerted action is needed.

Action which may be required can vary from simple precautions, such as ensuring smoke stop doors are not jammed open, to complex precautions involving structural alterations. More elaborate precautions may require a detailed knowledge of the construction of the building and service systems in the building so that, for example, concealed spaces and ductwork do not permit fire or smoke to spread, perhaps unseen, vertically or horizontally through a building.

If fire should occur it may spread beyond the control of any one individual. Laboratory staff should, therefore, know how to summon the fire brigade, raise the alarm and summon other help.

The nature and source of the fire risk

The potential risk arises from the presence of combustible solids, liquids or gases in conjunction with ignition sources. One or more class is generally found in most laboratories.

Solids
Most combustible solids will not present a great fire risk unless they are ground into powder. Powders of combustible solids can be explosive when dispersed in air and if large quantities are ground to a fine state then precautions may have to be taken against dust explosion. The main hazard will be those solids which are unstable and are likely to decompose explosively if they are heated or subjected to friction or even excessive light. Other solid materials may be hazardous due to their oxidising properties so that a hazardous situation can result if they should become contaminated with combustible material.

Another class of hazardous solids is those which will react spontaneously and exothermally with water or air. Obvious examples are alkali metals, metal hydrides and certain organometallic compounds. Even aluminium powder can react with water and although the reaction is not very vigorous hydrogen—which can be exploded by a small spark—may accumulate.

Special methods of storage of these materials are called for and periodical inspections are necessary to ensure that these conditions are being maintained. For example, some unstable materials are kept damped down with water or a high flash point liquid. During a long period of storage there may be evaporation of the liquid and drying out so that the unprotected solid is exposed. When attempting to scoop out some of the unstable material friction with the scoop may then ignite it, leading to fire or explosion.

Liquids
The most common of the hazardous materials to be found in laboratories are flammable liquids and it is essential to know their fire properties, such as flash point and ignition temperature. If the flash point is below room temperature then these liquids will always constitute a fire hazard and careful control

43

should be maintained over them. Liquids with flash points well above room temperature will support a fire if heated to a temperature exceeding their flash point.

Before ignition can occur, a flammable vapour has to be heated, at least locally, to a temperature exceeding its ignition temperature. Almost invariably the ignition temperatures are well below those of common igniting sources such as flames, sparks and incandescent surfaces. For some materials ignition temperatures can be extremely low. Carbon disulphide, for example, will be ignited by a source whose temperature just exceeds that of boiling water, while ethers and aldehydes usually have low ignition temperatures. Other properties of liquids have to be taken into account such as their propensity to form more hazardous substances after long periods. Ethers are one such class of materials, and form highly explosive peroxides after long standing. The possibility of mutual interaction between flammable liquids should be considered when keeping them in laboratories or stores. Only the minimal amount of flammable liquids should be kept in laboratories. One day's supply is often recommended but this may be inconvenient or impracticable. However, the requirements of the Highly Flammable Liquids and Liquefied Petroleum Gases Regulations 1972 may limit the total quantity of highly flammable liquids allowed to be stored in the laboratory to 50 litres within suitable containment. Nevertheless, the quantity of flammable liquid should be kept to a minimum.

Bottles containing flammable liquids should be positively identified by labels which can withstand the deleterious effects of any atmosphere likely to be present in the laboratory. Sand-blasted labels would make loss of labelling impossible. Large containers of flammable liquids should never be carried in the hand and in particular Winchester bottles should not be carried by the neck. Suitable Winchester carriers are available and should be used for the transfer of liquids from one place to another. Only trained staff should refill bottles with flammable liquids in a special room or in a laboratory where all ignition sources have been removed.

Damaged glassware should not be employed in experiments as it may crack and spill the contents during the experiment. Poorly assembled or unsuitable apparatus may introduce serious fire risks. Badly fitting corks and bungs will introduce a fire risk and ground glass joints are usually to be preferred. The breakage of equipment by localised overheating using direct gas flames is a hazard which can be avoided easily by using water baths, hot plates, heating mantles or sand baths.

Compressed and liquefied gases
Compressed or liquefied gases present hazards in the event of fire since heating will cause the pressure to rise and may rupture the container. Leakage or escape of flammable gases can produce an explosive atmosphere within the

laboratory which can be ignited and result in a devastating explosion. Cylinders of gases should be provided with a suitable pressure regulating valve. The pressure at which the gas is to be used should be determined by this valve and not by operating the needle valve, as this leads to erratic control and the application of full pressure on any tubing should the exit of the tubing become blocked. Gas cylinders should preferably be placed outside and the gas piped into the laboratory. Properly designed compartments and reinforced walls should be used to protect personnel against possible explosion.

Structural protection and segregation

The concept of structural protection and segregation is to ensure that if a fire occurs, the performance of the elements of building structure will not be sufficiently impaired to reduce their ability to act efficiently. Segregation is used to divide up or separate hazardous operations from one another so that any incident does not spread the conflagration but contains it in a known compartment. This makes fire-fighting operations easier by not having to tackle the blaze on many fronts. The elements of building structure include walls, floors, columns, beams, ceilings or roofs. Consideration should also be given to doors and the protection of lobbies and stairways.

The floor of the laboratory should be impervious to chemicals. It may be considered necessary to place a sill at the door which should be provided with a ramp which is not too steep. There should be at least two means of escape from the laboratory and benches should be laid out so that people can escape easily if a fire occurs on the bench. This means that blind aisles should be avoided and the bench should more or less run parallel with the line of the exits. A space of at least 1.2 m should be provided between benches for passage of personnel and equipment. Reagent shelves should be situated on the bench so that it is unnecessary to lean across experimental apparatus in order to reach reagent bottles. Similar consideration should apply to services such as gas, electricity supply and water supplies and drainage.

If laboratories are situated in single storey buildings, then it may be sufficient to construct the buildings of noncombustible material. If the laboratories are situated in multi-storey buildings then walls, floors and ceilings should be of fire resisting construction. Fire resistance should be of at least two hours but will vary according to the risk in adjoining compartments. Columns and beams should also be protected to the same standard of fire resistance. Openings made in walls for the passage of ducts, pipes and cables should be fire-stopped. Stairways should be enclosed by fire-resisting walls. Exit doors should be hung to swing outwards.

The quantities of flammable materials in the laboratory shall be kept as small as possible. Where it is possible a separate building should be used for the storage of flammable liquids and those laboratories subject to the HFL and LPG Regulations 1972 will have certain legal obligations regarding quantities stored. This should be a single storey building constructed from non-combustible materials. The roof should be of light construction and easily shattered or blows off in the event of an explosion. If a separate building cannot be provided the store should be on the ground floor and should be of fire-resisting construction. It should be well ventilated and unheated and have doors which open outwards. There should be a sill on the doorway in such a storage compartment so as to contain any spillages which may occur accidentally. As far as possible separate stores should also be used for materials of differing hazards, and materials which react vigorously with one another should not be stored together (*see* p. 76).

Pilot plant experiments are carried out principally because operation on this scale presents the possibility of unforeseen results. Laboratories which are used for pilot plant work are particularly hazardous as larger quantities of materials will be used and special buildings or areas should be set aside for these purposes.

It is inevitable that flammable vapour or gases will be produced in the laboratory atmosphere from time to time. The best general precaution is to ensure that the laboratory is well ventilated. Windows and doors cannot be relied upon to provide adequate ventilation as they may well be kept closed much of the time and mechanical ventilation should be used. Any chemical operation which involves the possible production of flammable vapours should be carried out in a fume cupboard. The apparatus should be set up over a large metal tray to catch any flammable liquids which may escape due to accidents, such as breakage of equipment. The recommended air-flow velocity through the face of a fume cupboard is 30 m min^{-1}. The air-flow should be maintained by means of a fan and the fan motor should be placed outside the duct serving the fume cupboard, driving the fan by means of a shaft. The extraction duct should be made of non-combustible material and it is desirable that it should pass directly to the outside of the building without passing through ceilings. If, however, it must pass through one or more floors, or there is intercommunication with other fume cupboards, then careful consideration must be given to the control of spread of fire and toxic gases within the building. Several alternatives are available, *viz*:

1 The extraction duct-work should be of a fire-resisting construction.
2 The extraction duct-work should be enclosed in a fire-resisting structure.
3 Fire dampers may be fitted at positions where the duct passes through ceilings or partition-walls. However, the complexity of the problems associated with ducts intercommunicating and/or passing through buildings requires expert advice.

More detailed advice can be found in a British Standard Code of Practice.[1]

Controlling ignition sources

For a fire or explosion to occur it is necessary for a flammable or explosive atmosphere to exist and to have sufficient energy added to it. In the laboratory, as in other places, electric equipment, open flames, static electricity, burning tobacco, lighted matches and hot surfaces can all cause ignition of flammable material.

Gas supplies to laboratory outlets should be by means of rigid permanent piping. A laboratory may have many outlets for gas supply and in this case a control valve should be placed just outside the laboratory so as to be able to cut supplies off in the case of an emergency.

Electric wiring should, as far as possible, be of a permanent nature and installed in accordance with the latest edition of the IEE regulations.[2] Switches, sockets and terminals should be placed where they are easily accessible and are safe from from accidental wetting by water or other liquids. Temporary wiring should be installed by a competent electrician or at least inspected by one before use. High-sensitivity current-operated earth-leakage circuit breakers can provide an excellent means of protection against fires originating from earth leakage faults, as well as protection against line-to-earth shocks.

Where it is likely that large amounts of flammable gases or vapours may be released into the atmosphere then it is necessary to install electrical equipment designed for use in explosible atmospheres. By various methods of design and construction electrical equipment suitable for employment in potentially flammable atmospheres can be manufactured. Probably the most well known of these are flameproof types for high power consumption equipment and intrinsically safe types where equipment requires only very low power for operation.

Flameproof equipment is not manufactured for all types of flammable vapours and gases and it may be necessary to use pressurised (purged) equipment. Care should be taken when using flameproof equipment that additions are not made which are below flameproof standard. A common mistake is to use domestic refrigerators for storage of materials and explosions have occurred when low flash point liquids have been placed in such refrigerators. Drying ovens are also a likely source of ignition and it should be remembered that, although the controls may be flameproof, the surface temperature may exceed the ignition temperature of some vapours. The many types of electrical equipment for use in potentially flammable atmospheres and their applicability to particular locations are summarised in the FPA Data Sheet[3] and given in detail in British Standard Specifications and Codes of Practice.[4,5] Thermostats on oil baths have often failed with consequent overheating and fire. Independent excess-temperature manual-reset cutouts should be incorporated to disconnect the supply if overheating should occur.

The use of organosilicon fluids in oil baths has much to commend it as they are highly stable and can have boiling points in excess of 400 °C. There are a number of general precautions which can be taken. Hotplates, furnaces and ovens should stand on heat resisting surfaces. Where the heated unit is on a wooden bench top there should be an air space between it and the bench to prevent charring. Gas jets not in use should be turned off or adjusted to give a small luminous flame as the pale blue flame cannot be seen easily in bright sunlight.

The discharge of accumulated static electricity can provide a spark which will ignite flammable vapours. Static electricity is created by the relative movement of two materials. Non-polar materials such as hydrocarbon solvents accumulate static charges readily as they have high insulation values and do not allow the charge to leak away. It should be noted that a dispersion of an immiscible polar liquid in a non-polar liquid can generate static charges even more rapidly than a non-polar liquid alone. Some improvement can be obtained by the use of a small proportion of a conductive additive where this would not affect the chemical properties of the fluid. Even crystallisation can produce static electricity, and the discharge of carbon dioxide has been known to produce static sparking. Laboratory coats and clothing made of certain synthetic fibres are prone to generate static electricity and should be avoided. Where possible all metalwork should be bonded and earthed when there is a static hazard. When no solution can be found to the static problem then all processes must be carried out as slowly as possible to give the accumulated charge time to disperse.

Space heating should be safe and adequate. Hot water or low pressure steam radiators are desirable and installation and maintenance should be first class so as to reduce the likelihood of the introduction of uncontrolled and dangerous forms of space heating. Radiators should not be used for drying materials and should be provided with sloping wire mesh screens over them to prevent such abuse.

Good housekeeping will help to reduce the number of fires in laboratories in addition to reducing other types of accidents. Cleaning up as soon as possible after an experiment has been completed will help and the provision of adequate storage space beneath the bench is an advantage. A clear bench will enable the experimenter to see any dangerous procedure without being distracted by unnecessary equipment. The provision of suitable waste bins, preferably of non-combustible material such as sheet iron, will help to encourage a positive attitude to housekeeping.

Emergency action

In the event of fire or explosion occurring there should be a prearranged plan of the necessary action to be taken. All personnel must be made aware of this and fire drills should be carried out at least twice a year in order to familiarise staff with these procedures. The essential elements which should be covered by instruction are:

1 Raising the alarm.
2 Summoning the local fire brigade.
3 First aid fire fighting (including practice in use of extinguishers).
4 Evacuation.

Raising the alarm
Personnel should be trained to recognise the severity of an outbreak of fire and the immediate danger presented by it. They should be able to report accurately on the fire situation, being able to discern whether a normal procedure or emergency procedure should be brought into operation. In the normal procedure the person discovering the fire may decide that it is relatively minor. He should report the fire immediately to the switchboard or instruct some other person to do so. A person in a position of responsibility should decide if it is necessary to evacuate the building. Meanwhile the switchboard operator should call the fire brigade and notify all persons who have been allocated special responsibility.

Emergency procedure is necessary when the fire has been found to have spread over a wide area or hazardous materials and processes are threatened. In such circumstances the premises must be evacuated immediately and the manual fire alarm system should be operated by the person discovering the fire. Then, if possible, the switchboard operator should be notified. The switchboard operator should have standing instructions to call the fire brigade on hearing the fire alarm unless the alarm is connected to the fire brigade *via* a central alarm depot, since there are virtually no direct connections.

Summoning the fire brigade
When the fire brigade is called by the switchboard operator it should be informed at the time which entrance is nearest to the fire. In all cases responsible persons should be sent to the entrance to give the brigade information it may need on the location and extent of the fire, water supplies, the nature of special risks and details of casualties and trapped persons. If any particularly hazardous materials or processes have been brought into the laboratory the fire brigade should be forewarned. Ideally the fire brigade should have prior knowledge of the hazards through a liaison or safety officer who should be responsible for informing the local fire authority on hazards at the time they are introduced into the laboratory (*see* Chapter 3, p. 22).

First aid fire fighting
Provided that no danger is involved, the fire should be attacked with first aid

fire fighting equipment as soon as it is discovered. It must be decided on relative merit whether attacking the fire takes precedence over reporting it. It can be dangerous to waste time trying to tackle a fire which cannot be controlled instead of immediately reporting the incident; on the other hand it is undesirable to allow a small fire to obtain a hold through spending time reporting it. In most circumstances there is no conflict, for there is normally more than one person near the scene of the outbreak, and one person can report the fire leaving the others to try to extinguish or contain it. On returning that person can aid the other personnel.

Fire fighters should withdraw from the scene if the heat and smoke threatens to overcome them or if the fire endangers their escape route. They should also retreat if the fire spreads towards explosive materials or gas cylinders. On withdrawing, windows and doors should be closed in an effort to contain the fire.

Evacuation of personnel

The essential features of evacuation are that it should proceed by a pre-arranged plan and that all personnel should be familiar with the escape routes to be used. It should be impressed on everybody that they should leave, without panic, immediately they have received instructions to evacuate the building. The assembly point should be fixed and known to all personnel, and the head of each department (or his deputy) should be responsible for ascertaining that all persons in his charge, including visitors, have been accounted for. To ensure that this is thoroughly carried out, one person must be delegated the responsibility of searching the department, including the lavatories and cloakrooms to ensure that nobody has been left behind. No person should attempt to re-enter the building without the expressed permission of the person taking the roll call. Escape routes should be clearly marked as such.

All responsibilities must have been allocated in advance in order to avoid delay and doubt. Moreover, each of the persons bearing responsibilities should designate deputies who could take over in the event of absence, illness or accident. This delegation of responsibilities could easily be made to coincide with that of the delegation of normal administrative duties.

Classes of fire and suitable extinguishing agents

It is standard international practice to classify fires according to their nature.

Class A fires are those involving solid materials, usually of a carbonaceous nature, but excluding materials that readily liquefy on burning.

Class B fires are those involving liquids or liquefiable solids.

Class C fires are those involving gases (but electrical equipment in the USA).

Class D fires are those involving burning metals, *e.g.* sodium, potassium, aluminium, titanium, magnesium and calcium.

For Class A and B fires portable hand extinguishers have been available for many years based on their content of suitable extinguishing agent, *e.g.* 9 l water and 0.9 kg dry powder. The current practice is to rate a particular extinguisher according to its ability to extinguish a standard fire under standard test conditions. Thus an extinguisher with a rating of 5A is capable of extinguishing a wood crib fire (Class A) having a cross-section of 0.5 m. A rating of 13B indicates that a Class B test fire of 13 litres of the test fuel has been extinguished.

The introduction of the new British Standard[6] in 1980 means that for some time acceptance of extinguishers will be based on contents until the older ones have obtained a rating or are given an assessed rating.

For a fire involving solid combustible material (Class A) such as wood, paper or textiles, water is the most effective extinguishing agent when used as a jet or spray. The cooling power of water is unsurpassed by other media and is especially useful on materials which are likely to re-ignite and for penetration into deep-seated fires. Portable water extinguishers are usually of 9 l capacity but smaller 4–6 l extinguishers which are particularly suitable for use by women are available. Small hose reels which should be at least 19 mm (3/4 in) inside diameter and capable of delivering at least 30 l min^{-1} give a practically inexhaustible supply.

Flammable liquid fires (Class B) cannot normally be extinguished with water. Dry powder, foam, carbon dioxide and vaporising liquids can be effective. Dry powder rapidly extinguishes the flames over burning liquid and gives a quick 'knock down'. It is particularly effective in the case of spill fires, but gives little protection against re-ignition. It can be safely used on electrical equipment as dry powder is non-conductive. The capacity of portable dry powder extinguishers ranges from about 1–12 kg.

Foam extinguishes a fire by forming a blanket which floats on the surface of the liquid preventing the access of air or the escape of vapour. The foam blanket remains in position for sufficient time to allow the liquid and surrounding to cool, so preventing re-ignition. Foam is particularly effective in dealing with fires in containers which have become overheated and is more effective than dry powder. It is impossible to form a foam blanket over liquids flowing down a vertical surface and difficult when liquids are flowing freely over a horizontal surface. Polar liquids break down the normal protein-based foam. An alcohol resistant foam has to be employed in fighting fires involving alcohol, acetone *etc.* Foam is electrically conductive and should not be used when electrical equipment is involved in the fire. Foam extinguishers are usually 9 l capacity which will produce approximately eight times this volume of foam.

Carbon dioxide smothers flames mainly by excluding oxygen. It is effective in situations where it is not easily dispersed, and so can be used on fires in refrigerators and ovens. It can also be used safely where there is danger of electric shock and where delicate equipment is involved since it leaves no residue. The extinguishers come in various sizes ranging from 1–6 kg in capacity.

Halons are vapourising liquids containing a halogen which acts by inhibiting the flame reactions. A halon is defined as a 'halogenated hydrocarbon used as an extinguishing medium' and is given a number where the first digit is the number of carbon atoms in the molecule, the second digit is the number of fluorine atoms in the molecule, the third digit is the number of chlorine atoms in the molecule and the fourth digit is the number of bromine atoms in the molecule. Halons work very effectively and are safe to use on electrical equipment, but some are very toxic and their use is not permitted. The principal ones in current use are:

Halon 1211 Bromochlorodifluoromethane (BCF)
Halon 1301 Bromotrifluoromethane (BTM) in fixed systems
Halon 1011 Chlorobromomethane (CBM)

All the halons produce toxic products on pyrolysis and they should not be used, or kept in confined spaces or any place where there is a risk that the vapours or their products could be inhaled. The capacity for hand halon extinguishers is 0·5–1 l.

Fires involving gases (Class C) may prove to be the most difficult to deal with, since the extinction of the burning gas, other than by cutting off the supply at source, will allow gas to build-up. If the gas is flammable a dangerous gas/air mixture may be produced which on reaching an ignition source will result in an explosion or re-ignition. A complete appraisal of the necessary actions to be taken in the event of a gas fire must be produced by the safety committee and the laboratory staff must be fully briefed on these actions.

Gas can be piped into a laboratory from a supply which is usually a town-main, bulk store or cylinders. Often individual cylinders are used within the laboratory. When a piped supply is used a shut-off valve, or valves, should be incorporated in the system so that isolation of the supply can be achieved safely. Shut-off valves in easily accessible positions on both sides of the wall through which the supply is piped are the simplest solution.

Where individual cylinders are involved no attempt to extinguish the burning gas should be made unless it is certain that the cylinder can be turned off safely. If not it is better to extinguish the surrounding fires caused by the burning gas and keep the cylinder cool by spraying water over it and the surrounding area. Cylinders of inert gases, as well as flammable ones, when heated strongly present a potential explosion hazard.

Class D fires deal with special risks. Some of the substances used in laboratories such as sodium, potassium and metal hydrides will react with extinguishing agents and it would be ineffectual or even dangerous to attempt to use these agents. Special dry powders are available to deal with such hazards, and manufacturers and suppliers of special risk chemicals will be able to assist in the selection of the appropriate dry powder for these substances.

Selection and distribution of extinguishers
A multiplicity of extinguishers is undesirable and probably all risks can be covered by two or three types. It is valuable to obtain extinguishers which are uniform in their method of operation; all personnel should be instructed in the use of these appliances and which type to use. Extinguishers should conform with British Standard Specifications[7] and be approved by the Fire Offices' Committee.[8] Further advice on choice of extinguishers may be obtained from the local fire authority or insurer.

The number and size of the extinguishers required will vary according to the risk but there should be at least one of each required type in each laboratory. In large laboratories there should be a minimum of one water-type extinguisher for every 200 m^2 of area and the total should be not less than 18 l capacity on every floor. Further information regarding the use of water and other types of extinguishers can be found in the FPA Data Sheets.[9, 10]

Fixed fire extinguishing systems
Large laboratories and pilot plant areas can be protected by means of fixed installations. The system used can be automatic or manually operated and any type of agent can be employed in the system.

Hydrants, hose reels or automatic sprinklers can provide general protection. Hydrants are advisable for premises covering a large area, remotely situated areas and tall buildings. Hose reels have already been mentioned. They are permanently connected to the water supply, simple to use and very effective for tackling fire at an early stage. Further details are given in the FPA Data Sheets.[11, 12] Although water is the extinguishing agent, automatic sprinklers are acceptable in many laboratories. It is advisable to link the automatic sprinkler alarm to the local fire brigade. As the sprinkler heads open automatically in response to elevated temperature the detection and operation is completely automatic. Automatic sprinklers are the most effective form of automatic protection (*see Rules of the Fire Offices' Committee for automatic sprinkler installations*, 29th edn).[13]

Protection for special risks can be given by water spray, foam, inert gas and dry powder. Water spray systems consist of pipes with high or medium pressure outlets projecting sprays of a predetermined droplet size. This system is suitable for protection of large scale flammable liquid and liquefied gas risks. Such a system is not only used to extinguish a fire but to protect plant, by cooling, from a nearby fire. Foam can be applied to a given risk by fixed pipework connected either to a self-contained foam generator or an inlet to which the fire brigade can connect their foam-producing equipment. Carbon dioxide or other inerting gases can be directed into plant or rooms by pipework. Most of the gases used are primarily suitable for flammable liquids, electrical equipment, valuable water-sensitive equipment and water reactive chemicals. The gas is delivered through pipes and the system involves automatically closing doors and ventilation ducts. Warning bells have to be sounded so as to warn occupants to make their escape as inert gases will asphyxiate them. If a total flooding system is contemplated, consultation at the design stage with the Health and Safety Executive and the insurers is strongly recommended.

A dry powder system consists of a dry powder container which is coupled to a gas cylinder and pipework and outlets. The pressurised gas drives the powder to the outlets. The system is suitable for flammable liquids, electrical equipment and materials where it is essential to avoid water contamination.

Working practice

Considerable amounts of heat are released during some reactions. This is obvious in acid-base neutralisation or oxidation reactions and can often be predicted when the mechanisms of the expected reaction are worked out beforehand. Such considerations may be second nature to experienced researchers but may not be so apparent to students and inexperienced laboratory assistants who should only undertake such work under the supervision of an experienced person.

When new work is being attempted, full consideration should be given to the possible dangers it presents. In these cases, before proceeding on the scale desired a very small scale experiment should be carried out to determine whether a large amount of heat is liberated or gases are produced or an unusually vigorous reaction occurs. These factors, although not serious on a small scale, may become extremely important in a large scale experiment. On scaling up a reaction heat loss will only increase according to a square law while the volume of reactants and heat produced will accord to a cube law. Additional means of cooling, such as cooling coils, may be required to take care of the extra heat produced.

The rate at which one reagent is added to another affects the rate at which heat is produced. In the case of a known exothermic reaction the reactants or reagents should be added as slowly as possible with adequate stirring to make sure that the heat is liberated slowly. One way in which too much reactant may be added at one time is when it is added to an overcooled mixture and accumulates in the reaction mix where the rate of reaction is artificially low. It then only requires a small increase in temperature, either due to the mixture warming up normally or to the gradual accumulation of heat of reaction, for a runaway condition to occur, leading to disastrous results. A better course is to allow the reaction to take place at a temperature where it can proceed at a suitable speed and add the reactant slowly with adequate stirring to make sure that it is consumed at the same rate at which it is added.

Flammable vapours can often escape into the atmosphere because they are not condensed as rapidly as they are produced. A balance must be struck between the rate at which the liquid is heated and the cooling provided by a water-cooled condenser.

Experiments should not be left unattended and if the person in charge of an experiment is called away he should ensure that whoever takes over is fully aware of the dangers involved and the precautions to be taken. No experiment should be left running at night unattended unless it can be ensured that somebody will come round and inspect it at regular intervals.

The disposal of waste materials calls for special attention. There have been fatal accidents during the disposal of waste materials and the disposer should be fully aware of the risks involved. The overriding consideration should be not to try to dispose of too much material at any one time. This is particularly important in the case of heavy metals such as silver and mercury which can form explosive compounds of which fulminates are a well-known example. Such by-products can be produced accidentally without the intentions of experimenters (*see* Chapter 5, p. 76) and full knowledge of the conditions under which explosive compounds can be produced should be available to such persons.

Untreated flammable liquids should not be disposed of down the drain, apart from *very* small quantities of water-soluble solvents. Limited spillage of flammable water insoluble solvents may be disposed of by the dispersion method described in Chapter 8. Larger quantities of flammable liquids should preferably be collected in sealed metal containers and recovered or disposed of in a safe fashion, for example by burning in a shallow metal tray in the open air. Drains should be properly trapped and vented and they should preferably discharge into an industrial waste sewer rather than a sanitary sewer. There are special designed incinerators for burning flammable liquids and if one attempts to employ this method a full investigation should be made of the types of material it can handle.

Materials which are known to decompose spontaneously and explosively should be kept in a safe manner until they can be disposed of safely. Waste material should be removed daily and destroyed or disposed of in a safe manner. Filter papers, residues and wiping cloths which have been in contact with unsaturated oils should be kept in covered receptacles of limited size to ensure frequent disposal of the contents. Bins used for disposal should be labelled appropriately and clearly so that no mistake is made in introducing the wrong type of material into the bin. Benches and glass apparatus should not be cleaned with flammable solvents after experiments.

Legislation

All statutory requirements cannot be covered in a survey of this nature, but it can show the range of legislation. The original legislation should be consulted by interested parties since no attempt is made here to interpret the requirements. An extensive guide to relevant legislation is contained in BS 5908.[1]

The legislation concerned with fire and explosion in laboratories is mainly contained in the Factories Act 1961 when laboratories are concerned with process control, and the Petroleum (Consolidation) Act, 1928 and Regulations and Byelaws made under the Acts. The Highly Flammable Liquids and Liquefied Petroleum Gases Regulations 1972 (SI 1972: No. 917) made under the Factories Act regulate the use and storage of such materials.

The Health and Safety at Work *etc* Act 1974 has far-reaching implications in that it enables regulations to be made to cover premises previously exempted from other legislation. It has transferred the implementation of the fire precautions requirements in the Factories Act, 1961 to the Fire Precautions Act, 1971 except for special premises, which are covered by the Fire Certificate (Special Premises) Regulations, 1976 (SI 1976: 2003). In particular the following sections of the Factories Act have been repealed and incorporated in the Fire Precautions Act:

Sections 40–47—Provision and maintenance of means of escape.
Section 48—Safety in case of fire.
Section 49—Escape instruction.
Section 50—Enabling special regulations for fire protection to be made.
Sections 51–52—Provision, maintenance and testing of fire fighting equipment.

In addition to statutory requirements, consideration has to be given to Common Law duties of occupier and employer.

There is a strict duty in Common Law in respect of injury to one's neighbour's property by fire and to employees in respect of the need to provide safe means of access, a safe place of work, a safe system of work and competent supervision; an obligation which almost certainly extends to fire protection measures. Common Law affects laboratories even if they are outside the scope of statutory laws.

The Factories Act, 1961

The Factories Act deals with many matters having a direct bearing on fire protection. These are fire drill, fire warning and means of escape as well as fire extinguishing equipment and fire prevention matters. The following sections of the Act are of special interest:

Section 31 Provisions for plant employing flammable gases, vapours and dusts
Section 54 Dangerous conditions and practices

Sections 80 and 81 Notification of accidents and dangerous occurrences
Section 148 Persons empowered to inspect premises
Sections 120, 155, 160 and Second Schedule Persons held responsible for compliance with the Act

Regulations made or deemed to have been made under the Factories Act, 1961
Certain processes and materials are considered to be extra hazardous and extra legislation has been introduced to deal with these. The requirements of such legislation can be considered supplementary to the requirements of the Factories Act.

The Chemical Works Regulations (SR & O 1922: No. 731)
In these comprehensive Regulations there are requirements for electrical installations and the prohibition of naked lights, matches *etc*, and special precautions for heating installations in any place where there is danger of an explosion from or ignition of flammable gas, vapour or dust. Notices prohibiting smoking and the presence of naked lights, matches *etc*, must be fixed at the entrance to any room or place where there is risk of explosion from flammable gas, vapour or dust.

The Electricity Regulations, 1908 (SR & O 1908: no. 1312) *as amended by the Electricity (Factories Act) Special Regulations* (SR & O 1944: No. 739).
The principal Regulation relating to fire is Regulation 27 of the Electricity Regulations, 1908, which requires that all conductors and apparatus exposed to the weather, wet, corrosion, flammable surroundings or explosive atmosphere, or used in any special process shall be so constructed or protected, and such precautions shall be taken as may be necessary adequately to prevent danger in view of such exposure or use.

The Highly Flammable Liquids and Liquefied Petroleum Gas Regulations 1972 (SI 1972: No. 917)
These Regulations impose requirements for the protection of persons employed in factories and other places to which the Factories Act 1961 applies, in which any highly flammable liquid or liquefied petroleum gas is present for the purpose of, or in connection with, any undertaking, trade or business.

Regarding highly flammable liquids, the Regulations contain requirements as to the manner of their storage, the marking of storage accommodation and vessels, the precautions to be taken for the prevention of fire and explosion, the provision in certain cases of fire-fighting apparatus and the securing in certain cases of means of escape in case of fire. Regarding liquefied petroleum gases, the Regulations contain requirements as to the manner of their storage and the marking of storage accommodation and vessels.

The Petroleum (Consolidation) Act, 1928
The Act requires premises used for the bulk storage of petroleum spirit, defined as petroleum which evolves a flammable vapour at less than 73 °F tested as prescribed, to be licensed and empowers licensing authorities to attach to licences such conditions as are necessary to ensure the safe keeping of the petroleum spirit.

The Secretary of State is empowered to make Regulations governing the conveyance of petroleum spirit by road, and the keeping, use and supply of petroleum for use in vehicles, motor boats and aircraft. In fact for the latter use certain exemptions are made by the Petroleum Spirit (Motor Vehicles) Regulations (SR & O 1929: No. 952). Provision is made to enable Regulations to be made under the Act which can extend the Act or portions of the Act to substances other than petroleum spirit.

Petroleum (Transfer of Licenses) Act, 1936
This Act empowers local authorities to transfer petroleum spirit licences granted under the Act of 1928.

Orders and Regulations made under the Petroleum (Consolidation) Act, 1928.
The Petroleum (Mixtures) Order, 1929 (SR & O 1929: No. 993)
This applies the whole of the Petroleum (Consolidation) Act to all mixtures of petroleum with any other substances, which possess a flash point below 73 °F.

The Petroleum (Carbon Disulphide) Order, 1958 (SI 1958: No. 257)
This applies the requirements of the 1928 Act in respect to labelling and conveyance by road to carbon disulphide. This Order is modified by The Petroleum (Carbon Disulphide) Order, 1968 (SI 1968: No 571).

The Petroleum (Compressed Gases) Order, 1930 (SR 1930: No. 34)
Air, argon, carbon monoxide, coal gas, hydrogen, methane, neon, nitrogen and oxygen when compressed into metal cylinders are brought under certain sections of the 1928 Act.
Other relevant regulations are:
The Gas Cylinder (Conveyance) Regulations, 1931 (SR & O: No. 679)
The Gas Cylinder (Conveyance) Regulations, 1947 (SR & O: No. 1594)
The Gas Cylinder (Conveyance) Regulations, 1959 (SI 1959: No. 1919)
However, compressed gases used for powering motor vehicles are subject to The Compressed Gas Cylinders (Fuel for Motor Vehicles) Regulations, 1940 (SR & O 1940: No. 2009).

The Petroleum (Organic Peroxides) Order, 1973 (SI 1973: No. 1897)
Specified organic peroxides and certain mixtures or solutions of them are brought under some sections of the 1928 Act.

The Petroleum (Inflammable Liquids) Order, 1971 (SI 1971: No. 1040)
This applies certain sections of the 1928 Act to a large number of flammable liquids, other than petroleum derivatives with a flash point below 73 °F, and also to certain other substances which are considered hazardous. The liquids and other substances are set out in Parts I and II of the Schedule attached to the Order.

Explosives Acts, 1875 and 1923
Explosive Substances Act, 1883
These Acts, and Orders and Regulations made under them, govern the manufacture, sale, importation and conveyance by road of all explosives.

Acetylene
    Order of the Secretary of State No. 5 dated 28.3.1898
    Order of the Secretary of State No. 5A dated 29.9.1905
These provide that under certain conditions compressed acetylene in admixture with oil–gas is not deemed to be an explosive within the meaning of the Act.
    Order of the Secretary of State No. 9 dated 23.6.1919
This provides that when acetylene is contained in a homogeneous porous substance with or without acetone or some other solvent and provided certain prescribed conditions are fulfilled shall not be deemed to be an explosive within the meaning of the Act.
    Order in Council No. 30 dated 2.2.1937 as amended by
    The Compressed Acetylene Order, 1947 (SR & O 1947: No. 805)
This prohibits the keeping, importation, conveyance and sale of acetylene compressed to over 9 psig. Certain exceptions are made among which is that acetylene at pressure not over 22 psig may be manufactured and kept under conditions approved by the Secretary of State. Subject to certain conditions, acetylene at pressures not exceeding 300 psig and not mixed with air or oxygen can be used in the production of organic compounds.
    The Compressed Acetylene (Importation) Regulations, 1978 (SI 1978: No. 1723).
These Regulations made under the Health and Safety at Work *etc* Act, indicate that acetylene at a pressure between 0.62 bar and 18.0 bar cannot be imported into the UK unless the HSE allows it under licence.

Other Acts and Regulations which may be applicable to laboratories in chemical works

Fire Precautions Act, 1971
Premises to which this Act can apply include those used for purposes of teaching, training or research.

The Public Health Act, 1961
Under this Act there have been made the very important Building Regulations
1972 (SI 1972: No. 317). These Regulations have far reaching provisions
determining height, floor area, cubic capacity and siting, and fire resistance of
structural elements of new buildings. Scotland and Inner London Area have
their own legal requirements.

Radioactive Substances Act, 1948
Radioactive Substances Act, 1960
The Ionizing Radiations (Sealed Sources) Regulations, 1969 (SI 1969: No. 808)

The Ionizing Radiations (Unsealed Radioactive Substances) Regulations, 1968
(SI 1968: No. 780)
The Ionizing Radiations Regulations (1985) and its accompanying Approved
Code of Practice (1985) now include the substance of the above 4 documents
and supercede them.

The Classification, Packaging and Labelling of Dangerous Substances Regula-
tions, 1984 (SI 1984: No. 1244).
The Regulations were introduced as the result of EEC directives. They are to
be applied in conjunction with the Authorised and Approved List: *Informa-*
tion approved for the classification, packaging and labelling of dangerous
substances for supply and conveyance by road. Health and Safety Commission.
London: HMSO, 1984. Approximately 2000 chemicals and mixtures appear
in the list. The Regulations recognise substances as being toxic, harmful,
explosive, corrosive, irritant, oxidising or highly flammable.

Notification of New Substances Regulations, 1982 (SI 1982: No. 1496).
These Regulations make provision for the notification of dangerous substances and their classification according to the type and degree of danger they
present.

Insurers' approach to fire protection

As previously advocated, consultation should take place with the fire insurers
at the design stage of the laboratory. Whilst appreciating that safety of life is
the prime object of fire protection, fire insurers are principally concerned with
material and consequential loss which may arise from damage to buildings
and contents. The fire authority approach will be to ensure that legal
requirements in relation to the safety of life are complied with, *e.g.* that there
are sufficient means of escape provided and that the construction is capable of
containing the fire long enough to allow the personnel to escape.

In some instances, insurers and brigades may put forward requirements in which the emphasis differs. In general, however, there is close liaison and difficulties are usually resolved and a solution is found acceptable to both. This removes the dilemma of management having to make a decision between two differing requirements. In most cases the fire insurer's requirements will be the more stringent, since they start from the basis that there is a statistical probability that a fire will occur at a particular risk some time. Their philosophy is to apply compartmentation by having the structure suitably divided by a fire resisting construction to contain a fire within a compartment to allow additional fire fighting equipment to be brought into use if necessary. By this means the premises at risk are sub-divided and they accept the possible loss of a portion of the premises, but seek to prevent any extension of the damage. In recognition of the fact that the extent of fire damage may be limited by the presence of suitably approved automatic sprinkler systems,[13] fire alarm systems,[14] or portable fire extinguishing equipment, the insurer may grant an allowance in the form of a reduced premium. The value fire insurers attach to automatic sprinkler systems is illustrated by the generous allowances which are given for this type of protection.

References

1 *Code of practice for fire precautions in chemical plant.* BS 5908. 1980.
2 *Regulations for electrical installations,* 15th edn. London: The Institution of Electrical Engineers, 1981 (revised 1986).
3 *Flammable liquids and gases: electrical equipment* (revised 1983). Fire Safety Data Sheet No. 6014. London: Fire Protection Association.
4 *Electrical apparatus for potentially explosive atmospheres.* BS 5501.
5 *Code of practice for the selection, installation and maintenance of electrical apparatus for use in potentially explosive atmospheres (other than mining applications or explosive processing and manufacturing).* BS 5345.
6 *Code of practice for fire extinguishing installations and equipment on premises,* Part 3. *Portable fire extinguishers.* BS 5306, 1980.
7 *Specification for portable fire extinguishers.* BS 5423, 1977.
8 *List of approved products and services.* London: Fire Offices' Committee.
   Volume 1 *Automatic sprinklers*
             Fire alarms   } May 1986
             Portable fire extinguishers
   Volume 2 *Approved firebreak doors*   March 1986.
             and shutters
9 *Portable fire extinguishers.* Data Sheet PE4. London: Fire Protection Association.
10 *First-aid fire fighting: training.* Data Sheet PE5. London: Fire Protection Association.
11 *Fixed fire-extinguishing equipment: hose reels.* Fire Safety Data Sheet No. PE7. London: Fire Protection Association.
12 *Fixed fire-extinguishing equipment: hydrant systems.* Fire Safety Data Sheet No. PE8. London: Fire Protection Association.
13 *Rules of the Fire Offices' Committee for automatic sprinkler installations* (29th Edn). London: Fire Offices' Committee, 1973.
14 *Rules of the Fire Offices' Committee for automatic fire alarm installations* (12th Edn). London: Fire Offices' Committee, 1985.

Bibliography

Fire and related properties of industrial chemicals. London: Fire Protection Association 1974.

N.I. Sax, *Dangerous properties of industrial materials*, 6th edn. New York: Reinhold, 1984.

R.R Young and P.J. Harrington, *Design and construction of laboratories.* (Lecture series 1962, No. 3). London: Royal Institute of Chemistry, 1962.

Guide to fire prevention in the chemical industry. London: Chemical Industries Association, 1968.

D.D. Libman, 'Safety in the chemical laboratory', *Laboratory Equipment Digest*, October, 1967.

Mathew M. Braidech, 'Fire and explosion problems in laboratories and pilot plants', *J. Chem. Educ.* 1967, **44**, A319.

N.V. Steere (ed.), *Safety in the chemical laboratory*, Vol. 1 1967, Vol. 2 1971, Vol. 3 1970, Vol. 4 1981. Easton, Penn: Chemical Education Publishing, 1967.

P.J. Gaston, *The care, handling and disposal of dangerous chemicals.* Aberdeen: Northern Publishers, 1970.

Safety measures in chemical laboratories, 4th edn. London: HMSO, 1981.

H.A.J. Pieters and J.W. Creyghton, *Safety in the chemical laboratory*, London: Butterworth, 2nd edn, 1957.

The General Safety Committee of the Manufacturing Chemists' Association Inc. *Guide for safety in the chemical laboratory*, New York: Van Nostrand, 2nd edn., 1972.

Fire Booklist. London: Fire Protection Association, 1985.

The storage of highly flammable liquids. HSE Guidance Note CS2. London: HMSO.

Chapter 5

Reactive Chemical Hazards

All chemical reactions involve changes in energy, usually evident as heat. This is normally released during exothermic reactions, but occasionally may be absorbed into the products in endothermic reactions, which are relatively few in number.

Reactive chemical hazards invariably involve the release of energy in a quantity or at a rate too great to be dissipated by the immediate environment of the reacting system, so that destructive effects appear. To try to eliminate such hazards from chemical laboratory operations, attempt to assess the likely degree of risk involved in a particular operation and then plan and execute the operation in a way which will minimise the risks foreseen. It may, of course, be necessary to accept that a certain degree of risk is likely to be attached to a particular course of action, but in this case, personal protection appropriate to the risk will be necessary.

This chapter is concerned with various aspects of the recognition and assessment of reactive hazards, of practical techniques for reaction control, and of personal protection where this is deemed necessary. Several references to sources of more detailed information are given by superscript numbers in the text, and where appropriate, page references in parentheses are also given to a recently updated monograph on reactive hazards.[1]

Physicochemical factors

Many of the underlying causes of incidents and accidents in laboratories which have involved unexpected violent chemical reactions are related to a lack of appreciation of the effects of simple physicochemical factors upon the kinetics of practical reaction systems. Probably the two most important of these factors are those governing the relationship of rate of reaction with concentration and with the rise in temperature during the reaction. The latter is the major factor.

Concentration of reagents

It follows from the law of mass action that the concentration of each reactant will directly influence the velocity of reaction and the rate of heat release.

It is, therefore, important not to use too-concentrated solutions of reagents, particularly when attempting previously untried reactions. In many preparations 10 per cent is a commonly used level of concentration where solubility and other considerations will allow, but when using reagents known to be vigorous in their action 5 per cent or 2 per cent may be more appropriate. Catalysts are commonly employed at these or even lower concentrations.

Of the many cases where increasing the concentration of a reagent, either accidentally or deliberately, has transformed a safe procedure into a hazardous event, three examples will suffice. When concentrated ammonia solution was used instead of the diluted solution to destroy dimethyl sulphate, explosive reaction occurred (297). Omission of most of the methanol during preparation of a warm mixture of nitrobenzene and sodium methoxide led to rupture of the containing vessel (580). Catalytic hydrogenation of *o*-nitroanisole at 34 bar (34×10^5 Pa) under excessively vigorous conditions (250 °C, 12 per cent catalyst, no solvent) ruptured the hydrogenation autoclave (678).

Reaction temperature

According to the Arrhenius equation, the rate of a reaction will increase exponentially with increase in temperature, and in practical terms an increase of 10 °C roughly doubles the reaction rate in many cases. This has often been the main contributory factor in cases where inadequate temperature control had caused exothermic reactions (normal, polymerisation, or decomposition) to run out of control.

An example relevant to the first two reaction types is the explosive decomposition which occurred when sulphuric acid was added to 2-cyano-2-propanol with inadequate cooling. Here, exothermic dehydration of the alcohol produced methacrylonitrile, acid-catalysed polymerisation of which accelerated to explosion, rupturing the vessel (1165). An example of an exothermic decomposition reaction is the violent explosion which occurred during storage of *m*-nitrobenzenesulphonic acid at ∼150 °C under virtually adiabatic conditions. It was subsequently found that exothermic decomposition of the solution set in at 145 °C, and the pure acid decomposed vigorously at 200 °C (582–3).

Many other examples of various types of hazardous and unexpected reactions have been collected and classified.[1,2]

Operational considerations

Effective control is essential to minimise possible hazards associated with a particular reaction system and, to allow you to achieve such control, relevant knowledge is necessary for you to assess potential hazards in the system.

Try to find what is already known about the particular procedure or reaction system (or a related one) from colleagues, or from existing literature.[1-4]

If no relevant information can be found, or the work proposed is known to be original, it will be necessary to conduct cautiously a very small scale preliminary experiment to assess the exothermic character and physical properties of the reaction system and its products.

When subsequently planning and setting up larger-scale reactions or preparations, attention to many practical details may be required to ensure safe working as far as possible. Relevant factors include:

- adequate control of temperature, with sufficient capacity for heating, and particularly cooling for both liquid and vapour phases;
- proportions of reactants and concentrations of reaction components or mixtures;
- purity of materials, absence of catalytic impurities;
- presence of solvents or diluents, viscosity of reaction medium;
- control of rates of addition (allowing for any induction period);
- degree of agitation;
- control of reaction atmosphere;
- control of reaction or distillation pressure;
- shielding from actinic radiation;
- avoiding mechanical friction or shock upon unstable or sensitive solids, and adequate personal protection if such materials will be isolated or dried (without heating).

Further details of specialised equipment, techniques and safety aspects are to be found in the publications devoted to preparative methods.[5-7] Information on recently published hazards is now readily accessible.[8]

Pressure systems

Some of the above considerations assume greater significance when conducting reactions in closed systems at relatively high pressures and/or temperatures. In high pressure autoclaves, for example, the thick vessel walls and generally heavy construction necessary to withstand the internal pressures implies high thermal capacity of the equipment, and really rapid cooling of such vessels to attempt to check an accelerating reaction is impracticable. This is why bursting discs or other devices must be fitted as pressure reliefs to high pressure equipment. A brief account of autoclave techniques in high-pressure hydrogenation is readily available.[9]

A further important point specific to closed systems which will be heated is to make adequate allowance for expansion of liquid contents. Several cases are recorded in which cylinders or pressure vessels partially filled with liquids at ambient temperature have burst under the hydraulic pressure generated when heating caused expansion of the liquid to fill completely the closed vessel. This is also likely to happen when cylinders of liquefied gases are exposed to fire conditions.

Adiabatic systems

If exothermic reactions proceed under conditions where heat cannot be lost (*i.e.* in adiabatic systems) they readily may accelerate out of control.

Although, in general, few laboratory situations will approach adiabatic conditions, occasionally the combination of a uniform heating system which is unusually well-insulated (such as a thick heating mantle with top jacket to surround completely a flask), or which is of unusually high thermal capacity and inertia (such as a deep, well-lagged oil-bath), coupled with a strongly exothermic reaction may approximate to an adiabatic system.

If it is really necessary for technical reasons to use such systems, provision must also be made for application of rapid cooling in the event of an untoward rise in internal temperature. This may be effected by lowering the heating mantle or bath, and application of an air blast, or of cooling to an internal coil.

Types of decomposition

There are three distinct types of chemical decomposition (fast reactions) and the violent effects manifest during deflagration, explosion or detonation are directly related to the rate of energy release, usually derived from combustion or similar processes. The two former types are most likely to be experienced in the laboratory, and usually arise from combustion of a flammable vapour or gas mixed with enough air to give a composition within the flammable limits.

If the fuel-air mixture is relatively small in volume and virtually unconfined when ignition occurs, a deflagration or 'soft' explosion will occur. Persons in the close vicinity may suffer flash burns from such an incident, but material damage other than scorching by the moving flame-front will be minimal. This is because the unconfined rapid combustion will give no significant pressure effects. Such a situation might arise from a small release of gas or vapour and its ignition in a relatively large room, or from a spill of flammable liquid under an open-fronted hood or lean-to building. There is, however, a volume effect, and larger-scale incidents of this type can produce significant destructive effects.

The effects of explosions under conditions of confinement are invariably more serious, and if the confinement is relatively close, and the fuel-air mixture is nearly stoicheiometric, instantaneous pressures several times that of the normal atmosphere may readily be produced. Such pressures are sufficient to demolish a laboratory building of normal construction. It is for this reason that operations of high potential hazard (involving highly energetic substances, and/or high temperature and pressure) are conducted in isolation cells of reinforced construction designed to withstand possible pressure effects.[10]

Explosion situations may also arise if a reaction accelerates out of control and to the point where the containing vessel fails and/or the vaporised contents reach their auto-ignition temperature. Fire will then definitely occur, but explosion may not.

Detonation is the name applied to a particularly severe form of explosion where the velocity of explosive propagation and the associated decomposition temperature and pressure are much higher (by up to two or three orders of magnitude) than in deflagrations. Under some circumstances, a gaseous deflagration can accelerate into detonation, but the necessary conditions (physically long vessels or pipelines) are seldom present in laboratories. However, explosive decomposition of unstable solids or liquids may occasionally involve detonation (during a few μs and with propagation velocities up to 8 km s^{-1}) with the associated violent shattering effects. A more detailed analysis of explosive phenomena is available.[11]

Chemical composition in relation to reactivity

In the last few years instrumental methods have been developed to determine the stability or otherwise of a compound or reaction mixture under a wide variety of possible processing conditions. These methods range from simple (and cheap) equipment which will usually give a general qualitative indication of likely instability, to complex and sophisticated sensitive calorimetric instrumentation of very high cost but which will determine the hazard potential quantitatively and with great precision. Such equipment is, however, unlikely to be found in a small or educational laboratory.

It may eventually be possible to calculate the hazards related to the stability and reactivity of chemical compounds and reaction mixtures in advance of experience. However, at the moment and in most cases, quantitative or qualitative assessments based on known examples represent the best practicable means of assessment of such hazards for small laboratories. Assessment may be based either on overall composition or on detailed structure of the materials involved.

Overall composition and oxygen balance

One of the fundamental factors which may determine the course of a reaction system is that of the overall elemental composition of the system. It is a fact that the majority of reactive chemical accidents or incidents have involved oxidation systems and, especially in organic systems, the oxygen balance is an important criterion.

Mixtures. Oxygen balance is the difference between the oxygen content of a system (a compound or a mixture) and that required to oxidise fully the carbon, hydrogen, and other oxidisable elements present to carbon dioxide, water *etc.* If there is a deficiency of oxygen, the balance is negative, and if a surplus, positive. Oxygen balance is often expressed as a weight percentage with appropriate sign.

In laboratory oxidation reaction systems, one should plan the operations to keep the negative oxygen balance at a maximum to minimise the potential energy release. This consideration will dictate, therefore, that wherever possible an oxidant will be added slowly (and with appropriate control of cooling, mixing *etc*) to the other reaction components, to maintain the minimal effective concentration of oxidant throughout the reaction. It is important to establish as early as possible, from the physical appearance or thermal behaviour of the system, that the desired exothermic reaction has become established. If this does not happen, relatively high concentrations of oxidant may accumulate before onset of reaction, which may then become uncontrollable.

Two relevant examples may be quoted. Mixtures of several water-soluble organic compounnds (ethanol, acetaldehyde, acetic acid, acetone *etc*) with aqueous hydrogen peroxide show clearly defined limits within which the mixtures are detonable (1161). Oxidation of 2,4,6-trimethyltrioxane ('paraldehyde') to glyoxal with nitric acid is subject to an induction period, and the reaction may become violent if addition is too fast. Presence of nitrous acid eliminates the induction period (1129).

In other cases it may be necessary for practical reasons to add one or more of the reaction components to the whole (or preferably part) of the oxidant, but the other considerations will still apply.

Compounds. The concept of oxygen balance has more usually been applied to isolated compounds rather than to reaction mixtures as mentioned above. A fairly rapid appreciation of any potential tendency towards explosive decomposition may be gained by inspection of the empirical formula of a particular compound.

If the oxygen content of a compound approaches that necessary to oxidise the other elements present (with the exceptions noted below) to their lowest state of valency, then the stability of that compound is doubtful. The exceptions are that nitrogen is excluded (it is usually liberated as the gaseous element), and halogen will go to halide if a metal or hydrogen is present. Sulphur, if present, counts as two atoms of oxygen.

This generalisation is related to the fact that most industrial high-explosives are well below zero oxygen balance, and some examples follow.

- Compounds of negative balance include:
  trinitrotoluene (661)
  $$C_7H_5N_3O_6 + 10.5O \rightarrow 7CO_2 + 2.5H_2O + 1.5N_2$$

  peracetic acid (271)
  $$C_2H_4O_3 + 3O \rightarrow 2CO_2 + 2H_2O$$
  Presence of an oxidant will decrease the negative balance
- Zero balance:
  performic acid (150)
  $$CH_2O_3 \rightarrow CO_2 + H_2O$$

  ammonium dichromate (1029)
  $$Cr_2H_8N_2O_7 \rightarrow Cr_2O_3 + 4H_2O + N_2$$
  Energy release is maximal at zero balance.
- Compounds of positive balance include:
  ammonium nitrate (1195)
  $$H_4N_2O_3 \rightarrow 2H_2O + N_2 + O$$

  glyceryl nitrate (360, 1437)
  $$C_3H_5N_3O_9 \rightarrow 3CO_2 + 2.5H_2O + 1.5N_2 + 0.5O$$

  dimanganese heptoxide (1277)
  $$Mn_2O_7 \rightarrow Mn_2O_3 + 4O$$

Presence of a fuel or reductant will increase the potential energy release.

Compounds with unusually high proportions of nitrogen and N—N bonds are also suspect (1541). Hydrazine (87.4 per cent nitrogen), hydrogen azide (97.6 per cent) are both explosively unstable, but not ammonia (82.2 per cent).

In practical terms, the margin between potential and actual hazard of explosive decomposition may be very narrow or quite wide, depending on the energy of activation necessary to initiate the decomposition. Performic acid is treacherously unstable (low energy of activation), and a sample at $-10\,°C$ exploded when moved (150). TNT, on the other hand, is relatively stable and will not detonate when burned, or under impact from incendiary bullets, but requires a powerful initiating explosive ('detonator') to trigger explosive decomposition.

Molecular structure
Instability and/or unusual reactivity in single compounds is often associated with a number of molecular structural features, which may include the specific bond systems given below.

$C{\equiv}C$	Acetylenes (1441), haloacetylenes (1526), metal acetylides (1563)
CN_2	Diazo compounds (1488)
$C{-}NO$	Nitroso compounds (1608)
$C{-}NO_2$	Nitro compounds (1599)
$C{-}(NO_2)_n$	*gem*-Polynitroalkyl compounds (1658)
$C{-}O{-}NO$	Alkyl or acyl nitrites (1438, 1424)
$C{-}O{-}NO_2$	Alkyl or acyl nitrates (1436, 1424)
$C{=}N{-}O$	Oximes (1633)
$C{\equiv}N{\rightarrow}O$	Metal fulminates (1574)
$N{-}NO$	*N*-Nitroso compounds (1608)
$N{-}NO_2$	*N*-Nitro compounds (1606)
$C{-}N{=}N{-}C$	Azo compounds (1457)
$C{-}N{=}N{-}O$	Arenediazoates (1451), bis(arenediazo) oxides (1459)
$C{-}N{=}N{-}S$	Arenediazo sulphides (1451), xanthates (1708), bis(arenediazo) sulphides (1460)
$C{-}N{=}N{-}N{-}C$	Triazenes (1703)
N_3	Azides (1456)
$N{=}N{-}NH{-}N$	Tetrazoles (1697)
$C{-}N_2^+$	Diazonium salts (1489–94)
$N{-}C(N^+H_2){-}N$	Guanidinium oxosalts (176–8)
$N^+{-}OH$	Hydroxylaminium salts (1546)
$N{-}Metal$	*N*—Metal derivatives (heavy metals) (1571)
$N{-}X$	*N*—Halogen compounds (1534), difluoroamino compounds (1499)
$O{-}X$	Hypohalites (1546)
$O{-}X{-}O$	Halites (1470), halogen oxides (1537)
$O{-}X{-}O_2$	Halates (1568, 1578, 1586)
$O{-}X{-}O_3$	Perhalates (1439, 1445, 1490), halogen oxides (1537)
$C{-}Cl{-}O_3$	Perchloryl compounds (1642)
$N{-}Cl{-}O_3$	Perchlorylamide salts (1642)

Xe—O$_n$	Xenon–oxygen compounds (1708)
O—O	Peroxides (1643)
O$_3$	Ozone (1359–64)

A further large group of compounds which cannot readily be represented by line formulae, and which contains a large number of unstable members is the amminemetal oxosalts (1446). These are compounds containing ammonia or an organic base coordinated to a metal, with coordinated or ionic chlorate, nitrate, nitrite, nitro, perchlorate, permanganate or other oxidising groups also present. Such compounds as dipyridinesilver perchlorate (766), tetra-amminecadmium permanganate (886), or bis-1,2-diaminoethanedinitro-cobalt(III) iodate (486) will decompose violently under various forms of initiation, such as heating, friction or impact.

An interesting application of computers to stability considerations is a program[12] which calculates the maximum possible energy release for a compound or mixture of compounds containing up to 23 elements. No information other than the chemical structure is necessary, and the result, which is semi-quantitative, is used as a screening guide to decide which reaction systems need more detailed and/or experimental investigation. A further program will calculate the maximum reaction heat possible from mixtures of 2 or 3 compounds and the composition for maximum heat release, and then assess the probability of ignition of that mixture.[13]

Redox compounds
When the coordinated base in the last described compounds is a reductant (hydrazine, hydroxylamine), decomposition is extremely violent. Examples are bishydrazinenickel(II) perchlorate (968) and hexahydroxylaminecobalt-(III) nitrate (1014), both of which have exploded while wet during preparation.

Other examples of highly energetic and potentially unstable redox compounds, in which reductant and oxidant functions are in close proximity in the same molecule, are salts of reductant bases with oxidant acids, such as hydroxylaminium nitrate (1200), hydrazinium chlorite or chlorate (922), or double salts such as potassium cyanide–potassium nitrite (reductant and oxidant respectively, 181).

Pyrophoric compounds
Materials which are so reactive that contact with air (and its moisture) causes oxidation and/or hydrolysis at a sufficiently high rate to cause ignition are termed pyrophoric compounds. These are found in many different classes of compounds, but a few types of structure are notable for this behaviour.

71

- Finely divided metals: calcium (870), titanium (1403)
- Metal hydrides: potassium hydride (1095), germane (1092)
- Partially- or fully-alkylated metal hydrides: diethylaluminium hydride (474), triethylbismuth (636)
- Alkylmetal derivatives: diethylethoxyaluminium (635), dimethylbismuth chloride (284)
- Analogous derivatives of non-metals: diborane (69), dimethylphosphine (303), triethylarsine (635)
- Carbonylmetals: pentacarbonyliron (492), octacarbonyldicobalt (696)

Many hydrogenation catalysts containing adsorbed hydrogen (before and after use) will also ignite on exposure to air.

Where such materials are to be used, an inert atmosphere and appropriate handling techniques and equipment are essential to avoid the distinct probability of fire or explosion.[14]

Peroxidisable compounds

A group of materials which react with air much more slowly and less spectacularly than pyrophoric compounds, but which give longer-term hazards, may now conveniently be described.

Peroxidation usually takes place slowly when the liquid materials are stored with limited access to air and exposure to light, and the hydroperoxides initially formed may subsequently react to form polymeric peroxides, many of which are dangerously unstable when concentrated and heated by distillation procedures.

The common structural feature in organic peroxidisable compounds is the presence of a hydrogen atom which is susceptible to autoxidative conversion to the hydroperoxy group —OOH. Some of the typical structures susceptible to peroxidation are:

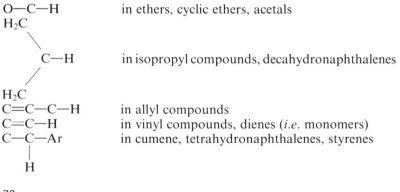

O—C—H	in ethers, cyclic ethers, acetals
C—H	in isopropyl compounds, decahydronaphthalenes
C=C—C—H	in allyl compounds
C=C—H	in vinyl compounds, dienes (*i.e.* monomers)
C—C—Ar	in cumene, tetrahydronaphthalenes, styrenes

Several commonly used organic solvents including diethyl ether, tetrahydrofuran, dioxan, 1,2-dimethoxyethane ('glyme'), bis-2-methoxyethyl ether ('diglyme') are often stored without inhibitors being present, and are therefore susceptible to peroxidation, and many accidents involving distillation in use of the peroxide-containing solvents have been reported. It is essential to test these solvents for peroxide (with acidified potassium iodide) before use and, if present, peroxides must be eliminated by suitable means[15] before proceeding.

Di-isopropyl ether must be mentioned as particularly dangerous for two reasons. Its structure is ideal for rapid peroxidation, and the peroxide separates from solution in the ether as a readily detonative crystalline solid. Several fatal accidents have occurred, and it should not be used.

If allyl and, particularly, vinyl monomers become peroxidised, they are potentially dangerous for two related reasons. The peroxides of some vinyl monomers, such as 1,1-dichloroethylene (231) or butadiene (419) separate from solution and are extremely explosive. Even when this does not happen, the peroxide present may initiate the exothermic and sometimes violent polymerisation of any vinyl monomer during storage. Reactive monomers must therefore be inhibited against oxidation and stored cool, with regular checks for presence of peroxide.[15] However, contained air should not be displaced by nitrogen, as some is essential for effective inhibition.

A few inorganic compounds, such as potassium and the higher alkali metals and sodium amide, are subject to autoxidation and production of hazardous peroxidic or similar products. Many organometallic compounds are also subject to autoxidation and require handling in the same way as pyrophoric compounds.[14]

Water-reactive compounds
Second to air (oxygen), water is the most common reagent likely to come into contact, deliberately or accidentally, with reactive chemical compounds.

Some of the classes of compounds which may react violently, particularly with a limited amount of water, are:

- alkali and alkaline-earth metals (potassium, 1240; calcium, 872)
- anhydrous metal halides (aluminium tribromide, 35; germanium tetrachloride, 1003)
- anhydrous metal oxides (calcium oxide, 881)
- non-metal halides (boron tribromide, 57; phosphorus pentachloride, 1009)
- non-metal halide oxides (*i.e.* inorganic acid halides, phosphoryl chloride, 997; sulphuryl chloride; chlorosulphuric acid, 906)
- non-metal oxides (acid anydrides, sulphur trioxide, 1366)

Concentrated solutions of some acids and bases also give an exotherm when diluted with water but this is a physical effect.

Endothermic compounds

Most chemical reactions are exothermic, but in the relatively few endothermic reactions heat is absorbed into the reaction product(s) which are thus endothermic (and energy-rich) compounds. These are thermodynamically unstable, because no energy would be required to decompose them into their elements, and heat would, in fact, be released.

There are a few endothermic compounds with moderately positive values of standard heat of formation (benzene, toluene, 1-octene) which are not usually considered to be unstable, but the majority of endothermic compounds do possess a tendency towards instability and possibly explosive decomposition under various circumstances.

Often the structure of endothermic compounds involves multiple bonding, as for example in acetylene, vinylacetylene, hydrogen cyanide, mercury(II) cyanide, dicyanogen, silver fulminate, cadmium azide or chlorine dioxide, and all these compounds have been involved in violent decompositions or explosions. Examples of explosively unstable endothermic compounds without multiple bonding are hydrazine and dichlorine monoxide.

In general terms, endothermic compounds may be considered suspect with regard to stability considerations (1505).

Hazardous reaction mixtures

Although the number of combinations of chemical compounds which may interact is virtually unlimited, the combinations which have been involved in hazardous incidents are limited to those which led to an exotherm too large or too fast for effective dissipation under the particular experimental conditions.

The exotherm may have arisen directly from the primary reaction, but a two-stage exotherm is also possible. This will arise when the undissipated primary exotherm leads to instability and subsequent further exothermic decomposition of a reaction intermediate or product.

The great majority of incidents of these types have involved a recognisable oxidant admixed with one or more oxidisable components (or fuels). Examples are the vigorous combustion of glycerol in contact with solid potassium permanganate (1245, the high viscosity of glycerol prevents effective heat transfer), and the violent or explosive oxidation of ethanol by excess concentrated nitric acid (involving formation of unstable fulminic acid, 1103). Some further common examples are given in the partial list of incompatible chemicals, and many others are available in classified form.[1, 2]

A smaller group of incidents has involved recognisable reductants admixed with materials capable of oxidation. Examples here are the explosive decomposition of a heated mixture of aluminium powder and a metal sulphate (29), the shock-sensitivity of sodium in contact with chlorinated solvents (1317–18), or the violent interaction of powdered magnesium with moist silica when heated (1271).

As might be expected, interaction of obvious oxidants and reductants is always potentially hazardous and must be conducted under closely controlled conditions with ample cooling capacity available. Attempted reduction of dibenzoyl peroxide by lithium tetrahydroaluminate led to a fairly violent explosion (829), and hydrazine is decomposed explosively in contact with chromium trioxide (1193). Rocket technology furnishes further examples of the extreme energy release possible in undiluted redox systems.

Potential storage hazards

Most of this chapter has been devoted to the possible outcome of the deliberate interaction of chemicals, but it must not be forgotten that hazardous reactions may occasionally arise from accidental contact of chemicals due to breakage, spillage, or more seriously, from fire in a chemical store. This possibility will dictate that a good deal of thought must be given to segregation of different materials in storage to minimise the effects of accidental contact.[18]

The lists overleaf give a selection of materials which need to be segregated on the grounds of potential reactive or toxic hazards.

Protection from reactive hazards
Where it has been decided after assessment of the various factors discussed in this chapter that a potential reactive hazard may exist in work being planned, consideration must also be given to the level of personal protection which may be required to allow the work to be executed safely or with minimum risk.

There is a considerable range of possibilities which may be involved, depending on the scale of operations and the type of assessed hazard, but in all cases eye-protection will be mandatory, and that from approved safety spectacles may be supplemented with a visor or full face mask.

A small-scale reaction not involving toxic hazards could be run on a wall-facing laboratory bench behind a portable safety screen, but a larger scale reaction, or one also involving toxic materials would best be run with the greater degree of protection afforded by a good fume cupboard. Where exceptionally reactive materials are involved, provision of a specific fire extinguisher additional to the general purpose laboratory extinguishers may be necessary. Analogously, where materials of high toxicity (especially cylinder gases) are involved, a specific respirator, or air breathing set may be required.

Where the possibility of explosive decomposition has been assessed, and in any case where high-pressure reactions are being used, an isolation cell of suitable design is appropriate,[10] and the Safety Code[17] should be applied.

In the absence of laboratory facilities appropriate to the degree of hazard assessed, operations must be deferred until such facilities can be made available.

Partial list of incompatible chemicals (reactive hazards)

Substances in the left hand column should be stored and handled so they cannot possibly accidentally contact corresponding substances in the right hand column under uncontrolled conditions, when violent reactions may occur.

Acetic acid	Chromic acid, nitric acid, peroxides and permanganates.
Acetic anhydride	Hydroxyl-containing compounds, ethylene glycol, perchloric acid.
Acetone	Concentrated nitric and sulphuric acid mixtures, hydrogen peroxide.
Acetylene	Chlorine, bromine, copper, silver, fluorine and mercury.
Alkali and alkaline earth metals, such as sodium, potassium, lithium, magnesium, calcium, powdered aluminium	Carbon dioxide, carbon tetrachloride and other chlorinated hydrocarbons. (Also prohibit water, foam, and dry chemical on fires involving these metals—dry sand should be available.)
Ammonia (anhyd.)	Mercury, chlorine, calcium hypochlorite, iodine, bromine and hydrogen fluoride.
Ammonium nitrate	Acids, metal powders, flammable liquids, chlorates, nitritres, sulphur, finely divided organics or combustibles.
Aniline	Nitric acid, hydrogen peroxide.
Bromine	Ammonia, acetylene, butadiene, butane and other petroleum gases, sodium carbide, turpentine, benzene and finely divided metals.
Calcium oxide	Water.
Carbon, activated	Calcium hypochlorite, other oxidants.
Chlorates	Ammonium salts, acids, metal powders, sulphur, finely divided organics or combustibles.
Chromic acid and chromium trioxide	Acetic acid, naphthalene, camphor, glycerol, turpentine, alcohol and other flammable liquids.
Chlorine	Ammonia, acetylene, butadiene, butane and other petroleum gases, hydrogen, sodium carbide, turpentine, benzene and finely divided metals.
Chlorine dioxide	Ammonia, methane, phosphine and hydrogen sulphide.
Copper	Acetylene, hydrogen peroxide.
Fluorine	Isolate from everything.
Hydrazine	Hydrogen peroxide, nitric acid, any other oxidant.
Hydrocarbons (benzene, butane, propane, gasoline, turpentine, *etc*)	Fluorine, chlorine, bromine, chromic acid, peroxides.
Hydrocyanic acid	Nitric acid, alkalis.
Hydrofluoric acid, anhyd. (hydrogen fluoride)	Ammonia, aqueous or anhydrous.
Hydrogen peroxide	Copper, chromium, iron, most metals or their salts, any flammable liquid, combustible materials, aniline, nitromethane.
Hydrogen sulphide	Fuming nitric acid, oxidising gases.
Iodine	Acetylene, ammonia (anhyd. or aqueous).

Partial list of incompatible chemicals (reactive hazards) (*continued*)

Substances in the left hand column should be stored and handled so they cannot possibly accidentally contact corresponding substances in the right hand column under uncontrolled conditions, when violent reactions may occur.

Mercury	Acetylene, fulminic acid,* ammonia.
Nitric acid (conc.)	Acetic acid, acetone, alcohol, aniline, chromic acid, hydrocyanic acid, hydrogen sulphide, flammable liquids, flammable gases and nitratable substances.
Nitroparaffins	Inorganic bases, amines.
Oxalic acid	Silver, mercury.
Oxygen	Oils, grease, hydrogen, flammable liquids, solids, or gases.
Perchloric acid	Acetic anhydride, bismuth and its alloys, alcohol, paper, wood, grease, oils.
Peroxides, organic	Acids (organic or mineral), avoid friction, store cold.
Phosphorus (white)	Air, oxygen.
Potassium chlorate	Acids (see also chlorates)
Potassium perchlorate	Acids (see also perchloric acid).
Potassium permanganate	Glycerol, ethylene glycol, benzaldehyde, sulphuric acid.
Silver	Acetylene, oxalic acid, tartaric acid, fulminic acid,* ammonium compounds.
Sodium	See alkali metals (above).
Sodium nitrite	Ammonium nitrate and other ammonium salts.
Sodium peroxide	Any oxidisable substance, such as ethanol, methanol, glacial acetic acid, acetic anhydride, benzaldehyde, carbon disulphide, glycerol, ethylene glycol, ethyl acetate, methyl acetate and furfural.
Sulphuric acid	Chlorates, perchlorates, permanganates.

* Produced in nitric acid–ethanol mixtures.

One topic related to several of the hazardous possibilities above is the use of powerful oxidants or mixtures (nitric acid, dichromate, or chromium trioxide with sulphuric acid) to clean glassware from traces of reaction residues, especially tars of organic derivation. In most cases this is not only potentially hazardous, particularly with large containers or residues, but unnecessary, as the specially formulated laboratory detergents which are available are just as effective.

Partial list of incompatible chemicals (toxic hazards)

Substances in the left hand column should be stored and handled so that they cannot possibly accidentally contact corresponding substances in the centre column, because toxic materials (right hand column) would be produced.

Arsenical materials	Any reducing agent*	Arsine
Azides	Acids	Hydrogen azide
Cyanides	Acids	Hydrogen cyanide
Hypochlorites	Acids	Chlorine or hypochlorous acid
Nitrates	Sulphuric acid	Nitrogen dioxide
Nitric acid	Copper, brass, any heavy metals	Nitrogen dioxide (nitrous fumes)
Nitrites	Acids	Nitrous fumes
Phosphorus	Caustic alkalies or reducing agents	Phosphine
Selenides	Reducing agents	Hydrogen selenide
Sulphides	Acids	Hydrogen sulphide
Tellurides	Reducing agents	Hydrogen telluride

---

* Arsine has been produced by putting an arsenical alloy into a wet galvanised bucket.

References

1 L. Bretherick, *Handbook of reactive chemical hazards*, London: Butterworth, 3rd Edn. 1985.

2 *Manual of hazardous chemical reactions*, 491M, 5th Edn. Boston: National Fire Protection Association, 1975.

3 A.I. Vogel, *A text-book of quantitative inorganic analysis*, 4th Edn. London: Longmans, 1978; A.I. Vogel, *A text-book of macro and semi-micro qualitative analysis*, 4th Edn. London: Longmans, 1976.

4 A.I. Vogel, *A text-book of practical organic chemistry*, 4th Edn. London: Longmans, 1978.

5 *Organic syntheses*, various eds, New York: Wiley, Coll. Vols 1–5, 1944–1973, annual volumes thereafter.

6 *Inorganic syntheses*, various eds, London: McGraw-Hill, Vols 1–20, 1939–1979, annual volumes thereafter.

7 G. Brauer, *Handbook of preparative inorganic chemistry*, 2nd Edn. London: Academic, Vols 1 and 2, 1963, 1965.

8 S. Templer (ed.), *Laboratory Hazards Bulletin*, Nottingham, Royal Society of Chemistry, issued monthly since July 1981.

9 Ref. 4, pp. 866–870.

10 W.G. High, 'The design of a cubicle for oxidation or high-pressure equipment', *Chem. Ind.*, 1967, 899–910. *Safety in the study of chemical reactions at high pressure*, Tech. Bull. No. 100, A.L. Glazebrook, Erie, Autoclave Engineers.

11 V.J. Clancey, 'Explosion hazards', *Protection*, 1971, **8** (9), 6.

12 *CHETAH, ASTM Chemical thermodynamics and energy release evaluation program*. Philadelphia: American Society for Testing and Materials, 1975.

13 T. Yoshida, *Chem. Abstr.*, 1983, **95**, 7293.

14 D.F. Shriver, *Manipulation of air-sensitive compounds*. London: McGraw-Hill, 1969.

15 H.L. Jackson, W.B. McCormack *et al.*, 'Control of peroxidisable compounds', *J. Chem. Educ.*, 1970, **47**, A175–188.
16 L. Bretherick, *Incompatible chemicals in the storeroom: identification and segregation*, in *Safe storage of laboratory chemicals*, D.A. Pipitone (ed.), Chichester, Wiley, 1984.
17 *High pressure safety code*. London: High Pressure Association, 1975.

Appeal for further information

The author of this chapter will be pleased to receive information on reactive chemical hazards from relatively inaccessible sources for a revised version of reference 1. Contributions will be acknowledged in print.

Warning on use of aluminium lithium hydride (LAH)

Following the investigation of a laboratory explosion, Dr M. L. H. Green of the University of Oxford has warned that the following precautions are essential for the safe use of lithium aluminium hydride (LAH). The measures are to prevent overheating of the hydride and dissociation to finely divided aluminium, which can then undergo thermite-type reactions with compounds (or solvents) containing combined oxygen or halogen.

1. All apparatus and reactants should be perfectly dry, and reactions should be run rigorously under nitrogen, with the reaction temperature below 60 °C at all times.
2. Order of addition is important. Always first add the hydride to the solvent in the nitrogen-purged apparatus, before adding the other reactant last.
3. The hydride should never be allowed to form a crust above the level of the liquid or to settle to the bottom, so efficient and gentle stirring is absolutely essential.
4. To prevent local overheating of the reaction vessel, heating mantles should never be used: always use an oil bath as heat source.
5. After reduction has been effected, destroy excess LAH by slow and careful addition of dry ethyl acetate (preferably diluted with inert solvent), again under nitrogen and keeping the temperature below 60 °C. All LAH reactions should be carried out behind suitable protective screens.

Chapter 6

Chemical Hazards and Toxicology

The basic rules of toxicology

a) How toxic is 'toxic'
In everyday language terms like poison, toxic or toxicity imply a self explanatory and absolute quality which separates one group of substances (the toxic) from another (the non-toxic). This is a misconception. Every chemical has the capacity to cause injury and that is why the rating of toxicity does not start with zero, but with the *practically* non-toxic category. Though there is no chemical which does not have the capacity to cause harmful effects, even the most toxic chemicals can be used harmlessly. Exposure to a chemical is harmless when it does not disturb the state of equilibrium of biological functions. To say that water is harmless is true only in the sense that water consumption to a certain limit has no adverse effect, but forced drinking may be lethal and in a hot environment uncontrolled consumption of non-salted water can cause severe muscle cramps. Drugs used for the treatment of diseases exert beneficial effects up to a certain dose level, but toxic effects become apparent in the case of overdose. Alkylmercury fungicides intentionally added to grain protect sown seeds against fungal infections, but the use of the same grain for making bread resulted in mass epidemics. In hazard

assessment toxicity is only one component, because hazard is the probability of a chemical to cause injury and is dependent on its handling and use.

The toxicity of one chemical relative to another may be different for single or repeated exposure, for oral or inhalation exposure, for the young or the old, for the mother or her foetus and naturally it can change from one species to another. Consequently, to say that a compound is toxic or non-toxic conveys no more information than to say that a chemical is soluble without defining its physical state, the solvent, temperature and pressure. It sounds even more absurd to make a general statement about solubility of a group of compounds, but it is not without precedent that no distinction is made between the toxicities of an element and its compounds. Statements made by a Professor of Chemistry in a scientific journal will help to pinpoint the most frequent mistakes and will illustrate the most basic rules of toxicology. He criticised the need for preventive measures aimed at decreasing exposure to methylmecury, a neurotoxic agent, which had already caused mass epidemics in Japan and Iraq.

Statements by the Professor:

1. He had suffered no adverse effects, with the exception of loose teeth, after a three hour exposure to butylmercury.

   Comment:
   Butylmercury is not neurotoxic and less toxic than methylmercury. Most people would find unacceptable an exposure which resulted in loose teeth.

2. He distilled dimethylmercury for two days and his technician worked with the same compound for 200 days without adverse effects. Rats exposed to dimethylmercury for 20 days in a static inhalation chamber of 4 m^3 where 25 g dimethylmercury was evaporated showed no visible signs of intoxication.

   Comment:
   Exposure was not measured and thus no judgement on toxicity is justified. Rats might have had co-ordination disorders only recognised by a trained observer.

3. During a mercury scare, which occurred 49 years ago, a test carried out on himself showed that he eliminated per day as much mercury as he absorbed, and therefore mercury does not accumulate.

   Comment:
   Supposing the balance study was carried out correctly, it indicated that his exposure to mercury was long enough to result in a steady state body burden and it did not prove that mercury does not accumulate.

4. The dominant use of organomercurial diuretics, before 1952, to combat

salt and fluid retention proves that the fuss about methylmercury as an environmental hazard is groundless.

Comment:

Organomercurial diuretics, unlike methylmercury, do not damage the sensory part of the nervous system, but frequently caused diarrhoea, vomiting, irritation, fever and heart injury. These side effects prompted a search for safer diuretics.

The errors in these statements are manifold. Firstly no distinction is made between the different forms of mercurial compounds, though their toxicity differs both quantitatively and qualitatively. Secondly a change in the molecular structure changes not only toxicity, but also physical and chemical properties which influence evaporation and absorption. Thirdly, the experience of occupational exposure cannot be extrapolated directly to the non-occupational exposure of a less homogeneous population partly because the route of intake may be different and partly because the general population includes children, pregnant women, *etc.* Fourthly, dose or exposure was never quantified.

b) Molecular structure and physico-chemical properties

Without proof one should never state that different compounds of an element have the same potential to be harmful or hazardous and that their toxicities are identical. In every statement concerning toxicity the exact molecular structure of a chemical must be identified. Change in molecular structure caused by oxidation, reduction, cleavage of a bond, introduction of a new ligand or replacement of one ligand with another can have profound effect on the potential of a compound to be hazardous. Some of these changes might occur during storage or by the use of a chemical and in these cases consideration must be given to the properties of the possible derivatives.

A change in the molecular structure can alter the physical properties. Thus dimethylmercury has a higher vapour pressure than methylmercury chloride. The volatility of methylmercury depends on the anion; methylmercury chloride is 350 times more volatile than is the corresponding dicyandiamide salt. The second important physical property is solubility. The solubility of mercury metal in water and biological fluids is very low, which explains why 2 lb of mercury taken orally in four divided doses by a young man caused no ill effects. The solubility of mercurous chloride is low compared with mercuric chloride, and in the case of oral administration the more soluble mercuric chloride is 60 times more toxic than the mercurous chloride. Other examples are: lipid solubility favours absorption through the outer layer of the skin and water solubility helps the transport through the lower layer. Consequently, of two lipid soluble compounds, like aniline and nitrobenzene, the absorption of aniline is somewhat faster due to its higher water solubility. The third physical property that can influence the potential hazard of a chemical is viscosity. Increasing viscosity would diminish the danger of swallowing the chemical when pipetting, although pipetting by mouth is undesirable.

Change in the molecular structure can affect the reactivity of a compound. Thus dimethylmercury having no charge, behaves like a neutral gas and becomes toxic only after decomposition to methylmercury. In mice within 2.5 hours of the intravenous administration of dimethylmercury 85 per cent of the dose was exhaled and only 10 per cent was converted to methylmercury, *i.e.* to the neurotoxic agent.

Dibutylmercury behaves in a similar way to dimethylmercury, but the metabolic product, butylmercury, is less toxic than methylmercury and seems to lack neurotoxic properties. Thus one cannot even make generalisations for the toxicity of alkylmercury compounds, let alone other organomercurials. Organomercurial diuretics are rapidly metabolised in the body and the role of the organic part of the molecule is to carry mercury and release it as mercuric ion at renal receptor sites involved in diuresis. This effect is qualitatively different from the neurotoxic effect of methylmercury which affects the sensory nerves or from the mercury vapour exposure which may result in psychotic disturbances. An extreme form of this type of intoxication was frequent in the hat industry when mercury was used for processing felt ('mad as a hatter').

c) Biological variation in response to chemicals
Not every species reacts with the same degree or type of response to equal doses of a chemical. Dimethylmercury might be quite innocuous to rodents as they are able to convert only a small proportion of it to methylmercury, but man might be more sensitive. In 1863 two laboratory technicians who worked with dimethylmercury died of methylmercury poisoning and in 1971 a chemist died of methylmercury poisoning in Czechoslovakia after synthesising 600 g of dimethylmercury.

The main difficulties in interpreting toxicological results are the extrapolation of data from one species to another and the biological variation in sensitivity within a species. Age and sex are the most important factors that influence sensitivity within a species, but there are more subtle causes such as nutrition or genetics. Thus male rats are more sensitive to the renotoxic effect of mercuric chloride than females, and female rats are more sensitive to the neurotoxic effect of methylmercury than males.

Epidemiological studies seem to indicate that some effects of lead on the central nervous system are age dependent and that children are at greater risk than adults. A selenium rich diet gives some protection against methylmercury toxicity in rats and susceptibility to lead depends on a wide variety of dietary factors like calcium, iron, protein, vitamin D, ascorbic acid, nicotinic acid, other metals and alcohol consumption. In rats methylmercury is liable to damage the kidney while in human cases of methylmercury intoxication renotoxicity is always absent. Different strains of rats have different sensitivities to the hepatotoxic effect of carbon tetrachloride; Fischer rats are more sensitive than are Wistar rats. The best known examples for the significance of genetic differences in humans are glucose-6-phosphatase deficiency resulting in hypersusceptibility to haemolytic agents like aniline, and serum antitrypsin deficiency which results in increased sensitivity to respiratory irritants. But even in a inbred group, which is homogenous in age, sex and nutrition, there will be some variation in the response to the same dose of the chemical. Thus individual observations, though they may be clinically interesting, do not satisfy the requirements of a responsible statement on the toxicity of a chemical. This needs greater numbers for suitable statistical analysis.

Kinetics and metabolism

a) Absorption

The toxic response in any cell, tissue or organ is a function of the concentration of the causative agent at the site of action. Chemical burns and irritation of the mucous membrane or skin are a result of external penetration into the layers of skin, but most chemicals act internally. This explains why knowledge on the absorption of toxic compounds from the site of administration, its subsequent distribution, metabolism and excretion are so important for the interpretation of data obtained from toxicity tests. Some conclusions can be drawn from single dose experiments: nearly identical LD_{50}s (dose which kills 50% of the group) for oral and parenteral administration indicate that the compound is almost completely absorbed from the gastrointestinal tract while incomplete gastrointestinal absorption requires substantially higher oral doses than the parenteral ones. Dermal application of the test material also gives some information on the ability of the tested chemical to pass through the skin. With increasing lipid solubility and decreasing molecular weight the absorption of non-polar substances increases through the outer layer of skin, though further transport into the blood vessels is promoted by water solubility. Abrasion of the skin increases absorption for both polar and non-polar compounds.

Lung offers a very much larger surface for absorption than either the gastronintestinal tract or skin as the thin walls of tiny air chambers, completely filled with a network of capillary blood vessels, cover an area of 90 m^2. The rate of absorption through these alveolar membranes is an equilibration process which depends on the solubility of the gas or vapour and the extraction of the chemical from blood by other tissues. Particles of 0.5 to 3μm diameter can reach the alveolar space, but their absorption involves a more complicated transport process. Respiratory bronchioles to a smaller extent also participate in the absorption of gases, vapours and particles. The rate of absorption through the lungs is influenced by the rate of respiration and blood circulation. As these are increased with work load, the respiratory intake of a toxic substance is higher in heavy than in light work.

b) Distribution
Once a chemical is absorbed, blood plays an important part in its distribution. In blood the chemical is either transported in the red blood cells or in the plasma. Thus methylmercury and atomic mercury are mainly carried by the cells and inorganic mercury (Hg^{2+}) by the plasma. Interestingly, the brain uptake of methylmercury and atomic mercury is relatively higher than inorganic mercury and it seems that the ability of a chemical to pass through the membrane of the red cells is related to its ability to cross other biological membranes. The easy passage of methylmercury and mercury vapour but not inorganic mercury salts through the so called blood–brain barrier explains why the central nervous system is only affected by the former forms of mercury. Passage of toxic compounds through the placenta is the prerequisite for direct toxic effects on the foetus, though the compound might interfere, without passing through the placenta, with the supply of essential nutrients to the foetus.

The passage of chemicals from the blood into organs is a complex process. An uncharged molecule can diffuse easily through membranes, but the transport of other molecules usually requires reaction with physiological carriers. Thus the transport of organic anions from plasma into the liver is facilitated by a transport protein called ligandin. Diffusion and metabolically-dependent transport are the two forms of transport. An example for diffusion is the distribution of hydrocarbons. There is some linear relationship between the anaesthetic potency of chlorinated hydrocarbons and their serum/air or oil/air partition coefficient. High lipid solubility increases uptake and decreases elimination from organs like the brain which has a high lipid content. Metabolically-dependent transport contributes to the accumulation of mercury in the kidneys. For example, dinitrophenol given before the administration of mercuric chloride decreases the kidney uptake of mercury.

c) Metabolism
The absorbed chemicals are in contact with highly reactive biological compounds and catalytic systems. Metallic mercury is mainly oxidised in blood, and depends on the rate of oxidation as to how much diffusable metallic mercury reaches the brain. However, the main organ for the metabolic conversion of chemicals is the liver which can oxidise, hydrolyse, reduce or conjugate many toxic compounds. The result of these metabolic

processes is usually a compound which can be easily excreted in bile or urine. This metabolic process can also change the toxic character of the compound.

The biotransformation might produce a more toxic derivative. Aniline is converted to *p*-aminophenol which is a potent methaemoglobin forming agent. Carbon disulphide reacts with amino compounds to give copper complexing dithiocarbamates. From carbon tetrachloride highly reactive free radicals are formed, which are at least partly responsible for its hepatotoxic effect. The toxicity of methanol resulting in the damage of the optic nerve is based on the formation of formaldehyde. Parathion is a weak acetylcholinesterase inhibitor but it is oxidised to a powerful inhibitor, paraoxon.

The metabolic process which increases the toxicity of one compound can decrease the toxicity of another. Butanol is oxidised through aldehyde to carbon dioxide. Cyanide reacts with sulphur donors to give the innocuous thiocyanate. The split of the C–Hg bond in phenylmercury or in alkoxyalkyl-mercury gives mercuric mercury which is less dangerous to health than a stable organomercury compound, like methylmercury.

Phenobarbitone potentiates the metabolism of many foreign chemicals. After phenobarbitone treatment carbon tetrachloride, carbon disulphide, chloroform or nitrosamine become more, but aflatoxin less toxic. Chemicals that inhibit the enzymes responsible for the formation of a more toxic compound have the opposite effect and they decrease the risk. Thus carbon disulphide or diethyldithiocarbamate decreases the toxicity of carbon tetrachloride.

There are other forms of interaction which do not fit into the normal biotransformation processes, such as the protective effect of methaemoglobin forming agents against cyanide. Interaction can change even the toxicity of metals, thus cadmium pretreatment has a protective effect against the renotoxic doses of mercuric chloride, and selenium protects against cadmium and mercury and *vice versa*.

As interaction can change the risk imposed by a certain level of exposure, from the theoretical point of view it would be justified to adopt TLVs (exposure limits) to fit the shift in the dose response curve. However, our knowledge of the mechanism of toxic effects, on the conditions of interactions and even the methods of measuring exposure is insufficient to put into practice such corrections.

d) Body burden and biological half time
The knowledge of half-time is extremely important when the development of a toxic effect depends on the accumulation of a toxic compound in the body. Half-time denotes the time that is needed to excrete half of the total amount of the compound from the body, *i.e.* half of the body burden when there is no further exposure. Accumulation of a toxic compound in the body also depends on the half-time as, in every day, a fixed proportion of the body

burden (and not of the daily dose) is eliminated. If the daily uptake (dose or exposure) remains constant, the body burden for this compound increases to the point at which absorption and elimination are equal. That condition is called the steady state. For example, methylmercury has an approximately 70 day half-time so if a person ingests 2 mg methylmercury per week of which 95 per cent is absorbed from the gastro-intestinal tract, one can calculate that after six months the body burden of mercury will be 24 mg, after one year 28 mg, and 28.9 mg at a steady state when exactly 2 mg of mercury will be excreted weekly.

As far as half-time is concerned, methylmercury belongs to the group of compounds with a single dominant half-time, and thus daily body burden plotted on a semilog paper against time gives a straight line. There are other compounds, such as lead or mercury vapour, that have more than one half-time, a short half-time followed by one or more longer half-times. The biological half-time of lead in bone is about 10–20 years, but in blood and soft tissues it is only 20–30 days.

This indicates that the half-time for different organs or tissues is also different and as time passes, tissues with a longer half-time dominate the elimination curve. Another complicating factor is the metabolism of the chemical yielding derivatives which might have a longer or shorter half-time than the parent compound. Thus the half-time of dimethylmercury is very short and the half-time of its derivative, methylmercury, very long. Phenyl-mercury, alkoxyalkyl mercury salts or organomercurial diuretics are rapidly metabolised to inorganic mercury and afterwards the half-time of the mercury component is identical with the half-time of mercuric mercury.

e) Excretion
The three main routes of excretion are exhalation, urinary and faecal excretion.

As a general rule every chemical that can be inhaled as vapour or gas can also be exhaled, and the concentration in the exhaled air depends on exposure and metabolism. However, not only inhaled chemicals can be exhaled. Exhalation is very noticeable if a solvent is given to an experimental animal either orally or by injection.

Excretion in the faeces or urine is the most important route for the majority of toxic chemicals and non-volatile compounds are excreted only by these routes, though small quantities might appear in tears, sweat or milk.

A proportion of an ingested chemical can be excreted with faeces without being absorbed. Less than 10% of toxic inorganic metallic compounds are absorbed, though with an organic radical their absorption (like that of methylmercury) might be nearly complete. The chemical which is absorbed into the blood stream might occur in the gastro-intestinal lumen through the normal shedding of the inner surface of the gut, but the most important source of faecal excretion is the bile. After oral exposure biliary secretion is helped by the fact that intestinal blood first passes through the liver before it reaches the general circulation. Biliary secretion is influenced by the molecular weight of the compound and by its polarity. The optimum molecular weight is 500–1000; between 300 and 500 mainly polar compounds are excreted. There are mechanisms in the liver which facilitate the biliary secretion of compounds by increasing their polarity and their molecular weight. A secreted compound might be excreted with the faeces or reabsorbed from the gut. The reabsorption of biliary methylmercury contributes to its long biological half-time. When biliary secretion is the major route of elimination, diseases of the liver which affect the function of liver cells or obstruct the bile flow, will increase the half-time of the chemical and increase its toxicity.

Extraction of chemicals by the kidneys is facilitated by the fact that one-quarter of the blood pumped into the circulation by the heart passes through the kidneys and about half of this is submitted to a filtration mechanism which is the first step in urine formation. Not all the chemicals that appear in the filtrate (tubular urine) are excreted, some are reabsorbed into the tubular cells. Urinary excretion depends on the ability of the kidney to extract the chemicals from the blood, and on the ability of tubular cells to accumulate or release the toxic compound. Chemicals with a high lipid/water partition coefficient are passively reabsorbed from the kidney tubules, while polar compounds and ions, if they are to remain in these forms in the tubular urine, are excreted. The availability of small molecules that form complexes with toxic chemicals can increase the urinary excretion because they compete for the chemical with protein binding sites in the cells. Thus the excretion of heavy metals can be increased very significantly by the administration of complexing agents, such as the excretion of inorganic mercury by BAL or D-penicillamine, or the excretion of lead by EDTA.

Dose and damage

a) Critical organ

After single or repeated administration the concentration of the toxic chemical in the various organs is usually different. The organ which first attains a concentration that affects its function is called the critical organ or, in other words, if the injury of the critical organ is prevented, intoxication is prevented.

For a toxic chemical there might be more than one critical organ. Acute exposure to cadmium fumes damages the lungs, whereas long-term exposure to the same metal results in kidney damage. The critical organ is not necessarily the organ which contains the highest amount or concentration of the chemical. The main storage organ for lead is the bone, but the critical organ is the haemopoetic system which forms the blood and, in infants, the critical organ can be the central nervous system. Irrespective of the chemical form of mercury the kidney is always the organ with the highest concentration, but the kidney is the critical organ only for mercuric mercury. For metallic mercury or methylmercury the critical organ is the nervous system.

The identification of the critical organ for a toxic chemical is based on the dose–effect relationship which gives the sequence of toxic effects in relation to dose.

b) Dose–effect relationship
Evaluation of a chemical hazard is based on two relationships, one is the dose–effect relationship, the other is the dose–response relationship. The methyl-mercury epidemic in Iraq, which involved many thousands of people, proved that the daily consumption of certain amounts of methylmercury-contaminated bread caused paraesthesia (sensory disturbances), higher levels of exposure caused ataxy (unco-ordinated movements) and even higher daily uptake resulted in the loss of vision, deafness, or even death. Thus methylmercury, depending on the exposure, is able to cause effects of varying severity. Another example is the narcotic and hepatotoxic effect of carbon tetrachloride. In the case of lead, porphyrin in the urine, anaemia, abdominal pain, palsy and, mainly in children, encephalopathy are the main constituents of the dose–effect relationship. Rats given 1–2 mg kg^{-1} amphetamine run around the walls of the cage, but when treated with 6 mg kg^{-1} amphetamine they soon become stationary and move their heads repetitively. Doses higher than this produce cardiovascular disorders. Listing effects against the corresponding dose gives the dose–effect relationship. If one effect, like anaemia or loss of nerve function, can be measured on a graded scale of severity, the gradation of this effect in relation to dose is also used as a dose effect relationship.

c) Dose–response relationship
If one effect is selected as a response, the percentage of animals giving this response in every dose group is used to produce a dose–response curve. From the dose–response curve one can calculate the dose that is able to produce a response in 50 per cent of the animals and this dose is called ED$_{50}$ (ED = effective dose). If the response is death the dose that killed 50 per cent of the animals is called LD$_{50}$ (LD = lethal dose). Both ED$_{50}$ and LD$_{50}$ can be calculated from single or multiple administration experiments, but in every case the route of administration, dose, treatment and observation period must be stated. In the case of inhalation exposure, the atmospheric concentration of the test compound and the length of the exposure are the essential data. Both ED$_{50}$ and LD$_{50}$ are valid only for the species tested with the qualification of age, sex and nutrition. Even when men or animals can be killed by a single dose, this does not necessarily mean that by lowering the dose, a dose–effect

relationship which includes all the possible effects can be established. In the case of methylmercury, the typical neurological symptoms in rats can be produced only after repeated administration. Many of the chemical carcinogens must be given for an extended period of time and the observation period might cover nearly the whole lifetime of the animal.

Toxicity Testing

a) Acute (single dose) toxicity tests
The most frequently used procedure in the assessment of acute toxicity is the LD_{50} test which is the estimation of the single dose lethal to 50% of the test animals. The chemical is most frequently administered orally, but occasionally it is given parenterally (intraperitoneally, intravenously or subcutaneously). For volatile compounds a more appropriate, and for gases the only, way of administration is by inhalation. The aim of the inhalation toxicity test is to calculate LC_{50}, *i.e.* the atmospheric concentration which results in 50% death. The aim of the acute dermal tests is usually restricted to the detection of local effects (*e.g.*, irritation) or whether by dermal application the compound can cause systemic toxic reactions. However, when the absorbed dose is measured, a dermal test can be used for the quantitation of acute toxicity. In the dermal toxicity test the compound is spread evenly on a skin area shaved 24 hours earlier. After treatment the area is bandaged for 24 hours with an aluminium foil or polythene lining. The dressing is removed 24 hours later and the treated area is washed and cleaned with soap and water. From the quantitative analysis of the compound in the washing liquid and bandage it is possible to calculate how much of the applied chemical penetrated the skin.

In acute toxicity tests the compound is given to two species at four dose levels. Each dose group should contain an equal number of males and females. The maximum dose usually does not exceed 2.0 g kg^{-1} orally or 2.8 g dermally. The maximum concentration for a six-hour inhalation exposure is approximately 8 mg l^{-1} when hazardous chemicals are screened for the possible consequences of occupational exposure. Animals are observed daily for 14 days. Some dying and some other less affected animals at day 14 are autopsied for gross pathological abnormalities and depending on the result of this examination organs are taken for histological examination. LD_{50} or LC_{50} are calculated for 24 hours, four days and 14 days. Some additional numbers of animals dosed at proven toxic levels are autopsied 2–3 days after dosing. The gross pathology and histology might reveal early or temporary changes not seen at 14 days.

Daily observations made on treated animals and the results of pathological and histological tests are of great importance for the design of further toxicity studies. They help to select more sensitive responses than death for ED_{50}

studies (effective doses which produce a certain response in 50% of the animals) and for subacute tests. They also help to narrow the dose range in these studies. Thus a well organized LD_{50} test can result in a significant saving in both the number of animals and work. When no substantial information is available on the toxicity of a chemical, in order to avoid death from causes other than cancer, the first step in the selection of realistic doses for long term carcinogenic tests is also an acute LD_{50} test. LD_{50} values are therefore the first estimate of toxicity which help to classify toxic compounds into broad categories. The categorization shown in Table 6.1 is widely used and indicates some relationship between a single oral dose which may cause death in man and the single oral LD_{50} for rats. Naturally if the toxic compound accumulates in the body, has a cumulative effect, is carcinogenic or can produce sensitisation, then the potential hazard bears no relationship to the above toxicity rating. Thus, based on a single dose, methylmercury is less toxic than mercuric chloride, but in the case of repeated administration animals develop a resistance to the renotoxic effect of mercuric chloride, but not to the neurotoxic effect of methylmercury. Di-isocyanates are only slightly toxic according to the toxicity rating, but it is not uncommon that following acute symptoms, an asthma-like syndrome is precipitated even by exposure to a minute amount of di-isocyanate. One of the difficulties when testing carcinogens is the absence of definite early signs before the occurrence of the tumour cells.

b) Sub-acute toxicity test
The acute toxicity test gives the relevant information for the organisation of sub-acute tests in which exposure lasts from a few days up to three months. In sub-acute or short-term tests the highest daily dose must be sufficiently below the single dose LD_{50} to avoid the early occurrence of severe toxicity, but must be high enough to cause intoxication. Usually three dose levels are used on

Table 6.1 Categories in relation to relative acute toxicities

Toxicity rating	Term of toxicity	Probable human lethal dose for a 70 kg man	Compound belonging to the group (oral LD_{50} for rats in mg kg^{-1})
1	practically non toxic	>15 g kg^{-1}	propylene glycol (26000)
2	slightly toxic	5–15 g kg^{-1}	sorbic acid (7400)
3	moderately toxic	0·5–5 g kg^{-1}	isopropanol (5800)
4	very toxic	50–500 mg kg^{-1}	hydroquinone (320)
5	extremely toxic	5–50 mg kg^{-1}	lead arsenate (100)
6	supertoxic	<5 mg kg^{-1}	nicotine (50)

male and female animals belonging to two species, and the route of administration should be relevant to the expected form of exposure. During the whole test time animals are regularly observed for their clinical conditions. Body weight and food intake are recorded and laboratory investigations are carried out for blood and urine chemistry and haematology. At the end of the test period animals are killed and subjected to full post-mortem and histological examination. Brain, heart, lungs, liver, kidneys, spleen, gonads, adrenals, thyroid, pituitary and pancreas are removed and weighed. Autopsy carried out at a later date gives information on the reversibility of damage. One of the objectives of the sub-acute toxicity test is to demonstrate some toxic response and the second objective is to provide an indication on the maximum dose which given for a certain time does not produce any adverse effects. The comparison of single dose LD_{50} with the repeated dose LD_{50} also helps to establish whether the compound has a cumulative effect. For example, if the LD_{50} of 28 daily doses is 28 times less than the single dose LD_{50}, this indicates that the compound is completely cumulative, and the closeness of the two LD_{50}s indicates that one daily dose and its effect is eliminated before the next treatment is given.

For many chemical agents, daily oral treatment or inhalation exposure lasting one-tenth of the lifespan of a species (90 days for rats) allows the setting of the non-adverse effect level for long term exposure. Of 122 toxic agents studied in rats only 3 (2.45%) and of 566 dosage levels only 15 (2.65%) induced effects after, but not before, 90 days exposure. Thus with the exception of carcinogenic effects, it is possible to predict from the negative outcome of a 90 day test on rats that even the life-time administration of the compound will not produce adverse effects.

c) Chronic toxicity test for carcinogenicity

The determination of carcinogenic potential requires chronic toxicity testing. The most widely used species are the rat and the mouse, but Syrian golden hamsters proved to be an excellent test animal to reveal bladder and lung cancer.

The route of administration is designed to follow the route of human exposure and can be oral (in food or drink), dermal or inhalation, though it can be administered also by gavage or by parenteral injection. Oral administration is carried out 7 days per week and the others 5 days per week. Treatment starts a few weeks after weaning and lasts for two years in rats and 18 months in mice.

The test is carried out on two species at two, but preferably more, dose levels. Doses must be selected on the basis that even the highest dose should not shorten the lifespan by non-tumourogenic lesions. Each group must consist of at least 30 male and 30 female animals, preferably more. For every group equal numbers of vehicle controls are included, that is animals which receive the vehicle without the test material. All animals in carcinogenicity tests are subjected to complete gross pathological and histological examination.

The sensitivity of the test depends on the frequency of tumours in the controls and on the number of animals per group. Thus when no tumour occurs in controls at a site, significant tumourogenic effect requires a more than 50% tumour frequency in 10 animals per group and slightly more than 10% in 75 animals per group. When in controls the probability of tumours is 20%, the corresponding threshold frequencies for detection are 87% and 43%. The predictive value of chronic tests for carcinogenicity is well established; 25 of 27 chemicals which are proved or probable human carcinogens produced cancer in at least one species.

d) In vitro *tests for carcinogenicity*
Chronic animal tests for cancer are extremely expensive (about £100000 per chemical). That is one of the reasons for interest in predictive *in vivo* tests. The most widely used predictive test is the Ames test named after its developer, Dr. Bruce Ames of the University of California. In this simple test, based on the view that carcinogens possess mutagenic activity, bacteria are used with mammalian microsomes to identify whether the chemical or its metabolite(s) is a mutagen, that is whether it is able to produce changes in the genetic material of a cell. So far about 85% of the known carcinogens gave a mutagenic response with the Ames test. Not every bacterium is equally sensitive and the best correlation between mutagenicity and carcinogenic potential was obtained with one particular strain of *S. typhimurium* (TA 1538); however, even with this strain some potent carcinogens produced weak mutagenic responses and *vice versa*.

Other less frequently used tests aim to detect gene mutation, chromosome damage or cell transformation in mammalian cells, primary DNA damage and repair in bacteria or mammalian cells and a wide variety of genetic damage in *Drosophila melanogaster*. These test systems are mainly used in addition to the Ames test.

e) Special tests
Special tests are those which measure skin irritation, eye irritation, sensitisation and teratogenicity or effects on reproduction. Irritation is the consequence of direct and injurious chemical interaction with the skin or eye, while in skin sensitisation an immunological process after various times of exposure results in reaction to concentrations that previously were innocuous.

Reproduction tests aim to estimate effects on male and female fertility, gestation and the offspring. The most frequently used procedure for the measurement of genetic damage in germ cells is the dominant lethal test. Treated male rats are mated with untreated virgin females. The failure of the fertilized eggs to implant in the uterine wall or to survive to midpregnancy is detected by the necropsy of females. The genetic damage of germ cells may cause malformations, but it is more typical that congenital malformation are caused by direct toxic action on the developing organs by treatment implemented during pregnancy. Only these malformations are called teratologic effects and thus in teratogenic tests the parenteral treatment is restricted to a short period: day 8 through 12 of gestation in rats. However, a toxic chemical may cause death and reabsorption before and after the end of organogenesis and degenerative changes and biochemical defects after organogenesis. Thus foetotoxicity is a wider concept than teratogenicity, that is teratogenicity is only a special case of foetotoxicity.

When teratogenicity tests are required, they are usually carried out on two species, one rodent and one non-rodent. The time of the administration of the toxic agent is very important. For example, 25 mg kg^{-1} lead nitrate given to pregnant rats on day 9 results in malformations in the foetuses, while on day 10 the same dose is lethal to them.

The predictive value of toxicity testing

a) Extrapolation from animal experiments to man

Extrapolation from animal experimental data to man is difficult for many reasons. Man has different body surface: body weight and respiratory minute volume: body weight ratios than the test animals. Also the metabolism of the compounds might be quite different. The extrapolation is relatively easy if a reference point is established both for the experimental animal and man. For example, if the dose required to elicit a well defined response is established for man and the experimental species, there is a scientific base for the extrapolation from the experimental dose response curve to safe exposure level in man. Data on the half-time and the metabolism of the toxic chemical in man and in experimental animals also helps extrapolation. Without parallel data, extrapolation from animal experiments is only guesswork based on analogies and probabilities with bias on the side of safety. Thus the threshold limit for uranium established on the basis of the urinary catalase activity of rabbits exposed to this metal was reduced four-fold when the same response had been established in hospitalised volunteers.

Zielhuis and Kreek proposed an extrapolation procedure that starts with a no-adverse effect level established in 90-day tests, and considers the difference between the respiratory volume of the test animal and the respiratory volume of man at work. Thus a male worker at rest has a respiratory volume of 55 l kg^{-1} 8 hr^{-1}, but at moderate physical activity this volume is increased to 130 l kg^{-1} 8 hr^{-1} (*i.e.* 9 m^3 8 hr^{-1} for a 70 kg man) which is 2.5 times less than the respiratory volume of a rat and twice as much as the respiratory volume of a cat. Thus a no-adverse effect level for the rat is multiplied by 2.5 and the no-adverse effect level of the cat must be divided by two to convert these values for man. A further division by a safety factor of 10 will give a tentative permitted atmospheric concentration for man.

The extrapolation is more complicated when only the oral or parenteral no-adverse effect level is known. In this case the oral no-adverse effect level must be multiplied by the LC_{50} and divided by the LD_{50} to calculate the approximate no-adverse effect level for inhalatory exposure in the test species and to calculate from this the permissible maximum atmospheric concentration as described.

This extrapolation method is not suitable for chemicals that are able to cause cancer. The toxicological evaluation of toxic compounds which might cause cancer requires longer tests than the 90-day test. These tests not only cover nearly the whole lifespan of the species, but their extrapolation to man requires a more cautious approach and depends on whether there is a threshold limit for a carcinogen at all. According to a booklet published by the New York Academy of Sciences most occupational cancer experts believe that there is no safe exposure to a cancer-causing substance. Others point out that there is ample evidence to show that a carcinogenic threshold exposure exists and such a threshold has been demonstrated for coal tar, β-napthylamine, bis-chloromethyl ether, 1,4-dioxane, vinyl chloride and dimethyl sulphate. They also point out that the carcinogens can be graded into high, middle and low potency carcinogens. This dependence on threshold explains why artificial sweeteners like cyclamate and saccharin produced tumours in mice only at doses which were unrealistically high when converted to human consumption. There are many chemicals with a long history of occupational exposure without any known case of cancer, though these compounds can be classified as carcinogens based on animal experiments. On the other hand Dr. Irving Selikoff of New York City's Mount Sinai School of Medicine found 3.5 times the expected rate of lung cancer in workers who were exposed to asbestos at least 25 years ago for three months or less. Though their short exposure might have been extremely high, this report calls attention to the irreversible consequences of any failure in keeping exposure below the threshold or minimal risk level.

b) Threshold and risk
The results of chronic experiments, like those of acute LD_{50} tests, are usually plotted to allow graphic evaluation. The choice of the graphic model partly depends on the dose range and partly on the acceptance of the threshold concept. Those who accept a threshold for carcinogens usually plot responses against log dose. When dose is increased by a constant factor this plotting method avoids the inconvenience of large gaps between neighbouring higher and overcrowding between neighbouring lower dose levels. Contrary to the sigmoid curve of similar data, plotted on a linear scale, the log dose model

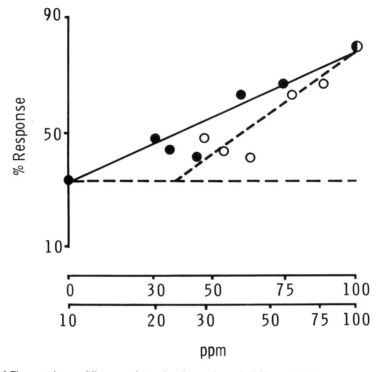

Fig. 6.1 The prevalence of liver neoplasms in mice at the end of 33 months dietary exposure to 2-acetylaminofluorene. Response is either plotted against linear dose (solid circles, upper abscissa) or against log dose (empty circles, lower abscissa). (Data of N.A. Littlefield *et al., J. Environ. Pathol. Toxicol.,* 1979, **3**, 17–34)

gives a straight line with a hockey stick end and the straight line intercepts the abscissa or the base line frequency at the threshold dose level (see right hand plot of Figure 6.1 or right hand corner on Figure 6.2). As the hockey stick model always gives a threshold dose for a response, those who reject the threshold concept will use other models. One possibility is to use a more compact dose range and plot response against dose on a linear scale. The smaller the dose range, the more likely that a straight or slightly curved line can be drawn through the data points. Moreover when the range of response is very much narrower than 0 to 100%, the intercept frequently moves from the abscissa to the origin, and when the background response is elevated, the dose–response line intercepts the ordinate at the background response level (indicating the absence of a threshold dose). However, when the background response is high, responses at low dose levels may fall below the upper

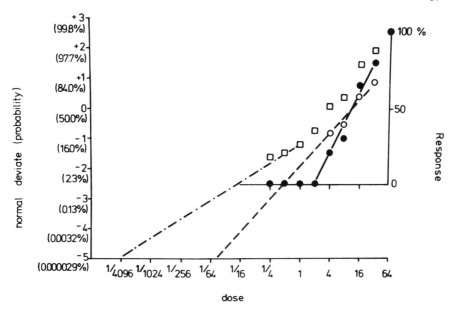

Fig. 6.2 The tumourogenic effect of a single injection of methylcholanthrene in mice. One dose unit equals 0.078 mg/mouse. Dose (log scale) is either plotted against linear frequency (solid symbols in insert where dose is lined up with the main abscissa) or against normal deviate (Probit) units without correction (empty circles) or corrected to the upper 99% confidence limit (empty squares). Note that without correction both zero and 100% responses are ignored and correction shifts zero responses upward. (Data of N. Mantel and W.R. Bryan, *J. Natl. Cancer Inst.*, 1961, 27, 455–470)

confidence limit of controls (*e.g.* in the experiment shown in Figure 6.1 the 95% upper confidence limit is 57.3% for eight liver cancers in 23 control mice) and the threshold concept cannot be rejected.

Another possibility is to retain the advantage of the log dose scale, but to change the response to Probit (or standard deviation) scale which has no zero response frequency. The Probit values are based on normal distribution and the middle of the highest point of the Gaussian bell shaped curve (50%) corresponds to the median lethal (effective) dose. Every percentage point left and right of 50% is converted to standard deviation units, that is $50 \pm 34\%$ to ± 1.0 S.D. and $50 \pm 47.7\%$ to ± 2.0 S.D and so on (see the left ordinate in Figure 6.2). The Probit plot gives a straight line without a threshold, and the straight line permits risk assessment at infinitely low responses (*e.g.*, 1 in a million).

The synergistic or additive effect of different carcinogens may disguise their individual threshold values by increasing the risk of cancer above the predicted risk for any of the two chemicals. Thus it has been documented that cigarette smoking increased lung cancer 10.8 times, exposure to asbestos by 5.15 times, but in the population of cigarette smoking asbestos workers the frequency of lung cancer increased by 53.24 times. Thus for those carcinogens which may promote the effects of each other, the setting of exposure limits, be they threshold or acceptable risk limits, always leaves a question mark.

A similar problem is presented by compounds which accelerate the ageing process. Ageing is a normal physiological process, but due to exposure to certain chemicals it becomes evident earlier than usual. For a long time it was believed that carbon monoxide has only acute effects which are present if at least 20 per cent of the haemoglobin is made unsuitable for oxygen transport. Recently it has become apparent that exposure resulting in less than this level is able to accelerate the ageing process of the cardiovascular system. A similar effect—though probably by a different mechanism—is exerted by another simple compound, carbon disulphide.

No model is available to predict the safe level of exposure if, as a result of antigen formation, antibodies are produced against the chemical. The reaction between the chemical and the antibodies in the sensitised individual can produce rashes (urticaria), allergic dermatitis or asthma. Sensitisation is quite different from hypersensitivity. In the case of sensitisation the response differs from the normal toxic effect of the chemical and its severity depends much less on the level of exposure than on the degree of sensitisation. In the case of hypersensitivity the reaction is the expected effect, but the dose response curve is shifted in the direction of lower doses.

Exposure as a measure of relative hazard

a) Toxicity and exposure
The main aim of toxicology testing is to establish how toxic the tested compound is in relation to other well known occupational chemical hazards. However, toxicity is only one of the factors that determine the degree of hazard for occupational exposure. Toxicity is a biological term and defines the capacity of a compound to interfere with normal functions or to cause injury. The probability that the toxic chemical will produce injury depends not only on its toxicity but also on the conditions of exposure. The results of the toxicity test already give the information as to whether the compound is irritative to skin or eye, whether it is a sensitiser and consequently if there is a need to protect the eyes with goggles and the hands with gloves. Toxic effects after the dermal application of the compound will reveal the hazard of absorption from the contaminated skin. The difference in oral and parenteral LD_{50} values will indicate how hazardous it is to risk ingestion by smoking or having a meal in a contaminated environment. It is necessary to know the physical form of the toxic substance, whether it is solid, liquid, gas or vapour.

Liquids can be splashed or sprayed, solids might form dust, and gas, vapour, spray and dust are the source of inhalation exposure.

b) The problem of complex exposure

Exposure to a single chemical is not the rule but an exception. Exposure to more than one compound might not change the risk presented by each of them, but there are examples of additive, antagonistic and synergistic effects. In the case of an additive effect, the critical organ or, in a broader sense, the critical mechanism is the same for the two compounds. Thus the effect of two narcotic solvents may add to each other. Contrary to additive effects, an antagonist decreases the toxicity of another compound. Thus methaemoglobin-forming agents act as antagonists against cyanide, azide or hydrogen sulphide. Another interaction is synergism. A synergist potentiates the effect of another compound either by interfering with its metabolism or by interfering with a basic biochemical process on which the toxic action depends. An example for the first mechanism is the potentiation of the toxicity of malathion by another pesticide, EPN, and for the second mechanism is the increase in the behavioural effect of amphetamine by exposure to CS_2.

c) Physical conditions and other factors in relation to exposure

The physical state, vapour pressure and solubility of a chemical are the factors which determine the type of exposure. The volume of the chemical, dilution, temperature and ventilation are the external factors which influence the degree of hazard. The knowledge of each of these factors is important in the assessment of possible danger to health.

If the toxic chemical is a gas, or a vapour, inhalation is the most important or possibly the only route of uptake; the exception is hydrogen cyanide. Fumes and solid particles floating in the air are also taken up by inhalation. Evaporation depends on the vapour pressure of the chemical and on its temperature. The higher the temperature the higher the rate of evaporation. Surface area is a very important factor—that is why spilled solvents evaporate so rapidly.

From the physical conditions of the chemical one cannot predict the volatility. Thus methylmercury chloride, which is a solid, is 5–7 times more volatile than the liquid metallic mercury. Compared with an undiluted chemical, evaporation is less from its solution, though in an enclosed space the final air concentration depends only on the vapour pressure at a given temperature. Fumes in the laboratory mainly derive from acids and their formation is also temperature dependent (external heat or reaction heat). The grinding of solid chemicals, and all solid chemicals in powder form, can form solid aerosols. Ventilation of the premises decreases inhalation exposure by the replacement of contaminated air by fresh air, though too strong a draught can increase evaporation or the formation of solid aerosols (dusting of the chemical).

d) Working processes as the source of exposure
A high proportion of exposure to toxic chemicals is of accidental origin. Fire promotes the evaporation of volatile chemicals and assists formation of toxic fumes (*e.g.*, oxides of nitrogen) and toxic gases (carbon monoxide, phosgene). Breakage of glass containers and spillage of volatile liquids also results in increased evaporation.

Pipetting by mouth can result in swallowing of the toxic chemical, but if the chemical is volatile, vapours dissolve in the saliva or can be inhaled. Grinding of solids or measuring out light powders in a draught increases the danger of the formation of solid aerosols. Inhalation exposure is a likely consequence when chemicals are heated outside a fume cupboard or in a fume cupboard with inadequate ventilation.

In analytical work chemicals are used in small quantities and thus the potential hazard is reduced, though for many purposes solvents, acids and alkalis are used in large quantities without dilution. If one considers only the health hazards, the preparation of a stock solution is always more dangerous than the final analytical reagent.

Distillation or other purification procedures and synthesis are the working processes where larger quantities of toxic chemicals might be handled. All the known laboratory methylmercury poisonings affected chemists who synthesised the compound.

e) Exposure and the route of uptake
Occupational exposure to toxic substances differs considerably from the intentional administration of a drug or a poison. In the latter case a known dose is selected to give the desired effect, in the case of occupational exposure the exact dose is not known and sometimes it needs a thorough examination even to identify the route of uptake. In occupational circumstances a toxic chemical can enter the body by inhalation, orally or through the skin, or by a combination of the routes. The respiratory tract is by far the most important route by which toxic chemicals enter the body. If atmospheric contamination is responsible for the exposure, the difference in the concentration of the chemical in the inhaled and exhaled air multiplied with the respiratory volume and time gives the dose. However, without any attempt to calculate the exact dose, the atmospheric concentration can be related to one or more responses. This is the basis for RL (recommended limit) or CL (control limit), TLV (threshold limit value) or MAC (maximum allowable concentration) figures.

It is more difficult to obtain the measure of exposure when oral or percutaneous absorption is the route of uptake or at least contributes to the total exposure. If the chemical is a gas which is absorbed through the skin and mucous membranes, air concentration is a measure of total exposure. Hydrogen cyanide—though a gas—can penetrate the skin to such an extent that without inhalation exposure the body can take up enough cyanide from the atmosphere to cause severe or even lethal intoxication.

In occupational circumstances oral or percutaneous exposure are usually interlinked. Contamination on the hand is the usual source of oral uptake through contamination of food and cigarettes. Absorption through the skin depends on the physical and chemical qualities of the toxic substance, but the surface area contaminated and time allowed between contamination and decontamination are equally important factors. Inside contamination of rubber gloves presents the most favourable conditions for absorption: increased temperature, vapour pressure, sweating and softening of the epidermis. Skin and the mucous membranes can be directly affected by the chemical. Chemical burns are usually caused by liquids.

In laboratories where the working procedures are alternating, it is unlikely that the occasional measurement of the atmospheric concentration of a chemical will give a truthful picture about the level of exposure. That is why it is important to judge whether appreciable exposure can exist at all. The awareness of the presence of toxic chemicals and of the conditions which might transform a potential hazard into an actual hazard is the best safeguard against harmful exposure. Nevertheless, in experimental plants or when a process is carried out for extended periods, *e.g.* synthesis, distillation, *etc*, it might be necessary to analyse the working environment for possible contamination.

There is no similar method of measuring absorption through the skin or through the gastrointestinal tract. If the chemical has a colour, discoloration of the skin can indicate the contact. Similarly the analysis of the skin wash may show how carefully the toxicants are handled. A very useful indicator of exposure is the concentration of a chemical or its metabolite in blood or urine samples. These biological tests give an overall picture of exposure independent of the route of entry.

f) The permissible limits of exposure
The permissible level of exposure has been defined by ILO and WHO experts as a quantitative hygienic standard of a level considered to be safe, expressed as a concentration with a defined average of time. The term permissible level for occupational exposure can mean recommended limit or control limit (RL or CL) in the UK, maximum allowable concentration (MAC) or threshold limit value (TLV). The American Conference of Governmental Industrial Hygienists (which is not an official governmental organisation) defined TLVs as conditions under which it is believed that nearly all workers may be repeatedly exposed, day after day, without adverse effects. Because of wide variation in individual susceptibility, exposure of an individual at or even

below the threshold limit may not prevent discomfort, aggravation of a pre-existing condition, or occupational illness. Contrary to this, the approach in the USSR is to set the maximum allowable concentration (MAC) not as a time weighted average but as the absolute upper limit of exposure which aims to prevent any deviation from normal (not only adverse health effects) detectable by the most modern research methods available. To define any deviation from the normal as unacceptable requires the definition of normal. The distribution of physiological functions within a 'normal' population, which being a mixture of healthy individuals and those affected by minor or major ailments, might not have a normal distribution at all. The suggestion that deviation from the normal must exceed twice the standard deviation is a compromise but it is unlikely to solve the problem.

This disagreement—according to a WHO document (1977)—has been a matter of concern for international organisations not only because of the difference in philosophy and terminology, but because of the experimental background on which these permissible limits are based. However, the problem of permissible limits is not only a philosophical one, but also a practical one. The frequency of control measurements, the sensitivity and accuracy of the method of measurement and the enforcement of permissible limits are the factors on which the value of TLVs or MACs depends.

g) Biological monitoring

The estimation of exposure frequently requires prolonged and complex environmental assessment. When intake of the toxic substance is not restricted to inhalation, biological monitoring, that is the estimation of the concentration of the toxic agent in biological samples, has many advantages over environmental tests because it uses the exposed person as a sampling device. Biological monitoring uses not only direct tests, that is the estimation of the toxic substance or its metabolite in blood, urine, hair or exhaled air, but also indirect tests which detect biochemical effects sensitive to exposure below or around the accepted TLV or MAC value. The demarcation line between direct and indirect tests is not always clear. For example, the estimation of carbon monoxide (CO) in blood is clearly a direct exposure test. However blood CO is associated with haemoglobin and the estimation of this abnormal blood pigment (COHb) fits into the category of indirect tests. In addition to being both a direct and indirect test, the estimation of COHb is also a diagnostic test because the impairment of oxygen transport through the formation of COHb is the essential step responsible for the toxic effects of CO. Though periodic medical examination does not fit into the category of biological monitoring, it may reveal the transgression of exposure above the acceptable limit, especially when employees are screened for the possible health effects of their particular exposure. Thus urinary porphyrin, an abnormal by-product of haemoglobin synthesis, and anaemia are indicators of toxic exposure to lead.

The consequences of toxic exposure

The symptoms and signs of intoxication are diverse and are neither always specific for, nor characteristic of, a toxic chemical. The following discussion might help to illustrate how the body responds to toxic chemicals. It also helps the understanding of the meaning of toxic effects in relation to the listed chemicals in Chapter 8.

a) The respiratory tract and the lungs
The upper respiratory tract through the bronchial system, which, like the branches of a tree, connects the trunk with the leaves, is connected with an intricate network of honeycomb-like airfilled spaces, the alveoli. Between the alveoli and the blood capillaries there is only a very thin membrane. This is the site of gas exchange between the inhaled air and the blood for the two physiological gases oxygen and carbon dioxide, but this is also the place for the uptake and release of toxic gases.

Any water soluble compound can be dissolved in the watery layer of mucous membrane of the bronchial system. However, high water solubility offers a protection as the dissolved compound is more easily cleared from the lungs. Though the immediate irritative effect of the water soluble sulphur dioxide might be more pronounced than that of the less water soluble nitrogen dioxide, the latter is more likely to produce pulmonary oedema (a retention of water in the lungs). Lung oedema and respiratory irritation have a pronounced detrimental effect on gas exchange. If the respiratory irritation is too severe, the lumen of the bronchial system constricts and increases the resistance to the respiratory airflow. Irritative chemicals such as sulphur dioxide, sulphuric acid mist, ozone, formaldehyde, phosgene, bromine, methyl bromide and nitrogen dioxide are all able to increase the respiratory resistance and many are able to cause oedematous changes in the lungs.

The bronchial system has a mechanism which removes particles from the lungs and any damage to this escalator promotes sensitivity to infections or dusts. These secondary effects with the primary irritant effect can lead to irreversible structural changes in the lungs manifested by a decrease in the surface area available for gaseous exchange.

b) Skin and eyes
In the laboratory environment contact of the skin with chemicals is a frequent though not an unavoidable risk. Compounds that irritate the skin or mucous membranes in the vapour form can cause more severe damage if splashed on the skin. Irritation is a general term which covers a multiplicity of processes. For example the removal of the lipid layer by solvents or alkaline detergents; dehydration of the skin by acids, anhydrides and alkalis; precipitation of proteins by heavy metals; oxidation by acids, peroxide or chlorine; the

dissolution of keratin by soap, alkalis or sulphides. If the contact lasts longer or is repeated frequently the injury called contact dermatitis becomes more and more severe.

There are so-called sensitisers which have the same effect as the primary irritants but the first signs appear only after 5–7 days' exposure. Chromate, formalin, phenylenediamine and turpentine belong to this group.

Halogenated compounds act on the skin where there is discontinuity: chloracne is composed of plugged sebaceous glands and suppurative inflammation of the hair follicles. There are substances such as coal tar, creosote, Rhodamine N and bergamot oil which sensitise the skin to sunlight.

Skin can be affected by a chemical taken orally or by other routes. Thus thallium results in the loss of hair, and arsenic the proliferation of the keratin layer.

Vapours that irritate the skin are more likely to irritate the conjunctiva or other mucous membranes. This effect is more pronounced when the irritant is rubbed into the eye with a dirty hand, or when acid, alkalis, lime, solvents or detergents are splashed into the eye. Chemical burns can affect the conjunctiva and the cornea. In the latter case scarring or vascularisation as a part of the healing process might affect the visual function.

Chemicals can affect not only the surface of the eye; chronic exposure to dinitrophenol is able to cause opacity in the lens and thallium or methanol can damage the optic nerve.

c) Gastrointestinal tract and liver

The gastrointestinal tract can be the route of uptake of a toxic chemical, but it can also be the site of action if corrosive chemicals are swallowed. Every chemical that can cause chemical burns on the skin, can injure the oesophagus and the stomach. Bleeding, perforation and deformation are the outcome of such a contact effect.

Other chemicals like arsenic, barium, chloromethane, fluorides or mercuric chloride lead to gastroenteritis and increased motility of the intestines.

The liver is the first organ exposed to a chemical absorbed from the gastrointestinal tract. However, its vulnerability is based more on the extraordinary metabolic activity of the liver cells which can structurally modify many foreign chemicals. If the chemical is detoxified, the liver is the last organ to be exposed to the more toxic parent compounds; if the change is in the opposite direction, the liver is the first organ to be exposed to the more toxic forms.

The uptake of non-polar chlorinated hydrocarbons by the liver is facilitated by their lipid solubility, but their toxicity mainly depends on the metabolic transformation with the formation of highly reactive metabolic products.

Chemicals can injure the liver in many ways; they can increase its water content, fat content and they can destroy the cells. The liver has an ability to regenerate, that is, to replace the dead cells, but repeated toxic effects result in a widespread scarring of the liver and it is therefore unable to carry out the normal physiological functions. These functions contribute to the general metabolic processes of the whole organism (lipid, protein and sugar metabolism), the formation of the bile and the extraction of chemicals from the plasma for bile secretion. Hepatotoxic agents, besides causing morphologic changes, can depress the metabolic activity of the liver, the bile flow and the excretion of bilirubin and other compounds. Bilirubin is a metabolic product of haem, the active part of haemoglobin and different haem enzymes. Approximately 300 mg bilirubin is formed per day (more when haemolysis occurs) and a failure of its biliary excretion due to 1,1,2,2-tetrachloroethane, carbon tetrachloride or arsenic can cause severe liver damage resulting in jaundice. Nearly all the solvents have the potential for causing some degree of liver injury which might be noticed only if enzymes, leaked from the liver cells into the blood, are estimated or some functional tests are carried out. If the liver once suffers chemical damage or an infective disease, its threshold to the effect of hepatotoxic agents is decreased.

d) Blood

The effect of a chemical on blood can be direct or indirect. In the former case the chemical acts on the blood itself and in the latter case on the formation of the formed elements or the plasma. The formed elements, red blood cells, white blood cells and platelets, are suspended in a volume ratio of 1:1 in the plasma, which is an aqueous solution of proteins, amino acids, salts and other small molecules. The main function of 4.5–5 million red blood cells per μl is the transport of oxygen from the lungs to the tissues. This function is carried out by a special protein called haemoglobin which makes up 97 per cent of the dry weight of these cells. The number of white cells is very much smaller, only 7000 per μl, their function is the defence against infection. The smaller platelets, which number 300000 per μl, contribute to another defensive mechanism: blood coagulation.

One of the most common effects of toxic chemicals on blood manifests itself in a change of the number of formed elements. There are so-called haemolytic poisons, like arsine, phenylhydrazine or a large group of aromatic amino and nitro compounds which are able to haemolyse the blood. In the case of arsine the effect can be so dramatic that the large quantity of haemoglobin released from the cell aggravates the effect of arsine on the liver and kidneys.

The aromatic amino and nitro compounds are able to oxidise the ferrous iron of the haem part of haemoglobin. This causes, through the formation of an inactive haemoglobin called methaemoglobin, a deterioration in the supply of oxygen to the tissues. Sometimes methaemoglobin formation is useful, as

105

methaemoglobin competes for cyanide with essential ferric enzymes engaged in the utilisation of oxygen.

One of the most common forms of inactive haemoglobin is carboxyhaemoglobin, the reaction product of carbon monoxide and haemoglobin. The inactivation is based on the affinity of carbon monoxide for haemoglobin which is approximately 240 times higher than that of oxygen. Carboxyhaemoglobin is fully dissociable and when the carbon monoxide concentration decreases in the inhaled air, carbon monoxide is replaced on the haemoglobin by oxygen. The reactivation of methaemoglobin needs the participation of reductive enzymes present in the red blood cells.

Anaemia caused by lead is the consequence of its interference with the synthesis of haem, *i.e.* the non-protein part of haemoglobin. The first block in the synthesis of haem results in the increased urinary excretion of δ-aminolaevulinic acid, the substrate of one of the inhibited enzymes; the other step of interference results in the excretion of a degradation product (coproporphyrin). The estimation of these two compounds is widely used in the clinical diagnosis of lead intoxication or in the measurement of exposure.

The best known toxic chemical affecting the white blood cells and platelets is benzene. Benzene can shift their numbers in two directions. At first the numbers of white blood cells and platelets decrease, as after ionising radiation. The most common manifestation at this stage is leakage of blood from the capillaries even after minor trauma. If the exposure is high and long lasting, the effect of benzene can move in the opposite direction resulting in the proliferation of the tissues which produce the white blood cells.

Cell death caused in any part of the body, but mostly in the liver, can leak enzyme proteins into the plasma, and thus their activity in blood can be used as a measure of tissue damage. Chronic liver damage caused by alcohol or other solvents might result in a disorder in the synthesis of plasma proteins with, as the first sign, a decrease in the albumin concentration. A general decrease in the concentration of plasma proteins might be the consequence of loss of proteins through the kidneys.

e) Nervous system

The nervous system is divided into two parts: the central and the peripheral. The central part is the brain and the spinal cord; the peripheral part consists of the motor and sensory nerves. The uptake of chemicals by the central nervous system depends on the vascularisation of the brain and the characteristics of the chemical. Non-polar, lipid soluble compounds, such as halogenated solvents easily penetrate the barrier between blood and brain tissue. Their effect, like that of chloroform, is anaesthetic-narcotic, that is they depress the function of the central nervous system. This, in severe cases, leads to respiratory and/or cardiac failure. Though the biochemical mechanism is quite different, the general effect of anoxaemic poisons is very similar to the narcotic ones. Chemical anoxaemia is caused either by the failure of the oxygen transport in the blood by the inactivation of haemoglobin (carbon monoxide, aromatic amino or nitro compounds) or by the inhibition of the utilisation of oxygen in the tissues (cyanide, azide).

Though it is a solvent, carbon disulphide has a more selective effect on the central nervous system than other solvents. It interferes with the metabolism of catecholamines, which belong to those compounds which participate in the transmission of an impulse from one nerve cell to the next. This effect is more pronounced in a certain part of the brain, which is affected also by manganese. Cerebral oedema caused by triethyltin, encephalopathy caused by lead, ataxy, blindness, deafness *etc* caused by methylmercury and behavioural disorders and tremor caused by mercury vapour are based on a more selective interaction between the toxic chemical and biochemical processes in the brain.

Many of the chemicals like carbon disulphide which affect the central nervous system also affect the peripheral nerves. Central nervous symptoms like drowsiness and headaches are the consequence of mild intoxication with anticholinesterase organophosphorus compounds, but some of them are able to cause delayed peripheral neuropathy in the form of paralysis. Tri-*o*-cresyl phosphate though it is not a potent anticholinesterase inhibitor caused a mass epidemic of paralysis in Morocco when engine-oil treated with tri-*o*-cresyl phosphate was sold and used as cooking oil. Acrylamide, a different type of compound, injures not only the motor but also the sensory nerves. Contamination of the skin with phenol results in a chemical burn which may go unnoticed because of the concurrent anaesthetic effect.

f) Kidneys

One of the main functions of the kidneys is the excretion of metabolic products (mainly from protein metabolism) and the regulation of the salt and water balance. Both processes are linked to the formation of urine which is also an important route for the clearance of toxic compounds and their metabolites. During the process of urinary excretion the chemicals filter, diffuse or are transported through cells in the kidneys, but some remain temporarily bound to cell components.

Heavy metals such as mercury, or chlorinated hydrocarbons like carbon tetrachloride, are the most prominent groups of compounds which can damage the kidney cells. The consequence is either a change in the composition or in the volume of urine or both. For example, in severe cases of mercuric chloride intoxication there might be no urine at all, in milder cases the occurrence of cell debris and protein in urine indicates renal damage.

Heavy metals like cadmium, mercury, uranium and lead are also able to increase the excretion of amino acids, glucose, or phosphate as a result of an impairment in the function of tubular cells to reabsorb these small molecules from the preformed urine. As the half time of cadmium is 10–40 years in the kidneys, even a short term exposure to cadmium will increase the kidney concentration for life. Without occupational exposure cadmium is taken up only from food and the cadmium content of the kidneys increases up to 50–60 years of age. Then the ageing process probably results in loss of kidney cells with their cadmium.

Excretion of toxic compounds with urine is a powerful protective mechanism. In some cases the concentration of toxic chemicals in the urine exposes the lower part of the urinary tract to higher concentrations than from the blood. 2-Naphthylamine and benzidine proved to be so potent in causing bladder tumours by this way that their use in the UK is now prohibited.

g) Cardiovascular system

Heart or respiratory failure is the cause of death in many acute intoxications. Chemicals such as carbon monoxide, phosgene, nitrogen dioxide, hydrogen sulphide, chlorinated compounds or solvents, which can injure cells in other organs can also injure the heart. In the not too severe cases of anoxaemia caused by carbon monoxide, cyanide or methaemoglobin-forming agents, the toxic effect is shown only by a decrease in blood pressure or an increase in the frequency of heart beat.

Ageing of the cardiovascular system is one of the cardinal processes of ageing in general. Lead, carbon monoxide or carbon disulphide are able to speed up atherosclerosis and increase the possibility and severity of coronary disease. Fluorocarbons, trichloroethylene and some other halogenated alkanes can produce, in addition to nausea and dizziness, irregular heartbeats. Heart failure may be precipitated after a symptom-free period by physical stress. Exposure to organic nitrates, like nitroglycol and nitroglycerine, results in skin flushes and throbbing headache. These vasodilatory effects subside and disappear when exposure is repeated regularly, but after a few years the interruption of the exposure, that is the daily supply of vasodilators, precipitates chest pain and even sudden death.

h) Bone and muscles

The whole body is supported by the skeletal system. Loss of calcium by exposure to cadmium, when aggravated by nutritional deficiency, can result in the softening and deformation of the bones. The compression of nerves by deformed bones can be painful. In a district in Japan where river water contaminated with cadmium was used for irrigating paddy fields this disease (called Ouch Ouch) reached epidemic proportions. In contrast to cadmium, fluoride and phosphorus can increase the fragility of bones. Interestingly lead, which is mainly stored in the bones, has no similar effects.

In severe carbon monoxide intoxication the loss of muscle power may trap the victim at the site of exposure, as he will be unable to move.

Bibliography

B. Ballantyne (ed). *Current approaches in toxicology*. Bristol: J. Wright and Sons, 1977.

V.K. Brown. *Acute toxicity in theory and practice*. John Wiley and Sons, Chichester, 1980.

L. Chiazza, Jr., F.E. Lundin and D. Watkins (eds). *Methods and issues in occupational and environmental epidemiology*. Ann Arbor Science, Ann Arbor, 1983.

J.F. Douglas (ed). *Carcinogenesis and mutagenesis testing*. Humana Press, Clifton, N.Y., 1984.

J. Doull, C.K. Klaassen and M.O. Amdur. *Casarett and Doull's Toxicology. The basic science of poisons*, 2nd Edn., Macmillan Publ. Inc., N.Y., 1980.

W.J. Hunter and J.G.P.M. Smeets (eds.). *The evaluation of toxicological data for the protection of public health*. Commission of European Communities, Oxford: Pergamon Press, 1977.

P. Lehman. *Cancer and the worker*, New York: New York Academy of Sciences, 1977.

R. Roi, W.G. Town, W.G. Hunter and L. Alessio. *Occupational health guidelines for chemical risk*. Commission of European Communities, Luxembourg, 1983.

H.E. Stokinger. *The case for carcinogen TLVs continues strong. Occup. Health and Safety*, 1977, **46**, 54–58.

E. Vigliani (ed). *Methods used in western European countries for establishing maximum permissible levels of harmful agents in the working environment*. Milan: Fondazione Carlo Erba, 1977.

World Health Organization. *Principles and methods for evaluating the toxicity of chemicals*. Geneva, 1978.

R.L. Zielhuis and F.W. Van der Kreek. *The use of safety factor in setting health-based permissible levels for occupational exposure. Int. Arch. Occup. Environ. Health*, 1979, **42**, 191–215.

Chapter 7

Health Care and First Aid

Throughout this book the emphasis is upon the importance of safety at work. To many people this conception still conjures up images of hard helmets and other devices to protect against physical hazards, while the more insidious and sometimes more dangerous effects upon health of exposure to toxic chemicals are often overlooked.

In the past there has been a neglect of these dangers in chemical laboratories where a great variety of reagents are in daily use, some toxic, others more innocuous. Furthermore, whenever such hazards have been recognised little guidance has been available to managers either on how to obtain medical advice or on what precautions are necessary to protect the health of laboratory staff.

The purpose of including a chapter on health care and first aid is to draw attention to some of the commoner health hazards that may threaten in a chemical laboratory and, more importantly perhaps, to give advice on where management may turn for help in protecting the health of their staff.

Medical services and poison information centres

The explosion of legislation on health and safety at work that has occurred during the past decade reflects the greatly increased concern felt by society for the welfare of those of its members who spend the greater part of their active life in employment. Much of this legislation and certainly the social obligations which have prompted it apply to the laboratory as to any other place of work. No longer can a laboratory manager afford to ignore possible hazards to the health of his staff.

In most industrialised countries there is an official labour inspectorate which usually includes a number of medical specialists; these doctors are normally available to advise on the precautions needed to protect against any health hazards that may be encountered in the course of laboratory operations.

A manager who has any doubts or queries either on health risks in his laboratory or on the first aid cover which he needs to provide, is well advised to contact a doctor in one of these services to enlist his aid. In the UK this advice is provided by the Employment Medical Advisory Service (EMAS) of the Health and Safety Executive. The head office address of EMAS and of comparable organisations in other European countries is given at the end of this chapter (*Appendix 1*). It is worth noting that in the UK and usually in other countries a government inspector may be approached directly by a staff association, trade union or indeed any single individual who is unsatisfied with the safety of his working conditions. Even if no medical service is available a non-medical member of a labour inspectorate may be able to offer useful advice on the general principles of maintaining a healthy working cnvironment. If a laboratory manager is unfortunate enough to have no local source of advice, then he can do no better than put into practice the precepts he will find in this book.

It goes without saying that a laboratory manager should be familiar with all the hazards that may be associated with the chemicals and reagents used in his laboratory. He should make it his duty to know which of the reagents used in the laboratory are toxic and to possess some knowledge on how their toxicity may manifest itself. More importantly, perhaps, he should understand through which route the chemical in question is likely to enter the human system—*via* the skin, by inhalation or by ingestion. Armed with this information he will be in a better position to institute appropriate precaution-ary measures. Chapter 8 sets out to provide exactly this sort of information. Managers who wish to know more about the toxic properties of chemicals and their effects in man should turn to one of the many text books on occupational toxicology.[1, 2]

Most countries maintain poison information centres which keep up-to-date information on the toxic properties of chemical substances and many common proprietary products. These centres are available for professional advice on the possible health hazards of chemicals and on the treatments recommended for dealing with poisoning incidents. In asking for assistance the inquirer should give an accurate description of the chemical responsible for the incident, the route of its entry into the body (mouth, skin or lungs) and the time elapsed between the incident occurring and the inquiry being made. Every prudent manager should keep handy the address and telephone number of the poison centre nearest to his laboratory. For convenience a list of addresses and telephone numbers of European poison information centres is given at the end of the chapter (*Appendix 2*). Like the Scouts Association, all well-conducted laboratories should adopt the motto: 'Be prepared'.

Exposure standards (*see also* p. 101)

There are a small number of chemicals that are so toxic that their effects can be rapidly lethal and those which are dangerous to this degree are usually well known to all laboratory personnel. In general, though, drama and health hazards are not common bed fellows. Managers (and doctors) should beware of jumping to the conclusion that a sudden medical emergency in the laboratory is necessarily due to the acute toxic effects of a chemical, however tempting the association may first appear. There are many other natural causes for a medical emergency which may have nothing to do with the chemical in use at the time of the incident. Common things happen most commonly; dramatic toxic effects are always uncommon.

Chemicals may have acute or chronic effects. The former are of rapid onset, quickly apparent, and for the most part reversible once further exposure is avoided. Chronic effects are likely to be insidious, delayed in their appearance, and may be irreversible. For these reasons they tend to be more dangerous and it is therefore against low level exposures that the manager and his staff need to guard most carefully. Many toxic chemicals have threshold limit values (TLV) recommended by the American Conference of Governmental Industrial Hygienists (ACGIH). These limits, expressed in ppm or mg m^{-3}, are updated annually and are published in a booklet.[3] This publication also contains valuable advice on precautions recommended for the handling and use of hazardous substances at work.

Many countries adopt the ACGIH TLVs for their own use, but others, such as the UK, GFR, the Netherlands and the Scandinavian countries, are developing their own limit values recognised by different abbreviations (*e.g.* RL—'recommended limits', CL—'control limits' in the UK; MAK or MAC values—'maximum acceptable concentrations'). In some countries (*e.g.* the US, the GFR, the Netherlands) these national exposure standards are legally enforceable; in others (*e.g.* the UK) they are advisory. Each laboratory should be aware of the permissible exposure levels and the definitions that are adopted by the country in which the laboratory operates. The Council of

Europe is also proposing to introduce a harmonised list of exposure standards for adoption by the EEC countries.

To observe workplace standards it is essential to measure the concentration of volatile or dusty substances in the working atmosphere. How this can be done is too complicated to describe in this chapter, but there are well recognised methods developed for the purpose which can be found in the published literature. Alternatively advice on such methods and their applications can be sought from labour inspectorates or other organisations expert in the field of occupational health and hygiene. It is important to stress, however, that no standard can guarantee freedom from health hazard. This principle is clearly stated in the ACGIH booklet which emphasises that although most persons may be repeatedly exposed without harm to the exposure limits (TLVs) recommended, there will be a small number who, for one reason or another, may experience adverse effects even at exposure levels below the TLV. Keeping to recommended exposures standards should not, therefore, lull laboratory staffs into a false sense of security. The overriding aim should always be to reduce any exposure from chemicals to the lowest reasonably achievable minimum below the recommended standard. Furthermore differences in permitted exposure levels must not be interpreted as indicators of relative toxicity between one chemical and another; many factors other than toxicity are taken into account before arriving at the recommended figure.

Carcinogenic and mutagenic chemicals (*see also* Chapter 8)

One of the most emotional health topics of which laboratory managers should take note is the problem of carcinogenic and mutagenic chemicals. They need to be aware of the background to the issue and be able to give reasonable answers to questions they may be asked by their staff or by critics outside the laboratory. It is not possible to discuss in this chapter complicated theories on the causation of cancer. On the other hand there are a number of practical remarks which may help laboratory managers to understand the background to the problem and to control the handling of carcinogenic substances in the laboratory.

In the first place it is important to recognise that hazards to health from carcinogens or mutagens are only likely to be of real significance if there is repeated and prolonged exposure over long periods of time. Thus, most attention is needed to prevent exposure of this type though precautions must also be taken to protect against short-term, intermittent exposures as far as is reasonably possible.

Mutagens
By definition a mutagen is an agent that changes the hereditary genetic

113

material contained in every living cell. There is good evidence to suggest that a mutation of genetic material is one of the early causes that trigger a sequence of events ultimately leading to the development of a cancer. This remote association has been misinterpreted in some quarters to mean that mutagenic chemicals must, *ipso facto*, be regarded as carcinogens.

There are many short term, *in vitro* laboratory tests to detect mutagenic properties of chemicals. The best known is the Ames test conducted on specially selected bacteria. A corollary of the misleading view that mutagens are carcinogens is that any chemical given a positive result in the Ames test (or any other single test for mutagenicity) is a human carcinogen. There is no scientific justification for this doom laden conclusion. However, an association between cancer and mutagenesis, distant though it may be, cannot be denied and any chemical which is proved to be mutagenic in more than two appropriate tests for mutagenesis should be handled with care. Unnecessary or uncontrolled exposure should be avoided.

Unfortunately there is no comprehensive list of proven mutagens so laboratory staffs must rely on sources of information such as professional journals and their contacts with experts in this subject to keep up to date in a rapidly advancing and bewildering field.

Carcinogens

There are only a limited number of proven occupational human carcinogens, proven in the sense that there is irrefutable evidence that they have caused cancer in persons exposed to them. A list of chemical substances and processes associated with cancer in man is reproduced in *Table 7.1*. It is taken from Supplement 4 of the International Agency for Research on Cancer (IARC) Monographs *Evaluation of the carcinogenic risk of chemicals to humans*.[4] Other sources of useful information on proven and suspect human carcinogens may be found in references 3 and 5. All three publications are strongly recommended for study by laboratory managers and their staffs who have responsibility for the handling and use of carcinogenic substances. As well as identifying carcinogenic substances the last two publications also give helpful advice on general principles to protect persons against exposure to carcinogens while at work. In nearly all industrial countries many carcinogens, proven or highly suspect, are covered by national legislation. Laboratory managements should be familiar with any legislation which applies to their own laboratory operations.

Safe handling precautions

It is not possible to lessen the inherent toxic nature of a chemical substance but, indirectly, the same result can be achieved by substitution with a less harmful one.

Table 7.1 Chemicals or industrial processes causally associated with cancer in humans†

1	4-Aminobiphenyl
2	Arsenic and arsenic compounds
3	Asbestos
4	Azothioprine*
5	Auramine (manufacture of)
6	Benzene
7	Benzidine
8	N,N-Bis(2-chloroethyl)-2-naphthylamine (chlornaphthazine)
9	Bis(chloromethyl) ether
10	Boot and shoe manufacture and repair (certain occupations)
11	1,4-Butanediol dimethanesulphonate (Myleran)*
12	Chlorambucil*
13	Chloromethyl methyl ether (possibly associated with bis(chloromethyl) ether)
14	Chromium and certain chromium compounds
15	Conjugated oestrogens*
16	Cyclophosphamide*
17	Diethylstilboestrol*
18	Furniture manufacture (hard woods)
19	Haematite mining (radon?)
20	Isopropyl oils (manufacture of isopropyl alcohol by the strong acid process)
21	Melphalan*
22	Mustard gas
23	2-Naphthylamine
24	Nickel (nickel refining)
25	Phenacetin*
26	Rubber industry (certain occupations)
27	Soot, tars, and unrefined oils
28	Treosulphan*
29	Vinyl chloride

* Medicinal drugs
† The list of chemicals or industrial processes given in this table should not be taken as a thorough compilation of chemicals known to induce cancer in humans. It reflects only those chemicals or industrial processes which have been evaluated in the IARC programme up to 1982 (IARC Monographs on the Evaluation of the carcinogenic risk of chemicals to humans. Volumes 1–29). Other agents have been designated as human carcinogens: tobacco smoking, betel-nut chewing, and alcoholic beverages, but have not yet been covered in the Monograph programme.

All chemists know that benzene is an ideal solvent. In terms of a health hazard, however, it is far from ideal and unless there are special reasons for its retention, benzene should be replaced by a less toxic solvent such as toluene or xylene. There are many other similar examples and wherever the principle of substitution can be adopted, it should be practised.

115

Chapter 3 provides a comprehensive account of the measures which can be taken to reduce the escape of hazardous fumes or dust into working atmosphere. The use of protective clothing to reinforce these precautions is in one sense an admission that a hazard still remains. There are, of course, many situations in which simple personal protective equipment such as goggles and gloves are an indispensable part of safe working practices, but protective clothes make uncomfortable wearing and sometimes poor dexterity. There is a well known aphorism in occupational health circles which says 'If you can bring an influence to bear external to the workman over which he can exercise no control you will be successful (in protecting him); if you do not or cannot you will never be wholly successful.' Protective clothing comes last on the list of effective safety precautions.

The following is a check list of precautions which have been found useful in protecting against unacceptable exposure to any toxic substance posing a health hazard:

- Substitution with a less harmful reagent.
- Avoidance of contact between employee and toxic chemicals by
    (i) a standard of laboratory design compatible with the need to avoid the escape of dangerous chemicals into the working environment at concentrations which could be hazardous to health;
    (ii) the application of sensitive techniques to detect small leaks of toxic chemicals into the working atmosphere; and
    (iii) insistence upon the proper use of personal protective clothing and equipment.
- Regular background and personal monitoring for levels of exposure to chemicals in the laboratory atmosphere.
- The dedication of special rooms or areas restricted to the handling of carcinogenic or other highly toxic substances.
- Conspicuous and intelligible labelling of carcinogenic and toxic reagents.
- Separate storage of carcinogenic and highly toxic substances away from other reagents and available only on request to a senior member of the laboratory staff.

In general the stringency with which laboratory safe working practices are applied should be geared to the degree of health hazard posed by the chemicals in use. The highest standards are necessary for protection against chemicals known to be human carcinogens. Training and a frank exchange of information should be an integral part of all precautions designed to reduce contact between chemicals and the laboratory staff.

Most of the advice given in this section may be summarised by the equation:

Toxicity × Exposure = Health Hazard

If one or both of the two left-hand factors can be reduced it follows that any potential health hazard will be smaller. If they can be reduced to zero then the hazard disappears. In practice zero values for either factor can rarely if ever be achieved, but if sufficient energy and resource is devoted to controlling them, a zero hazard can be approached. There is no human activity in which hazards to life or limb can be eliminated, but at least it should be possible to ensure that work in a laboratory is no more hazardous than life outside it.

Biological monitoring

Although the measures advised in Chapter 3 and in the preceding sections of this chapter should, if fully adopted, serve to protect most people working in the laboratory against the harmful effects of toxic chemicals, it has to be acknowledged that a degree of exposure or contact will inevitably take place, sometimes at unexpectedly high concentration. It is, for example, not possible to eliminate accidents or human error.

Despite the undoubted value of atmospheric monitoring to protect against harmful exposures, the best reassurance is a biological check to measure if unacceptable absorption into the body is taking place. The risk that this may be happening will partly depend upon the frequency and quantity with which chemicals are used in the course of the working day. When toxic chemicals are handled frequently it may be prudent for the laboratory management to ask advice from a medical doctor familiar with the practice of occupational health on whether biological monitoring should play a part. If biological checks are started they must of course be conducted under the supervision of a doctor who should treat individual results with the same confidentiality he would bring to the medical problems of any of his patients. Before disclosing his findings to a third person he should obtain the consent of the person providing the sample under examination.

A few chemicals which have important toxic properties and for which biological monitoring may be appropriate are listed below with their eight hour average threshold limit values as recommended by the ACGIH,[3] and with brief details of the biological checks which may detect excessive absorption. The last are no more than a very brief indication of procedures that may be useful. Expert advice is always needed on how to apply and interpret them.

Benzene
RL and ACGIH TLV 10 ppm (30 mg m^{-3})*
Benzene is known to cause leukaemia in some persons exposed regularly to

* There is a strong likelihood that in many countries this standard will be substantially reduced.

high concentrations of its vapour, but provided every effort is made to keep exposure to benzene vapour at the lowest reasonably achievable levels below the 10 ppm TLV, there is little likelihood that a health hazard will exist. Recent work has confirmed that absorption of liquid benzene through the skin is negligible.

Biological checks
1 Examination of the peripheral blood cells to detect benzene related abnormalities.
2 Measurements of the quantity of phenol conjugates excreted in the urine. Phenol is a metabolite of benzene.

Inorganic lead compounds
CL and ACGIH TLV 0.15 mg m^{-3}
If absorbed in excessive quantities, lead interferes with the normal production of red cells in the blood and also attacks the nervous system. Other less well defined, though more insidious, effects have been reported (but not confirmed) on the mental faculties of children exposed to lead.

Proposed Council of Europe standards (EEC countries)
Lead in air $\leqslant 150$ μg m^{-3} averaged over a 40 h working week.
Lead in blood $\leqslant 70$ μg 100 ml^{-1} blood.
ALA in urine $\leqslant 15$ mg l^{-1} urine.

Biological checks
1 Examination of the red cells in the blood.
2 Measurement of the level of lead in the blood. This test is difficult to conduct. It must be performed in a laboratory familiar with the technique or the results will be unreliable.
3 Measurements of coproporphyrins and/or amino laevulinic acid in the urine. Both may be excreted in increased quantities if excessive lead absorption has taken place.

Inorganic mercury
RL and ACGIH TLV $= 0.05$ mg m^{-3}
Excessive absorption of mercury may cause severe irreversible changes in the nervous system and kidney damage. The elusive nature of mercury droplets, combined with its volatility at room temperature, makes mercury a dangerous and insidious poison. Work with metallic mercury requires stringent precautions to prevent its escape from the laboratory bench.

Biological check
Measurement of the concentration of mercury excreted in the urine

Isocyanates
CL all isocyanates 0·02 mg m^{-3}; ACGIH TLV toluene di-isocyanate (TDI) $= 0·005$ ppm (0·04 mg m^{-3})— ceiling value.

ACGIH TLV methylenediphenyl isocyanate (MDI) = 0·02 ppm (0·2 mg m^{-3})—ceiling value.

The main target for both TDI and MDI is the lungs. The characteristic health effect is sensitisation manifested by the development of acute asthmatic symptoms in the presence of TDI or MDI. Repeated exposures may result in permanent lung damage. Sensitisation is an insidious and, at times, an unexpected complication. It takes time to develop and is always preceded by a period of symptomless exposure during which time the sensitisation process is maturing. Once established, severe symptoms may follow exposure to almost undetectable concentrations of the sensitising agent. If this condition has developed in a person exposed to isocyanates no further work with these chemicals should be permitted. In extreme cases this may mean the removal of the sensitised person from any employment in a laboratory in which isocyanates are present.

Biological check
Lung function tests. Both TDI and MDI are potent sensitisers, but of the two TDI is the more dangerous because of its much greater volatility; this is reflected in its much lower exposure limit standard (TLV). The main hazard from MDI is the possibility of its absorption by inhalation of aerosol droplets. Having entered the lungs both chemicals act in an identical way.

The chemicals that have been described in this section all present chronic health hazards which usually manifest themselves in a slowly progressive manner after exposure has taken place regularly over a relatively long period. Treatment, when necessary, always requires skilled medical attention and is usually a prolonged and tedious procedure in which emergency first aid measures have little or no part to play. A first essential must always be to remove the subject from any further exposure to the offending chemical until a full recovery has taken place. Even then careful thought is needed to decide whether work with the chemical responsible for the illness should ever again be allowed.

The development of any of these illnesses inevitably means that unacceptable exposures have been taking place. It calls for an urgent review of the general hygiene and safe working practices of the laboratory.

Dermatitis

In all fields of employment dermatitis present a problem of such magnitude and is the subject of so many misconceptions that it justifies an attempt to explain its causation in simple terms before outlining the principles of its prevention.

Laboratory technicians, exposed during their work to a great variety of chemicals each one potentially harmful to the skin, are especially vulnerable to occupational dermatitis which is the most frequent cause of medical disability arising out of their employment.

In the minds of many people dermatitis is synonymous with occupational skin disease. Inside medical circles however the terms means no more than 'inflammation of the skin'. It is not a disease entity in itself and its entry on medical certificates as a diagnosis without an indication of its cause is not helpful.

Many causes of dermatitis have no connection with occupation and it is worth observing that a skin disease that appears to be connected with work may have its origin in activities out of working hours or be a manifestation of some other more general medical condition.

In general terms, medical classification divides occupational skin disease into two main groups: contact and sensitisation dermatitis. Both are caused by contact with substances which for one reason or another interfere with normal skin physiology; the number which can be incriminated is almost infinite. Some are sufficiently powerful to damage the skin of all persons working with them if contact occurs. Others, usually regarded as innocuous, will give rise to skin disease only if handled by an individual wth an abnormal sensitivity.

Informed judgements, based upon past experience and a knowledge of its chemistry, may be offered on the likelihood of an unfamiliar substance inducing skin disease, but in the final analysis only contact between worker and substance will supply the answer.

Contact dermatitis

The diagnosis is applied to an inflammation of the skin caused by exposure to a substance that attacks its surface. Depending upon the degree of damage suffered the inflammation may be acute or chronic and will not heal until contact with the offending agent is stopped. The distribution of the skin affected will be largely determined by the physical nature of the dermatitis agent. Liquids and solids tend to affect the skin of the hands and forearms while fumes or dusts produce a much more diffuse pattern usually involving the face and neck. After an attack of contact dermatitis the behaviour of the skin is unpredictable. Whether it will respond in the same manner to a repetition of contact will depend upon the concentration of the substance to which it is exposed, the duration of that exposure and natural resistance to further attack. Small exposure may be tolerated without further trouble, and a return to work with a chemical known to have induced a previous attack of contact dermatitis is justified, provided it is done with caution and excessive or prolonged exposure is avoided. After two recurrences all further exposure should be avoided.

There are many well recognised skin irritants amongst the chemicals to be found in a laboratory though their action upon the skin of different individuals is not constant. Resistance of the skin to attack is the main factor in determining whether a contact dermatitis will or will not follow exposure. Because of their diversity and number it is impossible to give a comprehensive list of chemicals that are known to be irritants. Reference should be made to Chapter 8 where the irritant properties of individual chemicals are mentioned under toxic effects. They include all corrosive chemicals and most solvents because of their defatting and desiccating action on the skin. For this reason the practice of removing stains from the hands by cleaning with solvents is always to be deprecated. Other conditions that render the skin liable to attack are constant immersion in water, constant friction or pressure, and the 'soggy' skin which develops when natural perspiration is unable to escape during prolonged use of rubber gloves.

As a general rule people with fair skins or red hair are more vulnerable to contact dermatitis than their darker colleagues. Persons who suffer from any form of chronic non-occupational skin disease such as eczema should avoid employment in a laboratory where they may be exposed to chemicals whose actions could potentiate pre-existing disease.

Sensitisation dermatitis
Unlike contact dermatitis, sensitisation dermatitis develops following an activation of the natural immunity of the skin. Superficially the events that take place may be compared with the response made by the body to vaccination against smallpox. In the latter there is the deliberate introduction of minute quantities of an infectious virus so that the body will marshal and prepare its defences to meet a much more serious threat if and when exposure to the disease itself occurs. Similarly, in sensitisation dermatitis the offending compound must penetrate the skin where it becomes attached to one of its protein elements. This combination stimulates the immune defences present in the skin so that if further exposure to the same substance follows, a reaction, manifested by damage and inflammation of the skin, takes place between the foreign chemical and the reinforced defences. It is this sequence of events that is implied by the word sensitisation. A colourful analogy is the destruction inflicted upon a battlefield in the course of a battle.

Typically, a symptomless period varying from a few weeks to many months, during which time the immunological defences of the skin are being primed, follows first exposure to the sensitising compound. Once this priming has reached a certain stage of advancement, sensitisation dermatitis will break out if contact continues. It is not uncommon for a sensitising agent also to be a skin irritant producing a contact dermatitis by direct damage to the skin surface. When the two conditions coexist the diagnosis is difficult and treatment complicated.

Healing of the skin after sensitisation dermatitis is often delayed and irregular and will only follow complete removal of the sufferer from any further exposure. Unlike contact dermatitis the area of skin affected is usually extensive and on occasions may involve the whole body surface. If for any reason contact with the sensitising agent continues, the skin may become so responsive that it reacts violently to the presence of the agent at almost molecular concentrations, and in extreme cases the subject may not be able to enter a room in which the sensitising agent is present without developing a recurrence.

From this brief and simplified description it follows that a diagnosis of sensitisation dermatitis is more serious in its implications than is one of contact dermatitis. Once the former diagnosis has been made, no further exposure should be permitted to the sensitising agent when a return to work is allowed and, if no other alternative is available, the only solution may be a change of occupation.

Many substances are recognised to be skin sensitisers but because of individual susceptibilities it is not possible to predict who is liable to become sensitised or who will remain unaffected. Another difficulty that characterises sensitisation dermatitis is its appearance in an individual who may have worked harmlessly for many years with a substance to which he suddenly becomes intolerant. One authority quotes the five most common causes of sensitisation dermatitis in order of frequency as:

1 Hair dyes (*p*-phenylenediamine)
2 Nickel compounds
3 Rubber compounds
4 Epoxy- and phenol-formaldehyde plastics
5 Antihistamine skin medicaments derived from 1,2-diaminoethane (*see also* Chapter 6, p. 104)

Local preventive measures

It does not require much intelligence to appreciate that the most effective protection against dermatitis is to avoid all contact with potentially irritating or sensitising chemicals. This ideal cannot always be achieved, but the objective of preventive measures is to approach it as closely as possible. Clean working conditions, properly planned bench operations and careful attention to the familiar and simple principles of skin hygiene all contribute to reducing the incidence of skin disease. Cuts and abrasions should be treated without delay and kept covered with a clean dressing until fully healed. In all this the help of an efficient first-aider is invaluable.

For work with corrosives or with other known irritants, impermeable gloves should be worn and care must be taken to ensure that they are intact. It is not, however, good practice to wear such gloves for too long since the skin of the hands becomes soggy and vulnerable to damage. Nor is it wise to allow laboratory gloves to remain in use for long without replacement or to permit their indiscriminate use by all members of the laboratory staff. To do so will merely encourage contamination inside the glove and help to spread skin infections from one user to the next.

Barrier creams

There are many varieties of barrier cream on the market for which enthusiastic claims are made by the manufacturers. They are harmless in themselves and may help to keep the skin in good condition, but there is little convincing evidence that they provide effective protection against irritant dusts or vapours. Against sensitisation reactions they offer little protection.

Possibly their most useful function is to keep the skin clean and to help in removing grease and oil stains. Certainly they cannot replace more conventional measures such as good personal hygiene and careful working methods.

Treatment

Any persistent skin disease should be seen by a doctor and, if its cause is occupational, no resumption of contact with the causal agent should be allowed until healing is complete. If the condition proves to be a sensitisation dermatitis there should be a complete ban on any further work with the sensitiser unless or until medical approval to resume contact is obtained.

If an excessive incidence of dermatitis occurs in a laboratory and the cause is not apparent or easily eliminated the laboratory manager should seek the advice of a doctor familiar with the practice of occupational health to assist in detecting the source and controlling exposure to it.

Occasionally the agent responsible for an attack of dermatitis cannot be identified and this failure may result in mystifying recurrences. In such cases patch testing of the skin under medical supervision with a selection of substances normally encountered at work is advisable, but the physician conducting the tests needs special expertise in their technique and interpretation.

First aid

Before setting out to describe the arrangements for first aid that should be provided in any chemical laboratory it is as well to make a few introductory remarks which all persons responsible for organising first aid facilities and their application should consider.

In a working environment, wherever it may be or whatever the tasks performed inside it, there are broadly speaking two categories of incident for which first aid may be required:

1 Incidents arising from natural causes, and
2 Incidents arising as a direct result of the work performed.

Those in the first category are, by and large, unavoidable and are as likely to occur inside the workplace as outside it. Familiar examples range from a simple faint through the scale of common illnesses up to the serious acute emergency presented by a heart attack. For all of them there are appropriate first aid measures which in extreme cases may be life-saving. However, apart from ensuring that trained first aid assistance is readily available there is little else that can be done in the way of precautionary measures to prevent their occurrence. Their treatment is not the concern of this chapter and to learn about them the first aider must turn to such publications as *The First Aid Manual*, published jointly in the UK, by the St. John and St. Andrews Ambulance Associations and the Red Cross.

By contrast those incidents in the second category are avoidable and must be viewed as the direct result of a failure to observe the principles of safe working conditions. Since common things happen most commonly, these incidents are usually of such a trivial nature that no treatment is called for and they are dismissed as being of no significance; only rarely will they be of such a serious nature that urgent first aid is necessary. Nevertheless, whatever the outcome, such accidents always indicate a need to examine and perhaps correct the method of work from which they arose. Against incidents of this kind first aid is only a second line of defence and the provision of a first aid service, however comprehensive or well prepared, must never be allowed to supplant constant supervision and improvement of working methods by all members of a laboratory staff. The theme of working safely has already been the subject of preceding chapters of this book but its re-emphasis in the context of first aid can do no harm.

In the UK the 'Health and Safety (First Aid) Regulations 1981' require all employers (and self-employed persons) to provide '. . . such equipment and facilities as are adequate and appropriate in the circumstances for enabling first aid to be rendered to his employees if they are injured or become ill at work.'

The legal situation will vary in other countries, but each laboratory manager should be aware of how such regulations may affect his own obligations and act accordingly. The general principles, though, are the same.

Because of the tremendous variation in the size of laboratories and the multifarious nature of the work on which they are engaged it is not possible to offer in a single short chapter advice which can be universally applied without consideration of individual circumstances. For this reason any laboratory manager who is unsure of the scale of first aid cover appropriate to the operations for which he is responsible would be wise to invite a doctor experienced in occupational health to visit the laboratory and act upon the recommendations made.

The number of trained first aiders available in any single laboratory must obviously depend on the numbers of staff employed and upon the dangers presented by the work on which they are engaged, but as a basic rule at least one fully trained first aider, usually recruited from the staff, should be on duty at all times during working hours. It is equally important to ensure that everybody employed in the laboratory should know who the nominated first aider is and the site of his usual place of work. This information can be displayed on a board easily visible in every room where bench work is taking place.

In this context it should not be forgotten that there will be times when the nominated first aider will be away from work. For such absences a deputy needs to be appointed who should undergo the same training as the colleague for whom he is substituting. If a particularly hazardous operation such as work with hydrogen cyanide or hydrofluoric acid is to be undertaken the laboratory manager should consider whether to make it a rule that no work commences unless a first aider is on duty.

In the UK an approved code of practice has been issued[6] to complement the First Aid Regulations. It gives practical guidance on the application of the regulations, on the selection and training of occupational first aiders, on the equipment and facilities that need to be provided and, according to the degree of hazard that may be encountered, on the ratio of trained first aiders to number of employees.

Finally, it is worth pointing out that even in silent hours there may be cleaners at work who are not familiar with the chemicals by which they are surrounded and who are at special risk because of their isolation. Their safety remains the responsibility of the laboratory manager.

Training of the first aider

Like every other skill the practice of first aid can only be learned by training and practice. None of the many wall-charts and simple diagrams available for display illustrating the treatments recommended for first aid emergencies can in any way approach the level of knowledge and confidence acquired through a properly conducted course of training. In the UK the St. John Ambulance Association, St. Andrews Ambulance Association and the Red Cross offer such courses and there will certainly be an opportunity for any prospective first aider to attend one without having to travel far from his home or place of work. Sister organisations in most other industrialised countries arrange similar courses and information on them is usually available from local representatives of the training bodies.

Any laboratory manager is well advised to take advantage of these facilities in order to reassure both himself and his staff that when needed first aid will be given by a qualified person. Every first aider should be in possession of '*The First Aid Manual*' or comparable handbook.

The objective of this training is not to produce instant doctors, but rather to turn out a person able to assess the seriousness of an emergency rapidly and apply with confidence a few basic manoeuvres designed to prevent a deterioration in the casualty's condition until more professional medical attention can be obtained. The essence of its success is speed and simplicity of action combined with confidence and common sense. The measures to be learned are few, but the first aider must be familiar with them and be prepared to use them in difficult and sometimes cramped physical surroundings. The well-ordered conditions in which training sessions take place are rarely repeated on those occasions when the first aider is called upon to exercise his skills in earnest.

Equipment

In the UK, first aid boxes are part of the 'equipment and facilities' required by the Health and Safety (First Aid) Regulations. The Approved Code of Practice issued with the regulations recommends a list of contents for permanent first aid boxes. The list, which is reproduced below, has been selected on the basis of long experience and, although not obligatory, laboratory managers would be well advised to see that a similar range of items is quickly available.

- A card giving advice on first aid treatment available from HMSO (General first aid guidance for first aid boxes)

Sufficient numbers of:
- Individually wrapped sterile adhesive dressings
- Sterile eye pads
- Triangular bandages (if possible sterile)
- Safety pins
- A selection of medium, large and extra large sterile ummedicated wound dressings

First aid boxes should be placed in the care of a nominated first aider who should ensure that the contents are always kept up to standard. Often the latter may disappear because of unauthorised borrowings, but the remedy is not to keep the boxes locked or hidden in a safe place to preserve their inviolability.

In addition to first aid boxes there should always be one or more eyewash-bottles strategically placed about the laboratory and filled with at least 450 ml^3 (16 fl oz) of clean water. Many suitable varieties of bottles are available commercially or one can simply be constructed from standard laboratory equipment (*see Fig. 7.1*).

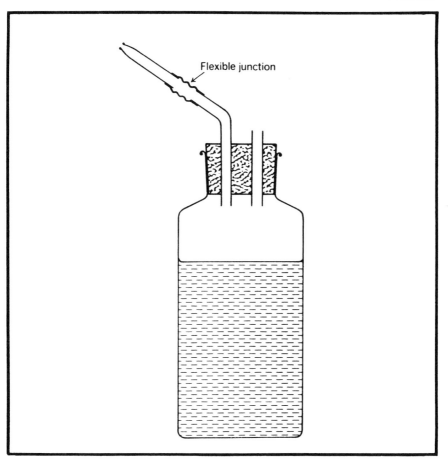

Flexible junction

Fig. 7.1 Eyewash-bottle. Note the use of a flexible dispensing nozzle. Over-enthusiastic irrigation in the heat of the moment with an unyielding nozzle can inflict added damage to the eye

Like the boxes, each bottle should be in the care of the nominated first aider who will be responsible for ensuring that the water is changed at least once a month. A number of sophisticated buffer solutions have been recommended for use in eyewash-bottles but they have no advantage over clean water and should be avoided. It is important that not only the first aider but all members of the laboratory staff are familiar with the proper use of eyewash-bottles since speed of action to wash out the eye is of overriding importance in the treatment of eye injuries occasioned by chemicals.

However, prevention is better than cure and eye protection in the form of goggles should always be used to protect against splashes entering the eyes of staff handling liquid reagents. Indeed, the laboratory manager would be wise to make a cardinal rule that no member of the staff should handle liquid chemicals, whatever their nature, without wearing some form of eye protection.

First aid boxes and eyewash-bottles are the only essential pieces of first aid equipment that should always be provided. There is on the market a large range of more sophisticated equipment such as oxygen-giving sets and bellows designed to administer artificial respiration. Whether these are added to the first aid armoury must depend largely on the size of the laboratory, the discretion of the laboratory manager and the opinion of his advisers, but the prudent director should be wary of purchasing expensive equipment from which there will be little or no practical benefit.

Last, but by no means least, it is incumbent upon every manager to ensure that all reagents in the laboratory are clearly labelled.

Treatments

It is not the purpose of this chapter to replace professional handbooks on first aid by giving instructions on methods of administering artificial respiration. These techniques are an essential part of the first aider's knowledge, but along with other basic practical measures can only be learned from an experienced instructor. It cannot be emphasised too strongly that every first aider should attend a suitable training course before he can be expected to exercise his skills with any degree of competence.

In addition to nominating and training individual members of the staff in the skills of first aid it should be impressed upon all members of the laboratory that each of them should also be prepared to provide first aid assistance if called upon to do so, and it is therefore wise for everyone to be familiar with a number of simple and effective treatments which can be applied without specialised training. Such measures are described below under the heading of *Standard treatments*. Each is designed to prevent a casualty's condition from worsening until more skilled aid is to hand and should be seen as a team responsibility in which every member of the laboratory staff has a part to play.

Most accidents with chemicals that are likely to occur in a laboratory will fall into one of the following categories:

splashes of the skin (including chemical burns)
splashes of the eyes
inhalation of gases or dusts
ingestion of chemicals
heat burns

Since there are few specific antidotes against the great majority of chemicals, the standard treatments described below are designed for the type of accident irrespective of the chemical involved. Thus all splashes of the skin should be treated in the same way. Similarly, each of the other four groups of accident calls for its own particular treatment. In Chapter 8 there are brief monographs of the hazardous properties and effects upon the body of many of the chemicals commonly used in laboratories. Under each one reference is made to first aid treatment. When only the entry *standard treatment* is made the reader should turn to the page indicated against it where a summary of the recommended first aid treatment will be found. It is essential that these simple standard treatments are applied with the minimum of delay.

Standard treatments

Splashes of the skin (for summary see p. 136)
In general, dangerous splashes are either corrosive (acids and alkalis) inflicting chemical burns, or are toxic by absorption through the skin (phenol, aniline). In all such serious emergencies continuous drenching with water and removal of all contaminated clothing is the essential treatment. All chemicals must be regarded as hazardous unless proved otherwise and if any doubt exists the drenching must be started immediately before further information is sought. In such circumstances it is better to err on the side of caution at the expense of the victim's comfort rather than allow a potentially hazardous chemical time to exert its action. Similarly, modesty in removing contaminated clothing must not be allowed to interfere with the treatment.

Drenching should continue for up to five minutes or until the first aider has satisfied himself that no further danger exists. This treatment must take priority over all other actions. Water is safe, normally available in large quantities and is simple to apply. In an emergency these features make it the treatment of choice over all other remedies that may be advocated. Gentle cleaning of the skin with soap while the drenching continues will help to remove the splash. If it is necessary to refer the casualty to hospital details of the treatment and the chemical responsible should be provided.

Splashes of the eye (for summary see p. 136)
Here again, the same principles apply as recommended under splashes of the skin. Flooding with a gentle stream of water is invariably the treatment of choice and should be continued for up to 10 minutes or longer if considered necessary. Strong alkalis are particularly damaging to the eye and must be treated with great care. A special feature of eye injuries is the spasm which may develop in the lids to keep them firmly shut. To overcome this the first aider must gently prise the lids apart with his fingers to ensure the water bathes the eyeball and continue to hold the lids apart until the treatment is completed. Flooding with water is best done by holding the casualty's head under a gently

129

running tap or by using one of the specially provided eyewash-bottles. Care must be taken not to use a powerful jet of water which may cause added damage to the eye. Regard for the casualty's clothing must not interfere with the thoroughness of the treatment. All injuries to the eye require medical attention and if removal to hospital is necessary information on the chemical responsible with brief details of the treatment given should accompany the casualty. As a precaution against this type of accident protective eye-shields or glasses should be used whenever liquid chemicals are in use.

Inhalation accidents (for summary see p. 136)
Of the routes by which toxic chemicals can enter the body, inhalation is the most dangerous and rapidly acting. Fortunately most toxic gases are either acutely irritating to the tissues of the respiratory tract or possess a warning odour which is detectable at concentrations well below the danger level. This last property however is an unreliable indicator of their presence since the nose quickly becomes insensitive to smell and a serious casualty may result if the initial warning is ignored. In all cases of gassing the essential treatment is to remove the subject into fresh air. Provided breathing has not stopped and the casualty is conscious there is little else that needs to be done except to maintain careful observation on his condition. If breathing has stopped artificial respiration should be administered until it resumes naturally or medical attention becomes available. In cases where hydrogen cyanide is the gas responsible, mouth-to-mouth artifical respiration must *not* be employed because of danger to the rescuer. In its place the Silvester method must be used (for details refer to a suitable first aid manual). Oxygen may be given to all serious cases of gassing if an oxygen set is available and the first aider trained in its use. If the victim is unconscious he must be placed in a face-down position in order to maintain a clear airway to his lungs and prevent the inhalation of vomit or foreign bodies such as false teeth into the breathing spaces. In all cases of serious gassing or if the first aider is not happy about a casualty's condition removal to hospital must be arranged as soon as possible. Information on the gas responsible and brief details of the treatment given should accompany the casualty.

Ingestion accidents (for summary see p. 137)
This route of entry presents the least practical hazard since it is unlikely that any signficant quantity of a foreign liquid or solid will be swallowed without deliberate intent. However, improbable accidents do occur though usually the offending chemical will be spat out and not swallowed. If it is corrosive, burns of the mouth may result and for such injuries repeated mouth-washes with water is the correct treatment. In this context it must be emphasised that pipetting by mouth is a practice that should be absolutely forbidden in every laboratory.

If a poisonous substance has been swallowed it should be diluted by giving about 250 ml of water to drink and the casualty moved to hospital as rapidly as possible. It is important in such cases to supply information on the chemical involved with brief details of the treatment given and an estimate of the quantity consumed.

The question of inducing vomiting is a difficult one. In general, it should not be attempted since it may result in further damage to the delicate tissues of the upper food passages if the chemical swallowed is corrosive and there is always a danger of vomit being inhaled. Tickling the back of the throat or drinking strong solutions of common salt are traditional methods recommended, but neither is reliable and the second may be dangerous. Present medical opinion is strongly against the use of salt water to induce vomiting. In general, attempts to induce vomiting should be deferred until the casualty reaches hospital where it can be done under skilled medical supervision.

No attempt should be made to induce vomiting in unconscious or semi-conscious persons.

Burns
Chemical burns from corrosive reagents should be treated as advised under the standard treatment for splashes of the skin.

Dry burns from bench fires and scalds from boiling water or hot chemicals are a familiar hazard even in the most elementary of laboratories. Depending upon the gravity of the burn the damage inflicted will vary from superficial reddening of the skin to extensive surface blistering and death of underlying tissues. However serious, the correct first aid treatment is no more than covering the burnt surface with loosely applied dry sterile dressings. These should form part of the standard contents of all first aid boxes. To reduce the dangers of infection, handling of the burnt area must be reduced to a minimum and any temptation to clean its surface must be resisted. All burns or scalds of more than a trivial nature should be referred to hospital.

If the victim's clothes are on fire the flames must be extinguished either by drenching with water or by throwing the victim to the ground and smothering with any garment that comes to hand such as the rescuer's lab coat. It is hardly necessary to emphasise that these moves must be resolute and swift.

Special treatments

There are a number of chemicals whose properties are so dangerous and rapidly acting that specific antidotes have been developed against them. Brief details of the treatments recommended are to be found under their respective monographs in Chapter 8. One or two of the chemicals however are in sufficiently common use to warrant an amplified description of the advance precautions needed and the treatment to be given if a casualty occurs. Where this is the case a reference to the appropriate page number giving the expanded account is made in the summarised entry to be found in Chapter 8 under the chemical in question.

In all laboratories where dangerous chemicals of this order are handled it is a wise precaution to liaise in advance with the nearest hospital and agree on a plan of action for the immediate admission and treatment of a casualty. The first aider should also make himself familiar beforehand with the measures he must take on the spot if the need arises.

*Hydrogen cyanide and other dangerous nitriles**

The immediate urgency is to remove the casualty well away from the neighbourhood of the gas. If there is time a breathing apparatus should be put on by the rescuer to ensure that he too does not fall victim to the gas and whenever possible a cord or similar device should be attached so that he can be dragged clear if he succumbs. Obviously circumstances will dictate the manner of the rescuer's actions, but it must be emphasised that inadequate preparations and unsuccessful heroics will worsen the emergency. If after reaching fresh air the casualty is still conscious and breathing he must be made to lie down quietly and await removal to hospital. No other treatment is required apart from keeping a close observation upon his general condition. Only if the breathing stops or consciousness is lost should the following measures be adopted:

Loss of consciousness

Place in the prone position with the mouth down. Ensure a clear breathing passage.

Breathing arrested

Apply artifical respiration by the Silvester method. Do not use the mouth-to-mouth method because of the danger of inhaling hydrogen cyanide gas from the victim's lungs. Cardiac message may be necessary if the casualty's heart has stopped beating. To learn the techniques of artificial respiration and cardiac massage reference should be made to a suitable first aid manual.

Amyl nitrite†

In cases of severe poisoning break two amyl nitrite capsules in quick succession beneath the victim's nose so that the vapour is inhaled. If the breathing has stopped this must be done by an assistant during the inhalation cycle of the artificial respiration. If these measures are ineffective and the patient's condition deteriorates, the Kelo-Cyanor resuscitation kit should be used.

---

* Acetonitrile, acrylonitrile, cyanogen bromide, cyanogen chloride, lactonitrile.

† Note: In the UK amyl nitrite capsules are available through retail chemists. They have a limited stock life and note must be taken of the expiry date for renewal. They should be stored at temperatures not exceeding 15 °C.

Kelo-Cyanor antidote (dicobalt ethylenediaminetetraacetate; EDTA)
Specially prepared packs of Kelo-Cyanor antidote* are available from most retail pharmacists. They include ampoules of Kelo-Cyanor ready for intravenous injection (20 ml ampoules containing 300 mg Kelo-Cyanor). One or more packs should be provided by the laboratory for on-the-spot use by trained personnel or to accompany the casualty to hospitals.

An intravenous injection of Kelo-Cyanor should only be given if both of the following criteria are satisfied.

1 The casualty is verified as a case of cyanide poisoning;
2 The casualty is unconscious or lapsing into unconsciousness.

The question of who is entitled to administer an intravenous injection in an emergency of this gravity is a contentious one. It may be possible for a nursing sister or an experienced first aider to receive proper training in this technique from a local doctor or the accident department of the nearest hospital. Whether it is necessary to prepare in this way will depend upon the degree of hazard present in a laboratory and the discretion of the laboratory manager, but any arrangements to do so must always be made in conjunction with a doctor who, for an emergency of this gravity, may be willing to accept responsibility in writing for the competence and judgement of the person selected to give the intravenous injection.

Soluble cyanides
Cyanide salts and solutions of cyanides present considerably less risk than does hydrogen cyanide gas since they are dangerous only by ingestion and this is unlikely to occur in significant quantities except by deliberate intention; poisoning by inhalation of these compounds is not a hazard. This, however, does not mean they should be treated without due circumspection. Solutions of over 1 per cent strength should always be kept under strict control; those under 1 per cent constitute, in practice, little danger, but always demand careful handling.

If ingestion of cyanide solution has taken place, amyl nitrite should be administered as for hydrogen cyanide and the casualty rapidly moved to hospital with details of the incident and treatment given.

In the past it has been the practice to prepare as cyanide antidotes solutions of ferrous sulphate (A)† and sodium carbonate (B)† to be taken by mouth as

---

* Distributed by Rona Laboratories Ltd., Cadwell Lane, Hitchin, Herts., UK.

† A. 158 g ferrous sulphate crystals ($FeSO_4 7H_2O$) and 3 g BP citric acid crystals in 1 l of cold, distilled water (the solution must be inspected regularly and replaced if any deterioration occurs). B. 60 g anhydrous sodium carbonate (Na_2CO_3) dissolved in 1 l of distilled water.
50 ml of solution A is placed in a 170 ml (6 oz) wide-necked bottle closed by a polythene-

antidotes to cyanide ingestion.

An attempt to estimate the quantity ingested should be made and its concentration noted to obtain some idea of the amount of cyanide consumed.

Only if the patient has collapsed and is in danger of dying should the question of an intravenous injection from the Kelo-Cyanor pack be considered.

Although cyanide can be absorbed through the skin, splashes of cyanide solutions do not constitute a major emergency; nevertheless they should be washed off immediately with large volumes of water.

The Health and Safety Executive has prepared a special wall-poster against cyanide poisoning (No. SHW 385). Copies are available through HMSO offices and give clear instructions on treatment. The Chemical Industries Association in the UK (CIA) has also prepared a label (No. 15), which gives details of first aid care and advice on medical treatment. It is intended for taking with the casualty to hospital. Wherever there is a significant risk of cyanide poisoning in a laboratory either by inhalation or ingestion these notices should be available and the former prominently displayed.

Hydrogen fluoride and hydrofluoric acid

Hydrogen fluoride and its solutions in water, (hydrofluoric acid) inflict destructive and extremely painful burns. With concentrated preparations these effects are rapidly apparent, but with dilute solutions they are often delayed and may not be noticed for a number of hours. All laboratory staff likely to handle hydrogen fluoride or hydrofluoric acid should be warned of this insidious hazard and instructed to take immediate action if they are splashed.

Any laboratory handling hydrogen fluoride or hydrofluoric acid regularly should make advance arrangements with a local hospital for the treatment and admission of casualties rather on the same lines as recommended for cyanide poisoning. The symptoms, which may start immediately or may be delayed, commence as a dull throbbing that builds up to become an acutely severe and persistent pain due to death of the underlying tissues. If not treated the condition can result in extensive and permanent damage which may involve the underlying bone. Accompanying the pain there may be a visible reddening of the damaged skin.

covered cork and labelled clearly CYANIDE ANTIDOTE A. 50 ml of solution B is similarly bottled and labelled CYANIDE ANTIDOTE B. Both bottles should bear the legend 'Mix the whole contents of bottles A and B and swallow the mixture'.

The merit of the basic ferrous hydroxide suspension that is swallowed is that it is likely to induce vomiting while at the same time forming insoluble non-toxic iron complexes with the cyanide.

The first aid treatment consists of removing all contaminated clothing and flooding the skin with large volumes of running water immediately. This should be continued for at least one minute. Thereafter 2 per cent calcium gluconate gel* should be applied in large quantities to the affected areas and continuously massaged into the skin for at least 15 minutes or until medical aid becomes available. Should the symptoms recur a further application of the gel should be made in the same way and continued until the pain subsides. If the acid has penetrated below the nails the ointment should be liberally applied over and around the nail and the area continuously massaged for at least 15 minutes.

All hydrogen fluoride or hydrofluoric acid splashes of any size must be referred to hospital after washing the skin and starting treatment with the ointment. The CIA has issued label No. 14, revised in 1975, which explains clearly and precisely how to treat hydrofluoric acid burns and all laboratories that handle the acid should have ready a special fluoride pack containing the ointment and a copy of the revised CIA label. A pack should always accompany the casualty to hospital. The Health and Safety Executive has produced a wall-poster (F2250) on the treatment of hydrofluoric acid burns. It is obtainable through any HMSO office and should be displayed prominently in the vicinity of any work with hydrogen fluoride or its acid.

Splashes of hydrogen fluoride or hydrofluoric acid in the eyes are particularly painful and dangerous; they must be treated immediately by flooding for at least 15 minutes with large volumes of running water from a tap (*for details of treatment* see *Splashes of the eye* p. 129).

The urgent removal of the patient to hospital is essential since failure to treat correctly may all too easily result in the loss of an eye.

Two compounds, phenol and hydrogen sulphide, commonly encountered in chemical laboratories deserve special mention because of their dangerous properties.

Phenol
Treatment for phenol splashes is the same as the standard treatment for *Splashes of the skin* given on p. 129. Phenol is caustic and rapidly absorbed through the skin. If a large surface is splashed there is a real danger of death from collapse and kidney damage. It is essential to remove all contaminated clothing and flood with water for at least 15 minutes. The casualty must be removed to hospital promptly and information given on the details of the accident.

---

* Obtainable from most retail pharmacists or suppliers of pharmaceutical products as 'H-F Antidote Gel'.

Hydrogen sulphide

Hydrogen sulphide gas is commonly encountered in many laboratories. In concentrations above a few hundred ppm it can be rapidly fatal. There is no specific antidote against it and first-aid treatment is the same as that recommended for inhalation accidents (*see* p. 130). Strict safety precautions must be in operation whenever it is generated.

There are, of course, other chemicals with equally dangerous properties but since they are not part of the normal stock-in-trade of chemical laboratories they have not been mentioned in this chapter. Appropriate warning against them are found under the individual chemicals given in Chapter 8.

Summary of standard treatments

Splashes of the skin

If the chemical responsible is hydrogen fluoride, hydrofluoric acid or a related compound turn to p. 134 for instructions.

1 Flood the splashed surface thoroughly with large quantities of running water and continue for at least 10 minutes, or until satisfied that no chemical remains in contact with the skin. Removal of agents insoluble in water will be facilitated by cleaning the contaminated skin area with soap.
2 Remove all contaminated clothing, taking care not to contaminate yourself in the process.
3 If the situation warrants it, arrange for transport to hospital or refer for medical advice to the nearest doctor. Provide information on the chemical responsible to accompany the casualty with brief details of the first aid treatment.

Splashes of the eye

If the chemical responsible is hydrogen fluoride, hydrofluoric acid or a related compound turn to p. 135 for instructions.

1 Flood the eye thoroughly with large quantities of gently running water either from a tap or from one of the eyewash-bottles provided and continue for at least 10 minutes.
2 Ensure the water bathes the eyeball by gently prising open the eyelids and keeping them apart until the treatment is completed.
3 All eye injuries from chemicals should be seen by a doctor. Supply information to accompany the casualty of the chemical responsible with brief details of the treatment already given.

Inhalation accidents

If the gas responsible is hydrogen cyanide turn to p. 132 for instructions.

1 Remove the casualty out of the danger area after first ensuring your own safety.

2 Loosen clothing.

3 If the casualty is unconscious place in a face-down position and watch to see if breathing stops.

4 If breathing has stopped apply artificial respiration by the mouth-to-mouth method. If the gas responsible is hydrogen cyanide *only use* the Silvester method.

5 If the emergency warrants it, remove the patient to hospital and provide information on the gas responsible with brief details of the first aid treatment given.

6 Administer oxygen if available and if the casualty's condition is serious.

Ingestion of poisonous chemicals

1 If the chemical has been confined to the mouth give large quantities of water as a mouth wash. Ensure the mouth wash is not swallowed.

2 If the chemical has been swallowed give about 250 ml of water to dilute it in the stomach.

3 Do not induce vomiting as a first aid procedure.

4 Arrange for transport to hospital. Provide information to accompany the casualty on the chemical swallowed with brief details of the treatment given and if possible an estimate of the quantity and concentration of the chemical consumed and the time elapsed since the emergency occurred.

References

1. G.D. Clayton and F.E. Clayton. *Patty's industrial hygiene and toxicology. Vol. II—Toxicology* (3rd revised edn.) New York/London: Wiley Interscience Publishers, 1982.
2. N.I. Sax. *Dangerous properties of industrial materials* (6th edn.). New York: Van Nostrand Reinhold, 1984.
3. *Threshold limit values for chemical substances and physical agents in the workroom environment.* Cincinnati: ACGIH (updated annually).
4. *IARC Monographs on the Evaluation of the Carcinogenic Risk of Chemicals to Humans.* Vols 1– 29. Supplement 4. Oct: 1982, Lyon, France.
5. *Occupational cancer—prevention and control.* Geneva: International Labour Office.
6. *First Aid at Work.* Health and Safety Series booklet HS(R)11, 1981, HMSO.

Appendix 1: Addresses of Principal labour inspectorates and occupational health authorities

Austria
Federal Ministry of Social
  Administration,
Central Labour Inspectorate
Stubenring 1
A1010 VIENNA

Belgium
Nationale Arbeidsraad
Blijde Inkomststraat 17–21
BRUSSELS
Tel: 02-7354000

Denmark
Directorate of the Danish
  Labour Inspection Services
Rosenvaengtallé 16–18
2100 COPENHAGEN Ø
Tel: 01-382800

Finland
National Board of Labour
  Protection
Hämeenkatu 13 bA
3310 TAMPERE 10
Tel: 31-37411

France
Inspection du Travail
391, Rue Vaugirard
5e Arondissement
PARIS
Tel: 01-8286311

GFR
Arbeitsambt Düsseldorf
Josef Gockelnstrasse 7
4000 DUSSELDORF
Tel: 211-43061

Greece
Section de la securite du Travail
Rue Piréos 40
ATHENS

Ireland
Industrial Inspectorate
Department of Labour
Ansley House
Mespil Road
DUBLIN 4
Tel: 01-765861

Italy
Inspectorat médical central du
  Travail
Via XX de Settembre
97/c ROME

Netherlands
General Directorate of Labour
Balen van Andelplein 2
VOORBURG
Tel: 070–694001

Norway
Directorate of Labour
  Inspection
P.O. Box 8103
OSLO 1
Tel: 02-469820

Spain
Plan Nacional de higiene y
  seguriad del Trabajo
Ministerio del Trabajo
Calle Torrelaguna
MADRID

Sweden
National Board of Occupational
  Safety and Health
S 17184 SOLNA
Tel: 8-7309000

Switzerland
Arbeitsanstalt Zürich
Pirmdurfstrasse 83
8000-ZURICH

United Kingdom
The Employment Medical
  Advisory Service
Health and Safety Executive
25 Chapel Street
Edgware Road
LONDON NW1 5TD
Tel: 01-262 3277

Appendix 2: Principal European and UK poison information centres

Austria
Med. Univ. Klinik
  Vergiftungsinformationszentral
  Lazarettgasse, 14
1090 VIENNA
Tel: 0222-434343
  0222-436898

Belgium
Centre National de Prévention et
  de Traitement des
  Intoxications
Rue Joseph Stallaer, 15
1060 BRUSSELS
Tel: 02-3454545
  02-3441515

Denmark
Giftinformationscentralen:
  Poisons Information Centre,
Rigshospitalet
Tagensvej 20
2100 COPENHAGEN N.
Tel: 01-386633

Finland
Poison Information Centre
Children's Hospital

University Central Hospital
Stenbäckinkatu, 11
SF-00290 HELSINKI 29
Tel: 0/4711: 2788

France
Centre National d'Informations
  Toxicologiques ASITEST
Hôpital Fernand Widal
200, rue du Faubourg St. Denis
75475 PARIS CEDEX 10
Tel: 01-205.63.29

GFR
Beratungsstelle für
  Vergiftungerscheinungen im
  Kindersalter
Universitätskinderklinik,
  Kaiserin Auguste Viktoria
  Haus
Heubnerweg 6
1000 BERLIN 19
Tel: 030/3023022

Greece
Poison Control Centre
Children's Hospital
'Aglaia Kyriakoy'

University of Athens
ATHENS 617
Tel: 01-7793777

Italy
Centro antiveleni
Universita di Roma
Policlinico Umberto I
Viale del Policlinico
00161 ROMA
Tel: 06/49.06.63

Netherlands
Nationaal Vergiftigingen
  Informatie Centrum
Rijks Instituut voor
  Volksgezondheid.
Antonie van
  leeuwenhoeklaan 9
Postbus 1
BILTHOVEN, UTRECHT
Tel: 030-742200
  030-742875

Norway
Giftkartoteket
Farmakologisk Instiutt
Universitet i Oslo
Odontologibygst
BLINDERN, OSLO 3
Tel: 02-465127

Spain
Servicio de Informaciòn
  Toxicologica
Farmacia 9
MADRID 4
Tel: 01-232.33.66

Sweden
Giftinformationscentralen
  Poison Information Centre
Karolinska Sjukhuset
Box 60, 500.
10401 STOCKHOLM 60
Tel: 08-340500

Switzerland
Centre Suisse d'Information
  Toxicologique
Klosbachstrasse 107
8030 ZURICH
Tel: 01-2515151
  01-2516666

United Kingdom
Poisons Unit
New Cross Hospital
Avonley Road
LONDON SE14 5ER
Tel: 01-4077600

Poisons Information Service
Royal Victoria Hospital
Grosvenor Road
BELFAST BT12 6BA
Tel: 0232-40503

Scottish Poisons Information
  Bureau
The Royal Infirmary
EDINBURGH EH3 9YW
Tel: 031-2292477
  ext Poisons Bureau

Chapter 8

Hazardous Chemicals

This chapter includes monographs describing briefly the hazardous properties and effects upon the human body of some 490 flammable, explosive, corrosive and/or toxic substances or groups of substances commonly used in chemical laboratories; it also recommends first aid and fire-fighting procedures in the case of accidents and suggests methods of dealing with spillages of these materials. Those materials which are known or are suspected to be carcinogenic in man are also identified.

Like the same chapter in the third edition, it incorporates brief references to dangerous reactions or preparations of many of these substances, and a large number of other materials that can be classed loosely as 'research chemicals', with other reagents or under specified physical or experimental circumstances. This was made possible in the second edition by the publication of *Handbook of reactive chemical hazards* by L. Bretherick (Butterworth, 1975). The publication of a third, much expanded edition of the latter work in 1985 has allowed a corresponding increase in the information on hazardous reactions in this chapter. The letter (B) followed by a page number, under the name of a substance, indicates a reference to the third edition *Handbook* page, and also to the page number of useful generalisations about the hazards common to groups of compounds (*e.g.*, acetylenic peroxides, halogen azides, tetrazoles) which are given in its second section. The page numbers given as references in

141

Chapter 5 'Reactive Chemical Hazards' also written by Bretherick, now refer to the third edition of the *Handbook*.

Monographs have been confined to chemicals in fairly general use in laboratories. Sophisticated chemicals important only to specific industries (*e.g.*, agriculture or plastics manufacture) are generally well catered for in the literature automatically supplied with these products by the manufacturers. The fact that other chemicals commonly used in laboratories are not included does not mean that they are harmless. To devote space to the large number of substances with low hazard properties would greatly increase the size of the chapter, thereby reducing the speed with which information can be found in it.

The nomenclature used is mainly based upon that currently recommended by the International Union of Pure and Applied Chemistry (IUPAC) and to a small extent on *Names for chemicals used in industry*, BS 2474: 1983. Cross-references are used extensively to cover the survival of many older but well-established names, and synonyms (some from ASE list) are also given after the main monograph title. Since all the chemicals are arranged in alphabetical order, their names do not appear in the Subject Index. The CAS Registry number now given opposite each bold title name affords access to a wide range of synonyms and further related information.

Properties

In a book of this kind, compromises must be made in the interests of clarity, conciseness and speed of reference. These characteristics are vital in an emergency. The temptation to list more of the physical properties of a toxic material than will assist rapid identification by the person confronted with a casualty has been resisted. Boiling points of liquids and melting points of low-melting solids have been included along with short qualitative descriptions of appearance, volatility and odour when these apply. Water plays a vital role in first aid measures and disposal procedures, and reference is therefore made to the solubility, miscibility or reactivity of the material with water.

Some changes in the method of presenting occupational exposure limits for volatile or dusty hazardous chemicals which it is planned to use on a more than occasional basis have become necessary in this edition because of a policy change by the HSE in 1984. Prior to that date, the annually revised Environmental Health Guidance Note EH 15 had reproduced entirely the 700-odd Threshold Limit Values (TLV) recommended by the American Conference of Governmental Industrial Hygienists in the previous year. These values for some 700 materials expressed as parts of gas or vapour per million parts (ppm) of contaminated air by volume at 25 °C and 1023 mbar (760 mmHg) pressure, and also as milligrams per cubic metre (mg m^{-3}) of air, and as mg m^{-3} for dusts, 'represent conditions under which it is believed that nearly all workers may be repeatedly exposed day after day, without adverse effect'. An eight-hour working day is implied. The word 'skin' in brackets after the TLV figure indicates that absorption by direct contact or absorption of airborne vapour or dust by skin, mucous membranes and eyes can occur, and that protection against poisoning through these routes has been assumed to have been provided in arriving at the figure. However, in 1984 the HSE departed from its previous practice and started publishing annual editions of *Occupational exposure limits* as Guidance Note EH 40, which differed in 2 ways from EH 15. First, only those limit values were included for materials available and used in the UK, and second, 2 types of limit were adopted. *Control limits* were set for those relatively few substances with unusually serious toxic effects and the subjects of specific Regulations or Codes of Practice, *etc*. These specific control limits should not normally be exceeded, but beyond that, additional action may be required, if it is reasonably practicable, to reduce the exposure further. *Recommended limits* are considered to represent good practice and realistic criteria for the control of exposure. The numerical values of the several hundred exposure limits are usually, though not invariably, identical with the TLVs, reflecting differences between UK and US situations and practice.

Like TLVs, both types of exposure limits are subject to annual review, though *control limits* may be modified at shorter notice. The abbreviations CL or RL are used in the following text to indicate these 2 types of UK exposure limits, but where the American value has not been adopted for UK use, the TLV figure is quoted in the same way as previously in the interests of completeness of information. Changes in exposure limits are often, but not invariably, to lower values as more information becomes available.

To some of the substances for which long-term exposure limits (LTEL, 8-hour time-weighted average values) are published, short-term exposure limits (STEL, usually 10-minute TWA) have also been assigned which are often in the range 1.25 to 3 times greater than the values quoted in the monographs. It is believed that exposure to this higher level for up to four 15-minute periods (provided they are minimally separated by one-hour periods in which the exposure is appropriately below the LTEL to avoid exceeding the latter) during an 8-hour day will not adversely affect the health or performance of those exposed. These STEL values have not been included in the monographs, but are readily available with all the adopted and intended LTEL, control and recommended limit values in *Occupational exposure limits 1985*. HSE Guidance Note EH 40/85, London, HMSO, 1985. The comparable list of TLVs is *Threshold limit values for chemical substances and physical agents in the work environment and biological exposure indices with intended changes for 1985–6*, Cincinnati, ACGIH, 1985.

Warnings

Capital letters are used to draw attention immediately to the main risks associated with the material being described. The risk phrases used in this edition are taken from the authorised list appended to *The Classification, Packaging and Labelling of Dangerous Substances Regulations 1984*, SI 1984 No. 1244. This brings into UK legislation the provisions of the EEC Directives and amendments which served as the source of risk phrases in the previous edition, and only minor changes have been necessary to the phrases used in the monographs to keep them strictly in line with the UK legislative requirements. In the previous edition, some monographs for chemicals not from the EEC list included risk phrases selected on the basis of experience and appropriate to the physical and chemical properties. In this edition the risk phrases in such monographs for chemicals not in the authorised list are retained, but in parentheses to indicate their non-official status.

While suppliers of chemicals in EEC countries may also have to adhere to the approved warnings system applying to sales in their domestic market, products exported to other EEC members by them, and by non-member countries, must have on each container a label bearing the name of the material, its origin, the danger symbols (*Fig. 8.1*), and the risk phrases prescribed in the Directive and now required in UK legislation. Safety advice phrases comparable with the injunctions used in this book are also prescribed in the legislation, but their inclusion on labels of small containers is still optional for low risk materials. For high risk materials, the risk and safety information may be provided as a separate label or sheet. If a chemical is covered by the UK Pharmacy and Poisons Act 1933, and later legislation, it is now exempted from the need to carry the warning 'POISON'.

Difficulties with the warning words used for flammable substances in the UK have not yet been completely resolved. Although it is widely agreed that 'flammable' is a better word than 'inflammable' to use for materials that burn readily in air, 'inflammable' remains the word used in the older legislation, *e.g.*, The Petroleum (Consolidation) Act 1928. The 1984 UK legislation seeks some reconciliation by specifying that hazardous materials correctly classified and labelled under its provisions will be deemed to comply with the requirements of the 1928 Act. The UK legislative wordings for degrees of flammability have been adopted in the monographs, as have the ranges of flash point defining those degrees. Thus, FLAMMABLE substances are those with a flash point between 21 °C and 55 °C, HIGHLY FLAMMABLE substances have flash points between 0 °C and 21 °C, and EXTREMELY FLAMMABLE substances have flash points below 0 °C and a boiling point below 35 °C. According to the Highly Flammable Liquids and Liquefied Gases Regulations 1972, liquids with a flash point below 32 °C are designated as HIGHLY FLAMMABLE in the UK. Again, the 1984 legislation specifies that correctly classified and labelled materials will be deemed to comply with the requirements of the 1972 Regulations. This unfortunate difference will have to be accepted until the issue is settled by international agreement.

The injunctions that follow the risk phrases do not always use the exact safety advice phraseology as it is thought useful to keep the distinction between the words 'avoid' and 'prevent', the latter being used to indicate that prevention of inhalation, and eye or hand contact is important.

Toxic effects

The revised chapter 'Chemical hazards and toxicology' provides a contempor-

Fig. 8.1 EEC Directive warning symbols

ary interpretation of the words 'toxic' and 'toxicity', and suggests a sensible attitude towards the risks of poisoning that may arise in a laboratory.

The use of less familiar medical terms has been avoided as far as possible in describing the toxic effects of harmful materials. Where the information is available and relevant to laboratory operations, chronic effects (those produced by prolonged exposure to small amounts of toxic materials) are included as well as the acute (short-term) effects. The effects are usually listed in each monograph in the order inhalation, eye contact, skin contact, mouth contact and ingestion.

Whereas the toxicologist uses six grades of toxicity (*see* Chapter 6 p. 91) the simple warning requirements of labelling call for a two-degree classification of the chemicals into 'toxic' or 'harmful' substances or alternatively into 'corrosive' or 'irritant' substances. According to the Council of Europe Yellow book and the derived UK legislation, toxic substances are those offering a serious risk of acute or chronic poisoning by any route while 'harmful' calls attention to a risk which, although minor is nevertheless real. Corrosive substances are those which destroy living tissues; irritant substances are those which can cause inflammation.

The literature drawn upon in trying to present a reasonable account of toxic effect does not always cover all the possible accidents that might occur. This is particularly true of mouth contact and ingestion and where evidence of such incidents is lacking certain effects are assumed to be probable. The rash use of an ordinary pipette to measure out a volume of a particular corrosive acyl chloride may not have occurred in the past, but one day it will — and it seems wise to anticipate probable effects and prescribe commonsense first aid treatment for a case of mouth contact and swallowing. A simple aspirator bulb must be used to avoid this particular risk in pipetting corrosive or poisonous liquids.

The use of protective clothing, gloves, goggles, breathing apparatus and so on is discussed in the chapter 'Safety planning and management' (p. 25). The level at which the quantity of a dangerous material that is being handled necessitates the wearing of protective clothing and equipment will always be a matter of debate. Although eye protection may be insisted upon, how many university laboratories would require a student to wear rubber gloves, let alone a respirator, when he is pouring a few cm^3 of aniline into a test-tube? Teaching laboratories are the schools for gaining first-hand experience in handling dangerous materials at a low level of risk.

Because the level of risk varies so widely in different laboratories, it is not possible to prescribe protective wear under each individual chemical; injunctions are qualified by using the word 'prevent' instead of 'avoid' when the danger from, say, skin absorption is of a high order, and the wearing of gloves becomes more important. Familiarity with the chapter 'Safety planning and management', together with the warnings and injunctions for the material to be handled, should tell the laboratory worker or his supervisor when and what protection is appropriate. However, in a few of the monographs reference is made to specific protection that is specially important.

Carcinogenic chemicals (*see also* Chapter 7)

One particular possible toxic effect of chronic exposure to various chemicals is that of induction of cancer (carcinogenesis), and much effort is now being devoted to understanding and overcoming this particular problem. It is an extremely complex problem, with several and often different factors affecting the risk of an individual developing the symptoms over a period, usually of many years. This overall difficulty in assessing risk realistically is reflected in the current uncertainty and diversity of regulatory activity designed to minimise carcinogenic risk when using chemicals.

In the United States, the Occupational Safety & Health Administration has moved towards a dual classification of 'potential suspected carcinogens'. Category 1 (where there is good scientific evidence of human carcinogenicity) with minimal feasible exposure or elimination from use if possible, and Category 2 (where there is suggestive evidence of carcinogenicity in man) with appropriate control of exposure levels. Some 150 and 350 candidate compounds respectively seem likely in these two categories. On the other hand the American Conference of Governmental Industrial Hygienists had developed the triple classification of 'human carcinogens', 'suspected carcinogens' and 'experimental carcinogens', and these were adopted by our Health & Safety Executive and in this chapter.

The policy decision to avoid completely the use of human carcinogens is desirable in teaching laboratories, while their presence, and that of suspected human carcinogens (below) in advanced laboratories must be under tight control in storage, use and disposal. The detailed code of practice *Guide to recognising and handling carcinogens in a veterinary laboratory*, Weybridge, MAFF Central Veterinary Laboratory, 1975 covers all aspects of recognition, laboratory hygiene and personal protection, control, accidents and disposal when carcinogenic materials are to be used.

Experimental animal carcinogens (those which have produced carcinogenic effects, often at rather high dose levels) are subdivided into those of high, medium or low potency. Less stringent precautions may be permissible when experimental carcinogens of medium or low potency are in use; a lack of precautions is not permissible when any potentially toxic material is in use.

All the currently proven and suspected human carcinogens are listed below, as well as appearing in the alphabetical sequence in this chapter, with a selection of the experimental animal carcinogenic substances, and an indication of the chemical classification of all of these.

147

Human carcinogens

2-Naphthylamine Benzidine 4-Aminobiphenyl 4-Nitrobiphenyl	Importation and use in manufacture of these are prohibited by legislation in the UK except if present at less than 1 per cent in another material.
1-Naphthylamine *o*-Tolidine Dianisidine 3,3′-Dichlorobenzidine	Use of these is controlled by legislation in the UK.
Chloromethyl methyl ether and four UK-prohibited compounds above	No Exposure Limit (CL, RL or TLV) has been assigned, no exposure permitted.
Acrylonitrile Asbestos Bis-chloromethyl ether Vinyl chloride	ELs have been assigned for these.

Suspected human carcinogens (most have ELs assigned)

3-Amino-1,2,4-triazole	Hexachlorobutadiene
Benzene	Hexamethylphosphoramide
Benz(a)pyrene	Hydrazine
Beryllium	4,4′-Methylenebis(2-chloroaniline)
Chloroform	Methylhydrazine
Chromates of Pb, Zn	2-Nitropropane
1,2-Dibromoethane	
Dimethylcarbamoyl chloride	*N*-Phenyl-2-naphthylamine
1,1-Dimethylhydrazine	1,3-Propanesultone
Dimethyl sulphate	1,3-Propiolactone
Epichlorhydrin	Vinylcyclohexene dioxide

Some experimental carcinogens

2-Acetylaminofluorene	Ethylenethiourea
Carbon tetrachloride	Ethyl *N*-nitrosocarbamate
Diazomethane	Hexamethylphosphoramide
1,4-Dichlorobutene	Iodomethane
4-Dimethylaminoazobenzene	3-Methylcholanthrene
7,12-Dimethylbenzanthracene	Methyl ethanesulphonate
Ethyl carbamate	*N*-Nitroso(methyl)urea
Ethyl diazoacetate	Propyleneimine
Ethyleneimine	

Chemical classification of carcinogens

Several examples of carcinogenic activity have been observed in each of the structural classes below.

- Biological alkylating agents
    Epoxides, chloromethyl ethers, bis(chloroethyl)-amines and sulphides, ethyleneimines, lower esters of hydriodic, sulphuric or alkanesulphonic acids, propiolactone, propanesultone.
- Polycyclic hydrocarbons or heterocycles (four rings and above)
    mono- and di-benzanthracenes, -pyrenes, -acridines; cholanthrenes.
- Aromatic amines, two rings or more
    naphthylamines amino-(or nitro-)biphenyls, acetylaminofluorene
- Nitroso compounds
    nitrosoamines, nitrosoamides.
- Azo-compounds and hydrazines
    azo-alkanes, azo-aromatics, aminoazobenzenes, diazonium salts, diazomethane, hydrazine and its methyl derivatives.
- Heavy metals and derivatives
    Be, Cr dusts, Ni subsulphide dusts, $CrOCl_2$

Closely related to carcinogens are those substances which have been observed to cause cell mutations (mutagens), and similar suspicions attach to those causing foetal deformities in experimental animals (teratogens). Many more structural classes of carcinogens, mutagens and teratogens are listed in *Safety in biological laboratories*, E. Hartree and V. Booth (ed.), London, The Biochemical Society, 1977.

Hazardous reactions

Normally, a brief indication of how an incident occurred is followed by a page reference to the third edition of Bretherick's *Handbook of reactive chemical hazards* — for example (B625) — which takes the reader back to the original report of the incident.

First aid

The general first aid principles that have been adopted, and the practice recommended in each monograph, have been described at some length in the preceding chapter which should be read and understood by those likely to use this book in an emergency. The emphasis that Chapter 7 rightly places on sound general first aid training for representatives of laboratory staff has led us to drop the repetitive presentation of first aid procedures in the monographs that were characteristic of the first edition. The time has come for the routine procedures for dealing with inhalation, eye, skin and ingestion casualties to be part of trained first aid knowledge, to be imparted to those in the laboratories for whom they act. A limited number of special procedures remain and are recorded.

Fire hazards

Chapter 4 has dealt fully with the nature of the risks involved in using flammable materials, precautions that can be taken to minimise these, and the methods of dealing with a fire once it has broken out. The short note **Fire hazard** in the monograph on each flammable substance provides three figures (if available):

(a) flash point (closed cup unless otherwise stated),

(b) explosive limits in terms of the range of percentage flammable component in a mixture with air that presents an explosion risk, and

(c) ignition temperature, which is the minimum temperature required to initiate or maintain self-sustained combustion independent of the source of heat.

The booklet *Laboratory waste disposal manual* formerly published by the Chemical Manufacturers Association has been the main source for the figures presented for these properties.

The note concludes with recommendations for the types of extinguishant that can be used in fighting a fire involving the substance concerned. Here we have drawn extensively from Fire Service Circular No 6/1970 issued by the Home Office.

Spillage disposal

Recommendations for dealing with spillages of hazardous materials have been based largely on the procedures suggested in the chart *How to deal with spillages of hazardous chemicals* published by BDH Chemicals Ltd and, to a lesser extent, on the Chemical Manufacturers Association's booklet referred to above. A more recent source which includes tested and environmentally acceptable disposal methods is *Hazardous chemicals information and disposal guide*, M.A. Armour, L.M. Browne and G.L. Weir, Edmonton, University of Alberta Chemistry Dept., 1984.

In interpreting these recommendations, the following points should be noted:

1 A suitable non-flammable dispersing agent and hand pump sprays for dealing with small spillages of water-insoluble liquids are commercially available. It is generally desirable to use one volume of the dispersing agent for every two volumes of flammable, water-insoluble liquid that has been spilt, together with 10 volumes of water. When this has been worked to an emulsion with a brush and run to waste, diluting greatly with running water, there is no risk of a flammable vapour mixture developing in a drainage system. Less dispersant is needed with non-flammable water-insoluble liquid spillages. This procedure lends itself to laboratory spillages of up to $2\frac{1}{2}$ litres (Winchester quart) proportions.

2 Generally speaking, local authorities will accept small quantities of various chemicals into their sewage systems, provided these are adequately diluted so as not to interfere with the purification system. It is important to know the acceptable dilutions in the case of poisonous materials such as arsenic, cadmium, chromium, lead, mercury, silver compounds, or sulphides. In no case should the amount of water-insoluble flammable liquid thus disposed of be sufficient to create an explosion hazard.

3 It is usually most convenient to allow volatile liquids of low or moderate toxicity to evaporate to atmosphere in an area where no nuisance can be caused by lachrymatory or foul-smelling materials. Lachrymatory materials should be either destroyed (alcoholic sodium or potassium hydroxide will often do this) or buried deeply. The great majority of harmful materials are rendered innocuous on burial by hydrolysis or slow dispersion in the soil by the action of rain water; these processes are assisted if the spillage has been absorbed on sand. However, it is important that the burial place selected should not be close to an area draining into a water supply system if large amounts of spillage are involved.

The monographs of the 42 compounds for which comprehensive Laboratory Hazard Data Sheets have now been prepared by the RSC Information Services Group at Nottingham end with a reference to the Data Sheet.

Abietates, metal

Hazardous reactions   Finely divided abietates of Al, Zn, Ca, Na, Co, Pb and Mn are subject to spontaneous heating and ignition (B1562).

Acetal —*see* 1,1-Diethoxyethane

Acetaldehyde (ethanal)   [75–07–0]

Colourless liquid with pungent fruity odour; miscible with water; bp 21 °C.

EXTREMELY FLAMMABLE
IRRITATING TO EYES AND RESPIRATORY SYSTEM

Avoid breathing vapour. Avoid contact with eyes. RL 100 ppm (180 mg m^{-3}). Readily forms peroxides.

Toxic effects   The vapour causes irritation of eyes, headache and drowsiness. The liquid, if swallowed, causes severe irritation of the digestive organs. **Chronic effects** Repeated inhalation of vapour can cause delirium, hallucinations, loss of intelligence, *etc*, as in chronic alcoholism.

Hazardous reactions   Extremely reactive with acid anhydrides, alcohols, halogens, ketones, phenols, amines, ammonia, hydrogen cyanide and hydrogen sulphide; exothermic polymerisation with acetic acid; fires and explosions with air, oxygen and hydrogen peroxide under various circumstances; exothermic polymerisation with trace metals; explosive products with $Hg(ClO_4)_2$; autoignition of vapour on corroded metals (B263–265).

First aid   *Vapour inhaled:* standard treatment (p. 136).
Affected eyes: standard treatment (p. 136).
If swallowed: standard treatment (p. 137).

Fire hazard   Flash point −38 °C; explosive limits 4–57%; Ignition temp. 185 °C. (130 °C at 56% in air). *Extinguish fire with* water spray, dry powder, carbon dioxide or vaporising liquids.

Spillage disposal   Shut off all possible sources of ignition. Instruct others to keep at a safe distance. Wear breathing apparatus and gloves. Mop up with plenty of water and run to waste diluting greatly with running water. Ventilate area well to evaporate remaining liquid and dispel vapour.

RSC *Lab. Haz. Data Sheet No. 38*, 1985 gives extended coverage.

Acetic acid (ethanoic acid) [64–19–7]

Colourless liquid with pungent acrid odour; bp 118 °C; glacial acetic acid freezes to a crystalline solid in cool weather, miscible with water.

FLAMMABLE
CAUSES SEVERE BURNS

Avoid breathing vapour. Avoid contact with eyes and skin. RL 10 ppm (25 mg m^{-3}).

Toxic effects The vapour irritates the respiratory system. The vapour irritates and the liquid burns the eyes severely. The liquid is very irritating to the skin and can cause burns and ulcers. If taken by mouth causes internal irritation and damage.

Hazardous reactions Causes exothermic polymerisation of acetaldehyde; violent or explosive reactions with oxidants BrF$_5$, CrO$_3$, KMnO$_4$ or Na$_2$O$_2$(B269–270).

First aid *Vapour inhaled:* standard treatment (p. 136).
Affected eyes: standard treatment (p. 136).
Skin contact: standard treatment (p. 136).
If swallowed: standard treatment (p. 137).

Fire hazard Flash point 43 °C; explosive limits 4–16%; ignition temp. 426 °C. *Extinguish fire with* water spray, dry powder, carbon dioxide or vaporising liquids.

Spillage disposal Shut off all possible sources of ignition. Wear face shield and gloves. Mop up with plenty of water and run to waste diluting greatly with running water. Ventilate area well to evaporate remaining liquid and dispel vapour.

Acetic anhydride [108–24–7]

Colourless liquid with strong acrid odour; bp 140 °C; reacts slowly with cold water to form acetic acid, but rapidly if hot and acid present.

FLAMMABLE
CAUSES BURNS

Avoid breathing vapour. Prevent contact with eyes and skin. RL 5 ppm (20 mg m^{-3}).

Toxic effects The vapour irritates the respiratory system. The vapour irritates and the liquid burns the eyes severely, with delayed damage. The liquid irritates and may burn the skin severely, with blistering and peeling. If swallowed, causes immediate irritation, pain and vomiting.

Hazardous reactions Vigorously oxidised by CrO$_3$, metal nitrates, nitric acid, KMnO$_4$; violent reactions with glycerol/POCl$_3$, perchloric acid/water, and with water when added slowly to mixture with acetic acid or tetrafluoroboric acid (B432–B435).

First aid	*Vapour inhaled:* standard treatment (p. 136).
	Affected eyes: standard treatment (p. 136).
	Skin contact: standard treatment (p. 136).
	If swallowed: standard treatment (p. 137).
Fire hazard	Flash point 54 °C; explosive limits 3–10%; ignition temp. 380 °C.
	Extinguish fire with water spray, dry powder, carbon dioxide or vaporising liquids.
Spillage disposal	Shut off all possible sources of ignition. Instruct others to keep at a safe distance. Wear breathing apparatus and gloves. Absorb on sand, shovel into bucket(s), transport to safe place and tip into large volume of water; leave to decompose before decanting water to waste, diluting greatly with running water. Site of spillage should be ventilated after washing thoroughly with water and soap or detergent.

Acetone (2-propanone) [67–64–1]

Colourless, mobile liquid with characteristic odour; bp 56 °C; miscible with water.

HIGHLY FLAMMABLE

Avoid breathing vapour. Prevent contact with eyes. CL 1000 ppm (2400 mg m^{-3}).

Toxic effects	Inhalation of vapour may cause dizziness, narcosis and coma. The liquid irritates the eyes and may cause severe damage. If swallowed may cause gastric irritation, narcosis and coma.
Hazardous reactions	Vigorously oxidised by air in the presence of active carbon, nitric/sulphuric acid mixtures, BrF_3, Br_2, nitrosyl chloride, nitrosyl perchlorate, nitryl perchlorate, chromyl chloride, CrO_3, F_2O_2, nitric acid, hydrogen peroxide, peroxomonosulphuric acid; violent reactions with bromoform or chloroform and base, SCl_2 (B365–366).
First aid	*Vapour inhaled:* standard treatment (p. 136).
	Affected eyes: standard treatment (p. 136).
	If swallowed: standard treatment (p. 137).
Fire hazard	Flash point −18 °C; explosive limits 3–13%; ignition temp. 538 °C. *Extinguish fire with* water spray, dry powder, carbon dioxide or vaporising liquids.
Spillage disposal	Shut off possible sources of ignition. Wear face shield and gloves. Mop up with plenty of water and run to waste diluting greatly with running water. Ventilate area well to evaporate remaining liquid and dispel vapour.

RSC *Lab. Haz. Data Sheet No. 21*, 1984 gives extended coverage.

Acetone cyanohydrin
—*see* 2-Hydroxy-2-methylpropiononitrile

Acetonitrile (methyl cyanide) [75–05–8]

Colourless, volatile liquid with similar odour to acetamide; bp 80 °C; miscible with water.

> HIGHLY FLAMMABLE
> TOXIC BY INHALATION, IN CONTACT WITH SKIN
> AND IF SWALLOWED

Do not breathe vapour. Prevent contact with skin and eyes. RL 40 ppm (70 mg m^{-3}).

Toxic effects Inhalation of vapour may cause fatigue, nausea, diarrhoea and abdominal pain; in severe cases there may be delirium, convulsions, paralysis and coma. Evidence is lacking on the effects of skin absorption and ingestion, but these may be similar to those resulting from inhalation.

Hazardous reactions Violent or explosive reactions with dinitrogen tetraoxide (in presence of indium), *N*-fluoro compounds, nitric acid, sulphuric acid; violent reaction with anhydrous $Fe(ClO_4)_3$; explosive mixture with H_2SO_4 (B245–246).

First aid *Vapour inhaled:* special treatment for hydrogen cyanide (p. 132).
Affected eyes: standard treatment (p. 136).
Skin contact: standard treatment (p. 136).
If swallowed: special treatment for hydrogen cyanide (p. 132).

Fire hazard Flash point 6 °C; explosive limits 4–16%; ignition temp. 524 °C. *Extinguish fire with* foam, dry powder, carbon dioxide or vaporising liquids.

Spillage disposal Shut off all possible sources of ignition. Instruct others to keep at safe distance. Wear breathing apparatus and gloves. Mop up with plenty of water and run to waste diluting greatly with running water. Ventilate area well to evaporate remaining liquid and dispel vapour.

RSC *Lab. Haz. Data Sheet No. 31*, 1985 gives extended coverage.

Acetylacetone —*see* Pentane-2,4-dione

Acetyl azide [24156–53–4]

Hazardous reactions Treacherously explosive (B250).

Acetyl bromide [506–96–7]

Colourless to yellow liquid; bp 77 °C; decomposed by water with formation of hydrobromic acid and acetic acid.

156

CAUSES BURNS
IRRITATING TO EYES, RESPIRATORY SYSTEM AND SKIN

Avoid breathing vapour. Prevent contact with skin and eyes.

Toxic effects	The vapour irritates all parts of the respiratory system. The vapour irritates the eyes severely. The liquid burns the skin and eyes. If taken by mouth, there is immediate and severe internal irritation and damage.
Hazardous reactions	Violent reaction with H_2O and hydroxylic compounds (B238).
First aid	*Vapour inhaled:* standard treatment (p. 136). *Affected eyes:* standard treatment (p. 136). *Skin contact:* standard treatment (p. 136). *If swallowed:* standard treatment (p. 137).
Spillage disposal	Instruct others to keep at a safe distance. Wear breathing apparatus and gloves. Spread soda ash liberally over the spillage and mop up cautiously with water — run to waste, diluting greatly with running water.

Acetyl chloride (ethanoyl chloride) [75-36-5]

Colourless, fuming, volatile liquid with a pungent odour; bp 51 °C; rapidly decomposed by water with formation of hydrochloric acid and acetic acid.

HIGHLY FLAMMABLE
REACTS VIOLENTLY WITH WATER
CAUSES BURNS

Avoid breathing vapour. Prevent contact with eyes and skin.

Toxic effects	The vapour irritates all parts of the respiratory system. The vapour irritates the eyes severely. The liquid burns the skin and the eyes. If taken by mouth, there is severe internal irritation and damage.
Hazardous reactions	Violent decomposition in preparation from PCl_3 and acetic acid; violent reactions with water, dimethyl sulphoxide (B240).
First aid	*Vapour inhaled:* standard treatment (p. 136). *Affected eyes:* standard treatment (p. 136). *Skin contact:* standard treatment (p. 136). *If swallowed:* standard treatment (p. 137).
Fire hazard	Flash point 4 °C; ignition temp. 390 °C. *Extinguish fire with* water spray, foam, dry powder, carbon dioxide or vaporising liquids.
Spillage disposal	Shut off all possible sources of ignition. Instruct others to keep at a safe distance. Wear breathing apparatus and gloves. Spread soda ash liberally over the spillage and mop up cautiously with water — run to waste, diluting greatly with running water.

157

Acetylene [74-86-2]

Colourless gas; bp −83 °C; commercial gas has garlic-like odour due to impurities; slightly soluble in water.

EXTREMELY FLAMMABLE
EXPLOSIVE WITH OR WITHOUT CONTACT WITH AIR
HEATING MAY CAUSE EXPLOSION

Avoid breathing gas. Forms sensitive explosive metallic compounds.

Toxic effects   Inhalation of gas may cause dizziness, headache, nausea. Mixed with oxygen it can have narcotic properties but it is primarily an asphyxiant.

Hazardous reactions   The endothermic gas explodes, alone or mixed with air, under various circumstances; contact with copper results in formation of explosive copper acetylide; reacts violently or explosively with Br_2 and Cl_2 and other oxidants; explodes on contact with potassium; contact with HNO_3 and $Hg(II)$ salts gives trinitromethane, and subsequent addition of H_2SO_4 gives tetranitromethane, both being oxidants and explosive (B226, 1117).

First aid   *Vapour inhaled:* standard treatment (p. 136).

Fire hazard   Explosive limits 3–82%; ignition temp. 335 °C. Since the gas is supplied in a cylinder, turning off the valve will reduce any fire involving it; if possible cylinders should be removed quickly from an area in which a fire has developed.

Acetylenebis(triethyltin) [994-99-0]

Hazardous reactions   Explodes on standing at room temperature (B398).

Acetylene dichloride
—*see* 1,2-Dichloroethylene

Acetylene tetrabromide
—*see* 1,1,2,2-Tetrabromoethane

Acetylene tetrachloride
—*see* 1,1,2,2-Tetrachloroethane

Acetylenic compounds Review of class (B1417).

Acetylenic peroxides Review of group (B1419).

Acetyl nitrate [591-09-3]

Hazardous reactions Thermally unstable and may decompose violently or explode under a variety of circumstances (B248).

Acetyl nitrite [5813-49-0]

Hazardous reactions Unstable and liable to explode (B248).

Acid anhydrides Review of group (B1420).

Acrylaldehyde (acrolein, propenal) [107-02-8]

Colourless to yellow, volatile liquid with pungent, choking odour; bp 53 °C; somewhat soluble in water.

HIGHLY FLAMMABLE
TOXIC BY INHALATION
IRRITATING TO EYES, RESPIRATORY SYSTEM AND SKIN

Avoid breathing vapour. Prevent contact with eyes and skin. RL 0.1 ppm (0.25 mg m^{-3}).

Toxic effects The vapour irritates all parts of the respiratory system and may cause unconsciousness. Short exposure may cause pain to the nose and eyes in addition to intense irritation. Assumed to have extremely poisonous and irritant action if taken by mouth.

Hazardous reactions Liable to polymerise violently, especially in contact with strong acid or basic catalysts (B349).

First aid *Vapour inhaled:* standard treatment (p. 136).
Affected eyes: standard treatment (p. 136).
Skin contact: standard treatment (p. 136).
If swallowed: standard treatment (p. 137).

Fire hazard Flash point −26 °C; explosive limits 3–31%; ignition temp. 278 °C. *Extinguish fire with* water spray, dry powder, carbon dioxide or vaporising liquids.

Spillage disposal	Shut off all possible sources of ignition. Instruct others to keep at a safe distance. Wear breathing apparatus and gloves. Mop up with plenty of water and run to waste, diluting greatly with running water. Ventilate area well to evaporate remaining liquid and dispel vapour.

Acrylamide (propenamide) [79-06-1]

White crystalline solid; soluble in water; may polymerise with violence on melting (85 °C).

TOXIC BY INHALATION, IN CONTACT WITH SKIN
AND IF SWALLOWED
DANGER OF CUMULATIVE EFFECTS

Avoid breathing dust. Avoid contact with skin and eyes. RL (skin) 0.3 mg m^{-3}.

Toxic effects	Irritates the skin and eyes. Expected, from animal experiments, to affect the central nervous system as a result of skin absorption. (Ch 6 p. 107)
First aid	*Affected eyes:* standard treatment (p. 136). *Skin contact:* standard treatment (p. 136). *If swallowed:* standard treatment (p. 137).
Spillage disposal	Wear face-shield or goggles, and gloves. Mop up with plenty of water and run to waste diluting greatly with running water.

Acrylic acid (propenoic acid) [79-10-7]

Colourless solid and liquid with acrid odour; mp 14 °C; bp 141 °C; miscible with water.

FLAMMABLE
CAUSES BURNS

Avoid breathing vapour. Prevent contact with skin and eyes. RL 10 ppm (30 mg m^{-3}).

Toxic effects	Severely irritates the skin, eyes and respiratory system in concentrated solutions or as a liquid; assumed to be extremely irritant if taken by mouth.
Hazardous reaction	Liable to polymerise violently. (B350—351).
First aid	*Affected eyes:* standard treatment (p. 136). *Vapour inhaled:* standard treatment (p. 136). *Skin contact:* standard treatment (p. 136). *If swallowed:* standard treatment (p. 137).
Fire hazard	Flash point 52 °C; ignition temp. 429 °C. *Extinguish fire with* water spray, dry powder, carbon dioxide or vaporising liquids.

160

Spillage disposal	Shut off all possible sources of ignition. Wear face-shield and gloves. Mop up with plenty of water and run to waste, diluting greatly with running water. Ventilate area well to evaporate remaining liquid and dispel vapour.

Acrylonitrile (propenonitrile, vinyl cyanide)   [107–13–1]

Colourless, volatile liquid with ethereal odour; bp 77 °C; 1 part dissolves in about 15 parts water at 25 °C.

HIGHLY FLAMMABLE
TOXIC BY INHALATION, IN CONTACT WITH SKIN
AND IF SWALLOWED

Prevent inhalation of vapour. Prevent contact with skin and eyes. Do not expose to flame or acids. Human carcinogen. CL (skin) 2 ppm (4.5 mg m^{-3}).

Toxic effects	Vapour may cause dizziness, nausea and unconsciousness. The liquid may cause blistering, dermatitis and acute effects by absorption. Poisonous if taken by mouth. Action similar to cyanides. **Chronic effects** Low concentrations of vapour inhaled over a long period may cause flushing of face, nausea, giddiness and jaundice. Suspected carcinogen.
Hazardous reactions	Violent or explosive polymerisation promoted by a variety of reagents including strong acids and bases, Br$_2$ and silver nitrate (B340–342).
First aid	Special treatment for hydrogen cyanide (p. 132).
Fire hazard	Flash point 0 °C (open cup); explosive limits 3–17%; ignition temp. 481 °C. *Extinguish fire with* water spray, dry powder, carbon dioxide or vaporising liquids.
Spillage disposal	Shut off all possible sources of ignition. Instruct others to keep at a safe distance. Wear breathing apparatus and gloves. Mop up with plenty of water and run to waste diluting greatly with running water. Ventilate area well to evaporate remaining liquid and dispel vapour.

Acyl azides Review of group (B1420).

Acyl halides Review of group (B1422).

Acyl hypohalites Review of group (B1423).

Acyl nitrates Review of group (B1424).

Acyl nitrites Review of group (B1424).

Alkali metal derivatives of hydrocarbons
Review of group (B1426).

Alkali metals Handling and uses (B1426).

Alkenes Reactions with oxides of nitrogen (B1427).

Alkylaluminium derivatives Review of class (B1429).

Alkylboranes Review of group (B1431).

Alkyl haloboranes Review of group (B1432).

Alkyl halophosphines Review of group (B1432).

Alkyl halosilanes Review of group (B1432).

Alkyl hydroperoxides Review of group (B1433).

Alkylmercury compounds
RL (skin, as Hg) 0.01 ppm (0.03 mg m^{-3}).

Alkylmetal halides Review of class (B1433–1434).

Alkylmetals Review of group (B1434).

Alkyl nitrates Review of group (B1436).

Alkyl nitrites Review of group (B1438).

Alkylnon-metal halides Review of class (B1438).

Alkylnon-metal hydrides Review of group (B1438).

Alkylnon-metals Review of group (B1439).

Alkyl perchlorates Review of group (B1438).

Allene —*see* Propadiene

Allyl alcohol (prop-2-en-1-ol, vinyl carbinol)   [107–18–6]

Colourless liquid with pungent odour; bp 97 °C; miscible with water.

HIGHLY FLAMMABLE
VERY TOXIC BY INHALATION
IRRITATING TO EYES, RESPIRATORY SYSTEM AND SKIN

Prevent breathing of vapour. Prevent contact with skin and eyes. RL (skin) 2 ppm (5 mg m^{-3}).

Toxic effects	The vapour irritates eyes and respiratory tract severely. The liquid irritates the eyes and skin. Causes severe irritation of digestive organs and kidneys if swallowed.
Hazardous reactions	CCl_4 gives unstable products; H_2SO_4 may polymerise it explosively (B366–367).
First aid	*Vapour inhaled:* standard treatment (p. 136). *Affected eyes:* standard treatment (p. 136). *Skin contact:* standard treatment (p. 136). *If swallowed:* standard treatment (p. 137).
Fire hazard	Flash point 21 °C; explosive limits 3–18%; ignition temperature 378 °C. *Extinguish fire with* water spray, dry powder, carbon dioxide or vaporising liquids.
Spillage disposal	Shut off all possible sources of ignition. Instruct others to keep at a safe distance. Wear breathing apparatus and gloves. Mop up with plenty of water and run to waste diluting greatly with running water. Ventilate area well to evaporate remaining liquid and dispel vapour.

Allyl benzenesulphonate   [7575–57–7]

Hazardous   Explosion of vacuum distillation residue (B747).
reaction

Allyl bromide (3-bromopropene)   [106–95–6]

Colourless liquid with an unpleasant smell; bp 71 °C; almost insoluble in water.

          HIGHLY FLAMMABLE
          VERY TOXIC BY INHALATION
          IRRITATING TO SKIN, EYES AND RESPIRATORY SYSTEM

Prevent breathing of vapour. Prevent contact with eyes and skin.

Toxic effects   The vapour irritates all parts of the respiratory system and may cause dizziness and headache. The vapour and liquid irritate the eyes. Assumed to be very irritant and poisonous if taken by mouth.

Hazardous   Contact with $AlCl_3$, BF_3, H_2SO_4 *etc* may cause exothermic violent polymeri-
reactions   sation (B352).

First aid   *Vapour inhaled:* standard treatment (p. 136).
           Affected eyes: standard treatment (p. 136).
           Skin contact: standard treatment (p. 136).
           If swallowed: standard treatment (p. 137).
Fire hazard   Flash point −1 °C; explosive limits 4–7%; ignition temp. 295 °C. *Extinguish fire with* water spray, foam, dry powder, carbon dioxide or vaporising liquids.
Spillage   Shut off all possible sources of ignition. Instruct others to keep at a safe
disposal   distance. Wear breathing apparatus and gloves. Apply non-flammable dispersing agent if available and work to an emulsion with brush and water — run this to waste, diluting greatly with running water. If dispersant not available, absorb on sand, shovel into bucket(s) and transport to safe open place for atmospheric evaporation or burial. Site of spillage should be washed thoroughly with water and soap or detergent.

Allyl chloride (3-chloropropene)   [107–05–1]

Colourless, volatile liquid with unpleasant, pungent odour; bp 45 °C; immiscible with water.

          HIGHLY FLAMMABLE
          VERY TOXIC BY INHALATION

Prevent breathing of vapour. Prevent contact with skin and eyes. RL 1 ppm (3 mg m^{-3}).

Toxic effects   The vapour irritates all parts of the respiratory system and inhalation may cause headache, dizziness and, in high concentrations, unconsciousness. The vapour and liquid irritate the eyes. Muscular pains may follow skin absorption. Assumed to be very irritant and poisonous if taken by mouth.

Hazardous reactions   Contact with $AlCl_3$, BF_3, H_2SO_4 *etc* may cause exothermic violent polymerisation (B352).

First aid   *Vapour inhaled:* standard treatment (p. 136).
Affected eyes: standard treatment (p. 136).
Skin contact: standard treatment (p. 136).
If swallowed: standard treatment (p. 137).

Fire hazard   Flash point $-32\,°C$; explosive limits 3–11%; ignition temp 392 °C. *Extinguish fire with* water spray, foam, dry powder, carbon dioxide or vaporising liquids.

Spillage disposal   Shut off all possible sources of ignition. Instruct others to keep at a safe distance. Wear breathing apparatus and gloves. Apply non-flammable dispersing agent if available and work to an emulsion with brush and water — run this to waste diluting greatly with running water. If dispersant not available, absorb on sand, shovel into bucket(s) and transport to safe open place for atmospheric evaporation or burial. Site of spillage should be washed thoroughly with water and soap or detergent.

RSC *Lab. Haz. Data Sheet No. 28*, 1984 gives extended coverage.

Allyl compounds Review of group (B1443).

Allyl ethyl ether [557–31–3]

Hazardous reaction   Explosion of peroxide at end of distillation (B521).

Allyl glycidyl ether [3-(2,3-epoxypropyl)propene] [106–92–3]

Colourless liquid; bp 154 °C; soluble in water.

HARMFUL BY INHALATION
MAY CAUSE SENSITISATION BY SKIN CONTACT

Avoid breathing vapour. Avoid contact with skin, eyes and clothing. RL (skin) 5 ppm (22 mg m^{-3}).

Toxic effects   The vapour irritates the respiratory system and has caused pulmonary oedema. The vapour irritates the eyes and skin. The liquid irritates the skin and alimentary system and may cause depression of the central nervous system and sensitisation.

First aid	*Vapour inhaled:* standard treatment (p. 136).
	Affected eyes: standard treatment (p. 136).
	Skin contact: standard treatment (p. 136).
	If swallowed: standard treatment (p. 137).
Spillage disposal	Shut off all possible sources of ignition. Wear face-shield or goggles, and gloves. Mop up with plenty of water and run to waste diluting greatly with running water.

Allyl iodide (3-iodopropene) [556-56-9]

Yellowish to brown liquid with unpleasant pungent odour; bp 102 °C; immiscible with water.

FLAMMABLE
CAUSES BURNS

Avoid breathing vapour. Prevent contact with skin and eyes.

Toxic effects	The vapour irritates the eyes and the respiratory system. The liquid irritates the eyes and skin. Assumed to be very irritant and poisonous if taken by mouth.
First aid	*Vapour inhaled:* standard treatment (p. 136).
	Affected eyes: standard treatment (p. 136).
	Skin contact: standard treatment (p. 136).
	If swallowed: standard treatment (p. 137).
Fire hazard	Flash point below 21 °C. *Extinguish fire with* water spray, foam, dry powder carbon dioxide or vaporising liquids.
Spillage disposal	Shut off all possible sources of ignition. Instruct others to keep at a safe distance. Wear breathing apparatus and gloves. Apply non-flammable dispersing agent if available and work to an emulsion with brush and water — run this to waste, diluting greatly with running water. If dispersant not available, absorb on sand, shovel into bucket(s) and transport to safe open place for burial. Site of spillage should be washed thoroughly with water and soap or detergent.

Allyl isothiocyanate [57–06–7]

Hazardous preparation	Explosion at end of reaction between allyl chloride and sodium thiocyanate in autoclave (B417).

Allyl trifluoromethanesulphonates Review of group (B1444).

Aluminium (metal) [7429-90-5]

FLAMMABLE
CONTACT WITH WATER LIBERATES HIGHLY FLAMMABLE GASES

Avoid breathing dust. RL (metal or oxide) 10 mg m^{-3}.

Hazardous reactions	Violent or explosive reactions with numerous oxidants including ammonium nitrate, ammonium peroxodisulphate, Na_2O_2, CuO, bromates; under certain conditions, the metal reacts violently with butanol, many halocarbons (including PTFE), halogens, HCl, iron/water, mercury salts, PCl_5, propan-2-ol, silver chloride, sulphur. Violent or explosive reactions with oxidants, including metal nitrates, oxides or sulphates; ignition in $SbCl_3$ or CS_2 vapours, or in hot CO_2 if aluminium halides present, or with picric acid on addition of water (B21-33).

Aluminium bromide anhydrous [7727-15-3]

Colourless to yellow or brown solid; violently decomposed by water with evolution of hydrogen bromide.

CAUSES BURNS

Avoid breathing dust. Prevent contact with skin and eyes. RL 2 mg m^{-3} (Al, soluble salts).

Toxic effects	Inhalation of the dust produces irritation or burns of the respiratory system. The dust will cause painful eye burns. Heat is produced on contact with moist skin resulting in thermal and/or acid burns. Severe internal burns and damage will result if taken by mouth.
Hazardous reaction	Very violent reaction with H_2O (B35).
First aid	*Dust inhaled:* standard treatment (p. 136). *Affected eyes:* standard treatment (p. 136). *Skin contact:* standard treatment (p. 136). *If swallowed:* standard treatment (p. 137).
Spillage disposal	Wear goggles and gloves. Mix with sand, shovel into dry bucket, transport to safe open area and add, a little at a time, to a large quantity of water; after reaction complete, run to waste diluting greatly with running water.

Aluminium chlorate [15477-33-5]

Hazardous reaction	Explosion when aqueous solution was evaporated (B39).

167

Aluminium chloride anhydrous [7446-70-0]

Yellow or off-white pieces, granules or powder; violently decomposed by water, with the formation of hydrogen chloride.

CAUSES BURNS

Avoid breathing dust. Prevent contact with eyes and skin. RL 2 mg m^{-3} (Al, soluble salts).

Toxic effects   Inhalation of dust produces irritation or burns of the mucous membranes. The material will cause painful eye burns. When moisture is present on the skin, heat is produced on contact, resulting in thermal and acid burns. If taken by mouth, the immediate local reaction causes severe burns.

Hazardous reactions   Violent reactions with water, ethylene oxide, nitrobenzene/phenol; alkenes polymerised exothermally by excess solid; mixture with PhNO$_2$ is thermally unstable; autoclave explosion during reaction of phenol with CO in MeNO$_2$–AlCl$_3$ (B36–38).

First aid   *Dust inhaled:* standard treatment (p. 136).
Affected eyes: standard treatment (p. 136).
Skin contact: standard treatment (p. 136).
If swallowed: standard treatment (p. 137).

Spillage disposal   Wear goggles and gloves. Mix with sand, shovel into dry bucket, transport to a safe open area and add, a little at a time, to a large quantity of water; after reaction complete, run to waste diluting greatly with running water.

Aluminium hydride [7784-21-6]

Hazardous reactions   May explode spontaneously at ambient temperatures; violent decomposition in certain methyl ethers in presence of carbon dioxide; forms explosive complexes with tetrazoles (B40–41).

Aluminium isopropoxide [555-31-7]

Hazardous reactions   Reaction with 6% H$_2$O$_2$ led to foaming and ignition (B1153).

Aluminium lithium hydride [16853-85-3]
(lithium tetrahydroaluminate)

White microcrystalline powder and lumps; decomposes above 125 °C; reacts rapidly with water with evolution of hydrogen.

CONTACT WITH WATER LIBERATES HIGHLY FLAMMABLE GASES

Prevent contact with skin, eyes and clothing. Keep container tightly closed. RL 2 mg m^{-3} (Al, soluble salts).

Toxic effects	Reaction with moisture forms corrosive lithium hydroxide which irritates the skin and eyes.
Hazardous reactions	May ignite when ground in mortar; use in dehydrating bis(2-methoxyethyl) ether has resulted in explosions; other experiments involving boron trifluoride diethyl etherate, dibenzoyl peroxide, 1,2-dimethoxyethane, ethyl acetate, fluoroamides also resulted in explosions; vigorous reactions with pyridine and tetrahydrofuran also reported. Use of excess to reduce alkyl benzoates is hazardous; hot residue from distillation of dioxane from the hydride exploded on admission of air; may ignite with limited water (B41– 44). (See special warning in Chapter 5, p. 79).
First aid	*Affected eyes:* standard treatment (p. 136). *Skin contact:* standard treatment (p. 136). *If swallowed:* standard treatment (p. 137).
Fire hazard	This arises usually by contact with small quantities of water. Such a fire is best extinguished by smothering with sand and disposing in the manner given below. Do not use extinguishers.
Spillage disposal	Instruct others to keep at a safe distance. Wear face-shield or goggles, and gloves. Cover with dry soda ash, shovel into dry bucket, transport to safe open area and add, a little at a time, to a large excess of dry propan-2-ol. Leave to stand for 24 hours and run to waste diluting greatly with running water.

RSC *Lab. Haz. Data Sheet No. 5,* 1982 gives extended coverage.

Aluminium sodium hydride   [13770–96–2]
(sodium tetrahydroaluminate)

White or grey crystalline powder; begins to melt at 183 °C and decomposes completely at 230–240 °C; decomposed by water with evolution of hydrogen.

CONTACT WITH WATER LIBERATES HIGHLY FLAMMABLE GASES

Protect eyes. Avoid contact with skin or clothing. Keep container tightly closed. RL 2 mg m^{-3} (Al, soluble salts).

Toxic effects	Reaction with moisture forms corrosive sodium hydroxide which irritates or burns the skin and eyes.
Hazardous preparation	Violent explosion when being synthesised from its elements in tetrahydro-furan (B45).

169

First aid	*Affected eyes:* standard treatment (p. 136).
	Skin contact: standard treatment (p. 136).
	If swallowed: standard treatment (p. 137).
Fire hazard	This arises usually from contact with small amount of water. Fire is best extinguished by smothering with dry sand and disposing in the manner given below. Do not use extinguishers.
Spillage disposal	Instruct others to keep at a safe distance. Wear face-shield or goggles and gloves. Cover with dry sand, shovel into dry bucket and add, a little at a time, to an excess of dry propan-2-ol. Leave to stand for 24 hours and run liquid to waste diluting greatly with running water.

Aluminium tetrahydroborate [16962-07-5]
(aluminium borohydride)

RL 2 mg m^{-3} (Al, soluble salts).

Hazardous reaction	Vapour is spontaneously flammable in air and explodes in oxygen (B35).

Aluminium triazide [39108-14-0]

Hazardous reaction	May detonate by shock (B46).

Aluminium triformate [7360-53-4]

Hazardous reaction	Explosion when aqueous solution was being evaporated (B336).

Amidosulphuric acid [5329-14-6]
(sulphamic acid; sulphamidic acid)

White crystals melting at about 205 °C with decomposition; one part dissolves in about six parts water at 0 °C, in about two parts at 80 °C.

IRRITATING TO EYES AND SKIN

Avoid contact with skin and eyes.

Toxic effects	The dust or solution irritates the eyes. Prolonged contact with the skin may cause irritation. It is not especially toxic when taken by mouth.

First aid	*Affected eyes:* standard treatment (p. 136).
	Skin contact: standard treatment (p. 136).
	If swallowed: standard treatment (p. 137).
Spillage disposal	Wear face-shield or goggles, and gloves. Clear up with dust pan and brush. May be disposed of after mixing with sand as normal refuse or flushed away to waste with water.

Aminium perchlorates Review of group (B1445).

4-Aminobiphenyl [92–67–1]

Human carcinogen, use prohibited in the United Kingdom under The Carcinogenic Substances Regulations 1967 (*see* p. 148). Inhalation or absorption through the skin of the dust has been recognised as a cause of bladder tumours. It is not therefore considered appropriate to deal with the hazards more fully in this book.

4-Amino-*N*,*N*-diethylaniline and salts —*see* *N*,*N*-Diethyl-*p*-phenylenediamine and salts

Aminodimethylbenzenes —*see* Xylidines

2-Aminoethanol [141–43–5]
(ethanolamine, 2-hydroxyethylamine)

Colourless, viscous liquid with ammoniacal smell; bp 170 °C; miscible with water.

> HARMFUL BY INHALATION
> IRRITATING TO EYES, RESPIRATORY SYSTEM AND SKIN

Avoid breathing vapour. Avoid contact with skin, eyes and clothing. RL 3 ppm (8 mg m^{-3}).

Toxic effects	As the vapour pressure is low, it is unlikely to cause irritation of respiratory system except when the liquid is hot. The liquid irritates the eyes and may irritate the skin and alimentary system if taken by mouth.
First aid	*Vapour inhaled:* standard treatment (p. 136).
	Affected eyes: standard treatment (p. 136).

171

	Skin contact: standard treatment (p. 136).
	If swallowed: standard treatment (p. 137).
Spillage	Wear face-shield or goggles, and gloves. Mop up with plenty of water and
disposal	run to waste diluting greatly with running water.

Aminoguanidinium nitrate [10308–82–4]

Hazardous
reaction
Violent explosion when aqueous solution was being evaporated on a steam bath (B178).

2-Aminophenol [95–55–6]

White to brown, light-sensitive crystals; sparingly soluble in water.

HARMFUL BY INHALATION, IN CONTACT WITH SKIN
AND IF SWALLOWED

Avoid contact with skin and eyes.

Toxic effects
May cause dermatitis and cyanosis by skin absorption. Assumed to be poisonous if taken by mouth.

First aid
Affected eyes: standard treatment (p. 136).
Skin contact: standard treatment (p. 136).
If swallowed: standard treatment (p. 137).

Spillage
disposal
Small amounts may be swept up and dispersed in a large volume of water which is then run to waste, diluting greatly with running water.

3-Aminopropiononitrile [151–18–8]

Hazardous
reaction
Stored material exploded after polymerisation to yellow solid (B364).

2-Aminopyridine (2-pyridineamine) [504–29–0]

White crystals or powder; mp 58 °C; soluble in water.

TOXIC BY INHALATION, IN CONTACT WITH SKIN
AND IF SWALLOWED

Avoid breathing dust or contact with skin or eyes. RL 0.5 ppm (2 mg m^{-3}).

Toxic effects
Headache, dizziness, flushing of skin, shortness of breath, nausea, collapse, convulsions (possibly fatal).

First aid   *Dust inhaled:* standard treatment (p. 136).
                       Affected eyes: standard treatment (p. 136).
                       Skin contact: standard treatment (p. 136).
                       If swallowed: standard treatment (p. 137).

Spillage   Instruct others to keep at a safe distance. Wear goggles or face mask and
disposal   gloves. Mop up with plenty of water — run to waste, diluting greatly with
                       running water.

2-Aminothiazole   [96–50–4]

Hazardous   Material ignited in drying oven; violent explosion when nitrated with nitric/
reactions   sulphuric acids (B347–348).

3-Amino-1,2,4-triazole   [61–82–5]

Suspected carcinogen (*see* p. 147).

Amminechromium peroxocomplexes Review of group (B1445).

Amminemetal oxosalts Review of this large class (B1446).

Ammonia (gas)   [7664–41–7]

Colourless gas with characteristic pungent odour. It is supplied to laboratories in cylinders of
various sizes in liquid (bp −33 °C) form.

            FLAMMABLE
            TOXIC BY INHALATION

Avoid breathing gas. RL 25 ppm (18 mg m^{-3}).

Toxic effects   The gas irritates all parts of the respiratory system. The gas irritates the eyes
                       severely.

Hazardous   Mixtures with air have exploded; violent reactions or explosive products
reactions   with halogens or interhalogens; violent reactions with boron halides;
                       explosive reaction with chlorine azide; causes explosive polymerisation of
                       ethylene oxide; forms explosive compounds with $AuCl_3$, Hg, NCl_3, silver
                       compounds, stibine, $TeCl_4$, tetramethylammonium amide, Ge derivatives or
                       Te halides (B1177–1181).

173

First aid	*Vapour inhaled:* standard treatment (p. 136).
	Affected eyes: standard treatment (p. 136).
Fire hazard	Explosive limits 16–25%; ignition temperature 651 °C. Since the gas is supplied in a cylinder, turning off the valve will reduce any fire involving it; if possible cylinders should be removed quickly from an area in which a fire has developed.
Disposal	Surplus gas or leaking cylinder can be vented slowly into water-fed scrubbing tower or column, or into a fume cupboard served by such a tower.

RSC *Lab. Haz. Data Sheet No. 37,* 1985 gives extended coverage.

Ammonia (solutions)

Ammonia solution is commonly supplied to laboratories as a 35% solution in water (0.88 specific gravity). In warm weather this strong solution develops pressure in its bottle and the cap must be released with care. The 25% solution (0.90) is free of this problem.

CAUSES BURNS
IRRITATING TO EYES, RESPIRATORY SYSTEM AND SKIN

Avoid breathing vapour. Prevent contact with eyes and skin. RL 25 ppm (18 mg m^{-3}).

Toxic effects	The vapour irritates all parts of the respiratory system. The solution causes severe eye burns. The solution burns the skin. If swallowed, the solution causes severe internal damage.
First aid	*Vapour inhaled:* standard treatment (p. 136).
	Affected eyes: standard treatment (p. 136).
	Skin contact: standard treatment (p. 136).
	If swallowed: standard treatment (p. 137).
Spillage disposal	Wear goggles and gloves (and rubber boots or overshoes if spillage is large). Mop up with plenty of water and run to waste diluting greatly with running water.

Ammonium amidosulphate [7773–06–0]
(ammonium sulphamate)

Internally harmful. RL 10 mg m^{-3}.

Hazardous reaction	Vigorous exothermic hydrolysis of 60% solution with acid (B1208).

Ammonium azide [12164–94–2]

Hazardous reaction	Explodes on rapid heating, friction or impact (B1201).

Ammonium bromate [13483–59–5]

Hazardous reaction Very friction-sensitive, may explode spontaneously (B95).

Ammonium chlorate [10192–29–7]

Hazardous reactions Occasionally explodes spontaneously, always above 100 °C; cold saturated solution may decompose explosively (B919).

Ammonium dichromate
—*see* Chromates and dichromates

Ammonium fluoride —*see* Fluorides (water-soluble)

Ammonium fluorosilicate
—*see* Hexafluorosilicic acid and salts

Ammonium hydrogen difluoride
—*see* Fluorides (water-soluble)

Ammonium iodate [13446–09–8]

Hazardous reaction Decomposed violently on touching with scoop (B1188).

Ammonium nitrate [6484–52–2]

Colourless crystals; mp 169 °C, decomposing at about 210 °C.

(CONTACT WITH COMBUSTIBLE MATERIAL MAY CAUSE FIRE)

Oxidant, keep out of contact with all combustible material.

Hazardous reactions Reviews of fire and explosion hazards of the salt; reactions with acetic acid, alkali metals, powdered Al, Sb, Bi, Cd, Cr, Co, Cu, Fe, Pb, Mg, Mn, Ni, Sn, Zn, brass and stainless steel, charcoal, organic fuels, potassium nitrite, potassium permanganate, sulphur and urea; decomposition temperature is markedly lowered by various impurities (B1195–1200).

175

Fire hazard	Mixtures of ammonium nitrate and combustible materials are readily ignited; mixtures with finely divided combustible materials can react explosively. *Extinguish fire with* water spray.
Spillage disposal	Mop up with plenty of water and run to waste diluting greatly with running water. Ensure that site of spillage is thoroughly washed down to eliminate future fire risks.

Ammonium oxalate —*see* Oxalates

Ammonium perchlorate —*see* Perchlorates

Ammonium periodate   [13446–09–8]

Hazardous reaction	Exploded while being transferred by scoop (B1188).

Ammonium permanganate   [13446–10–1]

Hazardous reactions	Friction sensitive when dry and explodes at 60 °C (B1190).

Ammonium peroxodisulphate   [7727–54–0]
(ammonium persulphate)

White crystals, mp 120 °C with decomposition; soluble in water.

(CONTACT WITH COMBUSTIBLE MATERIAL MAY CAUSE FIRE)

Keep out of contact with all combustible materials. Avoid contact with skin, eyes and clothing.

Hazardous reactions	Mixture with water and powdered Al may explode; in slightly acid concentrated solution, iron dissolved vigorously; mixture with sodium peroxide explodes on grinding in mortar (B1212–1213).
Fire hazard	Mixtures of ammonium persulphate with combustible materials are readily ignited. *Extinguish fire with* water spray.
Spillage disposal	Mop up with plenty of water and run to waste diluting greatly with running water.

Ammonium persulphate
—*see* Ammonium peroxodisulphate

Ammonium picrate   [58696–86–9]

Hazardous reactions   Explodes on heating or impact (B684).

Ammonium sulphamate
—*see* Ammonium amidosulphate

Ammonium sulphide solution   [12259–92–6]
(ammonium polysulphide)

Yellow liquid with offensive odour; contact with acid liberates poisonous hydrogen sulphide.

> CONTACT WITH ACID LIBERATES TOXIC GAS
> CAUSES BURNS

Avoid breathing vapour. Prevent contact with eyes and skin.

Toxic effects   Vapour inhaled in high concentration may cause unconsciousness. Lower concentrations cause headache, giddiness and loss of energy some time after exposure. The liquid severely irritates and may burn the eyes. The liquid irritates and may burn the skin. The liquid causes severe internal damage if taken by mouth.

First aid   *Vapour inhaled:* standard treatment (p. 136).
Affected eyes: standard treatment (p. 136).
Skin contact: standard treatment (p. 136).
If swallowed: standard treatment (p. 137).

Spillage disposal   Wear goggles and gloves. Mop up with plenty of water and run to waste diluting greatly with running water.

Ammonium vanadate
—*see* Vanadium compounds

Amyl acetates —*see* Pentyl acetate

Amyl alcohol (mixed isomers)   [71–41–0]

Colourless liquid. Ordinary amyl alcohol is mainly primary iso-amyl alcohol, bp 132 °C.

FLAMMABLE
HARMFUL BY INHALATION

Avoid contact with skin and eyes. RL 100 ppm (360 mg m^{-3}).

Toxic effects Vapour may irritate the eyes and respiratory system. Liquid irritates eyes severely and may irritate skin. The liquids, if swallowed, may cause headache, vertigo, nausea, vomiting, excitement and delirium followed by coma.

First aid *Vapour inhaled:* standard treatment (p. 136).
Affected eyes: standard treatment (p. 136).
Skin contact: standard treatment (p. 136).
If swallowed: standard treatment (p. 137).

Fire hazard Flash point 41 °C; explosive limits 1–9%. Ignition temp. about 350 °C. *Extinguish fire with* water spray, dry powder, carbon dioxide or vaporising liquids.

Spillage disposal Shut off all possible sources of ignition. Wear face-shield and gloves. Apply non-flammable dispersing agent if available and work to an emulsion with brush and water — run this to waste diluting greatly with running water. If dispersant not available absorb on sand, shovel into bucket(s) and transport to safe open area for atmospheric evaporation or burial. Ventilate area well to evaporate remaining liquid and dispel vapour.

t-Amyl alcohol —*see* 2-Methylbutan-2-ol

Amyl alcohols (n- and s-)—*see* Pentanols

Amyl nitrite (isopentyl nitrite) [110–46–3]

Colourless or pale yellow, highly volatile liquid with pungent fruity odour; bp 99 °C: advisable to store in refrigerator; immiscible with water.

HIGHLY FLAMMABLE
HARMFUL BY INHALATION

Avoid breathing vapour. Avoid contact with skin and eyes.

Toxic effects If inhaled may cause headache, flushing of face, weakness and collapse. If swallowed, similar effects may be expected.

First aid *Vapour inhaled:* standard treatment (p. 136).
Affected eyes: standard treatment (p. 136).
If swallowed: standard treatment (p. 137).

Fire hazard Flash point 10 °C; ignition temp. 209 °C. *Extinguish fire with* dry powder, carbon dioxide or vaporising liquids.

Spillage disposal Shut off all possible sources of ignition. Instruct others to keep at a safe distance. Wear breathing apparatus and gloves. Apply non-flammable dispersing agent if available and work to an emulsion with brush and water — run this to waste diluting greatly with running water. If dispersant not available, absorb on sand, shovel into bucket(s) and transport to safe open area for atmospheric evaporation or burial. Site of spillage should be washed thoroughly with water and soap or detergent.

Aniline (phenylamine)   [62–53–3]

Colourless to brown liquid; bp 185 °C; immiscible with water.

TOXIC BY INHALATION, IN CONTACT WITH SKIN
AND IF SWALLOWED
DANGER OF CUMULATIVE EFFECTS

Avoid breathing vapour. Avoid contact with eyes and skin. RL (skin) 2 ppm (10 mg m^{-3}).

Toxic effects Inhalation of the vapour or absorption through the skin causes headache, drowsiness, cyanosis, mental confusion and, in severe cases, convulsions. The liquid is dangerous to the eyes and the above effects are also experienced if it is swallowed. **Chronic effects** Prolonged exposure to the vapour, or slight skin exposure over a period, affects the nervous system and the blood, causing fatigue, loss of appetite, headache and dizziness. (*see* note in Ch 6 p. 106).

Hazardous reactions Vigorously oxidised by a number of oxidants including perchloric acid, fuming nitric acid, sodium peroxide and ozone. Violent reaction with BCl_3 (B601–602).

First aid *Vapour inhaled:* standard treatment (p. 136).
Affected eyes: standard treatment (p. 136).
Skin contact: standard treatment (p. 136).
If swallowed: standard treatment (p. 137).

Spillage disposal Wear breathing apparatus (or face-shield if amount is small) and gloves. Mix with sand and shovel mixture into a suitable vessel (glass, polythene or enamel) for dispersion in an excess of dilute hydrochloric acid (1 volume concentrated acid diluted with 2 volumes of water). Allow to stand, with occasional stirring, for 24 hours and then run acid extract to waste, diluting greatly with running water and washing the sand. Sand can be treated as normal waste.

RSC *Lab. Haz. Data Sheet No. 23,* 1984 gives extended coverage.

Anilinium salts

The commoner anilinium salts — the chloride and sulphate — are soluble in water. They are colourless to greyish-brown in colour.

TOXIC BY INHALATION, IN CONTACT WITH SKIN
AND IF SWALLOWED
DANGER OF CUMULATIVE EFFECTS

Avoid contact with eyes and skin.

Toxic effects   The acute and chronic effects of aniline poisoning are described under **Aniline**. Skin absorption does not occur so readily with the salts or their solutions as with aniline itself, but they are dangerous to the eyes, partly because of the acidity, and cause intense irritation. If taken by mouth, the effects of aniline poisoning (headache, drowsiness, cyanosis) will be apparent.

First aid   *Affected eyes:* standard treatment (p. 136).
Skin contact: standard treatment (p. 136).
If swallowed: standard treatment (p. 137).

Spillage disposal   Small amounts may be swept up and dispersed in a large volume of water which is then run to waste, diluting greatly with running water.

Anisidines (aminoanisoles)   *o*-[90–04–0] *p*-[104–94–9]

o-Anisidine is a pale yellow to orange liquid; bp 224 °C; immiscible with water. *p*-Anisidine is a pale yellow to brown solid; mp 59 °C; insoluble in water.

VERY TOXIC BY INHALATION, IN CONTACT WITH SKIN
AND IF SWALLOWED
DANGER OF CUMULATIVE EFFECTS

Avoid breathing vapour. Avoid contact with skin and eyes. RL (skin) 0.1 ppm (0.5 mg m^{-3}).

Toxic effects   These are not recorded but are assumed to be similar to those of aniline poisoning, *i.e.*, headache, drowsiness and cyanosis. *o*-Anisidine is a skin irritant.

First aid   *Vapour inhaled:* standard treatment (p. 136).
Affected eyes: standard treatment (p. 136).
Skin contact: standard treatment (p. 136).
If swallowed: standard treatment (p. 137).

Spillage disposal   Wear breathing apparatus (or face-shield if amount is small) and gloves. Mix with sand and shovel mixture into suitable glass, polythene or enamel vessel for dispersion in an excess of dilute hydrochloric acid (1 volume concentrated acid diluted with 2 volumes water). Allow to stand, with occasional stirring, for 24 hours and then run acid extract to waste, diluting greatly with running water and washing the sand. Sand can be treated as normal waste.

Antimony compounds (water-soluble)

Most soluble antimony compounds are colourless crystals or powder; the pentachloride is a reddish, fuming liquid with an offensive smell.

(TOXIC BY INHALATION, IN CONTACT WITH SKIN
AND IF SWALLOWED)

RL (as Sb) 0.5 mg m^{-3}.

Toxic effects All soluble antimony compounds must be considered to be poisonous when taken by mouth. Some compounds cause skin irritation and dermatitis. If taken by mouth, soluble antimony compounds may cause burning of the mouth and throat, choking, nausea and vomiting. Stibine, which may be formed by the action of acidic reducing agents on antimony-containing materials, is an extremely poisonous gas, causing blood destruction and damage to liver and kidneys. Insoluble antimony compounds, such as the oxide and sulphide, are not toxic.

First aid *Skin contact:* standard treatment (p. 136).
If swallowed: standard treatment (p. 137).

Spillage disposal The disposal of these in any quantity must be considered carefully in the light of local conditions and regulations. Burial in an isolated area can be considered as can gradual disposal at very high dilution into a sewage system permitting this. Both the chlorides $SbCl_3$ and $SbCl_5$ are readily hydrolysed by water.

Antimony(III) nitride   [12333–57–2]

Hazardous reactions Impure material explodes mildly on heating in air, or on contact with water or dilute acids (B1289).

Aprotic solvents Review of group (B1450)
(non-hydroxylic solvents)

Aqua regia (nitric acid/hydrochloric acid)   [8007–56–5]

Hazardous storage Pressure develops in screw-capped bottles (B23).

181

Arsenic compounds

Most arsenic compounds are colourless powders or crystals — they include arsenites and arsenates of many metals; syrupy arsenic acid and arsenic trichloride are liquids. All must be considered to be extremely poisonous. The metal itself 'has not been recognised as a noteworthy hazard': (*see* Ch 6 p. 104).

TOXIC BY INHALATION AND IF SWALLOWED

Do not inhale dust or fume. Prevent contact with skin and eyes. RL (as As) (0.2 mg m^{-3}).

Toxic effects	The inhalation of dust or fume irritates the mucous membranes and leads to arsenical poisoning. Certain compounds, especially the trichloride and arsenic acid, irritate the eyes and skin, and absorption causes poisoning. If swallowed, arsenic compounds irritate the stomach severely and affect the heart, liver and kidneys; nervousness, thirst, vomiting, diarrhoea, cyanosis and collapse may be symptoms. Arsine — *see* below. **Chronic effects** The inhalation of small concentrations of dust or fume over a long period will cause poisoning; skin contact over a long period may cause ulceration.
First aid	*Dust or fume inhaled:* standard treatment (p. 136). *Affected eyes:* standard treatment (p. 136). *Skin contact:* standard treatment (p. 136). *If swallowed:* standard treatment (p. 137).
Spillage disposal	The disposal of these in any quantity must be considered carefully in the light of local conditions and regulations. Deep burial mixed with sand in an isolated area or consignment to deep sea water in a heavy container can be considered as can gradual disposal at very high dilution into a sewage system permitting this.

Arsine (arsenic trihydride; hydrogen arsenide)   [7784-42-1]

Colourless gas with garlic odour; slightly soluble in water; formed whenever nascent hydrogen is in contact with an aqueous solution of arsenic.

VERY TOXIC BY INHALATION
(FLAMMABLE)

Prevent inhalation of gas. RL 0.05 ppm (0.2 mg m^{-3}).

Toxic effects	A few inhalations may be fatal, death resulting from anoxia or pulmonary oedema. Symptoms of poisoning include headache, weakness, vertigo and nausea. Damage is caused to kidneys and liver. (*see* note in Ch 6 p. 105.)
Hazardous reactions	Ignites in chlorine; explodes with fuming nitric acid; the endothermic compound is capable of detonation (B51).

First aid   *Gas inhaled:* standard treatment (p. 137); prompt medical attention vital.
Disposal   Leaking laboratory cylinders should be removed to open space, wearing
                  self-contained breathing apparatus, and allowed to discharge as slowly as
                  possible. Consult suppliers.

Asbestos   [1332-21-4]

Human carcinogen, use of all forms is controlled in the United Kingdom by the Asbestos
Regulations 1969. CL 0.2–2.0 fibres >5 μm long cc^{-1} depending on type of asbestos.

Auramine   [492-80-8]

Yellow flakes or powder. In the United Kingdom the manufacture (not use) of this substance is
controlled by The Carcinogenic Substances Regulations 1967. (*see* p. 148.)

> (TOXIC BY INHALATION, IN CONTACT WITH SKIN
> AND IF SWALLOWED
> DANGER OF VERY SERIOUS IRREVERSIBLE EFFECTS)

Prevent contact with skin, eyes and clothing.

Toxic effects   Absorption through the skin may result in dermatitis and burns, nausea and
                      vomiting.

First aid   *Skin contact:* standard treatment (p. 136).
                      If swallowed: standard treatment (p. 137).
Spillage   Wear face-shield or goggles, gloves and breathing apparatus. Mix with sand
disposal   and arrange for disposal by specialist contractor. Site of spillage should be
                      washed thoroughly with water and soap or detergent.

Azides Review of group (B1456).

Azidoacetic acid   [18523-48-3]

Hazardous   Destabilised by Fe salts and may explode at 90 °C (B250).
reaction

Azidoacetaldehyde   [67880-11-9]

Hazardous   Decomposed vigorously below 80 °C at 5 mbar (5×10^2 Pa) (B250).
reaction

Azidoacetone and oxime [4504–27–2] [101672–04–2]

Hazardous reactions Azidoacetone has exploded after 6 months' storage in the dark (B359), distillation residue of its oxime exploded violently (B364).

2-Azidocarbonyl compounds Review of group (B1456).

N-Azidodimethylamine [2156–66–3]

Hazardous reaction Rather explosive (B288).

Azidodimethylborane [2306–67–0]

Hazardous reaction Explodes on warming (B283).

Azidosilane [13807–60–4]

Hazardous preparation (B1183).

5-Azidotetrazole [35038–46–1]

Hazardous reactions The compound and its salts are explosive (B138–139).

Aziridine (ethyleneimine) [151–56–4]

HIGHLY FLAMMABLE
VERY TOXIC BY INHALATION, IN CONTACT WITH SKIN
AND IF SWALLOWED
POSSIBLE RISKS OF IRREVERSIBLE EFFECTS

Experimental carcinogen. RL (skin) 0.5 ppm (1 mg m^{-3}).

Hazardous reactions Erroneous preparative procedure — liable to polymerise explosively; reaction with hypochlorite gives explosive chloroaziridine; forms explosive silver derivatives (B278).

α-Azoisobutyronitrile [78–67–1]

Hazardous reactions   Decomposes when heated; explosive decomposition when technical material being recrystallised from acetone (B721).

Azo compounds Review of group (B1457).

Azo-*N*-nitroformamidine [53144–61–2]

Hazardous reaction   Decomposes explosively at 165 °C (B262).

Barium azide [18810–58–7]

Hazardous reaction   Impact sensitive when dry (B82–83]

Barium bromate [13967–90–3]

Hazardous reaction   Decomposes almost explosively at 300 °C; mixtures with S may ignite spontaneously (B79).

Barium compounds

Practically all barium compounds are colourless crystals or powders. All barium compounds, except the sulphate, must be considered to be poisonous when taken by mouth.

(POISON)
HARMFUL BY INHALATION AND IF SWALLOWED

RL (soluble compounds) 0.5 mg m^{-3}.

Toxic effects   If ingested, soluble barium compounds cause nausea, vomiting, stomach pains and diarrhoea. (*see* note in Ch 6 p. 104.)

Hazardous reactions   Dangerous reactions of the metal, hydride, hydroxide, nitrate, oxide, sulphate, sulphide are indicated (B78–85).

First aid   *If swallowed:* standard treatment (p. 137).
Spillage disposal   The sulphate may be brushed up and treated as normal refuse. Soluble barium salts should be mopped up with water and the solution run to waste, diluting greatly with running water.

Barium peroxide [1304–29–6]

CONTACT WITH COMBUSTIBLE MATERIAL MAY CAUSE FIRE
HARMFUL BY INHALATION OR IF SWALLOWED

Hazardous reactions
May ignite H_2S, hydroxylamine and organic materials especially in presence of water; mixtures with powdered metals ignite; explosions with acetic anhydride (B83–84).

Benzal chloride —*see* Benzylidene chloride

Benzaldehyde [100–52–7]

Hazardous reactions
Violent oxidation by 90% performic acid (B151).

Benzene (benzol; coal naphtha) [71–43–2]

Colourless, volatile liquid with characteristic odour; bp 80 °C; immiscible with water.

HIGHLY FLAMMABLE
TOXIC BY INHALATION AND CONTACT WITH SKIN
DANGER OF VERY SERIOUS IRREVERSIBLE EFFECTS

Avoid inhalation of vapour. Prevent contact with skin and eyes. RL (skin) 10 ppm (30 mg m^{-3}). Suspected carcinogen.

Toxic effects
Inhalation of the vapour causes dizziness, headache and excitement; high concentrations may cause unconsciousness. The vapour irritates the eyes and mucous membranes. The liquid is absorbed through the skin and poisoning may result from this. Assumed to be extremely poisonous if taken by mouth. **Chronic effects** Repeated inhalation of low concentrations over a considerable period may cause severe, even fatal, blood disease (leukaemia, *see* references to toxicity of benzene in Ch 6 and 7.)

Hazardous reactions
Complex with silver perchlorate exploded on crushing in mortar; certain mixtures with 84% of nitric acid are highly sensitive to detonation; mixture with liquid oxygen is explosive; benzene solution of rubber exploded when ozonised; reacts vigorously or explosively with other oxidants, inter-halogens, uranium hexafluoride (B587–588).

First aid
Vapour inhaled: standard treatment (p. 136).
Affected eyes: standard treatment (p. 136).
Skin contact: standard treatment (p. 136).
If swallowed: standard treatment (p. 137).

Fire hazard   Flash point −11 °C; explosive limits 1.4–8%; ignition temp 562 °C. *Extinguish fire with* foam, dry powder or vaporising liquids.

Spillage disposal   Shut off all possible sources of ignition. Instruct others to keep at a safe distance. Wear breathing apparatus and gloves. Apply non-flammable dispersing agent if available and work to emulsion with brush and water — run this to waste diluting greatly with running water. If dispersant not available, absorb on sand, shovel into bucket(s) and transport to safe open area for atmospheric evaporation. Site of spillage should be washed thoroughly with water and soap or detergent.

RSC *Lab. Haz. Data Sheet No. 20,* 1984 gives extended coverage.

Benzenediazonium-2-carboxylate   [17333–86–7]

Hazardous reactions   Internal salt explosive and reacts explosively with aniline and violently with aryl isocyanides (B654–655).

Benzenediazonium chloride   [100–34–4]

Hazardous reactions   Dry salt explosive (B573) as is zinc chloride complex (B799–800).

Benzenediazonium nitrate   [619–96–7]

Hazardous reactions   Highly sensitive to friction and impact and explodes at 90 °C (B585).

Benzenediazonium-4-sulphonate   [305–80–6]

Hazardous reaction   Dry internal salt exploded violently on touching (B566)

Benzene-disulphonic and -sulphonic acids
—*see* Sulphonic acids

Benzenesulphinyl chloride   [4972–29–6]

Hazardous reaction   Bottle, undisturbed for months, exploded (B574).

187

Benzenesulphonyl azide   [938–10–3]

Hazardous   Pure azide decomposed smoothly at 105 °C but crude exploded on heating
reaction   (B584).

Benzenesulphonyl chloride   [98–09–9]

Colourless to brown liquid; bp 251 °C with decomposition; reacts with water to form benzenesulphonic acid and hydrochloric acid.

> (IRRITATING TO EYES, RESPIRATORY SYSTEM AND SKIN
> CAUSES BURNS)

Prevent contact with skin and eyes.

Toxic effects   Irritates the eyes severely and causes skin burns. Causes severe internal irritation if taken by mouth.

Hazardous   Spontaneous decomposition in storage, violent reaction with dimethyl
reaction   sulphoxide (B574–575).

First aid   *Affected eyes:* standard treatment (p. 136).
Skin contact: standard treatment (p. 136).
If swallowed: standard treatment (p. 137).

Spillage   Wear face-shield or goggles, and gloves. Spread soda ash liberally over the
disposal   spillage and mop up cautiously with water — run to waste diluting greatly with running water.

Benzenethiol (phenyl mercaptan, thiophenol)   [108–98–5]

Colourless liquid, repulsive pungent smell; bp 168 °C; immiscible with water.

> (TOXIC BY INHALATION
> IRRITATING TO EYES, RESPIRATORY SYSTEM AND SKIN)

Avoid inhaling vapour or contact with skin or eyes. TLV 0.5 ppm (2 mg m^{-3}).

Toxic effects   Headache, dizziness: skin contact may lead to severe dermatitis.

Hazardous   Violent explosion during preparation from benzenediazonium chloride
preparation   (B99).

First aid   *Vapour inhaled:* standard treatment (p. 136).
Affected eyes: standard treatment (p. 136).
Skin contact: standard treatment (p. 137).

Spillage disposal   Instruct others to keep at safe distance. Wear breathing apparatus and gloves. Cover with plenty of soda-ash and shovel up into bucket(s), enclose in polythene bag and transport to safe open area for burial. Site of spillage should be washed thoroughly with dilute soda-ash solution and soap or detergent.

Benzidine and salts [92-87-5]

The use of these compounds in the United Kingdom is now prohibited under The Carcinogenic Substances Regulations 1967 (*see* p. 148). Inhalation or absorption through the skin of the dust has been recognised as a cause of bladder tumours. It is not therefore considered appropriate to deal with their hazards more fully in this book.

Benzonitrile [100-47-0]

Colourless liquid, bp 191 °C; sparingly soluble in water.

(HARMFUL BY INHALATION, IN CONTACT WITH SKIN AND IF SWALLOWED
IRRITATING TO EYES AND RESPIRATORY SYSTEM)

Avoid inhalation of vapour. Avoid contact with skin or eyes.

Toxic effects   No record has been found of instances of poisoning by this material, but its constitution is such that it must be assumed to be toxic by inhalation, skin and eye contact, and ingestion.

First aid   *Vapour inhaled:* standard treatment (p. 136).
Affected eyes: standard treatment (p. 136).
Skin contact: standard treatment (p. 136).
If swallowed: standard treatment (p. 137).

Spillage disposal   Instruct others to keep at a safe distance. Wear breathing apparatus and gloves. Apply dispersing agent if available and work to an emulsion with brush and water — run this to waste diluting greatly with running water. If dispersant not available absorb on sand, shovel into bucket(s) and transport to safe open area for burial. Site of spillage should be washed thoroughly with water and soap or detergent.

p-Benzoquinone (quinone) [106-51-4]

Yellow crystals, with characteristic, irritating odour. Slightly soluble in water.

TOXIC BY INHALATION AND IF SWALLOWED
IRRITATING TO EYES, RESPIRATORY SYSTEM AND SKIN

Avoid breathing dust. Avoid contact with skin and eyes. RL 0.1 ppm (0.4 mg m^{-3}).

189

Toxic effects   The dust irritates the respiratory system severely. Skin or eye contact is very irritating and can cause severe local damage. Must be considered highly irritant and dangerous if taken by mouth.

Hazardous reaction   Drums of moist material self-heated and decomposed (B570).

First aid   *Affected eyes:* standard treatment (p. 136).
Skin contact: standard treatment (p. 136).
If swallowed: standard treatment (p. 137).

Spillage disposal   Wear face-shield or goggles, and gloves. Mop up with plenty of water and run to waste diluting greatly with running water.

Benzotriazole   [95–14–7]

Hazardous reaction   Large batch exothermally decomposed and then detonated during distillation at 160 °C/2.5 mbar (2.5×10^2 Pa) (B583).

Benzotrichloride —*see* Benzylidyne chloride

Benzotrifluoride —*see* Benzylidyne fluoride

Benzoyl azide   [582–61–6]

Hazardous reaction   Crude material exploded violently between 120 °C and 165 °C (B661).

Benzoyl chloride (benzenecarbonyl chloride)   [98–88–4]

Colourless, fuming liquid with pungent smell; bp 197 °C; reacts with water forming benzoic acid and hydrochloric acid.

CAUSES BURNS
(IRRITATING TO EYES, RESPIRATORY SYSTEM AND SKIN)

Avoid breathing vapour. Prevent contact with eyes and skin.

Toxic effects   The vapour irritates the respiratory system. The vapour irritates and the liquid burns the eyes severely. The liquid is very irritating to the skin and can cause burns. If taken by mouth there is immediate irritation and damage.

Hazardous reaction	Violent reaction with dimethyl sulphoxide (B292).
First aid	*Vapour inhaled:* standard treatment (p. 136). *Affected eyes:* standard treatment (p. 136). *Skin contact:* standard treatment (p. 136). *If swallowed:* standard treatment (p. 137).
Spillage disposal	Wear goggles and gloves. Spread soda ash liberally over the spillage and mop up cautiously with plenty of water — run to waste diluting greatly with running water.

Benzoyl nitrate [6786–32–9]

Hazardous reactions	Unstable liquid which explodes on rapid heating and may also explode on exposure to light or contact with moist cellulose (B660).

Benzoyl peroxide —*see* Dibenzoyl peroxide

1,1-Benzoylphenyldiazomethane
—*see* 2-Diazo-2-phenylacetophenone

Benz(*a*)pyrene (3,4-benzpyrene) [50–32–8]

Suspected carcinogen. No TLV set.

Benzvalene (tricyclo[3.1.0.0^{2,6}]hex-3-ene) [659–85–8]

Hazardous reaction	Exploded violently when scratched (B579).

Benzylamine [100–46–9]

Colourless liquid; bp 185 °C; miscible with water.

CAUSES BURNS

Avoid contact with skin and eyes.

Toxic effects	These have not been recorded to any extent, but it has been found that benzylamine causes skin burns; by inference it must be assumed to damage the eyes and cause internal irritation and damage if taken by mouth.

191

First aid	*Affected eyes:* standard treatment (p. 136). *Skin contact:* standard treatment (p. 136). *If swallowed:* standard treatment (p. 137).
Spillage disposal	Instruct others to keep at a safe distance. Wear breathing apparatus and gloves. Mop up with plenty of water and run to waste diluting greatly with running water. Ventilate area well to evaporate remaining liquid and dispel vapour.

Benzyl bromide (α-bromotoluene) [100–39–0]

Colourless to pale yellow liquid; bp 198 °C; immiscible with water.

IRRITATING TO EYES, RESPIRATORY SYSTEM AND SKIN

Avoid breathing vapour. Prevent contact with skin and eyes.

Toxic effects	The vapour irritates the respiratory system. Low vapour concentrations cause lachrymation and severe irritation to the eyes. The vapour irritates the skin and the liquid causes burns. Can be assumed to cause severe internal irritation and damage if taken by mouth.
Hazardous reaction	Material drying over molecular sieve polymerized, evolving HBr (B670).
First aid	*Vapour inhaled:* standard treatment (p. 136). *Affected eyes:* standard treatment (p. 136). *Skin contact:* standard treatment (p. 136). *If swallowed:* standard treatment (p. 137).
Spillage disposal	Instruct others to keep at a safe distance. Wear breathing apparatus and gloves. Apply dispersing agent if available and work to an emulsion with brush and water — run this to waste diluting greatly with running water. If dispersant not available, absorb on sand, shovel into bucket(s) and transport to safe open area for burial. Site of spillage should be washed thoroughly with water and soap or detergent.

Benzyl chloride (α-chlorotoluene) [100–44–7]

Colourless to brown-yellow liquid with acrid smell; bp 179 °C; immiscible with water.

IRRITATING TO EYES, RESPIRATORY SYSTEM AND SKIN

Avoid breathing vapour Prevent contact with skin and eyes. RL 1 ppm (5 mg m^{-3}).

Toxic effects	The vapour irritates the respiratory system. Low vapour concentrations cause lachrymation and severe irritation to the eyes. The vapour irritates the skin and the liquid causes burns. Can be assumed to cause severe internal irritation and damage if taken by mouth.

Hazardous reactions	Absence of sufficient base to prevent acidity developing led to violent reaction or explosions (B671–672).
First aid	*Vapour inhaled:* standard treatment (p. 136). *Affected eyes:* standard treatment (p. 136). *Skin contact:* standard treatment (p. 136). *If swallowed:* standard treatment (p. 137).
Spillage disposal	Instruct others to keep at a safe distance. Wear breathing apparatus and gloves. Apply non-flammable dispersing agent and work to an emulsion with brush and water — run this to waste diluting greatly with running water. If dispersant not available, absorb on sand, shovel into bucket(s) and transport to safe open area for burial. Site of spillage should be washed thoroughly with water and soap or detergent.

Benzyl chloroformate [501–53–1]

Colourless or yellow, fuming, oily liquid; decomposes slowly at room temperature and needs refrigerated storage; bp 103 °C at 29 mbar (29×10^2 Pa); insoluble in water.

CAUSES BURNS
IRRITATING TO RESPIRATORY SYSTEM

Prevent contact with skin, eyes and clothing.

Toxic effects	Vapour causes severe irritation of eyes and respiratory system. Liquid blisters the skin and severely damages the eyes.
First aid	*Vapour inhaled:* standard treatment (p. 136). *Affected eyes:* standard treatment (p. 136). *Skin contact:* standard treatment washing with soap and water (p. 136). *If swallowed:* standard treatment (p. 137).
Spillage disposal	Instruct others to keep at a safe distance. Wear breathing apparatus and gloves. Apply non-flammable dispersing agent and work to an emulsion with brush and water — run this to waste diluting greatly. If dispersant not available, absorb on sand and transport to safe open area for burial.

Benzyl cyanide —*see* Phenylacetonitrile

Benzylidene chloride [98–87–3]
(benzal chloride, α,α,α-trichlorotoluene)

Colourless liquid; bp 205 °C: immiscible with water.

IRRITATING TO EYES, RESPIRATORY SYSTEM AND SKIN

Avoid breathing vapour. Avoid contact with eyes and skin.

Toxic effects	The vapour irritates the respiratory system. The vapour and liquid irritate the eyes and may cause conjunctivitis. The liquid may irritate the skin. If taken by mouth, internal irritation and damage must be assumed.
First aid	*Vapour inhaled:* standard treatment (p. 136). *Affected eyes:* standard treatment (p. 136). *Skin contact:* standard treatment (p. 136). *If swallowed:* standard treatment (p. 137).
Spillage disposal	Instruct others to keep at a safe distance. Wear breathing apparatus and gloves. Apply dispersing agent if available and work to an emulsion with brush and water—run this to waste diluting greatly with running water. If dispersant not available, absorb on sand, shovel into bucket(s) and transport to safe open area for burial. Site of spillage should be washed thoroughly with water and soap or detergent.

Benzylidyne chloride (benzotrichloride) [98–07–7]

Colourless to yellow, fuming liquid; bp 214 °C; immiscible with water.

HARMFUL BY INHALATION

Avoid breathing vapour. Avoid contact with eyes and skin.

Toxic effects	The vapour irritates all parts of the respiratory system. The vapour and liquid irritate the eyes and skin. If taken by mouth, internal irritation and damage must be assumed.
First aid	*Vapour inhaled:* standard treatment (p. 136). *Affected eyes:* standard treatment (p. 136). *Skin contact:* standard treatment (p. 136). *If swallowed:* standard treatment (p. 137).
Spillage disposal	Instruct others to keep at a safe distance. Wear breathing apparatus and gloves. Apply dispersing agent if available and work to an emulsion with brush and water—run this to waste diluting greatly with running water. If emulsifier not available, absorb on sand, shovel into bucket(s) and transport to safe open area for burial. Site of spillage should be washed thoroughly with water and soap or detergent.

Benzylidyne fluoride (benzotrifluoride) [98–08–8]

Colourless liquid with an aromatic odour; bp 101 °C; immiscible with water.

HIGHLY FLAMMABLE

Avoid breathing vapour. Avoid contact with skin and eyes.

Toxic effects	Animal experiments indicate the risk of central nervous system depression through inhalation, absorption and ingestion.
First aid	*Vapour inhaled:* standard treatment (p. 136). *Affected eyes:* standard treatment (p. 136). *Skin contact:* standard treatment (p. 136). *If swallowed:* standard treatment (p. 137).
Fire hazard	Flash point 12 °C. *Extinguish fire with* water spray, foam, dry powder, carbon dioxide or vaporising liquids.
Spillage disposal	Shut off all possible sources of ignition. Instruct others to keep at a safe distance. Wear breathing apparatus and gloves. Apply non-flammable dispersing agent and work to an emulsion with brush and water—run this to waste diluting greatly with running water. If dispersant not available, absorb on sand, shovel into bucket(s) and transport to safe open place for atmospheric evaporation or burial. Site of spillage should be washed thoroughly with water and soap or detergent.

Benzyloxyacetylene [40089–12–1]

Hazardous reaction Explosion if heated above 60 °C during vacuum distillation (B743).

Benzyltriethylammonium permanganate [68844–25–7]

Hazardous reaction The quaternary oxidant, previously reported as stable, is explosive under unexpectedly mild conditions. Related compounds are similarly unpredictable (B823).

Beryllium (metal) [7440–41–7]

VERY TOXIC BY INHALATION AND IN CONTACT WITH SKIN
IRRITATING TO RESPIRATORY SYSTEM
DANGER OF VERY SERIOUS IRREVERSIBLE EFFECTS

Suspected carcinogen. RL 0.002 mg m^{-3}.

Hazardous reactions Heavy impact flashes mixtures of powdered Be with CCl_4 or C_2HCl_3; incandescent reaction when heated with phosphorus (B86).

Beryllium compounds

White powder or crystals

195

VERY TOXIC BY INHALATION AND IN CONTACT WITH SKIN
IRRITATING TO RESPIRATORY SYSTEM
DANGER OF VERY SERIOUS IRREVERSIBLE EFFECTS

Prevent inhalation of dust. Prevent contact with eyes and skin. RL (as Be) 0.002 mg m^{-3}.

Toxic effects	Particles penetrating the skin through wounds and abrasions may cause local damage difficult to heal. Symptoms of poisoning, indicated by respiratory troubles or cyanosis, may develop within a week or after a latent period of even several years.
First aid	*Dust inhaled:* standard treatment (p. 136). *Affected eyes:* standard treatment (p. 136). *Skin contact:* standard treatment (p. 136). *If swallowed:* standard treatment (p. 137).
Spillage disposal	The disposal of these in even small quantity must be considered carefully in the light of local conditions and regulations. Deep burial mixed with sand in an isolated area or consignment to deep sea water in a heavy container can be considered as can gradual disposal at very high dilution into a sewage system permitting this. Larger amounts should involve either recycling or an approved contractor.

Biphenyl (phenylbenzene)   [92–52–4]

White crystals or powder, faint characteristic odour; mp 70 °C; insoluble in water.

HARMFUL BY INHALATION

Avoid inhalation of dust or eye contact. RL 0.2 ppm (1.5 mg m^{-3}).

Toxic effects	Irritation of respiratory tract and eyes, nausea, depressed appetite.
First aid	*Dust inhaled:* standard treatment (p. 136). *Affected eyes:* standard treatment (p. 136). *Skin contact:* standard treatment (p. 136).
Spillage disposal	Instruct others to keep at safe distance. Wear goggles or face mask and gloves. Mix with sand and transport to safe open area for burial.

Bis(4-aminophenyl)methane   [101–77–9]
(4,4'-diaminodiphenylmethane; *p,p'*-methylenebisaniline)

Light brown solid; sparingly soluble in water.

HARMFUL BY INHALATION, IN CONTACT WITH SKIN
AND IF SWALLOWED

Avoid inhalation of dust or contact with skin and eyes. RL 0.1 ppm (0.8 mg m^{-3}).

Toxic effects   Inhalation and/or ingestion of dust leads to liver damage.

First aid   *Affected eyes:* standard treatment (p. 136).
                       Skin contact: standard treatment (p. 136).
                       If swallowed: standard treatment (p. 137).

Spillage   Wear face-shield or goggles, and gloves. Mix with sand and shovel mixture
disposal   into a glass, enamel or polythene vessel for dispersion in dilute hydrochloric
                       acid (1 volume concentrated acid diluted with 2 volumes of water). Allow to
                       stand, with occasional stirring for 24 hours and then run extract to waste,
                       diluting greatly with running water and washing the sand. The residual sand
                       can be dealt with as normal refuse.

Bis-*o*-azidobenzoyl peroxide   [20442–99–3]

Hazardous   Exploded on touching with metal spatula (B825).
reaction

Bis(2-chloroethyl) ether   [111–44–4]
[di(2-chloroethyl) ether]

Colourless liquid with pungent odour; bp 178 °C; practically insoluble in water; liable to form
explosive peroxides on exposure to air and light, which must be decomposed before the ether is
distilled to small volume.

FLAMMABLE
VERY TOXIC BY INHALATION, IN CONTACT WITH SKIN
AND IF SWALLOWED
POSSIBLE RISKS OF IRREVERSIBLE EFFECTS

Avoid breathing vapour. Avoid contact with skin and eyes. TLV (skin) 5 ppm (30 mg m^{-3}).

Toxic effects   The vapour irritates the respiratory system and high concentrations may
                       result in lung damage after a latent period of some hours. The vapour and
                       liquid irritate the eyes and may cause conjunctivitis. Assumed to be
                       poisonous if taken by mouth.

First aid   *Vapour inhaled:* standard treatment (p. 136).
                       Affected eyes: standard treatment (p. 136).
                       Skin contact: standard treatment (p. 136).
                       If swallowed: standard treatment (p. 137).

Fire hazard   Flash point 55 °C; ignition temp. 369 °C. *Extinguish fire with* dry powder,
                       carbon dioxide or vaporising liquid.

Spillage   Shut off all possible sources of ignition. Instruct others to keep at a safe
disposal   distance. Wear breathing apparatus and gloves. Apply non-flammable
                       dispersing agent if available and work to an emulsion with brush and
                       water—run this to waste diluting greatly with running water. If dispersant
                       not available, absorb on sand, shovel into bucket(s) and transport to safe
                       open area for burial. Site of spillage should be washed thoroughly with
                       water and soap or detergent.

Bischloromethyl ether [542-88-1]

[di(chloromethyl) ether]

Colourless liquid, suffocating odour; bp 106 °C; decomposed by water.

VERY TOXIC BY INHALATION
DANGER OF VERY SERIOUS IRREVERSIBLE EFFECTS

Prevent inhalation of vapour (which is readily formed from vapours of formaldehyde and hydrochloric acid and at levels well above the RL) and contact with eyes or skin. Human carcinogen. RL 0.001 ppm.

Toxic effects	The vapour irritates the respiratory system and eyes, primary effects are probably due to rapid formation of hydrochloric acid and formaldehyde. Assumed to be toxic and corrosive if swallowed. **Chronic effects** A high incidence of lung cancer has been observed.
First aid	*Vapour inhaled:* standard treatment (p. 136). *Affected eyes:* standard treatment (p. 136). *Skin contact:* standard treatment (p. 136). *If swallowed:* standard treatment (p. 137).
Spillage disposal	Instruct others to keep at safe distance. Wear breathing apparatus and gloves. Spread soda-ash liberally over the spillage and mop up cautiously with plenty of water—run to waste diluting greatly with running water.

1,1-Bis(4-chlorophenyl)-2,2,2-trichloroethane
(DDT) [50-29-3]

RL 1 mg m^{-3}.

Bis(2-cyanoethyl)amine
—*see* 3,3'-Iminodipropiononitrile

Bis-2,3-epoxypropyl ether [2238-07-5]
(diglycidyl ether)

RL 0.5 ppm (3 mg m^{-3}).

Bis(4-isocyanatophenyl)methane [101-68-8]

[di(4-isocyanatophenyl)methane; 4,4'-methylenebisphenyl isocyanate]

Yellow crystals or fused solid with irritating smell; mp 37 °C; hydrolysed by water — by storing at 5 °C, a tendency to form polymeric solids is reduced to a minimum.

HARMFUL BY INHALATION

Avoid breathing vapour or dust. Avoid contact with skin, eyes and clothing. CL 0.02 ppm (0.2 mg m^{-3}).

Toxic effects	The vapour is irritating to the eyes and respiratory system, as is the dust. The solid or molten material irritates the eyes and skin and must be considered poisonous if taken by mouth.
First aid	*Vapour inhaled:* standard treatment (p. 136). *Affected eyes:* standard treatment (p. 136). *Skin contact:* standard treatment (p. 136). *If swallowed:* standard treatment (p. 137).
Spillage disposal	Wear face-shield or goggles, and gloves. Shovel into dry bucket(s), transport to safe open area and add, a little at a time, to a large quantity of water; after reaction is complete, run to waste, diluting greatly with running water.

Bis(2-methoxyethyl) ether [111–96–6]
(diethylene glycol dimethyl ether; diglyme)

Hazardous reactions	Forms peroxides on exposure to air and light, explosive reactions of peroxidised solvent with AlH_3, $LiAlH_4$ (B634).

Bismuth (metal) [7440–69–9]

Hazardous reactions	The finely divided metal reacts violently with BrF_5, and fuming nitric acid; violent or explosive reaction with fused ammonium nitrate; its reaction with perchloric acid may be explosive; that with nitrosyl fluoride or iodine pentafluoride may be accompanied by incandescence (B87).

Bismuth nitride [12232–97–2]

Hazardous reactions	Exploded on shaking, heating, or on contact with water or dilute acids (B88).

Bismuth pentafluoride [7787–62–4]

Hazardous reaction	Reacts vigorously with water and may ignite (B86).

199

Bismuth telluride [1304–82–1]

RL 10 mg m^{-3}; 5 mg m^{-3} if selenium doped.

Bistrichloroacetyl peroxide [2629–78–9]

Hazardous reaction   Explodes on standing at room temperature (B397).

Bistrifluoroacetyl peroxide [383–73–3]

Hazardous reaction   Explodes on standing at room temperature (B398).

Bleaching powder
—*see* Calcium hypochlorite

Borane — tetrahydrofuran complex [14044–65–6]

Hazardous reaction   May decompose in storage with liberation of hydrogen and bursting of bottle (B61).

Boranes Review of group (B1462).

Borazine [6569–51–3]

Hazardous reaction   Sealed ampoules exploded in daylight (B73).

Boron [7440–42–8]

Hazardous reactions   Ignites in Cl_2 or F_2 at ambient temperature. Reacts explosively when ground with silver fluoride at ambient temperature. Violence of interaction with fused metal nitrates increased by presence of nitrites. Violent reactions with other oxidants, *e.g.*, nitrosyl fluoride, Na_2O_2, PbO.
Many of the violent reactions of boron with reagents other than hot powerful oxidants now attributed to previous use of impure B samples (B55–56).

Boron oxide [1303–86–2]

RL 10 mg m^{-3}.

Boron tribromide [10294–33–4]

Colourless, fuming liquid with a pungent odour; bp 90 °C; reacts violently with water.

REACTS VIOLENTLY WITH WATER
VERY TOXIC BY INHALATION AND IF SWALLOWED
CAUSES SEVERE BURNS

Avoid breathing vapour. Prevent contact with skin and eyes. Do not put water into container. RL 1 ppm (10 mg m^{-3}).

Toxic effects	The vapour irritates all parts of the respiratory system. The vapour irritates the eyes. The liquid burns the skin and eyes. If taken by mouth, there would be severe internal burning.
Hazardous reactions	Reacts violently when poured into an excess of water and explosively when water is poured into it. Mixture with sodium metal explodes on impact and that with WO$_3$ may explode if not cooled effectively (B57).
First aid	*Vapour inhaled:* standard treatment (p. 136). *Affected eyes:* standard treatment (p. 136). *Skin contact:* standard treatment (p. 136). *If swallowed:* standard treatment (p. 137).
Spillage disposal	Instruct others to keep at a safe distance. Wear breathing apparatus and gloves. Spread soda ash liberally over the spillage and mop up cautiously with plenty of water — run to waste diluting greatly with running water.

Boron trichloride [10294–34–5]

Colourless fuming liquid or gas with a pungent odour; bp 12.5 °C; reacts rapidly with water forming boric and hydrochloric acids.

REACTS VIOLENTLY WITH WATER
VERY TOXIC BY INHALATION AND IF SWALLOWED
CAUSES BURNS

Avoid breathing gas. Prevent contact with skin and eyes. Do not put water into container.

Toxic effects	The gas irritates the eyes, skin and respiratory system. The liquid irritates or burns the skin and burns the eyes. If taken by mouth there would be severe internal burning.

201

Hazardous reaction	Reacts violently with aniline and phosphine (B58–59).
First aid	*Gas inhaled:* standard treatment (p. 136).
	Affected eyes: standard treatment (p. 136).
	Skin contact: standard treatment (p. 136).
	If swallowed: standard treatment (p. 137).
Spillage disposal	In warm weather it will exist as a gas, in which case instruct others to keep out of the affected area; wear breathing apparatus and organise adequate ventilation. In cool weather, a spillage of the liquid can be covered with excess of soda ash (wearing breathing apparatus and gloves) and then mopped up cautiously with water and run to waste.

Boron trifluoride [7637–07–2]

Colourless, fuming gas with pungent, suffocating odour; bp −100 °C. Decomposes in water forming fluoroboric and boric acids.

REACTS VIOLENTLY WITH WATER
VERY TOXIC BY INHALATION
CAUSES SEVERE BURNS

Prevent inhalation of gas. Prevent contact with skin and eyes. RL 1 ppm (3 mg m^{-3}).

Toxic effects	The gas irritates the skin, eyes and respiratory system; at high concentrations it may burn the skin.
Hazardous reaction	Reacts with hot alkali or alkaline earth (not Mg) metals with incandescence (B59).
First aid	*Gas inhaled:* standard treatment (p. 136).
	Affected eyes: standard treatment (p. 136).
	Skin contact: standard treatment (p. 136).
Disposal	Surplus gas or leaking cylinder can be vented slowly into water-fed scrubbing tower or column, or into a fume cupboard served by such a tower.

Boron trifluoride complexes

The liquid complexes formed between boron trifluoride and acetic acid, diethyl ether, methanol and propan-1-ol all display hazards and toxic effects associated with their constituents. All are readily hydrolysed by water, corrosive and, to some degree, flammable.

Boron triiodide [13517–10–7]

Hazardous reactions	Strongly exothermal reaction with ammonia. Incandescent reaction with warm red or white phosphorus. Violent reaction with limited amounts of water (B65–66).

Bromine [7726-95-6]

Dark reddish-brown fuming liquid; bp 59 °C; slightly soluble in water.

VERY TOXIC BY INHALATION
CAUSES SEVERE BURNS

Prevent breathing of vapour. Prevent contact with eyes and skin. RL 0.1 ppm (0.7 mg m^{-3}).

Toxic effects The vapour irritates all parts of the respiratory system. The vapour severely irritates the eyes and mucous membranes. The liquid burns the skin and eyes. If taken by mouth, severe local burns and internal damage would result.

Hazardous reactions Bromine reacts with varying degrees of violence with a large number of compounds and elements including acetone, acrylonitrile, ammonia, BrF_3, copper(I) hydride, diethyl ether, *N,N*-dimethylformamide, tetrahydrofuran, ethanol/phosphorus, fluorine, germane, hydrogen, metal acetylides and carbides, metal azides, Li, Na, K, Rb, Al, Hg, Ti, methanol, ozone, F_2O, P, tetracarbonylnickel, trialkylboranes, F_2O_3. Explosions or explosive products with Et_2Zn, GeH_4, Sb, Me_3N (B98–103).

First aid *Vapour inhaled:* standard treatment (p. 136).
Affected eyes: standard treatment (p. 136).
Skin contact: standard treatment (p. 136).
If swallowed: standard treatment (p. 137).

Spillage disposal Instruct others to keep at a safe distance. Wear breathing apparatus and gloves. Spread soda ash liberally over the spillage and mop up cautiously with plenty of water — run to waste diluting greatly with running water.

RSC *Lab. Haz. Data Sheet No. 24,* 1984 gives extended coverage.

Bromine azide [13973-87-0]

Hazardous reactions Very shock-sensitive in solid, liquid and vapour forms. Liquid explodes on contact with As, Na, Ag foil, P (B97).

Bromine pentafluoride [7789-30-2]

Pale yellow fuming liquid with pungent odour; bp 40 °C. It reacts vigorously and possibly explosively with water.
RL 0.1 ppm (0.7 mg m^{-3}).

and

Bromine trifluoride [7787–71–5]

Colourless to grey-yellow, fuming liquid with pungent choking smell; bp 127 °C; mp 8.8 °C; reacts vigorously with water; extremely reactive, etching glass and setting fire to paper, wood and other organic material.

(REACTS VIOLENTLY WITH WATER
CAUSES SEVERE BURNS)

Prevent inhalation of vapour. Prevent contact with skin, eyes, clothing and all combustible material.

Toxic effects	The vapours severely irritate and may burn the eyes, skin and respiratory system. The liquids burn all human tissue and cause severe damage.
Hazardous reactions	Review of hazards and necessary precautions in use of BrF_3 indicated; violent or explosive reactions occur with ammonium halides, antimony chloride oxide, CO, acetone, diethyl ether, toluene, uranium, water, silicone grease, CCl_4, benzene, other organic materials; incandescent reactions with Br_2, I_2, As, Sb — also with powdered Mo, Nb, Ta, Ti, V, B, C, P, S; BrF_5 reacts violently with strong nitric and sulphuric acids, Cl_2, I_2, ammonium chloride, KI, Sb, As, B, Se, Te, Al, Ba, Bi, Co, Cr, Ir, Fe, Li, Mn, C, P, S, As_2O_5, B_2O_3, CaO, CO, Cr_2O_3, I_2O_5, MgO, Mo_2O_3, P_2O_5, SO_2, W_2O_3; fire or explosion in contact with acetic acid, ammonia, benzene, ethanol, hydrogen, hydrogen sulphide, methane, cork, grease, paper, wax, *etc*; violent reaction or explosion with water; similar violent reactions occur with BrF_5 (B90–93).
First aid	*Vapour inhaled:* standard treatment (p. 136). *Affected eyes:* standard treatment (p. 136). *Skin contact:* standard treatment (p. 136). *Mouth contact:* standard treatment (p. 137).
Spillage disposal	Instruct others to keep at a safe distance. Wear breathing apparatus and gloves. Absorb on sand, shovel into bucket(s), transport to safe open area and tip slowly into large volume of water; when decomposed, decant to waste diluting greatly with running water. Site of spillage should be ventilated after washing thoroughly with water.

N-Bromoacetamide [79–15–2]

Hazardous reaction	Decomposes rapidly when hot in presence of moisture and light (B256).

Bromoacetic acid [79–08–3]

Colourless to pale brown solid; mp 50 °C; soluble in water.

TOXIC BY INHALATION, IN CONTACT WITH SKIN
AND IF SWALLOWED
CAUSES SEVERE BURNS

Prevent contact with skin and eyes.

Toxic effects   Contact of the solid or solution with the eyes causes severe burns. The effect on the skin is not immediate and blisters may not appear for 12 hours or more after contact. Can be assumed to cause severe internal irritation and damage if taken by mouth.

First aid   *Affected eyes:* standard treatment (p. 136).
Skin contact: standard treatment (p. 136).
If swallowed: standard treatment (p. 137).

**Spillage
disposal**   Wear goggles and gloves. Spread soda ash liberally over the spillage and mop up cautiously with plenty of water — run to waste diluting greatly with running water.

Bromoacetone oxime   [62116–25–0]

**Hazardous
reaction**   Explodes during distillation (B362).

α-Bromoacetophenone —*see* Phenacyl bromide

Bromoacetylene   [593–61–3]

**Hazardous
reaction**   Unstable — may burn or explode on contact with air (B217).

1-Bromoaziridine   [19816–89–8]

**Hazardous
reaction**   Unstable — explodes during or shortly after distillation (B255).

Bromobenzene   [106–86–1]

A colourless liquid with an aromatic smell; bp 156 °C: immiscible with water.

FLAMMABLE
IRRITATING TO SKIN

Avoid breathing vapour. Avoid contact with skin and eyes.

Toxic effects   Little is known about the toxic properties, but its relationship with benzene suggests caution in handling. The vapour may be narcotic in high concentrations. It should be assumed to be poisonous through skin absorption and if taken by mouth.

Hazardous reaction   May react violently with Na (B572).

First aid   *Vapour inhaled:* standard treatment (p. 136).
Affected eyes: standard treatment (p. 136).
Skin contact: standard treatment (p. 136).
If swallowed: standard treatment (p. 137).

Fire hazard   Flash point 51 °C; ignition temp. 566 °C. *Extinguish fire with* dry powder, carbon dioxide or vaporising liquids.

Spillage disposal   Shut off all possible sources of ignition. Instruct others to keep at a safe distance. Wear breathing apparatus and gloves. Apply non-flammable dispersing agent and work into an emulsion with brush and water — run this to waste diluting greatly with running water. If dispersant not available, absorb on sand, shovel into bucket(s) and transport to safe open place for burial. Site of spillage should be washed thoroughly with water and soap or detergent.

p-Bromobenzoyl azide   [14917-59-0]

Hazardous reaction   Explodes violently above mp 46 °C (B652).

Bromochloromethane   [74-97-5]
(methylene chlorobromide)

Colourless liquid with sweetish odour; bp 69 °C; insoluble in water.

(HARMFUL BY INHALATION)

Avoid breathing vapour. Avoid contact with skin and eyes. RL 200 ppm (1050 mg m^{-3}).

Toxic effects   The vapour irritates the respiratory system and the eyes. The liquid irritates the eyes severely. The liquid irritates the skin. Assumed to be irritant and narcotic if taken by mouth.

First aid   *Vapour inhaled:* standard treatment (p. 136).
Affected eyes: standard treatment (p. 136).
Skin contact: standard treatment (p. 136).
If swallowed: standard treatment (p. 137).

Spillage disposal   Instruct others to keep at a safe distance. Wear breathing apparatus and gloves. Apply dispersing agent if available and work to an emulsion with brush and water — run to waste diluting greatly with running water. If dispersant not available, absorb on sand, shovel into bucket and transport to safe open area for atmospheric evaporation. Site of spillage should be thoroughly ventilated.

4-Bromocyclopentene [1781–66–4]

Hazardous preparation In preparation from 3,5-dibromocyclopentene and lithium tetrahydroalumi-nate (B508).

p-Bromo-*N*,*N*-dimethylaniline [586–77–6]

Hazardous preparation Exploded during vacuum distillation (B714).

Bromoethane [74–96–4]
(ethyl bromide)

Colourless, volatile liquid with ethereal odour; bp 38 °C; sparingly soluble in water.

HARMFUL BY INHALATION, IN CONTACT WITH SKIN
AND IF SWALLOWED

Avoid breathing vapour. Avoid contact with skin and eyes. RL 200 ppm (890 mg m^{-3}).

Toxic effects The vapour irritates the respiratory system; it has anaesthetic and narcotic effects. The liquid irritates the eyes. The liquid is poisonous if taken by mouth, causing damage to the kidneys. **Chronic effects** Can produce damage to the nervous system.

Hazardous preparation In preparation from ethanol and bromine (B275).

First aid *Vapour inhaled:* standard treatment (p. 136).
Affected eyes: standard treatment (p. 136).
Skin contact: standard treatment (p. 136).
If swallowed: standard treatment (p. 137).

Spillage disposal Instruct others to keep at a safe distance. Wear breathing apparatus and gloves. Apply dispersing agent if available and work to an emulsion with brush and water — run this to waste diluting greatly with running water. If dispersant not available, absorb on sand, shovel into bucket(s) and transport to safe open area for atmospheric evaporation. Site of spillage should be thoroughly ventilated.

Bromoform (tribromomethane) [75–25–2]

Heavy, colourless liquid with smell like chloroform; bp 150 °C; insoluble in water.

TOXIC BY INHALATION
IRRITATING TO EYES AND SKIN

Avoid breathing vapour. Avoid contact with skin and eyes. RL (skin) 0.5 ppm (5 mg m^{-3}).

Toxic effects	The vapour is lachrymatory and irritates the respiratory system. The liquid irritates the skin. Ingestion can cause respiratory difficulties, tremors and loss of consciousness.
Hazardous reactions	Reacts violently with acetone if catalysed by powdered KOH or other bases, even in presence of diluting solvents (B131).
First aid	*Vapour inhaled:* standard treatment (p. 136). *Affected eyes:* standard treatment (p. 136). *Skin contact:* standard treatment (p. 136). *If swallowed:* standard treatment (p. 137).
Spillage disposal	Instruct others to keep at a safe distance. Wear breathing apparatus and gloves. Apply dispersing agent and work to an emulsion with brush and water — run this to waste diluting greatly with running water. If dispersant not available absorb on sand, shovel into bucket(s) and transport to safe open space for burial. Site of spillage should be washed thoroughly with water and soap or detergent.

Bromomethane [74-83-9]

(methyl bromide)

Colourless, volatile liquid (bp 4 °C) or gas with faint chloroform-like odour; liquid forms a crystalline hydrate with cold water and penetrates rubber.

VERY TOXIC BY INHALATION

Prevent inhalation of vapour. Prevent contact with skin and eyes. RL (skin) 5 ppm (60 mg m^{-3}).

Toxic effects	Short exposures to high concentrations of vapour cause headache, dizziness, nausea, vomiting and weakness; this may be followed by mental excitement, convulsions and even acute mania. The longer inhalation of lower concentrations may lead to bronchitis and pneumonia. Both the vapour and liquid cause severe damage to the eyes. The liquid burns the skin, blisters appearing several hours after contact; itching and reddening of the skin may precede this. Assumed to be very poisonous if taken by the mouth.
Hazardous reactions	Forms pyrophoric Grignard-type compounds with zinc, aluminium and magnesium, delayed explosion in reaction with dimethyl sulphoxide (B153).
First aid	*Vapour inhaled:* standard treatment (p. 136). *Affected eyes:* standard treatment (p. 136).

Skin contact: standard treatment (p. 136).
If swallowed: standard treatment (p. 137).

Spillage disposal If ampoule is broken, instruct others to keep at a safe distance. Wear breathing apparatus and gloves. Organise ventilation of area to dispel vapour completely. Leaking cylinder should be placed in a well ventilated fume cupboard and vented slowly until discharged.

2-Bromomethylfuran [4437–18–7]

Hazardous reaction Very unstable — will explode violently (B498).

3-Bromopropene —*see* Allyl bromide

3-Bromopropyne [106–96–7]
(1-bromoprop-2-yne; propargyl bromide)

Hazardous reactions Classed as extremely shock-sensitive. Also danger of explosion in contact with copper, high-copper alloys, mercury and silver (B336).

N-Bromosuccinimide [128–08–5]

White to pale buff crystalline solid smelling faintly of bromine; mp 177–181 °C with decomposition.

(CAUSES BURNS)

Avoid breathing dust. Prevent contact with skin or eyes.

Toxic effects Irritates or burns the skin, eyes or respiratory system. Strongly irritant if taken by mouth.

Hazardous reactions Reacts violently with aniline, diallyl sulphide, hydrazine hydrate (B409–410).

First aid *Dust inhaled:* standard treatment (p. 136).
Affected eyes: standard treatment (p. 136).
Skin contact: standard treatment (p. 136).

Spillage disposal Wear face-shield or goggles, and gloves. Mop up with plenty of water and run to waste, diluting greatly with running water.

N-Bromotetramethylguanidine [6926–40–5]

Hazardous   Unstable and explodes if heated above 50 °C (B527).
reaction

α-Bromotoluene *–see* Benzyl bromide

Bromotrifluoromethane [75–63–8]

RL 1000 ppm (6100 mg m^{-3}).

Buta-1,3-diene (vinylethylene) [106–99–0]

Colourless gas; bp −4.7 °C; insoluble in water.

EXTREMELY FLAMMABLE LIQUEFIED GAS

Avoid breathing gas. RL 1000 ppm (2200 mg m^{-3}).

Toxic effects   The gas is of low toxicity but has narcotic effects in high concentration and can irritate the skin.

Hazardous   May explode when heated under pressure. Peroxides formed on long
reactions   contact with air are explosive but may also initiate polymerisation. Violent or explosive reactions with crotonaldehyde, nitrogen oxide, sodium nitrite. Sealed glass or steel containers of the monomer may burst if 'popcorn polymerisation' occurs (B419–420).

First aid   *Gas inhaled:* standard treatment (p. 136).
Fire hazard   Flash point below −7 °C; explosive limits 2–11.5%; ignition temp. 429 °C. Since the gas is supplied in a cylinder, turning off the valve will reduce any fire involving it; if possible cylinders should be removed quickly from an area in which a fire has developed.
Disposal   Surplus gas or leaking cylinder can be vented slowly to air in a safe open area or gas burnt off in a suitable burner.

Buta-1,3-diyne [460–12–8]

Hazardous   Polymerises rapidly above 0 °C, a gas above 10 °C. Potentially very
reactions   explosive (B402).

Butane [106–97–8]

Colourless gas; bp −0.5 °C; sparingly soluble in water.

EXTREMELY FLAMMABLE LIQUEFIED GAS

Avoid breathing gas. RL 600 ppm (1430 mg m⁻³).

Toxic effects	The gas has an anaesthetic effect but is not toxic.
First aid	*Gas inhaled in quantity:* standard treatment (p. 136).
Fire hazard	Flash point −60 °C; explosive limits 1.9–8.5%; ignition temp. 405 °C. Since the gas is supplied in a cylinder, turning off the valve will reduce any fire involving it; if possible, cylinders should be removed quickly from an area in which a fire has developed.
Disposal	Surplus gas or leaking cylinder can be vented slowly to air in a safe open area or gas burnt off in a suitable burner.

iso-Butane —*see* Isobutane

Butane-2,3-dione monoxime [57–71–6]

Hazardous reaction	Has exploded during vacuum distillation (B441).

Butanethiol (butyl mercaptan) [109–79–5]

TLV 0.5 ppm (1.5 mg m⁻³).

Butan-1-ol (n-butyl alcohol; 1-butanol) [71–36–3]

Colourless liquid; bp 118 °C; 9 cm³ dissolves in about 100 cm³ water at 25 °C.

and

Butan-2-ol (s-butyl alcohol) [78–92–2]

Colourless liquid; bp 99.5 °C; one part dissolves in 12 parts of water at about 25 °C.

FLAMMABLE
HARMFUL BY INHALATION

211

Avoid breathing vapour. Avoid contact with skin and eyes. RL 50 ppm (150 mg m^{-3}) and 150 ppm (100) (450 mg m^{-3}) respectively.

Toxic effects	Vapour may irritate the respiratory system and the eyes. The liquid irritates the eyes and may irritate the skin causing dermatitis. If taken by mouth may cause headache, dizziness, drowsiness and narcosis.
Hazardous reaction	Old (peroxidised, up to 12% found) samples of butan-2-ol may explode on distillation (B504).
First aid	*Vapour inhaled:* standard treatment (p. 136). *Skin contact:* standard treatment (p. 136). *Affected eyes:* standard treatment (p. 136). *If swallowed:* standard treatment (p. 137).
Fire hazard	Flash points 24, 29 °C; explosive limits 1.4–11%; ignition temps. 365, 406 °C. *Extinguish fire with* water spray, dry powder, carbon dioxide or vaporising liquids.
Spillage disposal	Shut off all possible sources of ignition. Wear face-shield or goggles, and gloves. Mop up with plenty of water and run to waste, diluting greatly with running water. Ventilate area well to evaporate remaining liquid and dispel vapour.

Butanone (methyl ethyl ketone)   [78–93–3]

Colourless liquid with smell like acetone; bp 80 °C; one part dissolves in about 4 parts of water at 25 °C.

HIGHLY FLAMMABLE

Avoid breathing vapour. Avoid contact with eyes. RL 200 ppm (590 mg m^{-3}).

Toxic effects	Inhalation of vapour may cause dizziness, headache, nausea. The liquid irritates the eyes and may cause severe damage. If swallowed may cause gastric irritation and narcosis. Weak teratogen.
Hazardous reactions	Vigorous reaction with chloroform in presence of bases. Explosive peroxides formed by action of H_2O_2/HNO_3 (B446–447).
First aid	*Vapour inhaled:* standard treatment (p. 136). *Affected eyes:* standard treatment (p. 136). *Skin contact:* standard treatment (p. 136). *If swallowed:* standard treatment (p. 137).
Fire hazard	Flash point −7 °C; explosive limits 2–10%; ignition temp. 515 °C. *Extinguish fire with* water spray, dry powder, carbon dioxide or vaporising liquids.
Spillage disposal	Shut off all possible sources of ignition. Wear face-shield and gloves. Mop up with plenty of water and run to waste diluting greatly with running water. Ventilate area well to evaporate remaining liquid and dispel vapour.

RSC *Lab. Haz. Data Sheet No. 18,* 1983 gives extended coverage.

Butanone peroxide [1338–23–4]

('MEK peroxide', a mixture)

RL 0.2 ppm (1.5 mg m^{-3}).

Butenes 1-[106–98–9], 2-[107–01–7], iso-[115–11–7]

(butylenes)

Colourless gases; boil between −6 °C and 4 °C; all are insoluble in water.

EXTREMELY FLAMMABLE LIQUEFIED GASES

Avoid breathing gases.

Toxic effects	The butenes are generally regarded as simple asphyxiants with some anaesthetic properties
First aid	*Gas inhaled in quantity:* standard treatment (p. 136).
Fire hazards	Flash points are below −7 °C; explosive limits 1.6–9.7%; ignition temps. between 230 °C and 390 °C. Since the gases are supplied in cylinders, turning off the valve will reduce any fire involving them; if possible cylinders should be removed quickly from an area in which a fire has developed.
Disposal	Suplus gas or leaking cylinder can be vented slowly to air in a safe, open area or gas burnt off through a suitable burner.

But-1-en-3-yne [689–97–4]

Hazardous reaction	Forms explosive compounds on contact with air or AgNO$_3$, explodes on heating under pressure (B408–409).

Butoxyacetylene [3329–56–4]

Hazardous reaction	Explodes on heating in sealed tubes (B614).

2-Butoxyethanol [111–76–2]

(ethylene glycol monobutyl ether, butyl cellosolve)

Colourless liquid; bp 171 °C; 1 part dissolves in about 20 parts water at 25 °C.

213

HARMFUL BY INHALATION, IN CONTACT WITH SKIN OR IF SWALLOWED
IRRITATING TO RESPIRATORY SYSTEM

Avoid contact with skin and eyes. RL (skin) 50 ppm (240 mg m^{-3}).

Toxic effects	The liquid irritates the eyes and may irritate the skin. May have irritant and narcotic action if taken by mouth.
Hazardous reaction	Liable to form explosive peroxides on exposure to air and light which should be decomposed before the ether is distilled to small volume.
First aid	*Affected eyes:* standard treatment (p. 136). *Skin contact:* standard treatment (p. 136). *If swallowed:* standard treatment (p. 137).
Fire hazard	Flash point 61 °C; explosive limits 1.1–12.7%. *Extinguish fire with* water spray, dry powder, carbon dioxide or vaporising liquids.
Spillage disposal	Wear face-shield or goggles, and gloves. Mop up with plenty of water and run to waste diluting greatly with running water. Ventilate area well to evaporate remaining liquid and dispel vapour.

Butyl acetate [123–86–4]

Colourless liquid; bp 125 °C; slightly soluble in water.

FLAMMABLE

Avoid breathing vapour. Avoid contact with skin and eyes. RL 150 ppm (710 mg m^{-3}).

Toxic effects	The vapour may irritate the respiratory system and cause headache and nausea. The liquid will irritate the eyes and may cause conjunctivitis. The liquid may irritate the skin and cause dermatitis. If taken by mouth, the liquid will cause irritation and act as a depressant of the central nervous system.
First aid	*Vapour inhaled:* standard treatment (p. 136). *Affected eyes:* standard treatment (p. 136). *Skin contact:* standard treatment (p. 136). *If swallowed:* standard treatment (p. 137).
Fire hazard	Flash point 27 °C; explosive limits 1.4–7.6%; ignition temp. 399 °C. *Extinguish fire with* foam, dry powder, carbon dioxide or vaporising liquid.
Spillage disposal	Shut off all possible sources of ignition. Instruct others to keep at a safe distance. Wear goggles or face-shield, and gloves. Apply non-flammable dispersing agent if available and work to an emulsion with brush and water — run this to waste diluting greatly with running water. If dispersant not available, absorb on sand, shovel into bucket and transport to safe open area for atmospheric evaporation or burial. Ventilate area well to evaporate remaining liquid and dispel vapour.

Butyl acrylate [141–32–2]

A colourless liquid; bp 145 °C; immiscible with water.

IRRITATING TO EYES, RESPIRATORY SYSTEM AND SKIN
MAY CAUSE SENSITISATION BY SKIN CONTACT

Avoid breathing vapour. Avoid contact with skin and eyes. RL 10 ppm (55 mg m^{-3}).

Toxic effects	The liquid irritates the skin and eyes. Assumed to be poisonous if taken by mouth.
First aid	*Affected eyes:* standard treatment (p. 136). *Skin contact:* standard treatment (p. 136). *If swallowed:* standard treatment (p. 137).
Fire hazard	Flash point 49 °C. *Extinguish fire with* foam, dry powder, carbon dioxide or vaporising liquids.
Spillage disposal	Shut off all possible sources of ignition. Instruct others to keep at a safe distance. Wear breathing apparatus and gloves. Apply non-flammable dispersing agent and work to an emulsion with brush and water — run this to waste diluting greatly with running water. If dispersant not available, absorb on sand, shovel into bucket and transport to safe open area for burial. Site of spillage should be washed thoroughly with water and soap or detergent.

n-Butyl alcohol —*see* Butan-1-ol

iso-Butyl alcohol —*see* Isobutyl alcohol

s-Butyl alcohol —*see* Butan-2-ol

t-Butyl alcohol —*see* 2-Methylpropan-2-ol

Butylamines n-[109–73–9], iso-[78–81–9]

The butylamines (n-butylamine, isobutylamine, s-butylamine, t-butylamine, di-n-butylamine, di-isobutylamine, di-s-butylamine and tri-n-butylamine) are colourless liquids with an ammoniacal odour; miscible with water.

HIGHLY FLAMMABLE
IRRITATING TO EYES, RESPIRATORY SYSTEM AND SKIN

Avoid breathing vapour. Avoid contact with skin and eyes.
RL (n-, skin) 5 ppm (15 mg m^{-3}).

Toxic effects	The vapours irritate the respiratory system and eyes. The liquids may cause skin and eye burns. Assumed to be very irritant and poisonous if taken by mouth.
First aid	*Vapour inhaled:* standard treatment (p. 136). *Affected eyes:* standard treatment (p. 136). *Skin contact:* standard treatment (p. 136). *If swallowed:* standard treatment (p. 137).
Fire hazard	Flash points −12 °C (n-), −9 °C (iso-); explosive limits 1.7–9.8% (n-) 1.7–9.8% (t-); ignition temp. 312 °C (n-), 378 °C (iso-). *Extinguish fire with water spray, dry powder, carbon dioxide or vaporising liquid.*
Spillage disposal	Shut off all possible sources of ignition. Instruct others to keep at a safe distance. Wear breathing apparatus and gloves. Mop up with plenty of water and run to waste diluting greatly with running water. Ventilate area well to evaporate remaining liquid and dispel vapour.

t-Butyl azidoformate [1070–19–5]

Hazardous reaction	Has exploded during distillation at 74 °C/92mbar (92×10^2 Pa) (B517–518).

t-Butyl diazoacetate [35059–50–8]

Hazardous reaction	Distillation under vacuum potentially hazardous (B613).

Butyldichloroborane [14090–22–3]

Hazardous reactions	Ignites on prolonged exposure to air; hydrolysis may be explosive (B453).

iso-Butylene —*see* 2-Methylpropene

Butyl 2,3-epoxypropyl ether [2426–08–6]

(butyl glycidyl ether)

RL 50 (25) ppm (270 mg m^{-3}).

t-Butyl hydroperoxide [75–91–2]

Colourless liquid; stable below 75 °C; slightly soluble in water.

(FLAMMABLE
IRRITATING TO EYES AND SKIN)

Avoid contact with skin, eyes and clothing. Avoid contamination with other materials. Store in a cool place.

Toxic effects	The liquid irritates the eyes and skin. Assumed to be toxic if taken by mouth.
Hazardous reaction	Liable to explode when distilled (B469).
First aid	*Affected eyes:* standard treatment (p. 136). *Skin contact:* standard treatment (p. 136). *If swallowed:* standard treatment (p. 137).
Fire hazard	Flash point 27 °C. *Extinguish fire with* water spray, dry powder, carbon dioxide or vaporising liquids.
Spillage disposal	Wear face-shield or goggles, and gloves. Mix with sand and transport to safe open area for burial.

t-Butyl hypochlorite [507–40–4]

Hazardous storage	Ampoules liable to burst unless stored cool and in dark (B454–455).

Butyl lactate [138–22–7]

RL 5 ppm (25 mg m^{-3}).

Butyl mercaptan —*see* Butanethiol

217

Butyl methacrylate [97–88–1]

Colourless, mobile liquid, normally supplied containing a small amount of stabilising agent (*e.g.*, 0.01% quinol); bp 163 °C; slightly soluble in water.

IRRITATING TO EYES, RESPIRATORY SYSTEM AND SKIN
MAY CAUSE SENSITISATION BY SKIN CONTACT

Avoid breathing vapour. Avoid contact with eyes and skin.

Toxic effects	The vapour irritates the eyes and respiratory system. The liquid irritates the eyes and may irritate the skin. Considered moderately toxic if taken by mouth.
First aid	*Vapour inhaled:* standard treatment (p. 136). *Affected eyes:* standard treatment (p. 136). *Skin contact:* standard treatment (p. 136). *If swallowed:* standard treatment (p. 137).
Fire hazard	Flash point 52 °C. *Extinguish fire with* foam, dry powder, carbon dioxide or vaporising liquids.
Spillage disposal	Shut off all possible sources of ignition. Wear breathing apparatus and gloves. Apply non-flammable dispersing agent and work to an emulsion with brush and water — run this to waste diluting greatly with running water. If dispersant not available, absorb on sand, shovel into bucket(s) and transport to safe open area for burial. Site of spillage should be washed thoroughly with water and soap or detergent.

Butyl methyl ketone —*see* Hexan-2-one

iso-Butyl methyl ketone
—*see* 4-Methylpentan-2-one

t-Butyl peracetate [107–71–1]

Hazardous reaction	Explodes violently when rapidly heated (B627).

t-Butyl perbenzoate [614–45–9]

Hazardous reaction	Exploded during interrupted vacuum distillation (B788).

t-Butyl peroxophosphates Review of group (B1463).

2-s-Butylphenol [89-72-5]

RL (skin) 5 ppm (30 mg m^{-3}).

4-t-Butyltoluene [98-51-1]

TLV 10 ppm (60 mg m^{-3}).

But-1-yne (ethylacetylene) [107-00-6]

Colourless liquid and gas; bp 8.1 °C; insoluble in water.

EXTREMELY FLAMMABLE

Avoid breathing gas.

Toxic effects   The toxicity has not been fully investigated. It probably has some anaesthetic activity and can act as a simple asphyxiant.

First aid   *Gas inhaled in quantity:* standard treatment (p. 136).
Fire hazard   No figures on flash point *etc* available. Since the gas is supplied in a cylinder, turning off the valve will reduce any fire involving it. If possible cylinders should be removed quickly from an area in which a fire has developed.
Disposal   Surplus gas or leaking cylinder can be vented slowly to air in a safe open area or gas burnt off in a suitable burner.

But-2-ynedinitrile (dicyanoacetylene) [1071-98-3]

Hazardous reaction   Potentially explosive in pure state or in concentrated solutions (B489).

But-2-yne-1,4-diol [110-65-6]

Hazardous reactions   Explodes on distillation in presence of traces of alkali or alkaline earth hydroxides or halides (B429).

219

But-2-yne-1-thiol [101672–05–3]

Hazardous Exposure to air results in polymer which may explode on heating (B486).
reaction

Butyraldehyde (butanal) [123–72–8]

Colourless liquid; bp 76 °C; 7 parts dissolve in about 100 parts water at 25 °C.

HIGHLY FLAMMABLE

Avoid breathing vapour. Avoid contact with skin and eyes.

Toxic effects The vapour may irritate the eyes and respiratory system. The liquid will irritate the eyes and may irritate the skin. Assumed to be irritant and possibly narcotic if swallowed.

First aid *Vapour inhaled:* standard treatment (p. 136).
Affected eyes: standard treatment (p. 136).
Skin contact: standard treatment (p. 136).
If swallowed: standard treatment (p. 137).

Fire hazard Flash point −6.7 °C; ignition temp. 230 °C. *Extinguish fire with* water spray, foam, dry powder, carbon dioxide or vaporising liquids.

Spillage disposal Shut off all possible sources of ignition. Wear face-shield or goggles, and gloves. Mop up with plenty of water and run to waste, diluting greatly with running water. Ventilate area well to evaporate remaining liquid and dispel vapour.

Butyraldehyde oxime [151–00–8]

Hazardous Large batch exploded violently during vacuum distillation caused by metal-
reaction catalysed Beckmann rearrangement (B458).

Butyric acid (butanoic acid) [107–92–6]

Colourless, oily liquid with very pungent smell; bp 163.5 °C; miscible with water.

(CAUSES BURNS)

Prevent contact with skin, eyes and clothing.

Toxic effects Irritates or burns skin and eyes.

220

First aid   *Affected eyes:* standard treatment (p. 136).
                  Skin contact: standard treatment (p. 136).
                  If swallowed: standard treatment (p. 137).

Spillage   Wear face-shield or goggles and gloves. Mop up with plenty of water and
disposal   run to waste diluting greatly with water. Ventilate area well to evaporate
                  remaining liquid and dispel vapour.

iso-Butyric acid —*see* Isobutyric acid

Butyric *and* isobutyric anhydrides

n-[106–31–0]
iso-[97–72–3]

Colourless liquids with pungent odour (bp 200 °C and 182 °C respectively) reacting with water
to form the corresponding acids.

(CAUSE BURNS
IRRITATING TO EYES, RESPIRATORY SYSTEM AND SKIN)

Avoid contact with skin and eyes.

Toxic effects   The liquids burn the eyes and may burn the skin. Thay are irritant and
                     corrosive if taken by mouth.

First aid   *Affected eyes:* standard treatment (p. 136).
                     Skin contact: standard treatment (p. 136).
                     If swallowed: standard treatment (p. 137).

Spillage   Wear face-shield or goggles, and gloves. Spread soda ash liberally over the
disposal   spillage and mop up with plenty of water — run to waste, diluting greatly
                     with running water.

Butyronitrile (propyl cyanide)

[109–74–0]

Colourless liquid; bp 117 °C; immiscible with water but tending to break down to cyanide by
hydrolysis.

FLAMMABLE
TOXIC BY INHALATION, IN CONTACT WITH SKIN
AND IF SWALLOWED

Do not breathe vapour. Prevent contact with skin and eyes.

Toxic effects   Although there is no documented evidence of its toxicity to humans, it has
                     been stated that rats exposed to its vapour rapidly develop weakness,
                     laboured breathing and convulsions which usually result in death. Cases of
                     poisoning seem to warrant the same urgent attention as those caused by
                     hydrogen cyanide.

221

First aid	Special treatment of hydrogen cyanide (p. 132).
Fire hazard	Flash point 26 °C. *Extinguish fire with* dry powder, carbon dioxide or vaporising liquids.
Spillage disposal	Shut off all possible sources of ignition. Instruct others to keep at a safe distance. Wear breathing apparatus and gloves. Apply non-flammable dispersing agent if available and work to emulsion with brush and water — run to waste diluting greatly with running water. If dispersant not available, absorb on sand, shovel into bucket(s) and transport to safe open area for burial. Site of spillage should be washed thoroughly with water and soap or detergent.

Butyryl nitrate [101672–06–4]

Hazardous reaction	Detonates on heating (B442).

Cadmium (metal) [7440–43–9]

Hazardous reactions	The powdered metal reacts violently or explosively with fused ammonium nitrate; it reacts vigorously with Se or Te on warming (B885).

Cadmium compounds oxide [1306–19–0]

Chloride, nitrate and sulphate soluble in water; oxide and carbonate insoluble.

TOXIC BY INHALATION, IN CONTACT WITH SKIN
AND IF SWALLOWED
DANGER OF CUMULATIVE EFFECTS
POSSIBLE RISKS OF IRREVERSIBLE EFFECTS

Avoid inhaling dust. CL (cadmium oxide fume) (0.05 mg m^{-3}).

Toxic effects	The inhalation of dust (usually the metal or oxide) irritates the lungs. The compounds cause increased salivation, choking, vomiting, stomach pains and diarrhoea if taken by mouth. **Chronic effects** Prolonged exposure to dust may cause damage to the lungs and kidneys and discoloration of teeth. (See references to toxicity of cadmium compounds in Ch 6.)
First aid	*Dust inhaled:* standard treatment (p. 136). *If swallowed:* standard treatment (p. 137).
Spillage disposal	Cadmium compounds are not so toxic as to present serious disposal problems. The insoluble compounds can be mixed with wet sand, swept up and treated as normal waste. The soluble salts can be mopped up with water and run to waste, diluting greatly with water.

222

Cadmium diamide [22750–53–4]

| Hazardous reactions | May explode when heated rapidly. Reacts violently with water (B886). |

Cadmium diazide [14215–29–3]

| Hazardous reactions | Dry solid explodes on heating or light friction; preparative solution exploded after standing for some hours (B887). |

Cadmium propionate [16986–83–7]

| Hazardous reaction | The salt exploded during drying in oven (B612). |

Cadmium selenide [1306–24–7]

| Hazardous preparation | Mixtures of powdered metal and selenium may explode (B887). |

Caesium (metal) [7440–46–2]

| Hazardous reactions | Ignites immediately in air and oxygen and on contact with water, violent or incandescent reactions with halogens, P or S (B1033). |

Calcium (metal) [7440–70–2]

Grey metal with silver white surface when cut; mp 850 °C.

CONTACT WITH WATER LIBERATES HIGHLY FLAMMABLE GASES

Avoid contact with skin, eyes and clothing.

| Toxic effects | Reaction with moisture on skin and eyes may cause irritation. |

| Hazardous reactions | Pyrophoric when finely divided; reacts explosively with $PbCl_2$, P_2O_5, S; reaction with water or dilute acids may be violent; ignites in fluorine, warm NH_3 or in ClF_3 or ClF_5; explodes with N_2O_4 (B870–872). |

| First aid | Standard treatments for skin and eye contacts (p. 136). |
| Disposal | Allow to react in large excess of cold water and discharge to waste. |

Calcium arsenate
—*see* **Arsenic compounds**

Calcium bis-2-iodylbenzoate   [59643–77–5]

Hazardous   Overdried formulated granules exploded (B825).
reaction

Calcium carbide (calcium acetylide)   [75–20–7]

CONTACT WITH WATER LIBERATES HIGHLY FLAMMABLE
GASES

Hazardous   Incandesces with PbF_2 at room temperature, with HCl on warming, with Mg
reactions   when heated in air, with Cl_2, Br_2 and I_2 at temperatures over 245 °C; very
                       vigorous reaction with boiling methanol; forms highly sensitive explosive
                       with silver nitrate solution; a mixture with sodium peroxide is explosive
                       (B201–203).

Calcium cyanamide   [156–62–7]

Grey-black granules or powder; decomposed by water.

(HARMFUL BY INHALATION
IRRITATING TO EYES AND SKIN)

Avoid inhaling dust. Prevent contact with eyes and skin. RL 0.5 mg m^{-3}.

Toxic effects   Inhalation of dust can cause severe irritation of the mucous membranes. The
                       material irritates the eyes and can cause conjunctivitis. The material will
                       burn the skin. If taken by mouth, there is severe internal irritation and
                       damage which may result in death.

First aid   *Dust inhaled:* standard treatment (p. 136).
                       Affected eyes: standard treatment (p. 136).
                       Skin contact: standard treatment (p. 136).
                       If swallowed: standard treatment (p. 137).
Spillage   Wear face-shield or goggles, and gloves. Mop up with plenty of water and
disposal   run to waste diluting greatly with water.

Calcium diazide   [19465–88–8]

Hazardous   Explodes on heating at about 150 °C (B880).
reaction

Calcium dihydride   [7789-78-8]

CONTACT WITH WATER LIBERATES HIGHLY FLAMMABLE GASES

Hazardous reactions Mixtures with various bromates, chlorates, perchlorates explode on grinding; mixture with AgF becomes incandescent on grinding (B877–878).

Calcium disilicide   [12013-56-8]

Hazardous reactions Explodes when milled in CCl_4; ignites in close contact with alkali metal fluorides; mixture with potassium or sodium nitrate ignites readily (B884).

Calcium hydrogen di- and tri-fluorides
–*see* Fluorides

Calcium hydroxide   [1305-62-0]

RL 5 mg m^{-3}.

Calcium hypochlorite (pure,   [7778-54-3]
or as bleaching powder)

White powder, smelling of chlorine; absorbs water and is decomposed by it.

CONTACT WITH COMBUSTIBLE MATERIAL MAY CAUSE FIRE
CONTACT WITH ACIDS LIBERATES TOXIC GAS
CAUSES BURNS

Avoid breathing dust. Avoid contact with skin, eyes and clothing.

Toxic effects Dust irritates the respiratory system; irritates the skin, eyes and alimentary system.

Hazardous reactions Pure compound is powerful oxidant, 'burns' with *evolution* of oxygen; violent reactions with hydroxy-compounds, nitromethane, S compounds and many combustibles (B873–877).

First aid *Affected eyes:* standard treatment (p. 136).
Skin contact: standard treatment (p. 136).
If swallowed: standard treatment (p. 137).

Spillage disposal Wear face-shield or goggles, and gloves. Mop up with plenty of water and run to waste diluting greatly with running water.

225

Calcium oxalate —*see* Oxalates

Calcium oxide (lime) [1305–78–8]

White, amorphous lumps and powder; reacts vigorously with water forming calcium hydroxide.

(HARMFUL BY INHALATION
IRRITATING TO EYES, RESPIRATORY SYSTEM AND SKIN)

Avoid breathing dust. Avoid contact with skin and eyes. RL 2 mg m^{-3}.

Toxic effects	The dust irritates the skin, eyes and respiratory system.
Hazardous reactions	Incandesces in contact with liquid HF; mixture with P_2O_5 reacts violently if warmed or moistened; some mixtures with water develop enough heat to ignite combustible materials; glass bottles of the oxide may burst due to hydration expansion when the hydroxide is formed (B881–882).
First aid	*Affected eyes:* standard treatment (p. 136). *Skin contact:* standard treatment (p. 136). *If swallowed:* standard treatment (p. 137).
Spillage disposal	Wear face-shield or goggles, and gloves. Shovel into dry bucket, transport to safe area and add, a little at a time, to a large quantity of water; after reaction complete, run suspension to waste, diluting greatly with running water.

Calcium peroxide [1305–79–9]

Hazardous reaction	Grinding with oxidisable materials may cause fire (B882).

Calcium peroxodisulphate [13235–16–0]

Hazardous reactions	Shock-sensitive; explodes violently (B883).

Calcium silicide [12013–55–7]

Hazardous reaction	Reacts vigorously with acid; the silanes evolved ignite (B883).

Calcium sulphate [7778–18–9]

Hazardous reactions	Reduced violently or explosively by Al powder. Contact with diazomethane vapour may result in detonation (B882).

Calcium sulphide [20548–54–3]

CONTACT WITH ACIDS LIBERATES TOXIC GAS
IRRITATING TO EYES, RESPIRATORY SYSTEM AND SKIN

Hazardous reactions Reacts vigorously with chromyl chloride, lead dioxide; explodes with potassium chlorate and potassium nitrate (B883).

Camphor (synthetic) [76–22–2]

RL 2 ppm (12 mg m^{-3}).

Caproic acid —*see* Hexanoic acid

ε-Caprolactam [105–60–2]
(2-oxohexamethylenimine, hexanolactam)

Hygroscopic leaflets; mp 70 °C; freely soluble in water.

(CAUSES IRRITATION OF SKIN AND EYES)

Avoid contact with skin, eyes and clothing. RL (vapour) 5 ppm (20 mg m^{-3}), (dust) 1 mg m^{-3}.

Toxic effects Can cause local irritation.

First aid *Affected eyes:* standard treatment (p. 136).
Skin contact: standard treatment (p. 136).
If swallowed: standard treatment (p. 137).
Spillage disposal Wear face-shield or goggles, and gloves. Mop up with plenty of water and run to waste diluting greatly with running water.

Carbon [7440–44–0]

RL (carbon black) 3.5 mg m^{-3}.

Hazardous reactions Activated carbon is a potential fire hazard; contamination with drying oils or oxidising agents may ignite it spontaneously; numerous oxidants (O_2, oxides, peroxides, oxosalts, halogens, interhalogens, *etc.*) in intimate contact with carbon, may cause ignition or explosion (B111–113).

227

Carbon dioxide [124-38-9]

RL 5000 ppm (9000 mg m^{-3}).

Whereas carbon dioxide presents negligible hazards in the laboratory as a gas, the solidified gas presents the risk of unpleasant skin burns resulting from the handling of the material without using adequately thick gloves. Burns received in this way are akin to frostbite and require medical attention.

Hazardous reactions May react violently with various metal oxides or reducing metals (Al, Mg, Ti, Zr); mixtures with Na, K explode if shocked (B327–329).

Carbon disulphide (carbon bisulphide) [75-15-0]

Colourless to yellow liquid, with unpleasant odour; bp 46 °C; immiscible with water.

EXTREMELY FLAMMABLE
VERY TOXIC BY INHALATION

Avoid breathing vapour. CL (skin) 10 ppm (30 mg m^{-3}).

Toxic effects High concentrations when inhaled produce narcotic effects and may result in unconsciousness. The liquid and vapour irritate the eyes. The liquid is poisonous if taken by mouth. **Chronic effects** Repeated inhalation of the vapour over a period may cause severe damage to the nervous system, including failure of vision, mental disturbance and paralysis. (*See* references to toxicity of carbon disulphide in Ch 6.)

Hazardous reactions Many fires and explosions have been caused by the ignition of the vapour from liquid poured down laboratory sinks, and ignition was caused by high-intensity flash illumination; explosion may result from mixing with liquid chlorine in presence of iron; ignites on contact with fluorine; reacts with azide solutions to form explosive azidodithioformates; reacts with zinc dust with incandescence; violent or explosive reactions with Al, K, Na or alloys (B193–195).

First aid *Vapour inhaled:* standard treatment (p. 136).
Affected eyes: standard treatment (p. 136).
If swallowed: standard treatment (p. 137).

Fire hazard Flash point −30 °C; explosive limits 1–44%; ignition temp. 100 °C. *Extinguish fire with* foam, dry powder, carbon dioxide or vaporising liquids.

Spillage disposal Shut off all possible sources of ignition. Instruct others to keep at a safe distance. Wear breathing apparatus and gloves. Apply non-flammable dispersing agent if available and work to an emulsion with water and brush — run this to waste diluting greatly with running water. If dispersant not available, absorb on dry sand and transport to safe open area for atmospheric evaporation. Ventilate area of spillage thoroughly to dispel vapour.

RSC *Lab. Haz. Data Sheet No. 7,* 1982 gives extended coverage.

Carbon monoxide (exhaust or process gas constituent)   [630–08–0]

Colourless, odourless gas, only slightly soluble in water.

EXTREMELY FLAMMABLE
TOXIC BY INHALATION

Avoid breathing gas. RL 50 ppm (55 mg m^{-3}).

Toxic effects	Causes unconsciousness due to anoxia resulting from the combination of carbon monoxide with haemoglobin. Gas in lower concentrations causes headache, throbbing of temples, nausea, followed possibly by collapse. **Chronic effects** Headache, nausea and weakness. (*See* references to toxicity of carbon monoxide in Ch 6 p. 106).
Hazardous reactions	May react explosively with F_2 and O_2 in the preparation of $C_2F_2O_4$ — also with BrF_3 and other interhalogens; reacts readily with K to form explosive 'carbonylpotassium'; explosion occurred during reduction of Fe_2O_3 (B327).
First aid	*Vapour inhaled:* standard treatment (p. 136).
Fire hazard	Explosive limits 12.5–74%; ignition temp. 609 °C. Since the gas is supplied in a cylinder, turning off the valve will reduce any fire involving it; if possible cylinders should be removed quickly from an area in which a fire has developed.
Disposal	Surplus gas or a leaking cylinder can be vented slowly to air in a safe open area or burnt off in a suitable gas burner.

Carbon tetrabromide (tetrabromomethane)   [558–13–4]

Colourless crystals; insoluble in water.

(TOXIC BY INHALATION)

Avoid breathing vapour. Avoid contact with skin and eyes. RL 0.1 ppm (1.4 mg m^{-3}).

Toxic effects	The vapour is narcotic in high concentrations. Assumed that the solid is poisonous by skin absorption and if taken by mouth.
Hazardous reaction	Will react vigorously or explosively with alkali metals (B117).
First aid	*Vapour inhaled:* standard treatment (p. 136). *Affected eyes:* standard treatment (p. 136). *Skin contact:* standard treatment (p. 136). *If swallowed:* standard treatment (p. 137).
Spillage disposal	Wear face-shield or goggles, and gloves. Mix with sand and transport to safe open area for burial.

Carbon tetrachloride (tetrachloromethane) [56–23–5]

Heavy, colourless liquid with a characteristic odour; bp 77 °C; immiscible with water. Not now recommended for extinguishing fires, as phosgene (*qv*), which is very poisonous, is liable to be formed.

VERY TOXIC BY INHALATION AND IN CONTACT WITH SKIN

Avoid breathing vapour. Avoid contact with skin and eyes. RL 10 ppm (65 mg m^{-3}).

Toxic effects Inhalation of high concentrations of vapour can cause headache, mental confusion, depression, fatigue, loss of appetite, nausea, vomiting and coma, these symptoms sometimes taking many hours to appear. The vapour and liquid irritate the eyes. It causes internal irritation, nausea and vomiting if taken by mouth; there is damage to the liver, kidneys, heart and nervous system and small doses have caused death. **Chronic effects** Prolonged inhalation of low concentrations may cause headache, nausea, stupor, vomiting, bronchitis and jaundice. Dermatitis may follow repeated contact with the liquid. Experimental carcinogen, moderate teratogen. (*See* references in Ch 6 p. 105, 107).

Hazardous reactions Exploded when milled with calcium disilicide; CCl_4 solutions of ClF_3 may detonate; initiated by dibenzoyl peroxide, mixtures with ethylene may explode; may react violently with dimethylformamide in presence of Fe at well below 100 °C; may react violently or explosively with F_2, Al, Ba, Be, K, Na, Zn. (B121–123).

First aid *Vapour inhaled:* standard treatment (p. 136).
Affected eyes: standard treatment (p. 136).
Skin contact: standard treatment (p. 136).
If swallowed: standard treatment (p. 137).

Spillage disposal Instruct others to keep at a safe distance. Wear breathing apparatus and gloves. Apply dispersing agent if available and work to an emulsion with brush and water — run this to waste diluting greatly with running water. If dispersant not available, absorb on sand, shovel into bucket(s) and transport to safe open area for atmospheric evaporation. Ventilate area of spillage thoroughly to dispel vapour.

Carbonyl diazide [14435–92–8]

Hazardous reaction Violently explosive solid, usable only in solution (B188).

Carbonyl dichloride —*see* Phosgene

Carbonyl difluoride   [353–50–4]

TLV 2 ppm (10 mg m^{-3}).

'Carbonyllithium'
—*see* the hexameric **Lithium benzenehexoxide**

Carbonylmetals Review of group (B1464).

'Carbonylpotassium'
—*see* the hexameric **Potassium benzenehexoxide**

'Carbonylsodium'
—*see* the hexameric **Sodium benzenehexoxide**

Catechol (pyrocatechol, 1,2-dihydroxybenzene)   [120–80–9]

Colourless crystalline powder; soluble in water.

(HARMFUL IN CONTACT WITH SKIN
CAUSES BURNS)

Avoid contact with skin and eyes. RL 5 ppm (20 mg m^{-3}).

Toxic effects	Irritates the eyes severely, causing burns. Irritates the skin and causes poisoning by absorption. Assumed to be irritant and poisonous if taken by mouth.
Hazardous reaction	Explodes on contact with concentrated nitric acid (B596).
First aid	*Affected eyes:* standard treatment (p. 136). *Skin contact:* standard treatment (p. 136). *If swallowed:* standard treatment (p. 137).
Spillage disposal	Wear goggles and gloves. Mop up with plenty of water and run to waste diluting greatly with running water.

Caustic potash —*see* **Potassium hydroxide**

Caustic soda —*see* Sodium hydroxide

Cellulose [9004–34–6]

CL (fibre, nuisance particulate) 10 mg m^{-3}.

Hazardous reactions Reactions with calcium oxide and oxidants such as bleaching powder, perchlorates, perchloric acid, sodium chlorate, fluorine, nitric acid, $NaNO_2$, $NaNO_3$, Na_2O_2 are reviewed (B1467).

Cellulose nitrate (nitrocellulose) [9004–70–0]

White or yellowish-white amorphous powder or matted filaments; ignites at 160–170 °C; usually supplied moistened with alcohol.

HIGHLY FLAMMABLE

Hazardous reactions The combustion and explosion of cellulose nitrate are reviewed (B1467—1469). The hazards of cellulose nitrate (presented also as pyroxylin) centre on its flammability and explosive potential; the latter is classified as 'moderate' by one authority.

Spillage disposal Shut off all sources of ignition. After damping with water transfer the nitrocellulose to an iron, steel or tinned container and add to it an equal volume of 10% sodium hydroxide solution. Allow to stand for an hour and then pour to waste, diluting greatly with running water.

Cerium (metal) [7440–45–1]

Hazardous reactions Ignites and burns brightly at 160 °C; reaction with Zn is explosively violent, and with Sb or Bi very exothermic; Ce filings ignite in Cl_2 or Br_2 at about 215 °C; reacts violently with P above 400 °C, ignites in hot CO_2/N_2 mixtures (B888–889).

Cerium nitride [25764–08–3]

Hazardous reaction Contact with limited amount of water or dilute acid causes rapid incandescence with ignition (B889).

Cerium trihydride   [13864–02–3]

Hazardous   May ignite in moist air (B889).
reaction

Chloral (trichloroacetaldehyde)   [75–87–6]

Colourless, oily liquid with pungent, irritating odour; bp 98 °C; soluble in water forming chloral hydrate.

(TOXIC IF SWALLOWED
IRRITATING TO EYES AND SKIN)

Avoid contact with skin, eyes and clothing.

Toxic effects   The vapour irritates the respiratory system and eyes. The liquid irritates the eyes severely. The liquid irritates the skin. If taken by mouth it will show the effects of chloral hydrate, namely nausea, vomiting, coldness of extremities and unconsciousness.

First aid   *Vapour inhaled:* standard treatment (p. 136).
Affected eyes: standard treatment (p. 136).
Skin contact: standard treatment (p. 136).
If swallowed: standard treatment (p. 137).

Spillage   Wear face-shield or goggles, and gloves. Mop up with plenty of water, and
disposal   run to waste diluting greatly with water.

Chloral hydrate   [302–17–0]

Colourless crystals with acrid odour and bitter taste; soluble in water.

TOXIC IF SWALLOWED
IRRITATING TO EYES AND SKIN

Avoid contact with skin and eyes.

Toxic effects   Irritates the skin and eyes. If taken by mouth it may cause nausea, vomiting, coldness of extremities and unconsciousness.

First aid   *Affected eyes:* standard treatment (p. 136).
Skin contact: standard treatment (p. 136).
If swallowed: standard treatment (p. 137).

Spillage   Wear face-shield or goggles, and gloves. Mop up with plenty of water and
disposal   run to waste diluting greatly with running water.

Chloramine [10599-90-3]

Hazardous reaction	Hazardous preparation, stable in ethereal solution, but solvent-free material decomposes violently or explosively (B918).

Chloramine T
—see Sodium *N*-chloro-*p*-toluenesulphonamide

Chloric acid [7790-93-4]

Hazardous reactions	Aqueous solution explodes if evaporated too far; it ignites filter paper and explodes with copper sulphide if concentrated; reactions with other oxidisable substances similar to those of chlorates (B905).

Chlorinated diphenyl oxide [55720-99-5]

TLV 0.5 mg m^{-3}.

Chlorine [7782-50-5]

Greenish-yellow gas with irritating odour; soluble in water.

TOXIC BY INHALATION
IRRITATING TO EYES, RESPIRATORY SYSTEM AND SKIN

Avoid breathing gas. RL 1 ppm (3 mg m^{-3}).

Toxic effects	The gas causes severe lung irritation and damage. The gas irritates the eyes and can cause conjunctivitis. In high concentrations the gas irritates the skin.
Hazardous reactions	Numerous reports of violent or explosive reactions with alcohols, BrF_3, CS_2, dibutyl phthalate, Cs_2O, diethyl ether, F_2O_2, F_2, glycerol, hexachlorodisilane, hydrocarbons, H_2; metal acetylides, carbides, hydrides and phosphides; Al, Bi, Ca, Cu, Fe (also in chlorinated pyridine solvent), Ge, K, Mg, Mn, Na, Ni, Sb, Sn, Th, U, V, Zn; nitrogen compounds; AsH_3, PH_3, SiH_4, B_2H_6, SbH_3; P, B, C, As, Te; F_2O, P_2O_3, silicones, steel, metal sulphides, synthetic rubber, trialkylboranes (B949–961).

First aid   *Gas inhaled:* standard treatment (p. 136).
          Affected eyes: standard treatment (p. 136).
          Skin contact: standard treatment (p. 136).
Disposal   Surplus gas or leaking cylinder can be vented slowly into water-fed
          scrubbing tower or column, or into a fume cupboard served by such a tower.

RSC *Lab. Haz. Data Sheet No. 30*, 1984 gives extended coverage.

Chlorine azide   [13973–88–1]

Hazardous   Extremely unstable, usually exploding violently without cause (B936).
reactions

Chlorine dioxide   [10049–04–4]

RL 0.1 ppm (0.3 mg m^{-3}).

Hazardous   Explodes violently under slightest provocation (reference to guide on use
reactions   provided); explodes on mixing with CO, on shaking with Hg, in contact
          with solid KOH or concentrated solution; phosphorus, sulphur, sugar or
          combustible materials ignite on contact and may cause explosion (B946–
          947).

Chlorine fluoride   [7790–89–8]

Hazardous   Powerful oxidant reacting violently with a wide range of materials
reactions   (reference given to information sheet) (B890–892).

Chlorine nitrate —*see* Nitryl hypochlorite

Chlorine perchlorate   [27218–16–2]

Hazardous   Shock-sensitive and liable to explode, violent reactions with haloalkanes
reaction   and haloalkenes (B980–981).

Chlorine trifluoride   [7790–91–2]

Colourless gas or yellow-green liquid with somewhat sweet but highly irritant smell; bp
11.75 °C.

(CONTACT WITH COMBUSTIBLE MATERIAL MAY CAUSE FIRE
REACTS VIOLENTLY WITH WATER
VERY TOXIC BY INHALATION
CAUSES SEVERE BURNS)

Prevent inhalation of vapour. Prevent contact with skin, eyes and clothing. TLV 0.1 ppm (0.4 mg m^{-3}).

Toxic effects	The vapour severely irritates the eyes, skin and respiratory system, and may cause burns. The liquid severely burns all human tissue.
Hazardous reactions	Reacts violently or explosively with strong nitric and sulphuric acids, ammonium fluoride, carbon tetrachloride, ammonia, coal-gas, hydrogen, hydrogen sulphide, iodine; numerous metals and non-metals and their oxides; nitro compounds and organic materials generally; the reaction with water is violent and may be explosive even with ice. References to literature on the handling, *etc* are given (B896–900).
First aid	*Vapour inhaled:* standard treatment (p. 136). *Affected eyes:* standard treatment (p. 136). *Skin contact:* standard treatment (p. 136). *Mouth contact:* standard treatment (p. 137).
Spillage disposal	Instruct others to keep at a safe distance. Wear breathing apparatus and gloves. Spread soda ash liberally over the spillage and mop up cautiously with water — run to waste, diluting greatly with running water.

Chlorine pentafluoride [13637–63–3]

Hazardous reactions	Very vigorous reaction with water, anhydrous nitric acid (B901).

Chlorites Review of these unstable salts (B1470).

Chloroacetaldehyde [107–20–0]

Hazardous derivative	Oxime unstable above 60 °C (B256).

RL 1 ppm (3 mg m^{-3}).

Chloroacetamide [79–07–2]

Colourless solid; 1 part is soluble in about 10 parts of cold water.

(HARMFUL BY INHALATION
AND IN CONTACT WITH SKIN)

Avoid inhaling dust. Avoid contact with skin and eyes.

Toxic effects   The toxicity of chloroacetamide is not well documented but, in view of the presence of an active chlorine atom, it is assumed that it may have harmful effects.

First aid   *Affected eyes:* standard treatment (p. 136).
Skin contact: standard treatment (p. 136).
If swallowed: standard treatment (p. 137).

Spillage disposal   Wear face-shield or goggles, and gloves. Mop up with plenty of water and run to waste diluting greatly with running water.

N-Chloroacetamide   [598-49-2]

Hazardous reactions   Exploded during desiccation of solid and during concentration of chloroform solution (B256–257).

Chloroacetic acid   [79-43-6]

Colourless to pale brown crystals; soluble in water.

(TOXIC BY INHALATION, IN CONTACT WITH SKIN
AND IF SWALLOWED
CAUSES SEVERE BURNS)

Prevent contact with skin and eyes.

Toxic effects   The solid or its solutions severely irritate or burn the eyes. The solid or its solutions produce severe skin burns which may only be apparent several hours after contact. Assumed to cause severe internal irritation and damage if taken by mouth.

First aid   *Affected eyes:* standard treatment (p. 136).
Skin contact: standard treatment (p. 136).
If swallowed: standard treatment (p. 137).

Spillage disposal   Wear face-shield or goggles, and gloves. Spread soda ash liberally over the spillage and mop up cautiously with plenty of water — run to waste diluting greatly with running water.

Chloroacetone (lachrymator)   [78-95-5]

Hazardous reaction   May polymerise explosively on storage (B353).

ω-Chloroacetophenone
—*see* Phenacyl chloride

Chloroacetyl chloride [79–04–9]

Colourless to pale yellow liquid; bp 106 °C; reacts with water.

CAUSES BURNS
IRRITATING TO RESPIRATORY SYSTEM

Prevent inhalation of vapour. Prevent contact with skin and eyes. RL 0.05 ppm (0.2 mg m⁻³).

Toxic effects The vapour severely irritates all parts of the respiratory system. The vapour irritates and the liquid burns the eyes. The vapour irritates the skin and the liquid may produce blisters several hours after contact. Assumed to cause severe internal irritation and damage if taken by mouth.

First aid *Vapour inhaled:* standard treatment (p. 136).
Affected eyes: standard treatment (p. 136).
Skin contact: standard treatment (p. 136).
If swallowed: standard treatment (p. 137).

Spillage disposal Instruct others to keep at a safe distance. Wear breathing apparatus and gloves. Spread soda ash liberally over the spillage and mop up cautiously with water — run this to waste diluting greatly with running water.

Chloroacetylene [593–63–5]

Hazardous reaction Endothermic: may burn or explode in contact with air (B217–218).

Chloroanilines *o*-[95–51–2] *m*-[108–42–9] *p*-[106–47–8]

The *o*- and *m*-chloroanilines are yellow to brown liquids (bp 209 °C and 229 °C respectively); *p*-chloroaniline is an almost colourless crystalline solid or powder. All are insoluble in water.

TOXIC BY INHALATION, IN CONTACT WITH SKIN
AND IF SWALLOWED
DANGER OF CUMULATIVE EFFECTS

Avoid breathing vapour. Avoid contact with skin and eyes.

Toxic effects The inhalation of vapour and absorption through the skin may cause cyanosis, and damage to the liver and kidneys. Assumed that similar poisoning will result from ingestion.

Hazardous reactions	Subject to exothermic decomposition during high-temperature distillation (B590–591).
First aid	*Vapour inhaled:* standard treatment (p. 136). *Affected eyes:* standard treatment (p. 136). *Skin contact:* standard treatment (p. 136). *If swallowed:* standard treatment (p. 137).
Spillage disposal	Wear breathing apparatus (or face-shield if amount is small) and gloves. Mix with sand and shovel into suitable vessel (glass or enamel) for dispersion in an excess of dilute hydrochloric acid (1 volume concentrated hydrochloric acid in 2 volumes of water). Allow to stand, with occasional stirring, for 24 hours and then run acid extract to waste, diluting greatly with running water and washing the sand. The sand can be treated as normal refuse.

1-Chloroaziridine [25167–31–1]

Hazardous property	Liable to explode on long storage (B256).

2-Chlorobenzaldehyde [89–98–5]

Colourless liquid or crystals; mp 11 °C; very slightly soluble in water.

CAUSES BURNS

Avoid contact with skin, eyes and clothing.

Toxic effects	Contact with skin or eyes may cause irritation or burns. There may be severe irritation and damage if the substance is swallowed.
First aid	*Affected eyes:* standard treatment (p. 136). *Skin contact:* standard treatment (p. 136). *If swallowed:* standard treatment (p. 137).
Spillage disposal	Wear face-shield or goggles, and gloves. Apply dispersing agent if available and work to an emulsion with brush and water — run this to waste, diluting greatly with running water. If dispersant not available, absorb on sand, shovel into bucket(s) and transport to safe, open area for burial. Site of spillage should be washed thoroughly with water and soap or detergent.

Chlorobenzene (monochlorobenzene) [108–90–7]

Clear colourless liquid with a faint, not unpleasant almond-like odour; bp 132 °C; immiscible with water.

239

FLAMMABLE
HARMFUL BY INHALATION

Avoid breathing vapour. Avoid contact with skin, eyes and clothing. RL 75 ppm (350 mg m^{-3}).

Toxic effects	The vapour may cause drowsiness and unconsciousness. The liquid irritates the skin. The liquid may cause stupor and unconsciousness after a few hours if taken by mouth.
Hazardous reaction	Explodes with finely divided Na (B573).
First aid	*Vapour inhaled:* standard treatment (p. 136). *Affected eyes:* standard treatment (p. 136). *Skin contact:* standard treatment (p. 136). *If swallowed:* standard treatment (p. 137).
Fire hazard	Flash point 29 °C; explosive mixture 1.3–7.1%; ignition temp. 630 °C. *Extinguish fire with* foam, dry powder carbon dioxide or vaporising liquids.
Spillage disposal	Shut off all possible sources of ignition. Instruct others to keep at a safe distance. Wear breathing apparatus and gloves. Apply non-flammable dispersing agent if available and work to an emulsion with brush and water — run this to waste diluting greatly with running water. If dispersant not available, absorb on sand, shovel into bucket(s) and transport to safe, open area for atmospheric evaporation or burial. Site of spillage should be washed thoroughly with water and soap or detergent.

1-Chlorobenzotriazole [21050–95–3]

Hazardous reaction	May ignite spontaneously (B558).

Chlorobiphenyls a [53449–21–9] b [11097–69–1]

DANGER OF CUMULATIVE EFFECTS

a (42% chlorine) RL (skin) 1 mg m^{-3}.
b (54% chlorine) RL (skin) 0.5 mg m^{-3}.

Chlorobromomethane—*see* Bromochloromethane

2-Chloro-1,3-butadiene [126–99–8]

(β-chloroprene)

EXTREMELY FLAMMABLE
HARMFUL BY INHALATION

RL (skin) 10 ppm (36 mg m^{-3}).

Hazardous   Autoxidises very rapidly to produce unstable peroxide which will catalyse
reaction   exothermic polymerisation of monomer (B414).

1-Chlorobutan-2-one   [616–27–3]

Hazardous   Bottle of stabilised material exploded spontaneously (B439).
reaction

1-Chlorobut-1-en-3-one   [7119–27–9]

Hazardous   Liable to explode soon after preparation (B415).
reaction

Chlorocyanoacetylene
—see **3-Chloropropiolonitrile**

2-Chloro-1-cyanoethanol
—see **3-Chloro-2-hydroxypropiononitrile**

3-Chlorocyclopentene   [96–40–2]

Hazardous   Explosive decomposition after brief storage (B508).
reaction

Chlorodifluoromethane   [75–45–6]

RL 1000 ppm (3500 mg m^{-3}).

Hazardous   Reacts exothermically with Al (B25–27).
reaction

4-Chloro-2,6-dinitroaniline [5388–26–5]

Hazardous reactions   Dangerous decomposition set in during large-scale manufacture from 4-chloro-2-nitroaniline. Explosion occurred during large-scale diazotisation of the amine (B559).

1-Chloro-2,4-dinitrobenzene [97–00–7]
(2,4-dinitrochlorobenzene)

Pale yellow crystals; insoluble in water.

TOXIC BY INHALATION, IN CONTACT WITH SKIN
AND IF SWALLOWED
DANGER OF CUMULATIVE EFFECTS

Avoid breathing dust and vapour. Prevent contact with skin and eyes.

Toxic effects   The dust, or vapour from the molten compound, irritates the respiratory system. Irritates the skin and may cause dermatitis. Cyanosis and liver injury may follow inhalation or skin absorption. Assumed to be poisonous if taken by mouth.

Hazardous reaction   Has been used as an explosive; reaction with NH_3 under pressure may be violent (B548).

First aid   *Dust or vapour inhaled:* standard treatment (p. 136).
Affected eyes: standard treatment (p. 136).
Skin contact: standard treatment (p. 136).
If swallowed: standard treatment (p. 137).

Spillage disposal   Wear face-shield or goggles, and gloves. Mix with sand and transport to a safe open area for burial.

1-Chloro-2,3-epoxypropane [106–89–8]
(epichlorhydrin)

Colourless liquid with irritating chloroform-like odour; bp 118 °C; immiscible with water.

FLAMMABLE
VERY TOXIC BY INHALATION, IN CONTACT WITH SKIN
AND IF SWALLOWED
POSSIBLE RISK OF IRREVERSIBLE EFFECTS

Prevent inhalation of vapour. Prevent contact with skin and eyes. Suspected carcinogen RL (skin) 2 ppm (8 mg m^{-3}).

Toxic effects The vapour irritates the respiratory system and in severe cases can cause respiratory paralysis. The vapour and liquid irritate the eyes and may cause conjunctivitis. Poisonous by skin absorption and ingestion. Skin blistering and severe pain may develop after latent period, possibly with sensitisation and dermatitis **Chronic effects** Prolonged exposure to low concentrations of vapour may cause conjunctivitis, chronic weariness and stomach upset.

Hazardous reactions Reacts violently with aniline, isopropylamine, H_2SO_4 and also with trichloroethylene; polymerises exothermically (B354–355).

First aid *Vapour inhaled:* standard treatment (p. 136).
Affected eyes: standard treatment (p. 136).
Skin contact: standard treatment (p. 136).
If swallowed: standard treatment (p. 137).

Fire hazard Flash point 41 °C; *Extinguish fire with* water spray, dry powder, carbon dioxide or vaporising liquids.

Spillage disposal Shut off all possible sources of ignition. Instruct others to keep at a safe distance. Wear breathing apparatus and gloves. Apply non-flammable dispersing agent if available and work to an emulsion with brush and water — run this to waste diluting greatly with running water. If dispersant not available, absorb on sand, shovel into bucket(s) and transport to safe open area for burial. Site of spillage should be washed thoroughly with water and soap or detergent.

RSC *Lab. Haz. Data Sheet No. 9,* 1983 gives extended coverage.

Chloroethane (ethyl chloride) [75–00–3]

Colourless gas and liquid with pungent, ethereal odour, which is used as a local anaesthetic and refrigerant; bp 12.4 °C; sparingly soluble in water.

EXTREMELY FLAMMABLE LIQUEFIED GAS

Avoid breathing vapour. RL 1000 ppm (2600 mg m^{-3}).

Toxic effects The vapour is mildly irritating to the mucous membranes; at high concentrations it is narcotic.

Hazardous reaction Mixture with K shock-sensitive (B1236).

First aid *Vapour inhaled:* standard treatment (p. 000).

Fire hazard Flash point −50 °C; explosive limits 3.6–15.4%; ignition temp. 519 °C. Since the gas is supplied in a cylinder, turning off the valve will reduce any fire involving it; if possible, cylinders should be removed quickly from an area in which a fire has developed.

Disposal Surplus gas or leaking cylinder can be vented slowly to air in a safe open area or gas burnt off in a suitable burner.

2-Chloroethanol [107–07–3]

(ethylene chlorohydrin: chloroethyl alcohol)

Colourless liquid with a faint ethereal odour; bp 129 °C; miscible with water.

VERY TOXIC BY INHALATION, IN CONTACT WITH SKIN AND IF SWALLOWED

Prevent inhalation of vapour. Prevent contact with skin and eyes.
RL (skin) 1 ppm (3 mg m^{-3}).

Toxic effects The vapour causes nausea, headaches, vomiting, stupefaction and unconsciousness. It irritates the mucous membranes. The liquid is rapidly absorbed by the skin, producing similar effects to inhalation. Assumed to be extremely poisonous if taken by mouth.

First aid *Vapour inhaled:* standard treatment (p. 136).
Affected eyes: standard treatment (p. 136).
Skin contact: standard treatment (p. 136).
If swallowed: standard treatment (p. 137).

Spillage disposal Instruct others to keep at a safe distance. Wear breathing apparatus and gloves. Mop up with plenty of water and run to waste, diluting greatly with running water. Ventilate area well to evaporate remaining liquid and dispel vapour.

2-Chloroethylamine [689–98–5]

Hazardous reaction May polymerise explosively (B285).

Chlorofluoroalkanes *[75–68–3]

The commonly available compounds in this series, all of which are colourless gases or low-boiling liquids supplied in cylinders, are: chlorodifluoroethane*; chlorodifluoromethane; chloropentafluoroethane; chlorotrifluoromethane; dichlorodifluoromethane; dichlorofluoromethane; dichlorotetrafluoroethane; dichlorofluoroethane; trichlorofluoromethane. They are all relatively innocuous gases, their toxicity being of the same order as nitrogen or carbon dioxide. Only one, marked*, is flammable, the remainder are non-flammable.

Fire hazard (chlorodifluorethane) Explosive limits 9–14.8%; ignition temp. 632 °C. Since the gas is supplied in a cylinder, turning off the valve will reduce any fire involving it; cylinders should be removed quickly from an area in which a fire has developed.

Disposal Surplus gas or a leaking cylinder can be vented slowly in air in a safe open area or gas burnt off in a suitable burner.

Chloroform (trichloromethane) [67–66–3]

Colourless volatile liquid with a characteristic odour; bp 61 °C; immiscible with water.

HARMFUL BY INHALATION

Avoid breathing vapour. Avoid contact with eyes. Suspected carcinogen RL 10 ppm (50 mg m^{-3}).

Toxic effects The vapour has anaesthetic properties, causing drowsiness, giddiness, headache, nausea, vomiting and unconsciousness. The vapour and liquid irritate the eyes causing conjunctivitis. The liquid is poisonous if taken by mouth.

Hazardous reactions Vigorous reaction with acetone in the presence of KOH or Ca(OH)$_2$; may react explosively with fluorine, N$_2$O$_4$, Al, Li, Na, Na/methanol, NaOH/methanol, sodium methoxide (B132–134).

First aid *Vapour inhaled:* standard treatment (p. 136).
Affected eyes: standard treatment (p. 136).
If swallowed: standard treatment (p. 137).

Spillage disposal Instruct others to keep at a safe distance. Wear breathing apparatus and gloves. Apply dispersing agent if available and work to an emulsion with brush and water — run this to waste diluting greatly with running water. If dispersant not available, absorb on sand, shovel into bucket(s) and transport to safe, open area for atmospheric evaporation. Site of spillage should be washed thoroughly with water and soap or detergent.

Chlorogermane [13637–63–5]

Hazardous reaction Reacts with ammonia to form explosive product (B902).

3-Chloro-2-hydroxypropiononitrile [33965–80–9]
(2-chloro-1-cyanoethanol)

Hazardous reaction May explode during vacuum distillation (B346).

3-Chloro-2-hydroxypropyl perchlorate [101672–07–5]

Hazardous reaction Explodes violently on heating (B363).

Chloromethane (methyl chloride) [74–87–3]

Colourless gas; bp −24 °C; 2.2 volumes dissolve in 1 volume of water at 20 °C.

EXTREMELY FLAMMABLE LIQUEFIED GAS
HARMFUL BY INHALATION

Avoid breathing vapour. RL 100 (50) ppm (210 mg m^{-3}).

Toxic effects	Inhalation of vapour may cause dizziness, drowsiness, nausea, stomach pains, visual disturbances, mental confusion and unconsciousness; heavy exposure can be fatal and some symptoms may be delayed. (*See* note in Ch 6 p 104.)
Hazardous reactions	Ignites or explodes on contact with BrF$_3$ or BrF$_5$. May react explosively with Mg, K, Na and, probably, Zn (B154).
First aid	*Vapour inhaled:* standard treatment (p. 136).
Fire hazard	Flash point below 0 °C; explosive limits 10.7–17.4%; ignition temp. 632 °C. Since the gas is supplied in a cylinder, turning off the valve will reduce any fire involving it; if possible cylinders should be removed quickly from an area in which a fire has developed.
Disposal	Surplus gas or a leaking cylinder can be vented slowly to air in a safe open area or gas burnt off in a suitable burner.

2-Chloromethylfuran

Hazardous reaction	Liable to explode violently owing to polymerisation or decomposition (B524).

Chloromethyl methyl ether [107–30–2]
(chloromethoxymethane)

Colourless liquid, suffocating odour; bp 59 °C; decomposed by water.

(VERY TOXIC BY INHALATION, IN CONTACT WITH SKIN
AND IF SWALLOWED
DANGER OF VERY SERIOUS IRREVERSIBLE EFFECTS)

Prevent inhalation of vapour and contact with eyes or skin. Human carcinogen. No TLV set.

Toxic effects	Vapour irritates the respiratory system and eyes, primary effects probably being due to fairly rapid formation of hydrochloric acid, formaldehyde and methanol. **Chronic effects.** A high incidence of lung cancer has been observed.

First aid   *Vapour inhaled:* standard treatment (p. 136).
                   Affected eyes: standard treatment (p. 136).
                   Skin contact: standard treatment (p. 136).
                   If swallowed: standard treatment (p. 137).
Spillage   Instruct others to keep at safe distance. Wear breathing apparatus and
disposal   gloves. Spread soda-ash liberally over the spillage and mop up cautiously
                   with plenty of water — run to waste diluting greatly with running water.

4-Chloro-2-methylphenol [1570–64–5]

Hazardous   Vigorous reaction followed by explosion when large quantity was left in
reaction   contact with concentrated NaOH solution (B672).

2-Chloromethylthiophene [617–88–9]

Hazardous   Unstable and gradually decomposes: closed containers may explode
reactions   (B498).

N-Chloronitroamines Review of group (B1471).

Chloronitroanilines 2,4-[121–87–9] 4,2-[89–63–4]

2-Chloro-4-nitro- and 4-chloro-2-nitro-anilines are yellow to brownish-yellow powders or
crystals; insoluble in water.

VERY TOXIC BY INHALATION, IN CONTACT WITH SKIN
AND IF SWALLOWED
DANGER OF CUMULATIVE EFFECTS

Prevent contact with skin and eyes.

Toxic effects   Absorption through the skin may cause dermatitis, cyanosis and damage to
                    the liver and kidneys. Assumed that similar poisoning will result from
                    ingestion.

First aid   *Affected eyes:* standard treatment (p. 136).
                    Skin contact: standard treatment (p. 136).
                    If swallowed: standard treatment (p. 137).
Spillage   Wear face-shield or goggles, and gloves. Mix with sand and transport to a
disposal   safe open area for burial. Site of spillage should be washed thoroughly with
                    water and soap or detergent.

Chloronitrobenzenes   *o*-[88–73–3] *p*-[100–00–5]

o-, *m*- and *p*-Chloronitrobenzenes are yellow solids of low melting point; insoluble in water.

TOXIC BY INHALATION, IN CONTACT WITH SKIN
AND IF SWALLOWED
DANGER OF CUMULATIVE EFFECTS

Prevent contact with skin and eyes. Avoid breathing dust and vapour. RL (*p*-, skin) 1 mg m^{-3}.

Toxic effects   The dust or vapour from the molten compounds irritates the respiratory system. Contact with the skin may cause dermatitis. Cyanosis and liver injury may follow inhalation or skin absorption. Assumed to be poisonous if taken by mouth.

Hazardous reactions   Runaway reaction of *o*-chloronitrobenzene with ammonia under pressure involved 6 simultaneous fault conditions. *p*-Chloronitrobenzene reacted violently and finally explosively when added to a solution of sodium methoxide in methanol (B556–557).

First aid   *Dust or vapour inhaled:* standard treatment (p. 136).
Affected eyes: standard treatment (p. 136).
Skin contact: standard treatment (p. 136).
If swallowed: standard treatment (p. 137).

Spillage disposal   Wear face-shield or goggles, and gloves. Mix with sand and transport to a safe, open area for burial. Site of spillage should be washed thoroughly with water and soap or detergent.

Chloronitromethane   [1794–84–9]

Hazardous preparation   Product of chlorination of nitromethane decomposed explosively during vacuum distillation (B140).

1-Chloro-1-nitropropane   [600–25–9]

TLV 2 ppm (10 mg m^{-3})

Chloronitrotoluenes   2,4-[121–86–8] 4,3-[890–60–1]

Hazardous reactions   Residue from vacuum distillation of crude 2-chloro-4-nitrotoluene or isomers exploded; exothermic reaction of 4-chloro-3-nitrotoluene with CuCN in pyridine exploded (B664–665).

248

Chlorophenols
o-[95–57–8] p-[106–48–9]

o-Chlorophenol is a pale brown liquid; p-chlorophenol is a colourless to pale brown crystalline solid with a phenolic (carbolic) odour. Both compounds are sparingly soluble in water.

HARMFUL BY INHALATION, IN CONTACT WITH SKIN AND IF SWALLOWED

Avoid breathing vapour. Prevent contact with skin and eyes.

Toxic effects   The vapour when inhaled irritates the respiratory system. In contact with the eyes they cause irritation or burning. They irritate and may burn the skin and must be assumed to be very poisonous and irritant if taken by mouth.

First aid   *Vapour inhaled:* standard treatment (p. 136).
Affected eyes: standard treatment (p. 136).
Skin contact: standard treatment (p. 136). But see also note on phenol (p. 135).
If swallowed: standard treatment (p. 137).

Spillage disposal   Wear face-shield or goggles, and gloves. Mix with sand and transport to a safe open area for burial. Site of spillage should be washed thoroughly with water and soap or detergent.

Chlorophenyldiazirine
—*see* **Phenylchlorodiazirine**

m- or *p*-Chlorophenyllithium
m-[25077–87–6]
p-[14774–78–8]

Hazardous preparation   Absence of solvent, or presence of traces of O_2 in reaction atmosphere may cause explosion (B556).

Chloropicrin —*see* **Trichloronitromethane**

β-Chloroprene —*see* **2-Chloro-1,3-butadiene**

Chloropropanes
1-[540–54–5] 2-[75–29–6]

Colourless, volatile liquids; bps:1-isomer, 47 °C; 2-isomer, 35 °C; immiscible with water.

HIGHLY FLAMMABLE
HARMFUL BY INHALATION, IN CONTACT WITH SKIN
AND IF SWALLOWED

Avoid breathing vapour. Avoid contact with skin and eyes.

Toxic effects	Inhalation of 1-chloropropane irritates the respiratory system: high concentrations of both isomers cause narcosis. The liquids irritate the eyes. Assumed to be poisonous if taken by mouth.
First aid	*Vapour inhaled:* standard treatment (p. 136). *Affected eyes:* standard treatment (p. 136). *If swallowed:* standard treatment (p. 137).
Fire hazard	Flash point below −18 °C for 1-isomer, −32 °C for 2-; explosive limits 2.6–11.1% for 1-isomer, 2.8–10.7% for 2-; ignition temp. 520 °C for 1- isomer, 592 °C for 2-. *Extinguish fire with* water spray, foam, dry powder, carbon dioxide or vaporising liquids.
Spillage disposal	Shut off all possible sources of ignition. Instruct others to keep at a safe distance. Wear breathing apparatus and gloves. Apply non-flammable dispersing agent if available and work to an emulsion with brush and water — run this to waste, diluting greatly with running water. If dispersant not available, absorb on sand, shovel into bucket(s) and transport to safe open area for atmospheric evaporation. Site of spillage should be washed thoroughly with water and soap or detergent.

3-Chloropropene —*see* Allyl chloride

3-Chloropropiolonitrile [2003–31–8]

(chlorocyanoacetylene)

Hazardous reaction	Explosion hazard when heated in nearly closed vessel (B323–324).

3-Chloropropionyl chloride [625–36–5]

Colourless liquid with acrid odour; reacts with water, forming chloropropionic acid and hydrogen chloride.

(CAUSES BURNS
IRRITATING TO RESPIRATORY SYSTEM)

Prevent inhalation of vapour. Prevent contact with skin and eyes.

Toxic effects   The vapour irritates the eyes and respiratory system severely. The liquid
burns the eyes and skin. Causes severe internal irritation and damage if
taken by mouth.

First aid   *Vapour inhaled:* standard treatment (p. 136).
Affected eyes: standard treatment (p. 136).
Skin contact: standard treatment (p. 136).
If swallowed: standard treatment (p. 137).

Spillage   Shut off all possible sources of ignition. Instruct others to keep at a safe
disposal   distance. Wear breathing apparatus and gloves. Absorb on sand, shovel into
bucket(s), transport to safe, open area and tip into large volume of water;
leave to decompose before decanting the water to waste, diluting greatly
with running water. Site of spillage should be ventilated after washing
thoroughly with water and soap or detergent.

3-Chloropropyne (1-chloro-2-propyne)   [624–65–7]

Hazardous   Violent reaction with ammonia under pressure followed by explosion
reaction   (B337).

o-Chlorostyrene   [1331–28–8]

TLV 50 ppm (285 mg m^{-3}).

N-Chlorosuccinimide   [128–09–6]

Hazardous   Violent or explosive reaction with aliphatic alcohols, benzylamine, hydra-
reactions   zine hydrate (B410).

Chlorosulphuric acid   [7790–94–5]
(chlorosulphonic acid)

Colourless to brown fuming liquid; bp 151 °C; decomposing with explosive violence when
mixed with water.

REACTS VIOLENTLY WITH WATER
CAUSES SEVERE BURNS
IRRITATING TO RESPIRATORY SYSTEM

Prevent inhalation of vapour. Prevent contact with eyes and skin.

Toxic effects	The fumes are very irritant to the lungs and mucous membranes. The fumes irritate the eyes severely. The liquid burns the skin and eyes. If taken by mouth there would be severe local and internal corrosive effects.
Hazardous reactions	Powerful oxidising agent; explosive reaction with P; violent reaction with water or $AgNO_3$ (B905–906).
First aid	*Vapour inhaled:* standard treatment (p. 136). *Affected eyes:* standard treatment (p. 136). *Skin contact:* standard treatment (p. 136). *If swallowed:* standard treatment (p. 137).
Spillage disposal	Instruct others to keep at a safe distance. Wear breathing apparatus and gloves (and rubber boots or overshoes if spillage is large). Spread soda ash liberally over the spillage and mop up cautiously with plenty of water — run to waste diluting greatly with running water.

N-Chlorotetramethylguanidine [6926–39–2]

Hazardous reaction	Unstable; explodes if heated above 50 °C (B528).

Chlorotoluenes *o*-[95–49–8] *m*-[108–41–8] *p*-[106–43–4]

All three isomers are liquids at normal temperatures (*o*- bp 159 °C; *m*- bp 162 °C; *p*- bp 162 °C; mp 7 °C); they are insoluble in water.

HARMFUL BY INHALATION

Avoid breathing vapour. *o*- TLV (skin) 50 ppm (250 mg m^{-3}).

Toxic effects	The vapour of the chlorotoluenes is considered potentially toxic in moderate concentrations and is known to be narcotic at high concentrations in the case of the 3-isomer.
First aid	*Vapour inhaled:* standard treatment (p. 136). *Skin contact:* standard treatment (p. 136). *If swallowed:* standard treatment (p. 137).
Fire hazard	Flash point 47–50 °C. *Extinguish fire with* foam, dry powder, carbon dioxide or vaporising liquids.
Spillage disposal	Shut off all possible sources of ignition. Instruct others to keep at a safe distance. Wear breathing apparatus and gloves. Apply non-flammable dispersing agent if available and work to an emulsion with brush and water — run this to waste diluting greatly with running water. If dispersant not available, absorb on sand, shovel into bucket(s) and transport to safe, open area for burial. Site of spillage should be washed thoroughly with water and soap or detergent.

α-Chlorotoluene —*see* Benzyl chloride

Chloryl hypofluorite [101672–08–6]

Hazardous reaction   Explosive (B892).

Chloryl perchlorate [12442–63–6]

Hazardous reactions   Very powerful oxidant; reacts violently or explosively with ethanol, stopcock grease, wood and organic matter generally. Liable to explode on heating or contact with water, thionyl chloride, even at −70°C (B981–982).

Chromates and dichromates

Generally yellow or orange-red crystals or powder; usually soluble in water.

IRRITATING TO EYES, RESPIRATORY SYSTEM AND SKIN
MAY CAUSE SENSITISATION BY SKIN CONTACT

Avoid inhaling dust. Avoid contact with eyes and skin. RL (soluble, as Cr) 0.5 mg m^{-3}.

Toxic effects   The dust irritates the respiratory tract. The dust irritates the eyes severely. If taken by mouth there is irritation and internal damage. **Chronic effects** Lead and zinc chromates are suspected carcinogens. Frequent exposure of skin to dust can cause ulceration. Long-continued absorption can cause liver and kidney disease.

Hazardous reactions   Ammonium dichromate decomposes thermally at 190 °C, the flame spreading rapidly with emission of green Cr_2O_3; if confined it will explode; hydroxylamine reacts explosively with both potassium and sodium dichromates; the dihydrated sodium salt reacts violently and finally explosively with acetic anhydride (B1029–1030).

First aid   *Dust inhaled:* standard treatment (p. 136).
Affected eyes: standard treatment (p. 136).
Skin contact: standard treatment (p. 136).
If swallowed: standard treatment (p. 137).

Spillage disposal   Shovel into bucket of water and run solution or suspension to waste diluting greatly with running water. If amount is more than a few g, use a reduction procedure (*e.g.*, that given on p. 160 of ref. 20 in Ch 3). Site of spillage should be washed thoroughly to remove all oxidant, which is liable to render any organic matter (particularly wood, paper and textiles) with which it comes into contact, dangerously combustible when dry. Clothing wetted with the solution should be washed thoroughly.

Chromic acid —*see* Chromium trioxide

Chromium diacetate [628–52–4]

Hazardous reaction Anhydrous salt is pyrophoric in air (B423).

Chromium trioxide (chromic acid) [1333–82–0]

Dark red crystalline masses or flakes; soluble in water.

CONTACT WITH COMBUSTIBLE MATERIAL MAY CAUSE FIRE
CAUSES SEVERE BURNS
MAY CAUSE SENSITISATION BY SKIN CONTACT

Avoid inhaling dust. Prevent contact with eyes and skin. RL (as Cr) 0.05 mg m^{-3}.

Toxic effects The dust irritates all parts of the respiratory system. The solid and its solutions cause severe eye burns. The solid and its solutions burn the skin. If taken by mouth there would be severe internal irritation and damage. **Chronic effects** Frequent exposure of skin to the material may result in ulceration.

Hazardous reactions Very powerful oxidant; violent or explosive reactions with acetic acid, acetic anhydride, P, Se; reacts with incandescence with K, Na, NH_3, As, butyric acid, H_2S; may ignite acetone, methanol, ethanol, propan-2-ol, butanol, cyclohexanol, *N,N*-dimethylformamide, glycerol, pyridine, sulphur. Violent reactions or ignition with C_2H_2, CrS, hexamethylphosphoric triamide (complex formed) or organic solvents (B1022–1027).

First aid *Dust inhaled:* standard treatment (p. 136).
Affected eyes: standard treatment (p. 136).
Skin contact: standard treatment (p. 136).
If swallowed: standard treatment (p. 137).

Spillage disposal Wear face-shield or goggles, and gloves (and rubber boots or overshoes if spillage is large). Spread soda ash liberally over the spillage to neutralise and mop up cautiously with plenty of water — run to waste diluting greatly with running water. If amount is more than a few g, use a reduction procedure (*e.g.*, that given on p. 160 of ref. 20 in Ch 3).

Chromyl acetate [4112–22–5]

Hazardous preparation Explosion occurred when being prepared from chromium trioxide and acetic anhydride (B423).

Chromyl azide chloride 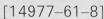 [14259–67–7]

Hazardous Explosive solid (B890).
reaction

Chromyl chloride . [14977–61–8]

Red fuming liquid with pungent musty odour; bp 117 °C; decomposed vigorously by water; can ignite organic matter on contact.

> CONTACT WITH COMBUSTIBLE MATERIAL MAY CAUSE FIRE
> CAUSES SEVERE BURNS

Avoid breathing vapour. Prevent contact with eyes and skin. RL 0.05 mg m^{-3} (as Cr).

Toxic effects The vapour irritates all parts of the respiratory system. The vapour irritates the eyes severely. The liquid burns the skin and eyes. If taken by mouth there would be severe local and internal irritation and damage. **Chronic effects** Frequent exposure of skin to the material may result in ulceration.

Hazardous reactions Explosions may occur with alkyl aromatics, liquid chlorine, PCl_3, sodium azide; ignites S_2Cl_2, acetone, ethanol, diethyl ether, turpentine, sulphur; incandescent reaction with ammonia, ignition with urea, H_2S, PH_3 or PBr_3 (B963–964).

First aid *Vapour inhaled:* standard treatment (p. 136).
Affected eyes: standard treatment (p. 136).
Skin contact: standard treatment (p. 136).
If swallowed: standard treatment (p. 137).

Spillage disposal Instruct others to keep at a safe distance. Wear breathing apparatus and gloves. Spread soda ash liberally over the spillage and mop up cautiously with water — run this to waste diluting greatly with running water.

Chromyl perchlorate [2597–99–3]

Hazardous reactions Powerful oxidant and nitrating agent which ignites many hydrocarbons and organic solvents on contact (B1021).

Chromyl perchlorate [62597–99–3]

Hazardous reactions Explodes violently above 80 °C; ignites organic solvents (B964).

Cinnamaldehyde [104–55–2]

Hazardous reaction   Rags soaked in NaOH solution and the aldehyde ignited in waste bin (B743).

Cleaning baths for glassware Review of hazards (B1471).

Coal gas —*see* Carbon monoxide

Cobalt, metal dust or fume [7440–48–4]

RL (as Co) 0.1 mg m^{-3}.

Cobalt nitride [12139–70–7]

Hazardous reaction   A pyrophoric powder (B1015).

Cobalt(II) perchlorate hydrates

Hazardous reactions   Partial dehydration of the stable hexahydrate to a trihydrate led to explosion (B962).

Cobalt trifluoride [10026–18–3]

Hazardous reactions   Fluorinating agent: reacts violently with hydrocarbons, water (B1012).

Complex acetylides Review of group (B1473).

Complex hydrides Review of group (B1474).

Copper [7440–50–8]

Metal fume RL (as Cu) 0.2 mg m^{-3}.
Dust or mist RL (as Cu) 1 mg m^{-3}.

Copper compounds

Blue or greenish-blue crystals or powder.

(HARMFUL IF TAKEN INTERNALLY)

Avoid inhaling dust. Avoid contact with eyes and skin.

Toxic effects	The dust irritates the mucous membranes. The dust and solutions of salts irritate the eyes. Ingestion may cause violent vomiting and diarrhoea with intense abdominal pain and collapse.
First aid	*Dust inhaled:* standard treatment (p. 136). *Affected eyes:* standard treatment (p. 136). *If swallowed:* standard treatment (p. 137).
Spillage disposal	Wear face-shield or goggles, and gloves. Mop up with plenty of water and run to waste, diluting greatly with running water.

Copper(I) hydride [13517–00–5]

Hazardous reaction   Impure material may decompose explosively on heating (B1042).

Copper(I) nitride [1308–80–1]

Hazardous reaction   May explode on heating in air (B1043).

Copper(II) phosphinate [34461–68–2]

Hazardous reaction   Explodes at about 90 °C (B1038).

Cresols (cresylic acid) [1319–77–3]

o-Cresol — colourless to pale brown crystals. *m*-Cresol — a colourless to yellow liquid. *p*-Cresol — colourless to pink crystals. Cresylic acid (mixed isomers) — colourless to brown liquid. All cresols have a phenolic (carbolic) odour and are sparingly soluble in water.

TOXIC IN CONTACT WITH SKIN AND IF SWALLOWED
CAUSES BURNS

Avoid breathing vapour. Prevent contact with skin and eyes. RL (skin) 5 ppm (22 mg m^{-3}).

Toxic effects   The vapour from heated cresols is irritant to the respiratory system. Cresols burn the eyes severely and irritate or burn the skin. Considerable absorption through the skin may give effects similar to those caused by ingestion, namely headache, dizziness, nausea, vomiting, stomach pain, exhaustion and possibly coma. **Chronic effects** Repeated inhalation or absorption of small amounts may cause damage to liver or kidneys; repeated contact with skin may cause dermatitis.

First aid   *Vapour inhaled:* standard treatment (p. 136).
Affected eyes: standard treatment (p. 136).
Skin contact: standard treatment (p. 136). But see also note on phenol (p. 135).
If swallowed: standard treatment (p. 137).

Spillage disposal   Wear face-shield or goggles, and gloves. Apply dispersing agent if available and work to an emulsion with brush and water — run this to waste, diluting greatly with running water. If dispersant not available, absorb on sand, shovel into bucket(s) and transport to safe, open area for burial. Site of spillage should be washed thoroughly with water and soap or detergent.

Crotonaldehyde   [123-73-9]

Colourless liquid with a pungent, suffocating odour; bp 104 °C; immiscible with water.

HIGHLY FLAMMABLE
TOXIC BY INHALATION
IRRITATING TO EYES, RESPIRATORY SYSTEM AND SKIN

Prevent inhalation of vapour. Prevent contact with skin and eyes. RL 2 ppm (6 mg m^{-3}).

Toxic effects   The vapour irritates the respiratory system. The vapour and liquid irritate the eyes severely, causing lachrymation and burns. The liquid irritates the skin. Assumed to be irritant and poisonous if taken by mouth.

Hazardous reactions   Exploded when heated with butadiene in autoclave at 180 °C; explodes on contact with concentrated nitric acid (B427).

First aid   *Vapour inhaled:* standard treatment (p. 136).
Affected eyes: standard treatment (p. 136).
Skin contact: standard treatment (p. 136).
If swallowed: standard treatment (p. 137).

Fire hazard   Flash point 13 °C; explosive limits 2.1–15.5%; ignition temp. 207°C. *Extinguish fire with* water spray, foam, dry powder, carbon dioxide or vaporising liquids.

Spillage disposal Shut off all possible sources of ignition. Instruct others to keep at a safe distance. Wear breathing apparatus and gloves. Apply non-flammable dispersing agent if available and work to an emulsion with brush and water — run this to waste diluting greatly with running water. If dispersant not available, absorb on sand, shovel into bucket(s) and transport to safe open area for atmospheric evaporation or burial. Site of spillage should be washed thoroughly with water and soap or detergent.

Cryogenic liquids
Review of two manuals and a recent bibliography on safe handling. Codes of Practice are defined in BS 5429:1976 (B1479).

Cumene (isopropylbenzene)   [98–82–8]

Colourless liquid; bp 152 °C; insoluble in water.

FLAMMABLE
VERY TOXIC IN CONTACT WITH SKIN

Avoid breathing vapour. Avoid contact with skin, eyes and clothing. RL (skin) 50 ppm (245 mg m^{-3}).

Toxic effects The vapour irritates the respiratory system. The vapour irritates the eyes and may cause conjunctivitis. The liquid irritates the skin and, by absorption and slow elimination, may cause serious poisoning. Assumed to be harmful if taken by mouth.

First aid *Vapour inhaled:* standard treatment (p. 136).
Affected eyes: standard treatment (p. 136).
Skin contact: standard treatment (p. 136).
If swallowed: standard treatment (p. 137).

Fire hazard Flash point 44 °C; explosive limits 0.9–6.5%; ignition temp. 424 °C. *Extinguish fire with* foam, dry powder, carbon dioxide or vaporising liquid.

Spillage disposal Shut off all possible sources of ignition. Wear face-shield or goggles, and gloves. Apply non-flammable dispersing agent if available and work to an emulsion with brush and water — run this to waste diluting greatly with running water. If dispersant not available, absorb on sand, shovel into bucket(s) and transport to safe, open area for burial. Ventilate site of spillage well to evaporate remaining liquid and dispel vapour.

Cyanamide   [420–04–2]

RL 2 mg m^{-3}.

Hazardous reactions Acids or alkalis and moisture speed decomposition which may become violent above 49 °C; explosive polymerisation may occur on evaporating aqueous solutions to dryness or storing unstabilised material (B143).

259

isoCyanatomethane —*see* Isocyanatomethane

Cyanides (water soluble)

Colourless crystals or powders which are soluble in water and react with acids to generate hydrogen cyanide.

(SERIOUS RISK OF POISONING BY INHALATION,
SKIN CONTACT AND IF SWALLOWED
CONTACT WITH ACIDS LIBERATES VERY TOXIC GAS)

RL (skin) (as CN) 5 mg m^{-3}.

Toxic effects   Cyanides and their solutions, and hydrogen cyanide liberated from these by the action of acids, are extremely poisonous. Both the cyanide solutions and the gas can be absorbed through the skin. Whatever the route of absorption, severe poisoning may result. The early warning symptoms of poisoning are general weakness and heaviness of the arms and legs, increased difficulty in breathing, headache, dizziness, nausea, vomiting. These may be rapidly followed by pallor, unconsciousness, cessation of breathing and death.

First aid   Special treatment for soluble cyanides (p. 133).
Spillage   Wear breathing apparatus and gloves. Instruct others to keep at a safe
disposal   distance. When cyanide solutions have been spilt, bleaching powder should be scattered liberally over the spillage, or an excess of sodium hypochlorite solution added. The treated spillage should then be mopped up into a bucket and allowed to stand for 24 hours before running to waste, diluting greatly with running water. Solid cyanides should be swept up and placed in a large volume of water in which they can be rendered innocuous by adding an excess of sodium hypochlorite solution and allowing to stand for 24 hours before running to waste, diluting greatly with running water.

Cyano compounds Review of class (B1480).

Cyanogen (dicyanogen)   [460–19–5]

Colourless gas with almond-like odour; bp −21 °C; 4 volumes dissolve in 1 volume of water at 15 °C.

HIGHLY FLAMMABLE
TOXIC BY INHALATION

Prevent inhalation of gas. RL 10 ppm (20 mg m^{-3}).

260

Toxic effects   The gas irritates the respiratory system, leading to headache, dizziness, rapid pulse, nausea, vomiting, unconsciousness, convulsions and death, depending upon exposure. The gas irritates the eyes causing lachrymation.

Hazardous reactions   Reaction of the endothermic compound with various oxidants may be explosive (B314).

First aid   Special treatment for hydrogen cyanide (p. 132).
Fire hazard   Explosive limits 6–32%. Since the gas is supplied in a cylinder, turning off the valve will reduce any fire involving it; if possible, cylinders should be removed quickly from an area in which a fire has developed.
Disposal   Surplus gas or leaking cylinder can be vented slowly in water-fed scrubbing tower or column, or into fume cupboard served by such a tower.

Cyanogen chloride   [506–77–4]

RL 0.3 ppm (0.6 mg m^{-6}).

Cyanogen *N,N*-dioxide   [4331–98–0]
(dicyanogen *N,N*-dioxide)

Hazardous reaction   Decomposes at −45 °C under vacuum before exploding (B315).

2-Cyanopropan-2-ol
—*see* **2-Hydroxy-2-methylpropiononitrile**

1-Cyano-2-propen-1-ol
—*see* **2-Hydroxybut-3-enonitrile**

3-Cyanotriazenes Review of group (B1483).

Cyanuric chloride
—*see* **2,4,6-Trichloro-*s*-triazine**

Cyclic peroxides Review of class (B1483).

261

Cyclohexa-1,3-diene [592-57-4]

Hazardous reaction   Slowly forms explosive peroxide on contact with air (B603).

Cyclohexane [110-82-7]

Colourless, mobile liquid with pungent odour when impure; bp 81 °C; insoluble in water.

HIGHLY FLAMMABLE

Avoid breathing vapour. Avoid contact with skin and eyes. RL 300 ppm (1050 mg m^{-3}).

Toxic effects   Irritates the eyes, skin and respiratory system; the inhalation of high concentrations may cause narcosis. Assumed to be irritant and narcotic if taken by mouth.

Hazardous reaction   Addition of liquid dinitrogen tetraoxide to hot cyclohexane caused explosion (B619).

First aid   *Vapour inhaled:* standard treatment (p. 136).
Affected eyes: standard treatment (p. 136).
Skin contact: standard treatment (p. 136).
If swallowed: standard treatment (p. 137).

Fire hazard   Flash point −20 °C; explosive limits 1.3–8.4%; ignition temp. 260 °C. *Extinguish fire with* foam, dry powder, carbon dioxide or vaporising liquid.

Spillage disposal   Shut off all possible sources of ignition. Instruct others to keep at a safe distance. Wear breathing apparatus and gloves. Apply non-flammable dispersing agent if available and work to an emulsion with brush and water — run this to waste, diluting greatly with running water. If dispersant not available, absorb on sand, shovel into bucket(s) and transport to safe open area for atmospheric evaporation. Site of spillage should be washed thoroughly with water and soap or detergent.

Cyclohexane-1,2-dione [765-87-2]

Hazardous preparation   Explosion when being prepared by HNO_3 oxidation of cyclohexanol (B607).

Cyclohexanol [108-93-0]

Colourless hygroscopic crystals or viscous liquid with camphorlike odour; bp 161 °C; mp 23–25 °C; sparingly soluble in water.

HARMFUL BY INHALATION AND IF SWALLOWED
IRRITATING TO RESPIRATORY SYSTEM AND SKIN

Avoid breathing vapour. Avoid contact with skin and eyes. RL 50 ppm (200 m^{-3}).

Toxic effects	The vapour may irritate the eyes, skin and respiratory system. The liquid irritates the eyes and may cause conjunctivitis and more serious damage. The liquid irritates the skin; major absorption may lead to tremors and kidney or liver damage. Assumed to be irritant and damaging to the alimentary system if taken by mouth.
Hazardous reactions	Ignited by CrO_3; explosion with HNO_3 (B624).
First aid	*Vapour inhaled:* standard treatment (p. 136). *Affected eyes:* standard treatment (p. 136). *Skin contact:* standard treatment (p. 136). *If swallowed:* standard treatment (p. 137).
Spillage disposal	Shut off all possible sources of ignition. Wear face-shield or goggles, and gloves. Apply non-flammable dispersing agent if available and work to an emulsion with brush and water — run this to waste, diluting greatly with running water. If dispersant not available, absorb on sand, shovel into bucket(s) and transport to safe open area for burial. Ventilate site of spillage well to evaporate remaining liquid and dispel vapour.

Cyclohexanone [108-93-0]

Colourless, oily liquid with odour somewhat similar to that of acetone; bp 156 °C; sparingly soluble in water.

FLAMMABLE
HARMFUL BY INHALATION

Avoid breathing vapour. Avoid contact with skin and eyes. RL 25 ppm (100 mg m^{-3}).

Toxic effects	The vapour may irritate the eyes, skin and respiratory system. The liquid irritates the eyes and may cause conjunctivitis. The liquid may irritate the skin. Assumed to be irritant and damaging to the alimentary system if taken by mouth.
Hazardous reactions	Forms explosive peroxide with H_2O_2; may explode when added to HNO_3 at about 75 °C (B614).
First aid	*Vapour inhaled:* standard treatment (p. 136). *Affected eyes:* standard treatment (p. 136). *Skin contact:* standard treatment (p. 136). *If swallowed:* standard treatment (p. 137).
Fire hazard	Flash point 44 °C; explosive limits 1.1–8.1%; ignition temp. 420 °C. *Extinguish fire with* water spray, foam, dry powder, carbon dioxide or vaporising liquid.

Spillage disposal	Shut off all possible sources of ignition. Wear face-shield or goggles, and gloves. Apply non-flammable dispersing agent if available and work to an emulsion with brush and water — run this to waste, diluting greatly with running water. If dispersant not available, absorb on sand, shovel into bucket(s) and transport to safe open area for burial. Ventilate site of spillage well to evaporate remaining liquid and dispel vapour.

Cyclohexanone oxime [100–64–1]

Hazardous reaction	Beckmann rearrangement with oleum to give caprolactam failed to start and led to explosion (B618).

Cyclohexene [110–83–8]

Colourless liquid; bp 83 °C; insoluble in water.

(HIGHLY FLAMMABLE
IRRITATING TO RESPIRATORY SYSTEM)

Avoid breathing vapour. RL 300 ppm (1015 mg m^{-3}).

Toxic effects	The vapour irritates the respiratory system and may irritate the eyes and skin. Assumed to be irritant if taken by mouth.
First aid	*Vapour inhaled:* standard treatment (p. 136). *Affected eyes:* standard treatment (p. 136). *Skin contact:* standard treatment (p. 136). *If swallowed:* standard treatment (p. 137).
Fire hazard	Flash point −60 °C. *Extinguish fire with* foam, dry powder, carbon dioxide or vaporising liquid.
Spillage disposal	Shut off all possible sources of ignition. Instruct others to keep at a safe distance. Wear breathing apparatus and gloves. Apply non-flammable dispersing agent if available and work to an emulsion with brush and water — run this to waste, diluting greatly with running water. If dispersant not available absorb on sand, shovel into bucket(s) and transport to safe open area for atmospheric evaporation. Site of spillage should be washed thoroughly with water and soap or detergent.

Cyclohexylamine (hexahydroaniline) [108–91–8]

Colourless liquid with an ammoniacal, fishy odour; bp 134 °C; miscible with water.

FLAMMABLE
HARMFUL IN CONTACT WITH SKIN AND IF SWALLOWED
CAUSES BURNS

Avoid breathing vapour. Prevent contact with skin and eyes. RL (skin) 10 ppm (40 mg m^{-3}).

Toxic effects The vapour may irritate the eyes and respiratory system, causing difficulty in breathing. The liquid can burn the eyes and skin; skin absorption may cause nausea and vomiting. Assumed to be poisonous if taken by mouth.

First aid *Vapour inhaled:* standard treatment (p. 136).
Affected eyes: standard treatment (p. 136).
Skin contact: standard treatment (p. 136).
If swallowed: standard treatment (p. 137).

Fire hazard Flash point 32 °C; ignition temp. 293 °C. *Extinguish fire with* water spray, foam, dry powder, carbon dioxide or vaporising liquids.

Spillage disposal Shut off all possible sources of ignition. Instruct others to keep at a safe distance. Wear breathing apparatus and gloves. Mop up with plenty of water and run to waste diluting greatly with running water. Ventilate area well to evaporate remaining liquid and dispel vapour.

Cyclopenta-1,3-diene [542–92–7]

Colourless liquid; bp 42 °C; insoluble in water.

(HIGHLY FLAMMABLE)

Avoid breathing vapour. Avoid contact with skin and eyes. TLV 75 ppm (200 mg m^{-3}).

Toxic effects The main toxic effects arise from inhalation which leads to depression of central nervous system, liver damage and a benzene-like effect upon the blood. **Chronic effects** include headache, abdominal pains, jaundice and anaemia. An acute effect may be narcosis.

Hazardous reactions Dimerisation is exothermic and may cause rupture of closed, uncooled containers; heat sensitive explosive peroxides formed on exposure to O_2; explosive reaction with N_2O_4 (B502–503).

First aid *Vapour inhaled:* standard treatment (p. 136).
Affected eyes: standard treatment (p. 136).
Skin contact: standard treatment (p. 136)
If swallowed: standard treatment (p. 137).

Fire hazard No data about flash point, *etc* was traced. *Extinguish fire with* foam, dry powder, carbon dioxide or vaporising liquid.

265

Spillage disposal   Shut off all possible sources of ignition. Instruct others to keep at a safe distance. Wear breathing apparatus and gloves. Apply non-flammable dispersing agent if available and work to an emulsion with brush and water — run this to waste, diluting greatly with running water. If dispersant not available, absorb on sand, shovel into bucket(s) and transport to safe open area for atmospheric evaporation. Site of spillage should be washed thoroughly with water and soap or detergent.

Cyclopentadienylmanganese tricarbonyl [12108–13–3]

RL (skin, as Mn) 0.1 mg m^{-3}.

Cyclopentadienylsodium [4984–82–1]

Hazardous reaction   Pyrophoric in air (B502).

Cyclopentane [287–92–3]

Colourless, mobile liquid; bp 49.3 °C; insoluble in water.

HIGHLY FLAMMABLE

Avoid breathing vapour.

Toxic effects   May act as mild narcotic in high concentrations.

First aid   *Vapour inhaled:* standard treatment (p. 136).
Fire hazard   Flash point below −6.7 °C *Extinguish fire with* foam, dry powder, carbon dioxide or vaporising liquid.
Spillage disposal   Shut off all possible sources of ignition. Instruct other to keep at a safe distance. Wear breathing apparatus and gloves. Apply non-flammable dispersing agent if available and work to an emulsion with brush and water — run this to waste, diluting greatly with running water. If dispersant not available, absorb on sand, shovel into bucket(s) and transport to safe open area for atmospheric evaporation. Site of spillage should be washed thoroughly with water and soap or detergent.

Cyclopentanone [120–92–3]

Hazardous reaction   Mixtures with nitric acid and hydrogen peroxide react vigorously and may become explosive (B512).

Cyclopentanone oxime

[1192–28–5]

Hazardous reaction Two instances of violet Beckmann rearrangement (B516).

Cyclopropane

[75–19–4]

Colourless gas with smell like that of petroleum spirit; bp −33 °C; 1 volume of gas dissolves in 2.7 volumes of water at 15 °C and 760 mmHg.

EXTREMELY FLAMMABLE LIQUEFIED GAS

Avoid breathing gas.

Toxic effects The gas is an anaesthetic and is employed for this purpose.
First aid *Gas inhaled:* remove from exposure, rest and keep warm.
Fire hazard Explosive limits 2.4–10.4%; ignition temp. 498 °C. Since the gas is supplied in a cylinder, turning off the valve will reduce any fire involving it; if possible, cylinders should be removed quickly from an area in which a fire has developed.
Disposal Surplus gas or a leaking cylinder can be vented slowly to air in a safe open area or burnt off in a suitable gas burner.

2,4-D – *see* 2,4-Dichlorophenoxyacetic acid

DDT – *see* 1,1-Bis(4-chlorophenyl)-2,2,2-trichloroethane

Decaborane

[17702–41–9]

TLV (skin) 0.05 ppm (0.3 mg m^{-3}).

Hazardous reactions Forms impact-sensitive mixtures with ethers and halocarbons; ignites in oxygen at 100 °C (B77)

Decanedioyl dichloride
— *see* Sebacoyl dichloride

267

Devarda's alloy Safe usage discussed (B1484).

Diacetone alcohol
—*see* 4-Hydroxy-4-methylpentan-2-one

Diacetyl peroxide [110–22–2]

Hazardous reaction When dry is a shock-sensitive explosive; use in ethereal solution (B436).

Diacyl peroxides Review of group (B1484).

Dialkyl hyponitrites Review of group (B1485).

Dialkylmagnesiums Review of group (B1486).

Dialkyl peroxides Review of group (B1486).

Dialkylzincs Review of group (B1487).

Diallyl ether [557–40–4]

Hazardous reaction Hazardous preparation, peroxidises readily in air and sunlight to explosive peroxide (B615).

Diallyl phosphite [23679–20–1]

Hazardous reaction Liable to explode during distillation (B619).

Diallyl phthalate [131–17–9]

Colourless, oily liquid; bp 157 °C: insoluble in water.

HARMFUL IF SWALLOWED

Avoid contact with skin, eyes and clothing.

Toxic effects The liquid irritates the eyes and skin and can cause internal disorders by continued skin absorption. Assumed to be harmful if taken by mouth.

First aid *Affected eyes:* standard treatment (p. 136).
Skin contact: standard treatment (p. 136).
If swallowed: standard treatment (p. 137).

Spillage disposal Wear face-shield or goggles, and gloves. Apply dispersing agent if available and work to an emulsion with brush and water — run this to waste, diluting greatly with running water. If dispersant not available, absorb on sand, shovel into bucket(s) and transport to safe open area for burial. Ventilate site of spillage well to evaporate remaining liquid and dispel vapour.

Diallyl sulphate [27063–40–7]

Hazardous reaction Exploded during distillation (B616)

4,4′-Diaminodiphenylmethane
—*see* Bis(4-aminophenyl)methane

1,2-Diaminoethane (ethylenediamine) [107–15–3]

Clear, colourless liquid with ammoniacal odour, bp 117 °C; miscible with water.

FLAMMABLE
HARMFUL IN CONTACT WITH SKIN AND IF SWALLOWED
CAUSES BURNS
MAY CAUSE SENSITISATION BY SKIN CONTACT

Avoid breathing vapour. Avoid contact with skin and eyes. RL 10 ppm (25 mg m^{-3}).

Toxic effects The vapour irritates the respiratory system. The liquid and vapour cause irritation of skin and eyes. If swallowed, may cause digestive disturbance and possibly damage to the kidneys. **Chronic effects** Repeated inhalation of vapour or skin contact may cause sensitisation of skin or respiratory system.

Hazardous reactions May ignite on contact with cellulose nitrate; dangerous reactions with nitromethane and diisopropyl peroxydicarbonate (B304).

First aid *Vapour inhaled:* standard treatment (p. 136).
Affected eyes: standard treatment (p. 136).
Skin contact: standard treatment (p. 136).
If swallowed: standard treatment (p. 137).

Fire hazard Flash point 43 °C. *Extinguish fire with* dry powder, carbon dioxide or vaporising liquid.

Spillage disposal Shut off all possible sources of ignition. Instruct others to keep at a safe distance. Wear breathing apparatus and gloves. Mop up with plenty of water and run to waste, diluting greatly with running water. Ventilate area well to evaporate remaining liquid and dispel vapour.

1,6-Diaminohexane (hexane-1,6-diamine) [124–09–4]

Colourless leaflets; mp 42 °C; freely soluble in water.

(IRRITATING TO EYES, RESPIRATORY SYSTEM AND SKIN)

Toxic effects Indicated in above warning.

First aid *Affected eyes:* standard treatment (p. 136)
Skin contact: standard treatment (p. 136).
If swallowed: standard treatment (p. 137).

Spillage disposal Wearing gloves, brush up and dissolve in bucket of water. Run solution to waste diluting with running water. Mop up residual amine with water.

Dianilinium dichromate [101672–09–7]

Hazardous reaction Unstable on storage (B809).

o-Dianisidine and hydrochloride
—*see* 3,3'-Dimethoxybenzidine and hydrochloride

Diazidodimethylsilane [4774–73–6]

Hazardous reaction Old sample exploded (B289).

Diazidoethanes   1,1-[67880–20–0] 1,2-[629–13–0]

Hazardous reaction   Explosive (B261–262).

Diazidomalononitrile   [67880–21–1]

Hazardous reaction   Shock-sensitive explosive but may explode without warning (B394).

1,3-Diazidopropene   [22750–69–2]

Hazardous reaction   Exploded while being weighed (B348).

Diazirines Review of group (B1488).

Diazoacetonitrile   [13138–21–1]

Hazardous reaction   Liable to explode through friction (B224).

Diazo compounds Review of group (B1488).

2-Diazocyclohexanone   [3242–56–6]

Hazardous reaction   May explode on heating (B605).

Diazocyclopentadiene   [1192–27–4]

Hazardous reaction   Exploded violently during vacuum distillation (B495–496).

Diazoindene   [35847–40–6]

Hazardous preparation   Distillation residue may explode (B739).

Diazomalononitrile (dicyanodiazomethane)   [1618–08–2]

Hazardous reaction   May explode at 75 °C (B394).

Diazomethane   [334–88–3]

A yellow gas generally employed for organic synthesis in chloroform or ethereal solution and prepared as part of the process.

> (EXTREME RISK OF EXPLOSION BY SHOCK, FRICTION, FIRE OR OTHER SOURCES OF IGNITION
> TOXIC BY INHALATION)

Prevent inhalation of vapour and contact with skin and eyes. Experimental carcinogen. TLV 0.2 ppm (0.4 mg m^{-3}).

Toxic effects   Irritates the respiratory system, eyes and skin. Inhalation may result in chest discomfort, headache, weakness and, in severe cases, collapse.

Hazardous reactions   Gaseous diazomethane may explode on ground glass surfaces and when heated to about 100 °C; concentrated solutions may also explode especially if impurities are present; explosions occur on contact with alkali metals and the exothermic reaction with calcium sulphate is also dangerous. Equipment for safe use of diazomethane has been described (B144–146).

First aid   *Vapour inhaled:* standard treatment (p. 136).
Affected eyes: standard treatment (p. 136).

RSC *Lab. Haz. Data Sheet No. 4,* 1982 gives extended coverage.

Diazonium salts Review of class (B1490).

2-Diazo-2-phenylacetophenone   [3469–17–8]
(1,1-benzoylphenyldiazomethane)

Hazardous reaction   May explode if heated above 40 °C (B827).

Dibenzoyl peroxide (benzoyl peroxide)   [94–36–0]

White granular crystals, normally supplied moistened with about 30% water; liable to explode when heated above melting point (mp 103–105 °C) or when subjected to friction or shock when dry. Insoluble in water.

EXTREME RISK OF EXPLOSION BY SHOCK, FRICTION, FIRE OR
OTHER SOURCE OF IGNITION
IRRITATING TO EYES, RESPIRATORY SYSTEM AND SKIN

Avoid contact with eyes and skin. RL 5 mg m^{-3}.

Toxic effects   Contact with skin or eyes causes irritation.

Hazardous reactions   Dry material burns and is sensitive to heat (explodes above mp), shock, friction or contact with combustible materials; has exploded on heating with ethylene and CCl_4 under pressure; explosions have resulted from contact with dimethylaniline or dimethyl sulphide; ignition occurs with methyl methacrylate; explosion with aniline; recrystallization from hot $CHCl_3$ leads to explosions (B827–829).

First aid   *Affected eyes:* standard treatment (p. 136).
Skin contact: standard treatment (p. 136).

Spillage disposal   Moisten well with water and mix with plenty of sand. Disposal in any quantity depends upon local conditions and regulations. Deep burial in an isolated area or consignment to deep sea water in a heavy container can be considered if quantities are large.

Dibenzylamine [103–49–1]

Colourless, oily liquid with ammoniacal odour; bp 300 °C with decomposition; immiscible with water.

CAUSES BURNS

Prevent contact with skin and eyes.

Toxic effects   The liquid irritates and burns the skin and eyes. Corrosive and highly irritant if taken by mouth.

First aid   *Affected eyes:* standard treatment (p. 136).
Skin contact: standard treatment (p. 136).
If swallowed: standard treatment (p. 137).

Spillage disposal   Wear face-shield or goggles, and gloves. Apply dispersing agent if available and work to an emulsion with brush and water — run this to waste, diluting greatly with running water. If dispersant not available, absorb on sand, shovel into bucket(s) and transport to safe open area for burial. Site of spillage should be washed thoroughly with water and soap or detergent.

Dibenzyl ether [103–50–4]

Hazardous reaction   Exploded with aluminium dichloride hydride diethyl etherate (B832).

Dibenzyl phosphite [17176–77–1]

Hazardous reaction
Decomposes at 160 °C (B833)

Diborane [19287–45–7]

RL 0.1 ppm (0.1 mg m^{-3}).

Hazardous reactions
Usually ignites in air and delayed ignition may be followed by violent explosions; reacts explosively with chlorine and forms explosive compound with dimethylsulphoxide; reacts violently with halocarbon liquids; explosive reaction with tetravinyllead, and with NF$_3$ in liquid phase (B69–70).

Diboron tetrachloride [13701–67–2]

Hazardous reaction
Exposure to air may cause explosion (B68).

Diboron tetrafluoride [13965–73–6]

Hazardous reactions
Extremely explosive in presence of oxygen; violent reaction or ignition with metal oxides (B68–69).

Dibromoacetylene [624–61–3]

Hazardous reactions
Ignites in air and explodes on heating (B200)

2,6-Dibromobenzoquinone- 4-chloroimide [537–45–1]

Hazardous reaction
Liable to decompose explosively on heating (B540).

Dibromodifluoromethane [75–61–6]

RL 100 pp, (860 mg m^{-3}).

1,2-Dibromoethane [106–93–4]
(ethylene dibromide)

Colourless liquid with sweetish chloroform-like odour, bp 131 °C; immiscible with water.

TOXIC BY INHALATION, IN CONTACT WITH SKIN
AND IF SWALLOWED

Avoid breathing vapour. Avoid contact with skin and eyes. Suspected carcinogen. TLV (skin) 20 ppm (145 mg m^{-3}).

Toxic effects	The vapour irritates the respiratory system and may have narcotic action. The vapour and liquid irritate the eyes. The liquid irritates the skin and may cause dermatitis. Poisonous by skin absorption and ingestion, effects being nausea, vomiting, pain and jaundice resulting from liver and kidney damage.
Hazardous reaction	Reaction with Mg may become violent (B256)
First aid	*Vapour inhaled:* standard treatment (p. 136). *Affected eyes:* standard treatment (p. 136). *Skin contact:* standard treatment (p. 136). *If swallowed:* standard treatment (p. 137).
Spillage disposal	Instruct others to keep at a safe distance. Wear breathing apparatus and gloves. Apply dispersing agent if available and work to an emulsion with brush and water — run this to waste, diluting greatly with running water. If dispersant not available, absorb on sand, shovel into bucket(s) and transport to safe open area for burial. Site of spillage should be washed thoroughly with water and soap or detergent.

Dibromomethane [74–95–3]

HARMFUL BY INHALATION

Hazardous reaction	Forms shock-sensitive explosive with potassium (B140)

Dibutylamines —*see* Butylamines

2-Dibutylaminoethanol [102–81–8]

TLV (skin) 2 ppm (14 mg m^{-3}).

Di-t-butyl chromate [1189-85-1]

Hazardous   Addition of t-butanol to CrO_3 resulted in explosion, explosive oxidation of
preparation   valencene (B730-731).

Dibutyl ether (n-butyl ether) [142-96-1]

Colourless liquid; bp 142 °C; insoluble in water. May form explosive peroxides on exposure to light and air which should be decomposed before distillation to small volume.

FLAMMABLE
IRRITATING TO EYES, RESPIRATORY SYSTEM AND SKIN

Avoid breathing vapour. Avoid contact with skin and eyes.

Toxic effects   The vapour is somewhat irritating to the respiratory system. The liquid irritates the eyes, and is considered to present some hazard by skin absorption.

First aid   *Vapour inhaled:* standard treatment (p. 136).
Affected eyes: standard treatment (p. 136).
Skin contact: standard treatment (p. 136).
If swallowed: standard treatment (p. 137).

Fire hazard   Flash point 25 °C; explosive limits 1.5-7.6%; ignition temp. 194 °C. *Extinguish fire with* dry powder, carbon dioxide, or vaporising liquid.

Spillage   Shut off all possible sources of ignition. Instruct others to keep at a safe
disposal   distance. Wear breathing apparatus and gloves. Apply non-flammable dispersing agent if available and work to an emulsion with brush and water — run this to waste diluting greatly with running water. If dispersant not available, absorb on sand, shovel into bucket(s) and transport to safe open area for burial. Ventilate site of spillage well to evaporate remaining liquid and dispel vapour.

Dibutyl hydrogen phosphate [107-66-4]

TLV 1 ppm (5 mg m^{-3}).

2,6-Di-t-butyl-4-nitrophenol [728-40-5]

Hazardous   Exploded on heating to 100 °C (B834).
reaction

Di-t-butyl peroxide [110–05–4]

HIGHLY FLAMMABLE
IRRITATING TO RESPIRATORY SYSTEM AND SKIN

Hazardous preparation   Addition of t-butanol to 50% H_2O_2/78% H_2SO_4 mixtures may result in explosions. The peroxide is thermally unstable (B732).

Dibutyl phthalate [84–74–2]

RL 5 mg m^{-3}.

Hazardous reaction   Mixture with liquid chlorine in S.S. bomb reacted explosively at 118 °C (B951).

Dichlorine oxide [7791–21–1]

Hazardous reactions   The liquid explodes on pouring, the gas on heating or sparking; explodes with charcoal, dicyanogen, NO, K, NH_3, As, Sb, S. PH_3, H_2S, paper, cork, rubber, turpentine and many other oxidisable materials; alcohols, CS_2, ethers and lower hydrocarbons are oxidised explosively (B975–977).

Dichlorine trioxide [17496–59–2]

Hazardous reaction   Vapour explodes well below 0 °C (B980).

Dichloroacetic acid [79–43–6]

Colourless liquid with pungent odour; bp 194 °C; miscible with water.

CAUSES SEVERE BURNS

Avoid breathing vapour. Prevent contact with skin and eyes.

Toxic effects   The vapour irritates the eyes and respiratory system. The liquid burns the skin and eyes. Assumed to cause severe irritation and damage if taken by mouth.
First aid   *Vapour inhaled:* standard treatment (p. 136).
Affected eyes: standard treatment (p. 136).
Skin contact: standard treatment (p. 136).
If swallowed: standard treatment (p. 137).

277

Spillage disposal Wear goggles and gloves (and rubber boots or overshoes if spillage is large). Spread soda ash liberally over the spillage and mop up cautiously with plenty of water – run this to waste, diluting greatly with running water.

Dichloroacetyl chloride [79–36–7]

Colourless, fuming liquid with acrid penetrating odour; bp 107 °C; reacts with water forming dichloroacetic and hydrochloric acids.

CAUSES SEVERE BURNS

Prevent inhalation of vapour. Prevent contact with skin and eyes.

Toxic effects The vapour irritates the eyes and respiratory system. The liquid burns the skin and eyes. Assumed to cause severe irritation and damae if taken by mouth.

First aid *Vapour inhaled:* standard treatment (p. 136).
Affected eyes: standard treatment (p. 136).
Skin contact: standard treatment (p. 136).
If swallowed: standard treatment (p. 137).

Spillage disposal Wear goggles or face-shield, and gloves (and rubber boots or overshoes if spillage is large). Spread soda ash liberally over the spillage to neutralise and mop up cautiously with plenty of water — run to waste diluting greatly with running water.

Dichloroacetylene [7572–29–4]

RL 0.1 ppm (0.4 mg m^{-3})'

Hazardous reaction Heat-sensitive explosive gas which ignites in contact with air (B206).

N,N-Dichloroaniline [70278–00–1]

Hazardous reaction Explosive (B576).

Dichlorobenzenes *o*-[95–50–1] *p*-[106–46–7]

o-Isomer is colourless liquid with a pleasant aromatic smell; bp 185 °C; *p*-isomer consists of colourless, volatile crystals with characteristic disinfectant smell, mp 53 °C; both are insoluble in water.

HARMFUL BY INHALATION (*o-*)
HARMFUL IF SWALLOWED (*p-*)

Avoid breathing vapour. Prevent contact with skin and eyes. RL *o-* 50 ppm (300 mg m^{-3}), *p-* 75 ppm (450 mg m^{-3}).

Toxic effects	Inhalation of vapours may cause drowsiness and irritation of nose; both isomers irritate the eyes; the *o*-isomer is more irritating to the skin and may cause dermatitis; long exposure to either isomer may result in liver damage.
First aid	*Vapour inhaled:* standard treatment (p. 136).
	Affected eyes: standard treatment (p. 136).
	Skin contact: standard treatment (p. 136).
	If swallowed: standard treatment (p. 137).
Spillage disposal	The solid *p*-isomer may be mixed with dry sand, swept up and placed in waste bin. The liquid *o*-isomer: wear breathing apparatus and gloves. Apply dispersing agent if available, and work to an emulsion with brush and water — run this to waste diluting greatly with running water. If dispersant not available, absorb on sand, shovel into bucket(s) and transport to approved open area for burial. Site of spillage should be washed thoroughly with water and soap or detergent.

3,3'-Dichlorobenzidine and dihydrochloride

[91–94–1]
[612–83–9]

Dichlorobenzidine is a colourless or pale crystalline solid, scarcely soluble in water, and the salt, of similar appearance, is slightly soluble in water. Use of both is controlled in the United Kingdom by The Carcinogenic Substances Regulations 1967 (see p. 148).

IRRITATING TO EYES AND SKIN
DANGER OF VERY SERIOUS IRREVERSIBLE EFFECTS

Prevent inhalation of dust. Prevent contact with skin and eyes. Suspected carcinogen. No TLV set.

Toxic effects	The dust irritates the eyes and skin and may lead to allergic reactions. **Chronic effects**. There is evidence that 3,3'-dichlorobenzidine through continued absorption may cause bladder tumours.
First aid	*Dust inhaled:* standard treatment (p. 136).
	Affected eyes: standard treatment (p. 136).
	Skin contact: standard treatment (p. 136).
	If swallowed: standard treatment (p. 137).
Spillage disposal	Mix with moist sand and shovel mixture into glass, polythene or enamelled vessel for dispersion in an excess of dilute hydrochloric acid (1 volume of concentrated acid diluted with 2 volumes of water). Allow to stand with occasional stirring for 24 hours and then run extract to waste, diluting greatly with running water and washing the sand. The residual sand can be treated as normal refuse. The site of the spillage should be washed with water and soap or detergent.

1,4-Dichlorobut-2-yne [821–10–3]

Hazardous preparation   A preferred method using dichloromethane as diluent is indicated; a distillation residue exploded (B410).

Dichlorodifluoromethane [75–71–8]

RL 1000 ppm (4950 mg m^{-3}).

Hazardous reaction   May react exothermically with Al. Mg dust ignited at 400 °C in vapour and suspension exploded violently on sparking (B119).

N,N'-Dichloro-5,5-dimethylhydantoin
—*see* 1,3-Dichloro-5,5-dimethylimidazolidindione

1,3-Dichloro-5,5-dimethylimidazolidindione
[118–52–5]

RL 0.2 mg m^{-3}.

Hazardous reaction   Violent explosion with xylene, but safe conditions devised (B504–505).

1,1-Dichloroethane [75–34–3]
(ethylidene chloride)

> EXTREMELY FLAMMABLE
> HARMFUL BY INHALATION

RL 200 ppm (810 mg m^{-3}).

1,2-Dichloroethane [107–06–2]
(ethylene dichloride)

Colourless liquid with a chloroform-like odour; bp 84 °C; immiscible with water.

> HIGHLY FLAMMABLE
> HARMFUL BY INHALATION

Avoid breathing vapour. Avoid contact with skin and eyes. RL 10 ppm (40 mg m^{-3}).

Toxic effects   In high concentrations, the vapour irritates the eyes and respiratory system; it may also cause drowsiness, hadache, vomiting and mental confusion. The liquid may cause serious damage to the eyes. Poisonous if taken by mouth. The liquid irritates the skin. **Chronic effects** Continued exposure to low concentrations may result in dizziness, nausea and abdominal pain, and there may be damage to the eyes and liver. Dermatitis may follow repeated contact with the skin.

Hazardous reactions   Mixtures with N_2O_4 or potassium are explosive when subjected to shock; reaction with aluminium powder may be violent or explosive; mixtures with HNO_3 easily detonate (B257, 1110).

First aid   *Vapour inhaled:* standard treatment (p. 136).
Affected eyes: standard treatment (p. 136).
Skin contact: standard treatment (p. 136).
If swallowed: standard treatment (p. 137).

Fire hazard   Flash point 13 °C; explosive limits 6.2–15.9%; ignition temp. 413 °C. *Extinguish fire with* water spray, foam, dry powder, carbon dioxide or vaporising liquid.

Spillage disposal   Shut off all possible sources of ignition. Instruct others to keep at a safe distance. Wear breathing apparatus and gloves. Apply non-flammable dispersing agent if available and work to an emulsion with brush and water — run this to waste, diluting greatly with running water. If dispersant not available, absorb on sand, shovel into buckets and transport to safe open area for atmospheric evaporation. Site of spillage should be washed thoroughly with water and soap or detergent.

2,2-Dichloroethylamine   [5960–88–3]

Hazardous reaction   Violent explosion when ethereal solution was being evaporated under vacuum (B276).

1,1-Dichloroethylene   [75–35–4]
(vinylidene chloride)

EXTREMELY FLAMMABLE
HARMFUL BY INHALATION
POSSIBLE RISKS OF IRREVERSIBLE EFFECTS

RL 10 ppm (40 mg m^{-3}).

Hazardous reactions   Rapidly absorbs oxygen forming a violently explosive peroxide; reaction products with ozone are particularly dangerous; reaction under pressure with chlorotrifluoroethylene may develop into explosive polymerisation (B231).

1,2-Dichloroethylene (acetylene dichloride)   *mixo-* [540–59–0]

Colourless liquid with a slight chloroform-like odour; bp 48–60 °C; immiscible with water.

HIGHLY FLAMMABLE
HARMFUL BY INHALATION

Avoid breathing vapour. Avoid contact with skin and eyes. TLV 200 ppm (790 mg m^{-3}).

Toxic effects   The vapour irritates the eyes and mucous membranes; in high concentrations it may cause drowsiness and unconsciousness. Poisonous if taken by mouth. **Chronic effects** Continued exposure to low concentrations of vapour may cause drowsiness and digestive disturbance.

Hazardous reactions   Contact with solid caustic alkalies or their concentrated solutions will form chloroacetylene which ignites in air; forms explosive mixtures with N_2O_4 (B231–232).

First aid   *Vapour inhaled:* standard treatment (p. 136).
Affected eyes: standard treatment (p. 136).
Skin contact: standard treatment (p. 136).
If swallowed: standard treatment (p. 137).

Fire hazard   Flash point 2–4 °C; explosive limits 9.7–12.8%. *Extinguish fire with* water spray, foam, dry powder, carbon dioxide or vaporising liquid.

Spillage disposal   Shut off all possible sources of ignition. Instruct others to keep at a safe distance. Wear breathing apparatus and gloves. Apply non-flammable dispersing agent if available and work to an emulsion with brush and water — run this to waste, diluting greatly with running water. If dispersant not available, absorb on sand, shovel into bucket(s) and transport to safe open area for atmospheric evaporation. Site of spillage should be washed thoroughly with water and soap or detergent.

Di(2-chloroethyl) ether
—*see* Bis(2-chloroethyl) ether

Dichlorofluoromethane   [75–43–4]

RL 10 ppm (40 mg m^{-3}).

1,6-Dichloro-2,4-hexadiyne   [16260–59–6]

Hazardous reaction   Extremely shock-sensitive explosive (B560).

1,3-Dichlorohydrin
— *see* **1,3-Dichloropropan-2-ol**

Dichloromethane (methylene chloride)   [75–09–2]

Colourless volatile liquid with chloroform-like odour; bp 40 °C; immiscible with water.

> HARMFUL BY INHALATION

Avoid breathing vapour. Avoid contact with skin and eyes. CL 100 ppm (350 mg m^{-3}).

Toxic effects	The vapour irritates the eyes and respiratory system and may cause headache and nausea; high concentrations may result in cyanosis and unconsciousness. The liquid irritates the eyes. Assumed to be poisonous if taken by mouth.
Hazardous reactions	Solution of dinitrogen pentaoxide in dichloromethane liable to explode; mixtures with Li, Na, N$_2$O$_4$, HNO$_3$ are liable to explode; violent reaction with hot Al, or potassium t-butoxide (B140–141).
First aid	*Vapour inhaled:* standard treatment (p. 136). *Affected eyes:* standard treatment (p. 136). *If swallowed:* standard treatment (p. 137).
Fire hazard	It has no measurable flash point, but is flammable from 12–19% in air with high energy ignition.
Spillage disposal	Wear face-shield or goggles, and gloves. Apply dispersing agent if available and work to an emulsion with brush and water — run this to waste, diluting greatly with running water. If dispersant not available, absorb on sand, shovel into bucket(s) and transport to safe open area for atmospheric evaporation. Site of spillage should be washed thoroughly with water and soap or detergent.

RSC *Lab. Haz. Data Sheet No. 3*, 1982 gives extended coverage.

N,N-Dichloromethylamine   [7651–91–4]

Hazardous reactions	Exploded on warming with water or on distillation over calcium hypochlorite; exploded on contact with solid sodium sulphide (B155).

1,4-Dichloro-5-nitrobenzene   [89–61–2]
(2,5-dichloronitrobenzene)

Colourless crystals; mp 33 °C; insoluble in water.

(HARMFUL VAPOUR
HARMFUL IN CONTACT WITH SKIN)

Avoid breathing vapour. Avoid contact with skin, eyes and clothing.

Toxic effects These are not well documented but it is assumed that, in common with other substituted benzene compounds of this type, it is irritant to the eyes, skin and respiratory system in vapour form. The liquid or solid would irritate the skin, eyes and respiratory system.

First aid *Vapour inhaled:* standard treatment (p. 136).
Affected eyes: standard treatment (p. 136).
Skin contact: standard treatment (p. 136).
If swallowed: standard treatment (p. 137).

Spillage disposal Wear face-shield or goggles, and gloves. Mix with sand, shovel into bucket(s) and transport to safe open area for burial. Site of spillage should be washed thoroughly with water and soap or detergent.

1,1-Dichloro-1-nitroethane [594–72–9]

(TOXIC BY INHALATION, IN CONTACT WITH SKIN AND IF SWALLOWED)

TLV 2 ppm (10 mg m^{-3}).

2,4-Dichlorophenol [120–83–2]

Colourless crystals; almost insoluble in water.

HARMFUL IF SWALLOWED
IRRITATING TO EYES AND SKIN

Toxic effects Causes severe irritation or burns in contact with the eyes and skin. Assumed

First aid *Affected eyes:* standard treatment (p. 136).
Skin contact: drench with water and swab contaminated skin with glycerol for at least 10 minutes (use water if glycerol is not available); remove and wash contaminated clothing before re-use; if contamination has been other than slight, obtain medical attention.
If swallowed: standard treatment (p. 137).

Spillage disposal Wear face-shield or goggles, and gloves. Mix with sand and transport to safe, open area for burial. Site of spillage should be washed thoroughly with water and soap or detergent.

2,4-Dichlorophenoxyacetic acid [94-75-7]
(2,4-D)

HARMFUL BY INHALATION, IN CONTACT WITH SKIN
AND IF SWALLOWED

RL 10 mg m^{-3}.

1,2-Dichloropropane (propylene dichloride) [78-87-5]

HIGHLY FLAMMABLE
HARMFUL BY INHALATION

TLV 75 ppm (350 mg m^{-3}).

1,3-Dichloropropan-2-ol [96-23-1]
(1,3-dichlorohydrin)

Colourless liquid with ethereal odour; bp 174 °C; sparingly soluble in water.

HARMFUL VAPOUR
IRRITATING TO SKIN AND EYES

Avoid breathing vapour. Avoid contact with skin, eyes and clothing.

Toxic effects The vapour irritates the eyes and respiratory system. Inhalation may cause headache, vertigo, nausea, vomiting and pulmonary oedema. The liquid irritates the skin and eyes. Nausea, vomiting, coma and liver damage may result if it is taken by mouth.

First aid *Vapour inhaled:* standard treatment (p. 136).
Affected eyes: standard treatment (p. 136).
Skin contact: standard treatment (p. 136).
If swallowed: standard treatment (p. 137).

Spillage disposal Wear face-shield or goggles, and gloves. Apply dispersing agent if available and work to an emulsion with brush and water — run this to waste, diluting greatly with running water. If dispersant not available absorb on sand, shovel into bucket(s) and transport to safe, open area for burial. Site of spillage should be washed thoroughly with water and soap or detergent.

Dichloropropene [542-75-6]

RL (skin) 1 ppm (5 mg m^{-3}).

2,2-Dichloropropionic acid [75–99–0]

TLV 1 ppm (6 mg m^{-3}).

Dichlorosilane [4109–96–0]

Hazardous reaction In spite of 70% chlorine content, it may ignite spontaneously in air (B967).

Dichlorotetrafluoroethane [76–14–2]

RL 1000 ppm (7000 mg m^{-3}).

Dicrotonoyl peroxide [93506–63–9]

Hazardous reaction Very shock-sensitive explosive (B716).

Dicyanoacetylene —*see* But-2-ynedinitrile

1,4-Dicyano-2-butene —*see* Hex-3-enedinitrile

Dicyanodiazomethane —*see* Diazomalononitrile

Dicyanomethane —*see* Malononitrile

Dicyclohexylcarbodiimide [538–75–0]

White powder; mp 34 °C; reacts with water to form the urea.

TOXIC BY INHALATION
IRRITATING TO EYES, RESPIRATORY SYSTEM AND SKIN
MAY CAUSE SENSITISATION BY INHALATION OR SKIN CONTACT

Prevent inhalation of vapour or contact of vapour or solid with eyes or skin.

Toxic effects   The vapour irritates the eyes severely, and prolonged use of the compound may lead to sensitisation and subsequent allergic reactions. Assumed to be harmful if taken by mouth, though probably fairly rapidly inactivated by water.

First aid   *Affected eyes:* standard treatment (p. 136).
Skin contact: standard treatment (p. 136).
If swallowed: standard treatment (p. 137).

Spillage disposal   Instruct others to keep at safe distance. Wear breathing apparatus and gloves. Cover liberally with moist sand, leave for several hours with occasional stirring. Shovel up and drop into polythene bucket of water. Allow to stand with occasional stirring for 24 hours, then decant off water, running to waste, diluting greatly with water and washing the sand. Bury the sand in safe open area.

Dicyclohexylcarbonyl peroxide   4904–55–4]

Hazardous reaction   In bulk, may explode without apparent reason (B834).

Dicyclopentadiene   [77–73–6]
(3a,4,7,7a-tetrahydro-4,7-methanoindene)

The commercial product that is commonly used is a colourless liquid with a camphor-like odour; bp about 167 °C; insoluble in water. The pure compound is in the form of colourless crystals, mp 33 °C.

(FLAMMABLE
IRRITATING TO SKIN AND EYES)

Avoid breathing vapour. Avoid contact with skin, eyes and clothing. RL 5 ppm (30 mg m^{-3}).

Toxic effects   These are not well documented, but suppliers advise against the inhalation of vapour and contact with the skin.

First aid   *Vapour inhaled:* standard treatment (p. 136).
Affected eyes: standard treatment (p. 136).
Skin contact: standard treatment (p. 136).
If swallowed: standard treatment (p. 137).

Fire hazard   Flash point 35 °C. *Extinguish fire with* foam, dry powder, carbon dioxide or vaporising liquid.

Spillage disposal   Shut off all possible sources of ignition. Wear face-shield or goggles, and gloves. Apply non-flammable dispersing agent if available and work to an emulsion with brush and water — run this to waste, diluting greatly with running water. If dispersant not available, absorb on sand, shovel into bucket(s) and transport to safe open area for burial. Ventilate site of spillage well to evaporate remaining liquid and dispel vapour.

Dicyclopentadienyliron (ferrocene) [102–54–5]

RL 10 mg m^{-3}.

Dienes Review of class (B1496–1497).

1,1-Diethoxyethane (acetal) [105–57–7]

Colourless, volatile liquid with a pleasant odour; bp 102 °C; sparingly soluble in water; liable to form explosive peroxides on exposure to light and air, which requires that these be decomposed before the ether is distilled to small volume.

HIGHLY FLAMMABLE
IRRITATING TO EYES AND SKIN

Avoid breathing vapour.

Toxic effects	These are not well documented. High concentrations of vapour are liable to cause narcosis when inhaled.
Hazardous reaction	Peroxidised material exploded during distillation (B633).
First aid	*Vapour inhaled in quantity:* standard treatment (p. 136). *Affected eyes:* standard treatment (p. 136). *Skin contact:* standard treatment (p. 136). *If swallowed:* standard treatment (p. 137).
Fire hazard	Flash point −20 °C; explosive limits 1.7–10.4%, ignition temp. 230 °C. *Extinguish fire with* water spray, foam, dry powder, carbon dioxide or vaporising liquid.
Spillage disposal	Shut off all possible sources of ignition. Wear face-shield or goggles, and gloves. Apply non-flammable dispersing agent if available and work to an emulsion with brush and water — run this to waste, diluting greatly with running water. If dispersant not available, absorb on sand, shovel into bucket(s) and transport to safe area for atmospheric evaporation or burial. Ventilate site of spillage well to evaporate remaining liquid and dispel vapour.

Diethylamine [109–89–7]

Colourless liquid with an ammoniacal odour; bp 56 °C; miscible with water.

HIGHLY FLAMMABLE
IRRITATING TO EYES AND RESPIRATORY SYSTEM

Avoid breathing vapour. Avoid contact with skin and eyes. RL 25 (10) ppm (75 mg m^{-3}).

Toxic effects   The vapour irritates the eyes and respiratory system. The liquid irritates the skin and eyes. Assumed to be poisonous if taken by mouth.

First aid   *Vapour inhaled:* standard treatment (p. 136).
Affected eyes: standard treatment (p. 136).
Skin contact: standard treatment (p. 136).
If swallowed: standard treatment (p. 137).

Fire hazard   Flash point below −26 °C; explosive limits 1.8–10.1%; ignition temp. 312 °C. *Extinguish fire with* water spray, foam, dry powder, carbon dioxide or vaporising liquid.

Spillage disposal   Shut off all possible sources of ignition. Instruct others to keep at a safe distance. Wear breathing apparatus and gloves. Mop up with plenty of water and run to waste, diluting greatly with running water. Ventilate area well to evaporate remaining liquid and dispel vapour.

2-Diethylaminoethanol   [100–37–8]

(*N,N*-diethylethanolamine)

Colourless hygroscopic liquid; bp 163 °C; miscible with water.

IRRITATING TO EYES, RESPIRATORY SYSTEM AND SKIN

Avoid breathing vapour. Avoid contact with skin, eyes and clothing. RL (skin) 10 ppm (50 mg m^{-3}).

Toxic effects   The vapour irritates the eyes and respiratory system. The liquid injures the eyes and is absorbed by the skin which may be irritated. The liquid is moderately toxic if taken by mouth.

First aid   *Vapour inhaled:* standard treatment (p. 136).
Affected eyes: standard treatment (p. 136).
Skin contact: standard treatment (p. 136).
If swallowed: standard treatment (p. 137).

Spillage disposal   Shut off all possible sources of ignition. Wear face-shield or goggles, and gloves. Mop up with plenty of water and run to waste, diluting greatly with running water. Ventilate area well to evaporate remaining liquid and dispel vapour.

N,N-Diethylaniline   [91–66–7]

Colourless to brown liquid; bp 216 °C; sparingly soluble in water.

TOXIC BY INHALATION, IN CONTACT WITH SKIN
AND IF SWALLOWED
DANGER OF CUMULATIVE EFFECTS

Avoid breathing vapour. Prevent contact with skin and eyes.

Toxic effects   Excessive breathing of the vapour or absorption of the liquid through the skin can cause headache, drowsiness, cyanosis, mental confusion and, in severe cases, convulsions. The liquid is dangerous to the eyes and the above effects can also be experienced if it is swallowed. **Chronic effects** Continued exposure to the vapour or slight skin exposure to the liquid over a period may affect the nervous system and the blood, causing fatigue, loss of appetite, headache and dizziness.

First aid   *Vapour inhaled:* standard treatment (p. 136).
Affected eyes: standard treatment (p. 136).
Skin contact: standard treatment (p. 136).
If swallowed: standard treatment (p. 137).

**Spillage
disposal**   Wear face-shield or goggles, and gloves. Mix with sand and shovel mixture into glass, enamel or polythene vessel for dispersion in an excess of dilute hydrochloric acid (1 volume of concentrated acid diluted with 2 volumes of water). Allow to stand, with occasional stirring, for 24 hours and then run acid extract to waste, diluting greatly with running water and washing the sand. The sand can be treated as normal refuse.

Diethylarsine   [692–42–4]

**Hazardous
reaction**   Inflames in air (B475).

Diethyl azoformate   [1972–28–7]

**Hazardous
reaction**   Shock-sensitive explosive (B613).

Diethylberyllium   [542–63–2]

**Hazardous
reactions**   Ignites in air; reacts explosively with water (B462).

Diethylcadmium   [592–02–9]

**Hazardous
reaction**   Liable to explode under differing circumstances (B463).

Diethyl carbonate   [105–58–8]

Colourless liquid with pleasant, ethereal smell; bp 126 °C; practically insoluble in water.

(FLAMMABLE
IRRITATING TO EYES AND RESPIRATORY SYSTEM)

Avoid breathing vapour. Avoid contact with eyes.

Toxic effects   The vapour irritates the eyes and respiratory system. The liquid irritates the eyes and is assumed to be irritant and harmful if taken by mouth.

First aid   *Vapour inhaled:* standard treatment (p. 136).
Affected eyes: standard treatment (p. 136).
Skin contact: standard treatment (p. 136).
If swallowed: standard treatment (p. 137).

Fire hazard   Flash point 25 °C. *Extinguish fire with* water spray, foam, dry powder, carbon dioxide or vaporising liquid.

Spillage disposal   Shut off all possible sources of ignition. Wear face-shield or goggles, and gloves. Apply non-flammable dispersing agent if available and work to an emulsion with brush and water — run this to waste, diluting greatly with running water. If dispersant not available, absorb on sand, shovel into bucket(s) and transport to safe open area for atmospheric evaporation or burial. Ventilate site of spillage well to evaporate remaining liquid and dispel vapour.

Diethylene dioxide —*see* Dioxan

Diethylene oximide —*see* Morpholine

Diethylenetriamine   [111–40–0]

Yellow, viscous liquid with ammoniacal smell; bp 208 °C; miscible with water.

HARMFUL IN CONTACT WITH SKIN OR IF SWALLOWED
CAUSES BURNS
MAY CAUSE SENSITISATION BY SKIN CONTACT

Prevent contact with skin and eyes. Avoid breathing vapour. RL (skin) 1 ppm (4 mg m^{-3}).

Toxic effects   As indicated in warning phrases.

First aid   *Vapour inhaled:* standard treatment (p. 136).
Affected eyes: standard treatment (p. 136).
Skin contact: standard treatment (p. 136).
If swallowed: standard treatment (p. 137).

Spillage disposal   Instruct others to keep at a safe distance. Wear breathing apparatus and gloves. Mop up with plenty of water and run to waste, diluting greatly with running water.

N,N-Diethylethanolamine
—*see* 2-Diethylaminoethanol

Diethyl ether (ether; ethyl ether; sulphuric ether)   [60–29–7]

Colourless, highly volatile liquid with characteristic odour, bp 34 °C; immiscible with water; liable to form explosive peroxides on exposure to air and light, which should be decomposed before the ether is distilled to small volume.

> EXTREMELY FLAMMABLE
> MAY FORM EXPLOSIVE PEROXIDES

Avoid breathing vapour. RL 400 ppm (1200 mg m^{-3}).

Toxic effects	Inhalation of vapour may cause drowsiness, dizziness, mental confusion, faintness and, in high concentrations, unconsciousness. Ingestion may also produce these effects. **Chronic effects** Continued inhalation of low concentrations may cause loss of appetite, dizziness, fatigue and nausea. Repeated inhalation or swallowing may lead to 'ether habit', with symptoms resembling chronic alcoholism.
Hazardous reactions	Peroxide formation and subsequent explosion extensively reviewed; powerful oxidants also produce explosive reactions readily; reacts vigorously with sulphuryl chloride (B467–469).
First aid	*Vapour inhaled:* standard treatment (p. 136). *Affected eyes:* standard treatment (p. 136). *If swallowed:* standard treatment (p. 137).
Fire hazard	Flash point −45 °C; explosive limits 1.85–48%; ignition temp. 180 °C. *Extinguish fire with* dry powder, carbon dioxide or vaporising liquid.
Spillage disposal	Shut off all possible sources of ignition. Instruct others to keep at a safe distance. Wear breathing apparatus and gloves. Apply non-flammable dispersing agent if available and work to an emulsion with brush and water — run this to waste, diluting greatly with running water. If dispersant not available, organise effective ventilation of area until the liquid and vapour have been dispersed.

RSC *Lab. Haz. Data Sheet No. 40,* 1985 gives extended coverage.

Di-2-ethyhexyl phthalate
—*see* Di-sec-octyl phthalate

Diethyl ketone —*see* Pentan-3-one

Diethylmagnesium [555-18-6]

Hazardous
reactions
Water ignites solid or ethereal solution, CO_2 ignites solid (B465).

Diethylmethylphosphine [1605-58-9]

Hazardous
reaction
May ignite on long exposure to air (B532).

Diethyl peroxide [628-37-5]

Hazardous
reaction
Explosive (B470).

N,N-Diethyl-*p*-phenylenediamine [93-05-0]
(4-amino-*N,N*-diethylaniline) **and salts**

The free base is a reddish-brown liquid which darkens on exposure to light and air; bp 261 °C; insoluble in water. The hydrochloride and sulphate are buff or grey crystalline powders which darken on exposure to light and air; they are soluble in water.

(HARMFUL IN CONTACT WITH SKIN)

Avoid contact with skin and eyes.

Toxic effects
These are not recorded but may be expected to be similar to those resulting from contact with phenylenediamine, namely eye and skin irritation; dermatitis and more serious eye injury may also result from major contact.

First aid
Affected eyes: standard treatment (p. 136).
Skin contact: standard treatment (p. 136).
If swallowed: standard treatment (p. 137).

Spillage
disposal
Wear face-shield or goggles, and gloves. Mix with sand and shovel mixture into a glass, polythene or enamel vessel for dispersion in an excess of dilute hydrochloric acid (1 volume of concentrated acid diluted with 2 volumes of water). Allow to stand, with occasional stirring, for 24 hours and then run acid extract to waste, diluting greatly with running water and washing the sand. The sand can be treated as normal refuse.

Diethylphosphine [627-49-6]

Hazardous
reaction
Readily ignites in air (B477).

Diethyl phosphite

[762–04–9]

Hazardous reaction   Violent reaction with *p*-nitrophenol in absence of solvent (B477).

Diethyl phthalate

[84–66–2]

RL 5 mg m^{-3}.

Diethyl sulphate (ethyl sulphate)

[64–67–5]

Colourless liquid with faint ethereal odour; bp 209 °C with decomposition; insoluble in water.

HARMFUL IF SWALLOWED
TOXIC BY INHALATION AND IN CONTACT WITH SKIN

Avoid breathing vapour. Avoid contact with skin and eyes.

Toxic effects   Its effect on humans has not been recorded, but authorities consider that animal experiments justify its classification as a dangerous chemical. It is assumed to be poisonous or irritant by inhalation, eye and skin contact and ingestion.

Hazardous reactions   Violent reactions with 2,7-dinitro-9-phenylphenanthridine/water and potassium t-butoxide (B472).

First aid   *Vapour inhaled:* standard treatment (p. 136).
Affected eyes: standard treatment (p. 136).
Skin contact: standard treatment (p. 136).
If swallowed: standard treatment (p. 137).

Spillage disposal   Instruct others to keep at a safe distance. Wear breathing apparatus and gloves. Apply dispersing agent if available and work to an emulsion with brush and water — run this to waste diluting greatly with running water. If dispersant not available, absorb on sand, shovel into bucket(s) and transport to safe open area for burial. Site of spillage should be washed thoroughly with water and soap or detergent.

Diethylzinc

[557–20–0]

Hazardous reactions   Pyrophoric in air; reacts violently with water; violent reaction with diiodomethane/alkene mixtures, methanol, halogens, $AsCl_3$, PCl_3, SO_2, nitro compounds (B473–474).

Difluoramine [10405–27–3]

Hazardous reaction A dangerous explosive (B1060).

Difluoroamino compounds Review of group (B1499).

Difluorodiazene [10578–16–2]

Hazardous reaction Reacts explosively with hydrogen above 90 °C. (B1061).

1,1-Difluoroethylene [75–38–7]

Colourless gas with faint ethereal odour; bp −83 °C; slightly soluble in water.

(FLAMMABLE)

Toxic effects The gas, according to current evidence, has no substantial toxicity but shows the asphyxiant properties of non-toxic gases such as nitrogen.

Hazardous reaction Reaction with HCl may become extremely violent (B233–234)

Fire hazard Explosive limits 5.5–21.3%. Since the gas is supplied in a cylinder, turning off the valve will reduce any fire involving it; if possible, cylinders should be removed quickly from an area in which a fire has developed.

Disposal Surplus gas or leaking cylinder can be vented slowly to air in an open area.

Di-2-furoyl peroxide [25639–45–6]

Hazardous reaction Explodes violently on heating and friction (B762).

Diglycidyl ether —*see* Bis-2,3-epoxypropyl ether

Dihexanoyl peroxide [2400–59–1]

Hazardous reaction Explodes at 85 °C (B812).

1,2-Dihydroxybenzene —*see* Catechol

1,4-Dihydroxybenzene —*see* Hydroquinone

Di-iodoacetylene [624–74–8]

Hazardous reaction   Explodes on friction, impact, and on heating to 84 °C (B311–312).

Di-iodamine [15587–44–7]

Hazardous reaction   Explosive (B1095).

Di-isobutylene —*see* 2,4,4-Trimethylpentene

Di-isobutyl ketone [108–83–8]
(2,6-dimethyl-4-heptanone)

Colourless liquid; bp 169 °C; sparingly soluble in water.

FLAMMABLE
IRRITATING TO RESPIRATORY SYSTEM

Avoid breathing vapour. Avoid contact with eyes. RL 25 ppm (150 mg m^{-3}).

Toxic effects   The vapour may irritate the respiratory system and is narcotic in high concentrations. The liquid irritates the eyes and may irritate the skin; it is assumed to be harmful if taken by mouth.

First aid   *Vapour inhaled in high concentrations:* standard treatment (p. 136).
Affected eyes: standard treatment (p. 136).
Skin contact: standard treatment (p. 136).
If swallowed: standard treatment (p. 137).

Spillage disposal   Shut off all possible sources of ignition. Wear face-shield or goggles, and gloves. Apply non-flammable dispersing agent if available and work to an emulsion with brush and water — run this to waste, diluting greatly with running water. If dispersant not available, absorb on sand, shovel into bucket(s) and transport to safe open area for burial. Ventilate site of spillage well to evaporate remaining liquid and dispel vapour.

Di-isobutyryl peroxide [3437–84–1]

Hazardous reactions	Explodes on standing at room temperature; solution in ether exploded during evaporation (B725).

Di(4-isocyanatophenyl)methane
—see **Bis(4-isocyanatophenyl)methane**

2,4-Di-isocyanatotoluene [584–84–9]
(toluene 2,4-di-isocyanate)

Pale yellow liquid with sharp pungent smell; bp 251 °C; reacts with water with evolution of carbon dioxide; commonly contains about 20% of the 2,6-isomer.

> VERY TOXIC BY INHALATION
> IRRITATING TO EYES, RESPIRATORY SYSTEM AND SKIN
> MAY CAUSE SENSITISATION BY INHALATION

Avoid breathing vapour. Avoid contact with skin, eyes and clothing. CL 0.02 mg m^{-3} (as –NCO).

Toxic effects	The vapour irritates the respiratory system and may cause bronchial asthma. The vapour and liquid are very irritating to the eyes. The liquid irritates the skin and may cause severe dermatitis. Assumed to be highly irritant and poisonous if taken by mouth.
Hazardous reactions	May polymerise vigorously on contact with bases and acyl chlorides, or slowly with gas evolution by diffusion of moisture into polythene containers (B740).
First aid	*Vapour inhaled:* standard treatment (p. 136). *Affected eyes:* standard treatment (p. 136). *Skin contact:* standard treatment (p. 136). *If swallowed:* standard treatment (p. 137).
Spillage disposal	Instruct others to keep at a safe distance. Wear breathing apparatus and gloves. Absorb on sand, shovel into bucket(s), transport to safe open area and tip into large volume of water; leave to decompose before decanting the water to waste, diluting with running water. Site of spillage should be ventilated after washing thoroughly with water and soap or detergent.

Di-isopropylamine [108–18–9]

Colourless, strongly alkaline liquid; bp 84 °C; miscible with water.

HIGHLY FLAMMABLE
IRRITATING TO EYES, RESPIRATORY SYSTEM AND SKIN

Avoid breathing vapor. Avoid contact with skin, eyes and clothing. RL (skin) 5 ppm (20 mg m^{-3}).

Toxic effects   The vapour irritates eyes and respiratory system. The liquid irritates the skin and eyes and may cause burns to the eyes. Assumed to be irritant and poisonous if taken internally.

First aid   *Vapour inhaled:* standard treatment (p. 136).
Affected eyes: standard treatment (p. 136).
Skin contact: standard treatment (p. 136).
If swallowed: standard treatment (p. 137).

Fire hazard   Flash point −1 °C. *Extinguish fire with* dry powder, carbon dioxide or vaporising liquid.

Spillage disposal   Shut off all possible sources of ignition. Instruct others to keep at a safe distance. Wear breathing apparatus and gloves. Mop up with plenty of water and run to waste diluting greatly with running water. Ventilate area well to evaporate remaining liquid and dispel vapour.

Di-isopropyl ether (isopropyl ether)   [108–20–3]

Colourless liquid with ethereal odour; bp 69 °C; sparingly soluble in water. The unstabilised ether readily forms crystalline explosive peroxides on exposure to light and air; these must be destroyed before the ether is distilled.

HIGHLY FLAMMABLE
MAY FORM EXPLOSIVE PEROXIDES

Avoid breathing vapour. Avoid contact with skin and eyes. RL 250 ppm (1050 mg m^{-3}).

Toxic effects   The vapour irritates the respiratory system and eyes and inhalation may lead to headache, dizziness, nausea, vomiting and narcosis. The liquid irritates the eyes causing conjunctivitis; it will defat the skin and may lead to dermatitis. If taken internally, it gives effects similar to those indicated for inhalation of the vapour.

Hazardous reaction   Formation of peroxides responsible for numerous explosions; methods of inhibiting this are reviewed. Vigorous reaction with propionyl chloride (B632–633).

First aid   *Vapour inhaled:* standard treatment (p. 136).
Affected eyes: standard treatment (p. 136).
Skin contact: standard treatment (p. 136).
If swallowed: standard treatment (p. 137).

Fire hazard   Flash point −28 °C; explosive limits 1.4–21%; ignition temp. 443 °C. *Extinguish fire with* dry powder, carbon dioxide or vaporising liquid.

Spillage disposal Shut off all possible sources of ignition. Instruct others to keep at a safe distance. Wear breathing apparatus and gloves. Apply non-flammable dispersing agent if available and work to an emulsion with brush and water — run this to waste, diluting greatly with running water. If dispersant not available, absorb on sand, shovel into bucket(s) and transport to safe open area for atmospheric evaporation. Site of spillage should be washed thoroughly with water and soap or detergent.

Di-isopropyl peroxydicarbonate [105–64–6]

Hazardous reaction Undergoes self-accelerating decomposition when warmed above its melting point (10 °C) which may become dangerously violent; ignites with dimethylaniline (B725–726).

Diketene (4-methyleneoxetan-2-one, diketen) [674–82–8]

Colourless liquid with pungent odour; bp 127 °C; decomposed by water; reacts violently with acids and alkalies.

FLAMMABLE
HARMFUL BY INHALATION

Prevent inhalation of vapour. Prevent contact with skin, eyes and clothing.

Toxic effects The vapour irritates the respiratory system and the eyes severely, causing lachrymation. The liquid irritates the skin and may cause burns; it is assumed to cause severe damage if taken internally.

Hazardous reactions Violent polymerisation is catalysed by acids or bases; residues decomposed on standing (B412–413).

First aid *Vapour inhaled:* standard treatment (p. 136).
Affected eyes: standard treatment (p. 136).
Skin contact: standard treatment (p. 136).
If swallowed: standard treatment (p. 137).

Fire hazard Flash point 46 °C (open cup). *Extinguish fire with* water spray, dry powder, carbon dioxide or vaporising liquids.

Spillage disposal Shut off all possible sources of ignition. Instruct others to keep at a safe distance. Wear breathing apparatus and gloves. Apply non-flammable dispersing agent if available and work to an emulsion with brush and water—run this to waste, diluting greatly with running water. If dispersant not available, absorb on sand, shovel into bucket(s) and transport to safe open area for atmospheric evaporation or burial. Site of spillage should be washed thoroughly with water and soap or detergent.

Dilithium acetylide   [1070–75–3]

Hazardous   Burns brilliantly in chlorine or fluorine; burns vigorously in P, S and Se
reactions   vapours (B313).

Dimercury dicyanide oxide   [1335–31–5]

Hazardous   Heat- and impact-sensitive explosive (B310–311).
reaction

3,3'-Dimethoxybenzidine (*o*-dianisidine)   [119–90–4]
and its dihydrochloride   [20325–40–0]

o-Dianisidine is a colourless to grey-mauve powder; insoluble in water. Its dihydrochloride is a
colourless to grey powder, sparingly soluble in water. The use of *o*-dianisidine and its salts is
controlled in the United Kingdom by The Carcinogenic Substances Regulations 1967. (p. 148)

> VERY TOXIC BY INHALATION, IN CONTACT WITH SKIN
> AND IF SWALLOWED
> DANGER OF CUMULATIVE EFFECTS

Prevent inhalation of dust. Prevent contact with skin and eyes. Human carcinogen.

Toxic effects   The dust irritates the nose severely, causing sneezing. Solutions of the
                       dihydrochloride irritate the eyes. The effects of ingestion are not recorded.
                       Chronic effects There is evidence that *o*-dianisidine, through continued
                       absorption, can cause cancer of the bladder.

First aid   *Dust inhaled:* standard treatment (p. 136).
                       Affected eyes: standard treatment (p. 136).
                       Skin contact: standard treatment (p. 136).
                       If swallowed: standard treatment (p. 137).
Spillage   Wear breathing apparatus and gloves. Mix spillage with moist sand and
disposal   shovel mixture into a glass, enamel or polythene vessel for dispersion in an
                       excess of dilute hydrochloric acid (1 volume of concentrated acid diluted
                       with 2 volumes of water). Allow to stand, with occasional stirring, for 24
                       hours and then run extract to waste, diluting greatly with running water and
                       washing the sand. The residual sand can be treated as normal refuse. The
                       site of the spillage should be washed with water and soap or detergent.

1,2-Dimethoxyethane   [110–71–4]
(ethylene glycol dimethyl ether)

Colourless liquid with sharp, ethereal odour; bp 85 °C; miscible with water; liable to form
explosive peroxides on exposure to air and light which should be decomposed before the ether is
distilled to small volume.

FLAMMABLE
MAY FORM EXPLOSIVE PEROXIDES
HARMFUL BY INHALATION

Avoid breathing vapour. Avoid contact with skin and eyes.

Toxic effects These are not well documented apart from animal experiments which indicate that the vapour is irritant to the respiratory system and that skin and eyes are liable to be affected by contact with the liquid.

First aid *Vapour inhaled:* standard treatment (p. 136).
Affected eyes: standard treatment (p. 136).
Skin contact: standard treatment (p. 136).
If swallowed: standard treatment (p. 137).

Fire hazard Flash point 4.5 °C; *Extinguish fire with* water spray, foam, dry powder, carbon dioxide or vaporising liquids.

Spillage disposal Shut off all possible sources of ignition. Instruct others to keep at a safe distance. Wear breathing apparatus and gloves. Mop up with plenty of water and run to waste diluting greatly with running water. Ventilate area well to evaporate remaining liquid and dispel vapour.

Dimethoxymethane (methylal) [109–87–5]

Colourless, volatile liquid; bp 42 °C; 1 part dissolves in 3 parts water at about 25 °C; may form explosive peroxides on exposure to air and light which should be decomposed before the ether is distilled to small volume.

(EXTREMELY FLAMMABLE
MAY FORM EXPLOSIVE PEROXIDES)

Avoid breathing vapour. Avoid contact with eyes. RL 1000 ppm (3100 mg m^{-3}).

Toxic effects Considered to be of low toxicity though high concentrations may cause narcosis. It has produced injury to lungs, liver, kidneys and heart in experiments on animals.

First aid *Vapour inhaled:* standard treatment (p. 136).
Affected eyes: standard treatment (p. 136).
Skin contact: standard treatment (p. 136).
If swallowed: standard treatment (p. 137).

Fire hazard Flash point −18 °C (open cup); ignition temp. 237 °C. *Extinguish fire with* dry powder, carbon dioxide or vaporising liquid.

Spillage disposal Shut off all possible sources of ignition. Instruct others to keep at a safe distance. Wear breathing apparatus and gloves. Mop up with plenty of water and run to waste, diluting greatly with running water. Ventilate area well to evaporate remaining liquid and dispel vapour.

2,2-Dimethoxypropane [77–76–9]

Hazardous reactions	Violent explosions when dehydration of hydrated manganese and nickel perchlorates was attempted using the ether. As a *gem*-diether it is subject to cool flame behaviour and subsequent explosion (B530).

N,N-Dimethylacetamide [127–19–5]

Colourless liquid, slight basic odour; bp 165 °C; miscible with water.

HARMFUL BY INHALATION AND IN CONTACT WITH SKIN
IRRITATING TO EYES

Avoid contact with eyes and skin. Prevent contact during pregnancy, experimental teratogen. RL (skin) 10 ppm (35 mg m^{-3}).

Toxic effects	Prolonged contact with the skin may lead to absorption and liver damage.
Hazardous reactions	Exothermic reactions with CCl_4, hexachlorocyclohexane, violent in presence of iron (B459).
First aid	*Affected eyes:* standard treatment (p. 136). *Skin contact:* standard treatment (p. 136).
Spillage disposal	Wear face shield or goggles, and gloves. Mop up with pleny of water and run to waste, diluting greatly with running water.

Dimethylamine and solutions [124–40–3]

Colourless gas at ordinary temperatures (bp 7 °C); readily soluble in water. Commonly available in aqueous and ethanolic solution.

EXTREMELY FLAMMABLE
IRRITATING TO EYES AND RESPIRATORY SYSTEM

Avoid breathing vapour. Avoid contact with skin and eyes. RL 10 ppm (18 mg m^{-3}).

Toxic effects	The vapour irritates the mucous membranes and respiratory system; in high concentrations it may affect the nervous system. The vapour and solutions irritate the eyes. The solutions may irritate the skin. Assumed to be poisonous if taken by mouth.
Hazardous reactions	The gas incandesces on contact with fluorine; causes maleic anhydride to decompose exothermically above 150 °C (B301).

First aid	*Vapour inhaled:* standard treatment (p. 136). *Affected eyes:* standard treatment (p. 136). *Skin contact:* standard treatment (p. 136). *If swallowed:* standard treatment (p. 137).
Fire hazard (Gas)	Flash point −50 °C; explosive limits 2.8–14.4%; ignition temp. 402 °C. *Extinguish fire with* water spray, foam, dry powder, carbon dioxide or vaporising liquid. Latter applies also to solutions in water or alcohol.
Spillage disposal	(Liquid and solutions) Shut off all possible sources of ignition. Instruct others to keep at a safe distance. Wear breathing apparatus and gloves. Mop up with plenty of water and run to waste, diluting greatly with running water. Ventilate area well to evaporate remaining liquid and dispel vapour.

Dimethylaminobenzenes —*see* Xylidines

2-Dimethylaminoethanol   [108–01–0]
(*N,N*-dimethylethanolamine)

Colourless liquid; bp 135 °C; miscible with water.

> FLAMMABLE
> IRRITATING TO EYES, RESPIRATORY SYSTEM AND SKIN

Avoid contact with skin, eyes and clothing.

Toxic effects   Little is recorded about these, but it is suggested that splashing of the eyes could cause serious injury.

First aid	*Affected eyes:* standard treatment (p. 136). *Skin contact:* standard treatment (p. 136). *If swallowed:* standard treatment (p. 137).
Fire hazard	Flash point 41 °C (open cup). *Extinguish fire with* water spray, foam, dry powder, carbon dioxide or vaporising liquid.
Spillage disposal	Shut off all possible sources of ignition. Wear face-shield or goggles, and gloves. Mop up with plenty of water and run to waste diluting greatly with running water. Ventilate area well to evaporate remaining liquid and dispel vapour.

N,N-Dimethylaniline   [121–69–7]

Colourless to brown liquid; bp 193 °C; sparingly soluble in water.

> TOXIC BY INHALATION, IN CONTACT WITH SKIN
> AND IF SWALLOWED
> DANGER OF CUMULATIVE EFFECTS

Avoid breathing vapour. Prevent contact with skin and eyes. RL (skin) 5 ppm (25 mg m^{-3}).

303

Toxic effects   Excessive breathing of the vapour or absorption of the liquid through the skin can cause headache, drowsiness, cyanosis, mental confusion and, in severe cases, convulsions. The liquid is dangerous to the eyes and the above effects can also be experienced if it is swallowed. **Chronic effects** Continued exposure to the vapour or slight skin exposure to the liquid over a period may affect the nervous system and the blood causing fatigue, loss of appetite, headache and dizziness.

Hazardous reaction   Contact with a drop causes dibenzoyl peroxide or diisopropyl peroxydicarbonate to explode (B718).

First aid   *Vapour inhaled:* standard treatment (p. 136).
Affected eyes: standard treatment (p. 136).
Skin contact: standard treatment (p. 136).
If swallowed: standard treatment (p. 137).

Spillage disposal   Wear face-shield or goggles, and gloves. Mix with sand and shovel mixture into glass, enamel or polythene vessel for dispersion into an excess of dilute hydrochloric acid (1 volume of concentrated acid diluted with 2 volumes of water). Allow to stand, with occasional stirring, for 24 hours and then run acid extract to waste, diluting greatly with running water and washing the sand. The sand can be treated as normal refuse.

Dimethylantimony chloride [18380–68–2]

Hazardous reaction   Ignites in air at 40 °C (B286).

Dimethylarsine [593–57–7]

Hazardous reaction   Inflames in air (B300).

Dimethylbenzenes —*see* Xylenes

2,2'-Dimethylbenzidine —*see* *o*-Tolidine

3,5-Dimethylbenzoic acid [499–06–9]

Hazardous preparation   Explosion when mesitylene being oxidised with nitric acid in autoclave (B746).

Dimethylberyllium   [506–63–8]

**Hazardous
reactions**   Ignites in moist air or in CO_2; reacts explosively with water (B284).

2,3-Dimethylbuta-1,3-diene   [513–81–5]

**Hazardous
reaction**   Polymeric peroxide autoxidation residue exploded violently on ignition;
reacts explosively with FN=S (B610).

Dimethylcadmium   [506–82–1]

**Hazardous
reaction**   Peroxide formed on exposure to air is explosive (B284–285).

Dimethylcarbamoyl chloride   [79–44–7]
(dimethylcarbamyl chloride)

Colourless lachrymatory liquid; bp 165 °C; decomposed by water.

                 (VERY TOXIC BY INHALATION, IN CONTACT WITH SKIN
                 AND IF SWALLOWED
                 IRRITATING TO EYES AND SKIN)

Prevent inhalation of vapour. Prevent contact with skin and eyes. Suspected carcinogen, no TLV
set.

Toxic effects   The vapour irritates the eyes and respiratory system severely. The liquid
burns the eyes and skin. Assumed to cause severe internal irritation and
damage if taken by mouth.

First aid   *Vapour inhaled:* standard treatment (p. 136).
Affected eyes: standard treatment (p. 136).
Skin contact: standard treatment (p. 136).
If swallowed: standard treatment (p. 137).

**Spillage
disposal**   Instruct others to keep at safe distance. Wear breathing apparatus and
gloves. Mop up cautiously with water and run to waste diluting greatly with
water.

Dimethyl carbonate (methyl carbonate)   [616–38–6]

Colourless liquid with pleasant odour; bp 90 °C; insoluble in water.

305

HIGHLY FLAMMABLE
HARMFUL BY INHALATION, IN CONTACT WITH SKIN
AND IF SWALLOWED

Avoid breathing vapour. Avoid contact with skin, eyes and clothing.

Toxic effects	The vapour irritates the eyes and respiratory system. The liquid irritates the eyes and is assumed to be poisonous if taken internally.
First aid	*Vapour inhaled:* standard treatment (p. 136). *Affected eyes:* standard treatment (p. 136). *Skin contact:* standard treatment (p. 136). *If swallowed:* standard treatment (p. 137).
Fire hazard	Flash point 19 °C (open cup). *Extinguish fire with* water spray, foam, dry powder, carbon dioxide or vaporising liquid.
Spillage disposal	Shut off all possible sources of ignition. Instruct others to keep at a safe distance. Wear breathing apparatus and gloves. Apply non-flammable dispersing agent if available and work to an emulsion with brush and water — run this to waste, diluting greatly with running water. If dispersant not available, absorb on sand, shovel into bucket(s) and transport to safe open area for atmospheric evaporation. Site of spillage should be washed thoroughly with water and soap or detergent.

Dimethyldichlorosilane [75–78–5]

Colourless, fuming liquid which reacts violently with water; bp 70 °C.

HIGHLY FLAMMABLE
IRRITATING TO EYES, RESPIRATORY SYSTEM AND SKIN

Avoid breathing vapour. Avoid contact with skin, eyes and clothing.

Toxic effects	The vapour irritates the eyes and respiratory system. The liquid burns the skin and eyes. The liquid will burn the mouth and alimentary system if taken by mouth.
First aid	*Vapour inhaled:* standard treatment (p. 136). *Affected eyes:* standard treatment (p. 136). *Skin contact:* standard treatment (p. 136). *If swallowed:* standard treatment (p. 137).
Fire hazard	Flash point −9 °C; explosive limits 3.4–9.5%. *Extinguish fire with* dry powder, carbon dioxide, dry sand or earth.
Spillage disposal	Shut off all possible sources of ignition. Instruct others to keep at safe distance. Wear breathing apparatus and gloves. Absorb on sand, shovel into bucket(s), transport to safe open area and tip into large volume of water; leave to decompose before decanting the water to waste, diluting greatly with running water. Site of spillage should be ventilated after washing thoroughly with water and soap or detergent.

N,N-Dimethylethanolamine
—*see* 2-Dimethylaminoethanol

Dimethyl ether (methyl ether)   [115–10–6]

Colourless gas with slight ethereal odour; bp −25 °C; slightly soluble (7% by weight) in water.

EXTREMELY FLAMMABLE LIQUEFIED GAS

Avoid breathing gas.

Toxic effects	The gas is about one fourth as potent as diethyl ether as an anaesthetic but is not used for this purpose because of other toxic effects, notably rushing of blood through the head and sickness.
First aid	*Vapour inhaled:* standard treatment (p. 136).
Fire hazard	Flash point −41 °C; explosive limits 3.4–18%; ignition temp. 350 °C. Since the gas is supplied in a cylinder, turning off the valve will reduce any fire involving it; if possible, cylinders should be removed quickly from an area in which a fire has developed.
Disposal	Surplus gas or leaking cylinder can be vented slowly to air in a safe open area or gas burnt off in a suitable burner.

Dimethylformamide (formdimethylamide)   [68–12–2]

Colourless liquid with faint amine-like odour; bp 153 °C; miscible with water.

HARMFUL IN CONTACT WITH SKIN
IRRITATING TO EYES

Avoid breathing vapour. Avoid contact with skin and eyes. RL (skin) 10 ppm (30 mg m^{-3}).

Toxic effects	The vapour from the hot liquid irritates the eyes and respiratory system. The liquid irritates the skin and eyes. Assumed to be poisonous if taken by mouth. **Chronic effects** Prolonged inhalation of vapour has resulted in liver damage in experimental animals.
Hazardous reactions	Reacts vigorously or violently with a range of materials including Br_2, CCl_4, Cl_2, CrO_3, Na, magnesium nitrate, $KMnO_4$ (B374–375).
First aid	*Vapour inhaled:* standard treatment (p. 136). *Affected eyes:* standard treatment (p. 136). *Skin contact:* standard treatment (p. 136). *If swallowed:* standard treatment (p. 137).

| **Spillage disposal** | Shut off all possible sources of ignition. Wear face-shield or goggles, and gloves. Mop up with plenty of water and run to waste, diluting greatly with running water. |

RSC *Lab. Haz. Data Sheet No. 17*, 1983 gives extended coverage.

2,6-Dimethyl-4-heptanone
—*see* Diisobutyl ketone

1,1-Dimethylhydrazine [57–14–7]

Suspected carcinogen, TLV (skin) 0.5 ppm (1 mg m^{-3}).

| **Hazardous reactions** | Ignites violently with oxidants such as N_2O_4, H_2O_2, HNO_3 (B305). |

Dimethylketen [598–26–5]

| **Hazardous reaction** | Forms extremely explosive peroxide when exposed to air (B427). |

Dimethylmagnesium [2999–74–8]

| **Hazardous reaction** | Water ignites the solid or its ethereal solution (B287). |

Dimethylmercury [593–74–8]

| **Hazardous reaction** | Reacts explosively with diboron tetrachloride (B287). |

N,N-Dimethyl-*p*-nitrosoaniline [138–89–6]
(nitrosodimethylaniline)

Green powder, insoluble in water.

(HARMFUL BY INHALATION
IRRITATING TO EYES, RESPIRATORY SYSTEM AND SKIN)

Avoid breathing dust. Avoid contact with skin and eyes.

Toxic effects   The dust irritates the respiratory system. The dust irritates the eyes and skin and may cause dermatitis. Assumed to be irritant and poisonous if taken by mouth.

Hazardous reaction   Delayed violent reaction with acetic anhydride (B715–716).

First aid   *Dust inhaled:* standard treatment (p. 136).
Affected eyes: standard treatment (p. 136).
Skin contact: standard treatment (p. 136).
If swallowed: standard treatment (p. 137).

Spillage disposal   Wear face-shield or goggles, and gloves. Mix with sand and transport to a safe, open area for burial. Site of spillage should be washed thoroughly with water and soap or detergent.

1,2-Dimethyl-1-nitrosohydrazine   [101672–10–0]

Hazardous reaction   Deflagrates on heating (B302).

Dimethyl peroxide   [690–02–8]

Hazardous reaction   Heat- and shock-sensitive explosive as liquid or vapour (B295).

3,3-Dimethyl-1-phenyltriazene   [7227–91–0]

Hazardous reaction   Exploded on attempted distillation at atmospheric pressure (B718).

Dimethylphosphine   [676–59–5]

Hazardous reaction   Readily ignites in air (B303).

Dimethyl phthalate   [131–11–3]

RL 5 mg m^{-3}.

2,2-Dimethylpropane (neopentane)   [463–82–1]

Colourless liquid or gas; bp 9.5 °C; insoluble in water.

EXTREMELY FLAMMABLE LIQUEFIED GAS

Avoid breathing gas.

Toxic effects	This is classed as a simple asphyxiant, anaesthetic gas of low toxicity which may show irritant and narcotic effects in high concentrations.
First aid	*Vapour inhaled:* standard treatment (p. 136).
Fire hazard	Flash point below −7 °C; explosive limits 1.4–7.5%; ignition temp. 450 °C. Since the gas is supplied in a cylinder, turning off the valve will reduce any fire involving it; if possible, cylinders should be removed from an area in which a fire has developed.
Disposal	Surplus gas or leaking cylinder can be vented slowly to air in a safe, open area or gas burnt off in a suitable gas burner.

Dimethyl selenate   [6918–51–0]

Hazardous reaction	Explodes at about 150 °C when distilled at atmospheric pressure (B298).

Dimethyl sulphate (methyl sulphate)   [77–78–1]

Colourless, odourless liquid; bp 189 °C with decomposition; somewhat soluble in water.

VERY TOXIC BY INHALATION AND IN CONTACT WITH SKIN
POSSIBLE RISKS OF IRREVERSIBLE EFFECTS

Prevent inhalation of vapour. Prevent contact with skin and eyes. Suspected carcinogen. RL (skin) 0.1 ppm (0.5 mg m^{-3}).

Toxic effects	Vapour causes severe irritation of respiratory system, with possible severe lung injury after a latent period. Vapour and liquid irritate or burn the eyes severely after a latent period, resulting in temporary or permanent dimming of vision. The vapour or liquid may blister the skin and skin absorption may result in severe poisoning after a latent period. Extremely poisonous and irritant if taken by mouth.
Hazardous reaction	Reacts violently with concentrated aqueous ammonia; ignites with barium chlorite (B297–298).
First aid	*Vapour inhaled:* standard treatment (p. 136). *Affected eyes:* standard treatment (p. 136). *Skin contact:* standard treatment (p. 136). *If swallowed:* standard treatment (p. 137).

Spillage disposal   Instruct others to keep at a safe distance. Wear breathing apparatus and gloves. Apply dispersing agent if available and work to an emulsion with brush and water — run this to waste, diluting greatly with running water. If dispersant not available, absorb on sand, shovel into bucket(s) and transport to safe open area for burial. Site of spillage should be washed thoroughly with water and soap or detergent.

Dimethyl sulphoxide   [69–68–5]

Colourless, hygroscopic liquid; bp 189 °C; miscible with water.

(HARMFUL SUBSTANCE IF TAKEN INTERNALLY
IRRITATING TO EYES)

Avoid contact with skin or eyes. Solutions of toxic solutes in dimethyl sulphoxide or its mixtures with water may penetrate some synthetic rubber gloves (and the skin) and so promote the toxic effect.

Toxic effects   May cause redness, itching and scaling of skin and damage to eyes. Absorbed readily by skin and volunteers have reported nausea, vomiting, cramps, chills and drowsiness from applications.

Hazardous reactions   Subject to thermal decomposition just above b.p. Reacts violently or explosively with acetyl chloride, benzenesulphonyl chloride, cyanuric chloride, PCl_3, $POCl_3$, $SiCl_4$, SCl_2, S_2Cl_2, $SOCl_2$, N_2O_4, IF_5, magnesium perchlorate, $HClO_4$, HIO_4, sodium hydride, SO_3 or AgF_2; violent reactions or ignition with P_2O_3, K powder, K t-butoxide, Na isopropoxide or trifluoroacetic anhydride; delayed explosion with bromomethane under pressure; complexes with metal nitrates, perchlorates explosive (B291–295).

First aid   *Affected eyes:* standard treatment (p. 136).
Skin contact: standard treatment (p. 136).
If swallowed: standard treatment (p. 137).

Spillage disposal   Wear goggles and gloves. Mop up with plenty of water and run to waste diluting greatly with running water.

RSC *Lab. Haz. Data Sheet No. 11,* 1983 gives extended coverage.

Dimethylzinc   [544–97–8]

Hazardous reactions   Ignites in air and explodes in oxygen (B299).

Di-1-naphthoyl peroxide   [29903–04–6]

Hazardous   Explodes on friction (B858).
reaction

2,4-Dinitroaniline   [97–02–9]

Yellow granules or powder; mp 188 °C; insoluble in water.

> VERY TOXIC BY INHALATION, IN CONTACT WITH SKIN
> AND IF SWALLOWED
> DANGER OF CUMULATIVE EFFECTS

Prevent contact with skin and eyes.

Toxic effects   Records have not been traced, but it can be assumed that skin absorption is
                      liable to cause dermatitis and cyanosis, and that the eyes would be damaged
                      by contact. It must also be assumed to be poisonous if taken by mouth.

Hazardous   Hazardous preparation; a high fire and explosion risk with fast flame
reactions   propagation (B585).

First aid   *Affected eyes:* standard treatment (p. 136).
                      Skin contact: standard treatment (p. 136).
                      If swallowed: standard treatment (p. 137).

Spillage   Wear face-shield or goggles, and gloves. Mix with sand and shovel mixture
disposal   into glass, enamel or polythene vessel for dispersion in an excess of dilute
                      hydrochloric acid (1 volume of concentrated acid diluted with 2 volumes of
                      water). Allow to stand, with occasional stirring, for 24 hours and then run
                      acid extract to waste, diluting greatly with running water and washing the
                      sand. The sand can be disposed of as normal refuse.

1,3-Dinitrobenzene   [99–65–0]

Colourless to yellow crystals; mp 80–90 °C; insoluble in water.

> VERY TOXIC BY INHALATION, IN CONTACT WITH SKIN
> AND IF SWALLOWED
> DANGER OF CUMULATIVE EFFECTS

Avoid breathing vapour. Prevent contact with skin and eyes. RL (skin) 0.15 ppm (1 mg m^{-3}).

Toxic effects   The vapour may cause headache, vertigo and vomiting; in severe cases this
                      may be followed by exhaustion, cyanosis, drowsiness, and unconscious-
                      ness. Contact will damage the eyes, and skin absorption may lead to the
                      above symptoms. Assumed to be poisonous if taken by mouth.

First aid   *Vapour inhaled:* standard treatment (p. 136).
                  Affected eyes: standard treatment (p. 136).
                  Skin contact: standard treatment (p. 136).
                  If swallowed: standard treatment (p. 137).

Spillage   Wear face-shield or goggles, and gloves. Mix with sand and transport to a
disposal   safe, open area for burial. Site of spillage should be washed thoroughly with
                  water and soap or detergent.

Dinitrobenzenes   [528-29-0]

RL (all isomers, skin) 0.15 ppm (1 mg m^{-3}).

Hazardous   Mixtures with concentrated nitric acid possess high explosive properties
reaction   (B1122).

2,4-Dinitrobenzenesulphenyl chloride   [528-76-7]

Hazardous   Must not be overheated as it may explode (B548).
preparation

Dinitrobutenes   1,1,3-[10229-09-1] 2,3,2-[28103-63-6]

Hazardous   Liable to violent decomposition or explosion when heated (B425).
reaction

2,4-Dinitrochlorobenzene
—*see* 1-Chloro-2,4-dinitrobenzene

Dinitro-*o*-cresol   [534-52-1]

Yellow crystals or powder; mp 85 °C almost insoluble in water.

          VERY TOXIC BY INHALATION, IN CONTACT WITH SKIN
          AND IF SWALLOWED
          DANGER OF CUMULATIVE EFFECTS

Prevent inhalation of dust. Prevent contact with skin and eyes. RL (skin) (0.2 mg m^{-3}).

Toxic effects   The inhalation of dust may cause profuse sweating, fever, shortness of
                     breath and yellow coloration of skin of hands and feet. Similar symptoms
                     may follow ingestion and absorption through the skin. Skin contact may
                     cause dermatitis.

First aid *Dust inhaled:* standard treatment (p. 136).
Affected eyes: standard treatment (p. 136).
Skin contact: standard treatment (p. 136).
If swallowed: standard treatment (p. 137).

Spillage disposal Wear face-shield or goggles, and gloves. Mix with sand and transport to a safe, open area for burial. Site of spillage should be washed thoroughly with water and soap or detergent.

Dinitrogen oxide (nitrous oxide) [10204-97-2]

Hazardous reactions Endothermic, with O content 1.7 times that of air. Amorphous boron ignites when heated in the gas; a mixture of phosphine with excess of the oxide can be exploded by sparking; decomposes exothermally if heated locally; explosive mixtures formed with NH_3, CO, H_2, H_2S, PH_3 or other fuels (B1293-1294).

Dinitrogen pentaoxide [10102-03-1]

Hazardous reactions K and Na burn in gas, Hg and As are vigorously oxidised; explodes with naphthalene, other organic materials react vigorously; reacts explosively with sulphur dichloride and sulphuryl chloride (B1301).

Dinitrogen tetraoxide [10102-44-0]
(nitrogen dioxide; nitrous fumes)

Nitrogen dioxide is a red-brown gas which is a common by-product of the reaction of nitric acid with metals and organic materials; it is also available in cylinders as the pure liquefied gas (bp 20 °C).

VERY TOXIC BY INHALATION
IRRITATING TO RESPIRATORY SYSTEM

Avoid breathing gas. RL 3 ppm (5 mg m^{-3}).

Toxic effects Although nitrogen dioxide has some irritant effect upon the respiratory system, its danger lies in the delay before its full effects upon the lungs are shown by feelings of weakness and coldness, headache, nausea, dizziness, abdominal pain and cyanosis; in severe cases, convulsions and asphyxia may follow. (*See* note in Ch 6 p. 103.)

Hazardous reactions Reacts violently or explosively with wide range of materials including acetonitrile, alcohols, liquid ammonia, carbonylmetals, dimethyl sulphoxide, halocarbons, hydrazine and derivatives, hydrocarbons, nitrobenzene, organic compounds, Mg filings, potassium, reduced iron, pyrophoric manganese; violent or explosive reactions or products with hydrocarbons (especially if unsaturated), Me_4Sn, $Ni(CO)_4$ (B1295-1300).

First aid   It is advisable in all cases where appreciable inhalation of nitrogen dioxide is believed to have occurred to obtain medical attention immediately, even if the person exposed is not complaining of discomfort; removal from exposure, followed by rest and warmth are essential until under professional care. In severe cases, administer oxygen through a face mask. If breathing has stopped, apply artificial respiration.

Disposal   Surplus gas or leaking cylinder can be vented slowly into a water-fed scrubbing tower or column in a fume cupboard, or into a fume cupboard served by such a tower.

Dinitrophenols   2,4-[51–28–5]

Yellow crystals or powder; sparingly soluble in water.

> TOXIC BY INHALATION, IN CONTACT WITH SKIN
> AND IF SWALLOWED
> DANGER OF CUMULATIVE EFFECTS

Prevent inhalation of dust. Prevent contact with skin and eyes.

Toxic effects   The inhalation of dust may cause profuse sweating, fever, shortness of breath and yellow coloration of skin of hands and feet. Similar symptoms may follow ingestion and absorption through the skin. Skin contact may cause dermatitis. (*See* note in Ch 6 p. 104.)

First aid   *Dust inhaled:* standard treatment (p. 136).
Affected eyes: standard treatment (p. 136).
Skin contact: drench with water and swab contaminated skin with glycerol for at least 10 minutes (use water if glycerol is not available); remove and wash contaminated clothing before re-use; if contamination has been other than slight, obtain medical attention.
If swallowed: standard treatment (p. 137).

Spillage disposal   Wear face-shield or goggles, and gloves. Mix with sand and transport to a safe, open area for burial. Site of spillage should be washed thoroughly with water and soap or detergent.

2,4-Dinitrophenylhydrazine   [119–26–6]

Red crystalline powder; mp about 200 °C; slightly soluble in water which is normally added in storage to reduce explosion risk.

> (RISK OF EXPLOSION BY SHOCK, FRICTION, HEAT OR OTHER
> SOURCES OF IGNITION
> HARMFUL BY INHALATION, IN CONTACT WITH SKIN AND IF
> SWALLOWED)

Avoid inhalation of dust. Avoid contact with skin and eyes.

Toxic effects No effects have been recorded, nor testing carried out, but the relationship with phenylhydrazine and presence of two nitro groups indicate that the above warning is likely to be justified.

First aid *Dust inhaled:* standard treatment (p. 136).
Affected eyes: standard treatment (p. 136).
Skin contact: standard treatment (p. 136).
If swallowed: standard treatment (p. 137).

Spillage disposal Wear face-shield or goggles and gloves. Mix with sand and transport to a safe, open area for burial. Site of spillage should be washed thoroughly with water and detergent.

3,5-Dinitro-*o*-toluamide [148–01–6]

TLV 5 mg m^{-3}.

2,4-Dinitrotoluene [121–14–2]

Yellow crystals; insoluble in water.

TOXIC BY INHALATION, IN CONTACT WITH SKIN
AND IF SWALLOWED
DANGER OF CUMULATIVE EFFECTS

Prevent contact with skin and eyes. RL (skin) 1.5 mg m^{-3}.

Toxic effects The vapour and crystals irritate the eyes. Contact with the skin may cause dermatitis. Assumed to be poisonous if taken by mouth.

Hazardous reaction Prolonged heating with alkali at >150 °C may cause explosion (B668).

First aid *Affected eyes:* standard treatment (p. 136).
Skin contact: standard treatment (p. 136).
If swallowed: standard treatment (p. 137).

Spillage disposal Wear face-shield or goggles, and gloves. Mix with sand and transport to a safe, open area for burial. Site of spillage should be washed thoroughly with water and soap or detergent.

Di-sec-octyl phthalate [111–81–7]
(di-2-ethylhexyl phthalate)

RL 5 mg m^{-3}.

Dioxan (diethylene dioxide; diethylene oxide; dioxane) [123-91-1]

Colourless, almost odourless, liquid; bp 101 °C; miscible with water; liable to form explosive peroxides on exposure to light and air which should be decomposed before the ether is distilled to small volume.

HIGHLY FLAMMABLE
MAY FORM EXPLOSIVE PEROXIDES
HARMFUL BY INHALATION

Avoid breathing vapour. RL (skin) 50 (25) ppm (180 mg m^{-3}).

Toxic effects The vapour irritates nose and eyes and this may be followed by headache and drowsiness. High concentrations may also cause nausea and vomiting, while injury to the kidney and liver are possible. Shows some of the above effects when taken by mouth.

Hazardous reactions Forms explosive peroxides on exposure to air; reacts almost explosively with Raney nickel above 210 °C; addition complex with sulphur trioxide decomposes violently on storage; explosion during distillation from LiAlH$_4$ and during reaction with HClO$_4$/HNO$_3$ (B450–451).

First aid *Vapour inhaled:* standard treatment (p. 136).
Affected eyes: standard treatment (p. 136).
If swallowed: standard treatment (p. 137).

Fire hazard Flash point 12 °C; explosive limits 2–22%; ignition temp. 180 °C. *Extinguish fire with* water spray, dry powder, carbon dioxide or vaporising liquid.

Spillage disposal Shut off all possible sources of ignition. Instruct others to keep at a safe distance. Wear breathing apparatus and gloves. Mop up with plenty of water and run to waste, diluting greatly with running water. Ventilate area well to evaporate remaining liquid and dispel vapour.

RSC *Lab. Haz. Data Sheet No. 29,* 1984 gives extended coverage.

Dioxygen difluoride [7782-44-0]

Hazardous reactions Very powerful oxidant reacting with many materials at low temperatures (B1064).

Dioxygenyl polyfluorosalts Review of group (B1500).

Dipentene (DL-limonene; *p*-mentha-1,8-diene) [138-86-3]

Colourless liquid with pleasant, lemon-like odour; bp 178 °C; insoluble in water.

FLAMMABLE
IRRITATING TO SKIN

Avoid contact with skin, eyes and clothing.

Toxic effects	It is considered to be a moderate skin irritant and sensitiser, but its effects are not well documented.
First aid	*Affected eyes:* standard treatment (p. 136).
	Skin contact: standard treatment (p. 136).
	If swallowed: standard treatment (p. 137).
Fire hazard	Flash point 45 °C; ignition temp. 237 °C. *Extinguish fire with* foam, dry powder, carbon dioxide or vaporising liquid.
Spillage disposal	Shut off all possible sources of ignition. Wear face-shield or goggles, and gloves. Apply non-flammable dispersing agent if available and work to an emulsion with brush and water — run this to waste diluting greatly with running water. If dispersant not available, absorb on sand, shovel into bucket(s) and transport to safe open area for burial. Ventilate site of spillage well to evaporate remaining liquid and dispel vapour.

Diphenyl —*see* Biphenyl

Diphenylamine [122–39–4]

Colourless or grey crystals, mp 54 °C; insoluble in water.

TOXIC BY INHALATION, IN CONTACT WITH SKIN
AND IF SWALLOWED
DANGER OF CUMULATIVE EFFECTS

RL 10 mg m^{-1}

Toxic effects	May irritate mucous membranes or skin. Toxic action similar to aniline, but less severe.
First aid	*Skin contact:* standard treatment (p. 136).
	If swallowed: standard treatment (p. 137).
Spillage disposal	Instruct others to keep at safe distance. Wear face mask or goggles, and gloves. Mix with sand and shovel up mixture into glass, enamel or polythene vessel for dispersion into an excess of dilute HCl (1 volume of concentrated acid diluted with 2 volumes of water). Allow to stand, with occasional stirring, for 24 hours and then run the acid extract to waste, diluting greatly with running water and washing the sand. The sand can be treated as normal refuse.

Diphenyldistibene [5702–61–4]

Hazardous reactions Ignites in air and is oxidised explosively by nitric acid (B804).

Diphenyl ether [101–84–8]

TLV (vapour) 1 ppm (7 mg m^{-3}).

1,1-Diphenylethylene [530–48–3]

Hazardous reaction Forms explosive peroxide with oxygen (B830).

Diphenylmagnesium

Hazardous reactions Ignites in moist air; reacts violently with water (B801).

Diphenylmercury [587–85–9]

Hazardous reactions Reacts violently with sulphur trioxide or dichlorine monoxide (B800).

Diphenylmethane diisocyanate
—*see* Methylenebis-4-phenyl isocyanate

Diphenyltin [6381–06–2]

Hazardous reaction Ignites on contact with fuming HNO$_3$ (B804).

1,3-Diphenyltriazene [136–35–6]

Hazardous reactions Decomposes explosively at mp 98 °C; mixture with acetic anhydride exploded on warming (B805).

Diphosphane [13445-50-6]

Hazardous reactions   Ignites in air; when present 0.2% v/v causes other flammable gases to ignite in air (B1203).

Dipotassium acetylide [22754-96-7]

Hazardous reaction   Contact with water may cause ignition and explosion of evolved acetylene (B312).

Dipropionyl peroxide [3248-28-0]

Hazardous reaction   Explodes on standing at room temperature (B616).

Dipropylene glycol methyl ether [34590-94-8]

TLV (skin) 100 ppm (600 mg m^{-3}).

Dipyridinesodium [101697-88-5]

Hazardous reaction   This addition product of sodium and pyridine ignites in air (B769).

Dirubidium acetylide [22754-97-8]

Hazardous reactions   Ignited by concentrated hydrochloric acid; burns in halogens; may react violently or explosively with some metal oxides; ignites on warming in CO_2, NO or SO_2; ignites with As, burns in S or Se vapour; reacts vigorously with B or Si (B321).

Disilane [1590-87-0]

Hazardous reactions   Explodes on contact with Br_2, CCl_4, or SF_6; ignites spontaneously in air (B1211).

Disilver acetylide [7659-31-6]

Hazardous reaction   Powerful detonator which will initiate explosive acetylene-containing gas mixtures (B196).

Disodium acetylide [2881-62-1]

Hazardous reactions Burns in chlorine and probably in fluorine, and in contact with bromine and iodine on warming; violent reactions when ground with finely divided lead, aluminium, iron or mercury; incandesces in CO_2, or SO_2; ignites on warming in oxygen; burns vigorously in P vapour; rubbing in mortar with some salts (chlorides, iodides and sulphates — and probably nitrates) results in vigorous reactions or explosions (B319–320).

Disulphur dichloride [10025-67-9]

Yellow-brown, fuming liquid with a suffocating odour; bp 136 °C; decomposed by water with formation of hydrochloric acid, sulphur dioxide and sulphur.

> REACTS VIOLENTLY WITH WATER
> CAUSES BURNS
> IRRITATING TO RESPIRATORY SYSTEM

Avoid breathing vapour. Prevent contact with skin and eyes. RL 1 ppm (6 mg m^{-3}).

Toxic effects The vapour irritates the respiratory system. The vapour and liquid irritate the eyes and skin severely and the liquid may cause burns. Its ingestion would result in severe internal irritation and damage.

Hazardous reactions Vigorous or violent reactions with Sb, Sb or As sulphides, chromyl chloride, dimethyl sulphoxide, P_2O_3, HgO, Na_2O_2, water and a number of organic compounds; a mixture with potassium is shock-sensitive and explodes on heating (B985).

First aid *Vapour inhaled:* standard treatment (p. 136).
Affected eyes: standard treatment (p. 136).
Skin contact: standard treatment (p. 136).
If swallowed: standard treatment (p. 137).

Spillage disposal Instruct others to keep at a safe distance. Wear breathing apparatus and gloves. Spread soda ash liberally over the spillage and mop up cautiously with water — run this to waste, diluting greatly with running water.

Disulphur dinitride [25474-92-4]

Hazardous reactions Explosion initiated by shock, friction, pressure and heat (B1303).

Disulphur heptaoxide [12065-85-9]

Hazardous reaction Explodes in moist air (B1374).

Disulphuryl diazide   [73506–23–7]

Hazardous   Explodes below 80 °C; with dilute alkali at 0 °C an explosive deposit is
reaction   formed (B1311).

Disulphuryl dichloride   [7791–27–7]

Hazardous   Reacts vigorously with red phosphorus; reaction with water can be violent
reactions   (B981).

Disulphuryl difluoride   [13036–75–4]

Hazardous   Reacts violently with ethanol (B1066).
reaction

Divinylbenzene   [108–57–6]

RL 10 ppm (50 mg m^{-3}).

Dodecanoyl peroxide (lauroyl peroxide)   [105–74–8]

White powder; mp 55 °C; decomposes slowly above 60 °C, especially in sunlight; insoluble in
water.

HIGHLY FLAMMABLE
IRRITATING TO EYES, RESPIRATORY SYSTEM AND SKIN

Avoid contact with skin, eyes and clothing.

Toxic effects   It irritates and may burn the skin and eyes; assumed to be irritant and harmful
if taken by mouth.

Hazardous   Becomes shock-sensitive on heating (B862).
reaction

First aid   *Affected eyes:* standard treatment (p. 136).
Skin contact: standard treatment (p. 136).
If swallowed: standard treatment (p. 137).

Spillage   Wear face-shield or goggles, and gloves. Mix with sand and transport to a
disposal   safe, open area for burial. Site of spillage should be washed thoroughly with
water.

Endothermic compounds Review of group (B1505–1508).

Epichlorohydrin
—*see* 1-Chloro-2,3-epoxypropane

1-Epoxyethyl-3,4-epoxycyclohexane [106–87–6]
(vinylcyclohexene dioxide)

Suspected carcinogen RL 10 ppm (60 mg m^{-3}).

1,2-Epoxypropane —*see* Propylene oxide

2,3-Epoxy-1-propanol (glycidol) [556–52–5]

TLV 25 ppm (75 mg m^{-3}).

Hazardous reaction   May polymerize explosively if stored inappropriately (B369).

2,3-Epoxypropyl isopropyl ether [4016–14–2]
(isopropyl glycidyl ether, IGE)

RL 50 ppm (240 mg m^{-3}).

2,3-Epoxypropyl phenyl ether [122–60–1]
(phenyl glycidyl ether, PGE)

TLV 1 ppm (6 mg m^{-3}).

Ethane [74–84–0]

Colourless gas; bp −89 °C; insoluble in water.

EXTREMELY FLAMMABLE

Avoid breathing gas.

Toxic effects   Considered to be a simple asphyxiant gas which can act as an anaesthetic at high concentrations.

First aid   *Gas inhaled in quantity:* standard treatment (p. 136).
Fire hazard   Explosive limits 3–12.5%; ignition temp. 515 °C. Since the gas is supplied in a cylinder, turning off the valve will reduce any fire involving it; if possible, cylinders should be removed quickly from an area in which a fire has developed.
Disposal   Surplus gas or leaking cylinder can be vented slowly to air in a safe, open area or gas burnt off through a suitable burner.

Ethane-1,2-diol (ethylene glycol) [107–21–1]

Colourless syrupy liquid with sweetish taste; miscible with water; bp 198 °C. It is the main constituent of 'anti-freeze' mixtures.

HARMFUL IF SWALLOWED

RL (vapour) 60 mg m^{-3}.
RL (mist) 10 mg m^{-3}.

Toxic effects   Death has followed the drinking of ethanediol as a substitute for spirits; 100 cm^3 may prove fatal. Smaller quantities may result in restlessness, unsteady gait, drowsiness, coma, and injury to the kidneys.

Hazardous reactions   Product of reaction with perchloric acid explodes on addition of water; explosion occurred when it was heated with P_2S_5 in hexane; may ignite in contact with silvered-copper d.c. conductors (B296).

First aid   *If swallowed:* standard treatment (p. 136).
Spillage disposal   Mop up with plenty of water and run to waste, diluting greatly with running water.

Ethane-1,2-diol dinitrate [628–96–6]
(1,2-ethylidene nitrate, nitroglycol)

RL (skin) 0.2 ppm (1.2 mg m^{-3}).

Hazardous reaction   It is an explosive, and a sample of the isomeric 1,1-dinitrate [55044–04–7] detonated during analytical combustion (B261).

Ethanethiol (ethyl mercaptan) [75–08–1]

Colourless liquid with penetrating garlic-like odour; bp 35 °C; sparingly soluble in water (6.8 g in one litre at 20 °C).

HIGHLY FLAMMABLE
HARMFUL BY INHALATION

Avoid breathing vapour. Avoid contact with eyes and skin. RL 0.5 ppm (1 mg m^{-3}).

Toxic effects The vapour irritates the respiratory system and may be narcotic in high concentrations. The liquid irritates the eyes and mucous membranes; it is assumed to be poisonous if taken by mouth.

First aid *Vapour inhaled:* standard treatment (p. 136).
Affected eyes: standard treatment (p. 136).
Skin contact: standard treatment (p. 136).
If swallowed: standard treatment (p. 137).

Fire hazard Flash point below 21 °C; explosive limits 2.8–18%; ignition temp. 299 °C. *Extinguish fire with* water spray, foam, dry powder, carbon dioxide or vaporising liquid.

Spillage disposal Shut off all possible sources of ignition. Instruct others to keep at a safe distance. Wear breathing apparatus and gloves. Apply non-flammable dispersing agent if available and work to an emulsion with brush and water — run this to waste, diluting greatly with running water. If dispersant not available, absorb on sand, shovel into bucket(s) and transport to safe open area for atmospheric evaporation or burial. Site of spillage should be washed thoroughly with water and soap or detergent.

Ethanol (ethyl alcohol) [46–17–5]

Colourless, mobile liquid with characteristic smell, bp 79 °C; miscible with water. Ethanol is supplied to laboratories in a variety of forms mainly because of Excise control. In the United Kingdom, 'duty paid' material comes as 90% (rectified spirit), 95% and 99/100% (absolute) grades; denatured ethanol is sold as industrial methylated spirit (supplied against Excise requisitions) and methylated spirit (mineralised). The materials used for denaturing ethanol add substantially to its toxicity and are not taken into account in this monograph which deals with pure (99/100%) material.

HIGHLY FLAMMABLE

Avoid breathing vapour in high concentrations. RL 1000 ppm (1900 mg m^{-3}).

Toxic effects The intoxicating qualities of ethanol are so well appreciated that a stark summary of them is superfluous.

Hazardous reactions Reacts with varying degrees of violence with a wide range of oxidants, including silver nitrate with which an explosion was reported (B289–291).

First aid *Affected eyes:* standard treatment (p. 136).

Fire hazard Flash point 12 °C; explosive limits 3.3–19%; ignition temp. 423 °C. *Extinguish fire with* water spray, dry powder, carbon dioxide or vaporising liquid.

Spillage **disposal**	Shut off all possible sources of ignition. Wear face-shield or goggles, and gloves. Mop up with plenty of water and run to waste diluting greatly with running water. Ventilate area well to evaporate remaining liquid and dispel vapour.

Ethanolamine —*see* 2-Aminoethanol

Ethene —*see* Ethylene

Ether —*see* Diethyl ether

Ethers Review of peroxidation hazards (B1509].

Ethoxyacetylene   [927–80–0]

Hazardous **reaction**	Explodes when heated at around 100 °C in sealed tubes (B428).

Ethoxyanilines (phenetidines)   *o*-[94–70–2] *p*-[156–43–4]

o- and *p*-Phenetidines are red to brown liquids; bp 229 °C and 254 °C respectively; immiscible with water.

TOXIC BY INHALATION, IN CONTACT WITH SKIN
AND IF SWALLOWED
DANGER OF CUMULATIVE EFFECTS

Avoid breathing vapour. Avoid contact with skin and eyes.

Toxic effects	These are not recorded, but are assumed to be similar to those of aniline poisoning, *i.e.* headache, drowsiness and cyanosis.
First aid	*Vapour inhaled:* standard treatment (p. 136). *Affected eyes:* standard treatment (p. 136). *Skin contact:* standard treatment (p. 136). *If swallowed:* standard treatment (p. 137).
Spillage **disposal**	Wear face-shield or goggles, and gloves. Mix with sand and shovel mixture into glass, enamel or polythene vessel for dispersion in an excess of dilute hydrochloric acid (1 volume of concentrated acid diluted with 2 volumes of water). Allow to stand, with occasional stirring, for 24 hours and then run acid extract to waste, diluting greatly with running water and washing the sand. The sand can be disposed of as normal refuse.

2-Ethoxyethanol [110–80–5]
(ethylene glycol monoethyl ether; cellosolve)

Colourless liquid; bp 135 °C; miscible with water; liable to form explosive peroxides on exposure to light and air which should be decomposed before the ether is distilled to small volume.

FLAMMABLE
IRRITATING TO EYES

Avoid breathing vapour. Avoid contact with skin, eyes and clothing. RL (skin) 100 ppm (370 mg m^{-3}).

Toxic effects	The vapour irritates the eyes and respiratory system when in high concentrations. The liquid irritates the eyes. The liquid is poisonous if taken by mouth and may cause blood and kidney damage.
First aid	*Vapour inhaled:* standard treatment (p. 136). *Affected eyes:* standard treatment (p. 136). *Skin contact:* standard treatment (p. 136). *If swallowed:* standard treatment (p. 137).
Fire hazard	Flash point 40 °C; explosive limits 1.7–15.6%; ignition temp. 238 °C. *Extinguish fire with* water spray, foam, dry powder, carbon dioxide or vaporising liquid.
Spillage disposal	Shut off all possible sources of ignition. Wear face-shield or goggles, and gloves. Mop up with plenty of water and run to waste, diluting greatly with running water. Ventilate area well to evaporate remaining liquid and dispel vapour.

2-Ethoxyethyl acetate [111–15–9]
(ethylene glycol monoethyl ether monoacetate)

Colourless liquid with pleasant ester-like odour; bp 156 °C; sparingly soluble in water; liable to form explosive peroxides on exposure to air and light which should be decomposed before the ether ester is distilled to small volume.

FLAMMABLE
HARMFUL BY INHALATION AND IN CONTACT WITH SKIN

Avoid breathing vapour. Avoid contact with skin, eyes and clothing. RL (skin) 100 ppm (540 mg m^{-3}).

Toxic effects	The vapour irritates the eyes and respiratory system in high concentrations. The liquid irritates the eyes. The liquid is poisonous if taken by mouth and may cause blood and kidney damage.
First aid	*Vapour inhaled:* standard treatment (p. 136). *Affected eyes:* standard treatment (p. 136). *Skin contact:* standard treatment (p. 136). *If swallowed:* standard treatment (p. 137).

Fire hazard　Flash point 51 °C; explosive limits 1.2–13%; ignition temp. 379 °C. *Extinguish fire with* foam, dry powder, carbon dioxide or vaporising liquid.

Spillage disposal　Shut off all possible sources of ignition. Wear face-shield or goggles, and gloves. Apply non-flammable dispersant and work to an emulsion with brush and water — run this to waste, diluting greatly with running water. If dispersant not available, absorb on sand, shovel into bucket(s) and transport to safe, open area for atmospheric evaporation or burial.

3-Ethoxypropyne (1-ethoxy-2-propyne) [628-33-1]

Hazardous reaction　Exploded during distillation under vacuum (B512).

Ethyl acetate (ethyl ethanoate) [141-78-6]

Colourless, volatile liquid with fragrant odour; bp 77 °C; 1 part soluble in about 35 parts water at 25 °C.

HIGHLY FLAMMABLE

Avoid breathing vapour. Avoid contact with eyes. RL 400 ppm (1400 mg m^{-3}).

Toxic effects　The vapour may irritate the eyes and respiratory system. The liquid irritates the eyes and mucous surfaces. Prolonged inhalation may cause kidney and liver damage.

Hazardous reaction　May ignite or explode with solid LiAlH$_4$ or K t-butoxide (B451).

First aid　*Vapour inhaled:* standard treatment (p. 136).
Affected eyes: standard treatment (p. 136).
Skin contact: standard treatment (p. 136).
If swallowed: standard treatment (p. 000).

Fire hazard　Flash point −4.4 °C; explosive limits 2.5–11.5%; ignition temp. 427 °C. *Extinguish fire with* water spray, foam, dry powder, carbon dioxide or vaporising liquid.

Spillage disposal　Shut off all possible sources of ignition. Wear face-shield or goggles, and gloves. Apply non-flammable dispersing agent if available and work to an emulsion with brush and water — run this to waste, diluting greatly with running water. If dispersant not available, absorb on sand, shovel into bucket(s) and transport to safe open area for atmospheric evaporation. Ventilate site of spillage well to evaporate remaining liquid and dispel vapour.

RSC *Lab. Haz. Data Sheet No. 41,* 1985 gives extended coverage.

Ethylacetylene —*see* But-1-yne

Ethyl acrylate [140–88–5]

Colourless liquid with acrid, penetrating odour; bp 100 °C; immiscible with water; normally supplied containing stabiliser.

HIGHLY FLAMMABLE
HARMFUL BY INHALATION AND IF SWALLOWED
IRRITATING TO EYES, RESPIRATORY SYSTEM AND SKIN
MAY CAUSE SENSITISATION BY SKIN CONTACT

Avoid breathing vapour. Avoid contact with skin and eyes. RL (skin) 25 (5) ppm (100 mg m^{-3}).

Toxic effects The vapour irritates the eyes and respiratory system. High concentrations may result in lethargy and convulsions. The liquid irritates the skin and eyes. Animal experiments indicate high degree of toxicity if taken by mouth.

Hazardous reaction Polymerisation in a large clear glass bottle burst it (B516).

First aid *Vapour inhaled:* standard treatment (p. 136).
Affected eyes: standard treatment (p. 136).
Skin contact: standard treatment (p. 136).
If swallowed: standard treatment (p. 137).

Fire hazard Flash point 16 °C; ignition temp. 273 °C. *Extinguish fire with* water spray, foam, dry powder, carbon dioxide or vaporising liquid.

Spillage disposal Shut off all possible sources of ignition. Instruct others to keep at a safe distance. Wear breathing apparatus and gloves. Apply non-flammable dispersing agent if available and work to an emulsion with brush and water — run this to waste diluting greatly with running water. If dispersant not available, absorb on sand, shovel into bucket(s) and transport to safe, open area for burial. Site of spillage should be washed thoroughly with water and soap or detergent.

Ethyl alcohol —*see* Ethanol

Ethylamine and solutions [75–04–7]

Colourless volatile liquid with a fishy ammoniacal odour; bp 17 °C. The solutions in water and ethanol are commonly used.

EXTREMELY FLAMMABLE LIQUEFIED GAS
IRRITATING TO EYES AND RESPIRATORY SYSTEM

Avoid breathing vapour. Avoid contact with skin and eyes. RL 10 ppm (18 mg m^{-3}).

Toxic effects The vapour irritates the mucous membranes and respiratory system; in high concentrations it may affect the nervous system. The vapour and solutions irritate the eyes. The solutions may irritate the skin. Assumed to be irritant and poisonous if taken by mouth.

329

First aid	*Vapour inhaled:* standard treatment (p. 136).
	Affected eyes: standard treatment (p. 136).
	Skin contact: standard treatment (p. 136).
	If swallowed: standard treatment (p. 137).
Fire hazard	Flash point below −18 °C; explosive limits 3.5–14%; ignition temp. 384 °C.
	Extinguish fire with water spray or dry powder.
Spillage disposal	Shut off all possible sources of ignition. Instruct others to keep at a safe distance. Wear breathing apparatus and gloves. Mop up with plenty of water and run to waste, diluting greatly with running water. Ventilate area well to evaporate remaining liquid and dispel vapour.

Ethyl amyl ketone — *see* 5-Methyl-3-heptanone

N-Ethylaniline [103–69–5]

Colourless to brown liquid; bp 204 °C; insoluble in water.

TOXIC BY INHALATION, IN CONTACT WITH SKIN
AND IF SWALLOWED
DANGER OF CUMULATIVE EFFECTS

Avoid breathing vapour. Prevent contact with skin and eyes.

Toxic effects	Excessive breathing of the vapour or absorption of the liquid through the skin can cause headache, drowsiness, cyanosis, mental confusion and, in severe cases, convulsions. The liquid damages the eyes, and the above effects can be expected if it is swallowed. **Chronic effects** Continued exposure to the vapour, or slight skin exposure to the liquid, over a period may affect the nervous system and the blood, causing fatigue, loss of appetite, headache and dizziness.
First aid	*Vapour inhaled:* standard treatment (p. 136).
	Affected eyes: standard treatment (p. 136).
	Skin contact: standard treatment (p. 136).
	If swallowed: standard treatment (p. 137).
Spillage disposal	Wear face-shield or goggles, and gloves. Mix with sand and shovel mixture into glass, enamel or polythene vessel for dispersion in an excess of dilute hydrochloric acid (1 volume of concentrated acid diluted with 2 volumes of water). Allow to stand, with occasional stirring, for 24 hours and then run acid extract to waste, diluting greatly with running water and washing the sand. The sand can be disposed of as normal refuse.

Ethyl azide [871–31–8]

Hazardous reaction	May detonate on rapid heating (B280).

Ethyl azidoformate   [817–87–8]

**Hazardous   Liable to explode at 114 °C (B360).
reaction**

Ethylbenzene   [100–41–4]

Colourless liquid with aromatic odour; bp 136 °C; immiscible with water.

HIGHLY FLAMMABLE
IRRITATING TO RESPIRATORY SYSTEM

Avoid breathing vapour. Avoid contact with skin and eyes. RL 100 ppm (435 mg m^{-3}).

Toxic effects   The vapour irritates the eyes and respiratory system and may cause
                       dizziness. The liquid irritates the skin and eyes. Assumed to be poisonous if
                       taken by mouth.

First aid   *Vapour inhaled:* standard treatment (p. 136).
                       Affected eyes: standard treatment (p. 136).
                       Skin contact: standard treatment (p. 136).
                       If swallowed: standard treatment (p. 137).

Fire hazard   Flash point 15 °C; explosive limits 1–6.7%; ignition temp. 432 °C; *Extinguish
                       fire with* foam, dry powder, carbon dioxide or vaporising liquid.

Spillage   Shut off all possible sources of ignition. Instruct others to keep at a safe
disposal   distance. Wear breathing apparatus and gloves. Apply non-flammable
                       dispersing agent if available and work to an emulsion with brush and
                       water — run this to waste, diluting greatly with running water. If dispersant
                       not available, absorb on sand, shovel into bucket(s) and transport to safe,
                       open area for burial. Site of spillage should be washed thoroughly with
                       water and soap or detergent.

Ethyl bromide —*see* Bromoethane

Ethyl butyl ketone —*see* Heptan-3-one

Ethyl butyrate   [105–54–4]

Colourless liquid with pineapple-like odour; bp 121 °C; sparingly soluble in water.

(FLAMMABLE)

Avoid contact with skin and eyes.

Toxic effects	In high concentrations the vapour irritates the eyes and respiratory system, and is narcotic.
First aid	*Vapour inhaled in high concentrations:* standard treatment (p. 136).
	Affected eyes: standard treatment (p. 136).
	Skin contact: standard treatment (p. 136).
	If swallowed: standard treatment (p. 137).
Fire hazard	Flash point 26 °C; ignition temp. 463 °C; *Extinguish fire with* water spray, foam, dry powder, carbon dioxide or vaporising liquid.
Spillage disposal	Shut off all possible sources of ignition. Wear face-shield or goggles, and gloves. Apply non-flammable dispersing agent if available and work to an emulsion with brush and water — run this to waste, diluting greatly with running water. If dispersant not available, absorb on sand, shovel into bucket(s) and transport to safe, open area for atmospheric evaporation or burial. Ventilate site of spillage well to evaporate remaining liquid and dispel vapour.

Ethyl carbonate —*se* Diethyl carbonate

Ethyl chloride —*see* Chloroethane

Ethyl chloroacetate [105–39–5]

Colourless liquid with fruity odour; bp 144 °C: immiscible with water.

TOXIC BY INHALATION, IN CONTACT WITH SKIN AND IF SWALLOWED

Avoid breathing vapour. Avoid contact with skin and eyes.

Toxic effects	The vapour irritates the respiratory system. The vapour and liquid irritate the eyes severely. Repeated or prolonged contact between the liquid and skin causes irritation. Assumed to be irritant if taken by mouth.
First aid	*Vapour inhaled:* standard treatment (p. 136).
	Affected eyes: standard treatment (p. 136).
	Skin contact: standard treatment (p. 136).
	If swallowed: standard treatment (p. 137).
Fire hazard	Flash point 54 °C. *Extinguish fire with* water spray, foam, dry powder, carbon dioxide or vaporising liquid.
Spillage disposal	Shut off all possible sources of ignition. Instruct others to keep at a safe distance. Wear breathing apparatus and gloves. Apply non-flammable dispersing agent if available and work to an emulsion with brush and water — run this to waste, diluting greatly with running water. If dispersant not available absorb on sand, shovel into bucket(s) and transport to safe, open area for burial. Site of spillage should be washed thoroughly with water and soap or detergent.

Ethyl chloroformate (ethyl chlorocarbonate) [541–41–3]

Colourless liquid with pungent odour; bp 95 °C; immiscible with water.

> HIGHLY FLAMMABLE
> TOXIC BY INHALATION
> IRRITATING TO EYES, RESPIRATORY SYSTEM AND SKIN

Prevent inhalation of vapour. Prevent contact with skin and eyes.

Toxic effects The vapour irritates the respiratory system. The vapour and liquid irritate the eyes severely. The liquid irritates the skin. Assumed to be irritant and poisonous if taken by mouth.

First aid *Vapour inhaled:* standard treatment (p. 136).
Affected eyes: standard treatment (p. 136).
Skin contact: standard treatment (p. 136).
If swallowed: standard treatment (p. 137).

Fire hazard Flash point 16 °C; *Extinguish fire with* water spray, foam, dry powder, carbon dioxide or vaporising liquid.

Spillage disposal Shut off all possible sources of ignition. Instruct others to keep at a safe distance. Wear breathing apparatus and gloves. Apply non-flammable dispersing agent if available and work to an emulsion with brush and water — run this to waste, diluting greatly with running water. If dispersant not available, absorb on sand, shovel into bucket(s) and transport to safe, open area for burial. Site of spillage should be washed thoroughly with water and soap or detergent.

Ethyl diazoacetate [623–73–4]

Experimental carcinogen.

Hazardous reactions Explosive, and distillation may be dangerous (B424).

Ethylene (ethene) [74–85–1]

Colourless gas with sweetish smell; bp −103.7 °C; 1 volume dissolves in about 9 volumes of water at 25 °C.

> EXTREMELY FLAMMABLE LIQUEFIED GAS

Avoid breathing gas.

Toxic effects Considered to be a simple asphyxiant gas which can act as an anaesthetic at high concentrations.

333

Hardous reactions	Mixtures of ethylene and AlCl$_3$ are liable to explode in presence of nickel and other catalysts; mixtures with CCl$_4$ are liable to explode as are mixtures with chlorine and other gaseous oxidants; violent explosion with tetrafluoroethylene; highly compressed gas may decompose explosively if sparked or in presence of Cu; explosions with bromotrichloromethane or chlorodifluoroethylene (B252–255).

First aid	*Gas inhaled in quantity:* standard treatment (p. 136).
Fire hazard	Explosive limits, 3.1–32 °C; ignition temp. 450 °C. Since the gas is supplied in a cylinder, turning off the valve will reduce any fire involving it; if possible, cylinders should be removed quickly from an area in which a fire has developed.
Disposal	Surplus gas or leaking cylinder can be vented slowly to air in a safe, open area or gas burnt off through a suitable burner.

Ethylene chlorohydrin —*see* 2-Chloroethanol

Ethylenediamine —*see* 1,2-Diaminoethane

Ethylene dibromide —*see* 1,2-Dibromoethane

Ethylene dichloride —*see* 1,2-Dichloroethane

Ethylene dinitrate—*see* Ethane-1,2-diol dinitrate

Ethylene diperchlorate [52936–25–1]

Hazardous reaction	Highly sensitive, violently explosive material (B258).

Ethylene glycol —*see* Ethane-1,2-diol

Ethylene glycol monoethyl ether —*see* 2-Ethoxyethanol

Ethylene glycol monomethyl ether
—*see* 2-Methoxyethanol

Ethylene glycol monomethyl ether acetate
—*see* 2-Methoxyethyl acetate

Ethyleneimine —*see* Aziridine

Ethylene oxide (oxirane) [75–21–8]

Colourless liquid or gas with no very distinctive smell; bp 11 °C; soluble in water.

EXTREMELY FLAMMABLE LIQUEFIED GAS
VERY TOXIC BY INHALATION

Avoid breathing vapour. Avoid contact with eyes and skin. CL 5 ppm (10 mg m^{-3}).

Toxic effects   If inhaled, the vapour irritates the respiratory tract and may give rise to bronchitis or pulmonary oedema; other possible effects include nausea, vomiting, convulsions and coma. The vapour irritates the eyes and may cause conjunctivitis and corneal damage. Excessive contact with liquid or solution can cause delayed burns and blistering.

Hazardous reactions   Vapour readily initiated into explosive decomposition (detailed recommendations on usage); liable to explode in autoclave reactions with some alkanethiols or alcohols; ammonia and other contaminants may cause explosive polymerisation. Iron blue pigment reacts to give a pyrophoric solid (B265–269).

First aid   *Vapour inhaled:* standard treatment (p. 136).
Affected eyes: standard treatment (p. 136).
Skin contact: standard treatment (p. 136).
If swallowed: standard treatment (p. 137).

Disposal   The contents of a broken ampoule of ethylene oxide are best dispersed by thorough ventilation of the area concerned; care must be taken to shut off all possible sources of ignition and breathing apparatus must be worn. A leaking cylinder of the substance can be vented slowly to air in a safe, open area.

RSC *Lab. Haz. Data Sheet No. 2,* 1982 gives extended coverage.

Ethylene ozonide   [289-14-5]
(1,2,4-trioxolane)

Hazardous   Explodes very violently on heating, friction or shock (B271).
reaction

Ethyl ether —*see* Diethyl ether

Ethyl formate (ethyl methanoate)   [109-94-4]

Colourless, volatile liquid with a pleasant odour; bp 54 °C; somewhat soluble in water.

HIGHLY FLAMMABLE

Avoid breathing vapour. Avoid contact with skin and eyes. RL 100 ppm (300 mg m^{-3}).

Toxic effects   The vapour irritates the respiratory system. Large concentrations affect the
nervous system and can cause unconsciousness. The vapour and liquid
irritate the eyes severely. Assumed to be irritant and poisonous if taken by
mouth.

First aid   *Vapour inhaled:* standard treatment (p. 136).
Affected eyes: standard treatment (p. 136).
If swallowed: standard treatment (p. 137).

Fire hazard   Flash point −20 °C; explosive limits 2.7–13.5%; ignition temp. 455 °C.
Extinguish fire with water spray, foam, dry powder, carbon dioxide or
vaporising liquid.

Spillage   Shut off all possible sources of ignition. Instruct others to keep at a safe
disposal   distance. Wear breathing apparatus and gloves. Apply non-flammable
dispersing agent if available and work to an emulsion with brush and
water — run this to waste, diluting greatly with running water. If dispersant
not available, absorb on sand, shovel into bucket(s) and transport to safe
open area for atmospheric evaporation. Site of spillage should be washed
thoroughly with water and soap or detergent.

Ethyl hypochlorite   [624-85-1]

Hazardous   Ignition or rapid heating of vapour causes explosion (B275).
reaction

Ethylidene chloride
—*see* 1,1-Dichloroethane

1,2-Ethylidene nitrate
—*see* Ethane-1,2-diol dinitrate

Ethylidene norbornene [16219–75–3]

TLV 5 ppm (25 mg m^{-3}).

Ethyl iodide —*see* Iodoethane

Ethyl isocyanide [624–79–3]

Hazardous reaction Liable to explode (B357).

Ethyl lactate [97–64–3]

Colourless liquid; bp 154 °C; soluble in water with partial hydrolysis.

FLAMMABLE

Toxic effects No evidence was found that ethyl lactate has significant toxic properties.

Fire hazard Flash point 46 °C; explosive limits 1.5–30%; ignition temp. 400 °C; *Extinguish fire with* water spray, foam, dry powder, carbon dioxide or vaporising liquid.

Spillage disposal Shut off all possible sources of ignition. Mop up with plenty of water and run to waste, diluting greatly with running water.

Ethyl mercaptan —*see* Ethanethiol

Ethyl methacrylate [97–63–2]

Clear, colourless liquid; bp 119 °C; insoluble in water.

HIGHLY FLAMMABLE
IRRITATING TO EYES, RESPIRATORY SYSTEM AND SKIN
MAY CAUSE SENSITISATION BY SKIN CONTACT

Avoid contact with skin and eyes. Avoid breathing vapour.

Toxic effects   The vapour irritates the eyes and respiratory system. The liquid irritates the skin and eyes. Less toxic when taken by mouth than ethyl acrylate.

First aid   *Vapour inhaled:* standard treatment (p. 136).
Affected eyes: standard treatment (p. 136).
Skin contact: standard treatment (p. 136).
If swallowed: standard treatment (p. 137).

Fire hazard   Flash point 20 °C. *Extinguish fire with* water spray, foam, dry powder, carbon dioxide or vaporising liquid.

Spillage disposal   Shut off all possible sources of ignition. Instruct others to keep at a safe distance. Wear breathing apparatus and gloves. Apply non-flammable dispersing agent if available and work to an emulsion with brush and water. If dispersant not available, absorb on sand, shovel into bucket(s) and transport to safe open area for burial. Site of spillage should be washed thoroughly with water and detergent.

Ethyl methyl ketone —*see* Butanone

Ethyl methyl peroxide   [70299–48–8]

Hazardous reactions   Shock-sensitive as liquid or vapour; explodes violently when heated strongly (B381).

N-Ethylmorpholine   [100–74–3]

Colourless liquid; bp 138 °C; miscible with water.

> (FLAMMABLE
> IRRITATING TO EYES)

Avoid breathing vapour. Avoid contact with skin, eyes and clothing. RL (skin) 5 ppm (23 mg m^{-3}).

Toxic effects   These are not well documented, but it is considered prudent to treat it with the same caution as its parent, morpholine, regarding it as hazardous by skin absorption, inhalation (irritant) and ingestion.

First aid   *Vapour inhaled:* standard treatment (p. 136).
Affected eyes: standard treatment (p. 136).
Skin contact: standard treatment (p. 136).

Fire hazard   Flash point 32 °C. *Extinguish fire with* water spray, foam, dry powder, carbon dioxide or vaporising liquid.

Spillage disposal   Shut off all possible sources of ignition. Wear face-shield or goggles, and gloves. Mop up with plenty of water and run to waste diluting greatly with running water. Ventilate area well to evaporate remaining liquid and dispel vapour.

Ethyl nitrate   [625–58–1]

Colourless liquid with pleasant odour; bp 89 °C; insoluble in water.

RISK OF EXPLOSION BY SHOCK, FRICTION, FIRE OR OTHER
SOURCES OF IGNITION

Avoid breathing vapour. Avoid contact with skin, eyes or clothing.

Toxic effects	Acute local effects are not documented but it is stated to be moderately toxic systemically by inhalation and ingestion.
Hazardous reaction	Explodes at 85 °C (B280).
First aid	*Vapour inhaled:* standard treatment (p. 136). *Affected eyes:* standard treatment (p. 136). *Skin contact:* standard treatment (p. 136). *If swallowed:* standard treatment (p. 137).
Fire hazard	Flash point 10 °C; in view of the probability of explosion shortly after ignition, the area involving an ethyl nitrate fire (or the risk of it occurring) should be evacuated and the fire brigade informed at once of the situation.
Spillage disposal	Wear face-shield or goggles, and gloves. Mix with sand and transport to a safe, open area for burial. Site of spillage should be washed thoroughly with water and soap or detergent.

Ethyl nitrite   [109–95–5]

RISK OF EXPLOSION BY SHOCK, FRICTION, FIRE
OR OTHER SOURCES OF IGNITION
HARMFUL BY INHALATION, IN CONTACT WITH SKIN
AND IF SWALLOWED

Hazardous reaction	Explodes above 90 °C (B279).

Ethyl perchlorate   (22750–93–2]

Hazardous reaction	Reputedly most explosive substance known — sensitive to impact, friction and heat (B276).

Ethylphosphine   [593–68–0]

Hazardous reactions	Ignites in air; explodes on contact with Cl_2, Br_2, or fuming nitric acid (B303).

N-Ethylpiperidine [766–09–6]

Colourless liquid; bp 131 °C; miscible with water.

(HIGHLY FLAMMABLE
IRRITATING TO EYES)

Avoid breathing vapour. Avoid contact with skin, eyes and clothing.

Toxic effects	These are not well documented, but the substance is considered to be harmful by vapour inhalation, skin absorption and ingestion.
First aid	*Vapour inhaled:* standard treatment (p. 136). *Affected eyes:* standard treatment (p. 136). *Skin contact:* standard treatment (p. 136). *If swallowed:* standard treatment (p. 137).
Fire hazard	Flash point 19 °C. *Extinguish fire with* water spray, foam, dry powder, carbon dioxide or vaporising liquid.
Spillage disposal	Shut off all possible sources of ignition. Wear face-shield or goggles, and gloves. Mop up with plenty of water and run to waste diluting greatly with running water. Ventilate area well to evaporate remaining liquid and dispel vapour.

Ethyl silicate —*see* Tetraethyl silicate

Ethyl sulphate —*see* Diethyl sulphate

Ethyl vinyl ether [109–92–2]

Hazardous reaction	Methanesulphonic acid catalysed explosive polymerisation of the ether (B447–448).

Ethynediylbis(triethyltin)
—*see* Acetylene bis(triethyltin)

2-Ethynylfuran [18649–64–4]

Hazardous reactions	Explodes on heating, or on contact with concentrated nitric acid (B569–570).

Ethynyl vinyl selenide   [101672–11–1]

Hazardous Explodes on heating (B414).
reaction

Ferric chloride anhydrous —*see* Iron(III) chloride

Fluoramine   [15861–05–9]

Hazardous Impure material is very explosive (B1046).
reaction

Fluorides (water soluble)

Normally colourless crystals or powders; soluble in water.

TOXIC BY INHALATION, IN CONTACT WITH SKIN
AND IF SWALLOWED

Avoid inhalation of dust. Prevent contact with eyes and skin. RL (as F) 2.5 mg m^{-3}.

Toxic effects The dust irritates all parts of the respiratory tract. The dust irritates the eyes and skin. Nausea, vomiting, diarrhoea and abdominal pains follow ingestion. **Chronic effect** Shortness of breath, cough, elevated temperature and cyanosis. (*See* note in Ch 6 p. 104.)

First aid *Dust inhaled:* standard treatment (p. 136).
Affected eyes: standard treatment (p. 136).
Skin contact: standard treatment (p. 136).
If swallowed: standard treatment (p. 137).
Spillage Wear face-shield or goggles. Mop up with plenty of water and run to waste,
disposal diluting greatly with running water.

Fluorinated cyclopropenyl methyl ethers
Review of group (B1517–1518).

Fluorine   [7782–41–4]

Pale yellow gas with sharp penetrating odour; bp −188 °C; reacts vigorously with water and may cause organic materials to inflame

341

MAY CAUSE FIRE
VERY TOXIC BY INHALATION
CAUSES SEVERE BURNS

Prevent inhalation of vapour. Prevent contact with skin and eyes. RL 1 ppm (2 mg m^{-3}).

Toxic effects   The gas is highly irritant to the eyes and respiratory system; high concentrations may produce thermal-type burns on the skin and chemical-type burns may result from lower concentrations.

Hazardous reactions   Anyone working with fluorine should study the whole section devoted to its dangerous reactions in Bretherick's *Handbook of reactive chemical hazards* (B1050–1060).

First aid   *Gas inhaled:* standard treatment (p. 136).
Affected eyes: standard treatment (p. 136).
Skin contact: standard treatment (p. 136).

Disposal   Instruct others to keep at a safe distance. Wear breathing apparatus and gloves. Organise ventilation of the area to dispel the gas completely. Leaking cylinder should be vented slowly into a water-fed scrubbing tower or column in a fume cupboard, or into a fume cupboard served by such a tower.

Fluorine azide   [14986–60–8]

Hazardous reaction   Unstable — explodes on vaporisation (at −82 °C) (B1049).

Fluorine nitrate — *see* Nitryl hypofluorite

Fluorine perchlorate   [10049–03–3]

Hazardous reactions   Liquid explodes on freezing at −167 °C; explosion of gas initiated by sparks, flame or contact with grease, dust or rubber tube; ignition occurs in excess of hydrogen (B895).

Fluoroacetylene   [2713–09–9]

Hazardous reactions   Liquid explodes close to its bp, −80 °C; ignition occurs in contact with solution of Br_2 in CCl_4; mercury salt decomposes but silver salt explodes on heating (B221).

Fluoroboric acid and salts
—*see* Tetrafluoroboric acid and salts

1-Fluoro-2,4-dinitrobenzene   [70-34-8]
(2,4-dinitrofluorobenzene)

Pale yellow crystals; mp 26 °C; insoluble in water.

> (TOXIC BY INHALATION, SKIN CONTACT
> AND IF SWALLOWED
> MAY CAUSE SENSITISATION BY INHALATION
> OR SKIN CONTACT)

Avoid breathing dust, vapour or aerosols. Prevent contact with skin or eyes. Use only nitrile gloves.

Toxic effects The dust, vapour from the molten compound or sprayed aerosols irritate the respiratory system. These also irritate the skin and may cause blistering, dermatitis and severe allergic reactions. Assumed to be poisonous if taken by mouth.

Hazardous preparation Explosions with distillation residues (B549–550).

First aid *Dust or vapour inhaled:* standard treatment (p. 136).
Affected eyes: standard treatment (p. 136).
Skin contact: wash with dilute sodium carbonate or ammonia solutions, then standard treatment (p. 136).
If swallowed: standard treatment (p. 137).

Spillage disposal Wear face-shield or goggles and gloves. Mix with soda ash and transport to a safe open area for burial. Wash spill site thoroughly with soda ash solution, then with soap or detergent.

Fluorodinitro compounds Review of group (B1518–1519).

Fluoroethylene —*see* Vinyl fluoride

Fluorosilicic acid and salts
—*see* Hexafluorosilicic acid and salts

Fluorosulphuric acid (fluorosulphonic acid) [7789–21–1]

Colourless or yellow fuming liquid; soluble in water.

> HARMFUL BY INHALATION
> CAUSES SEVERE BURNS

Avoid inhaling vapour. Prevent contact with eyes and skin.

Toxic effects The vapour irritates all parts of the respiratory system. The vapour irritates and the liquid burns the eyes severely. The liquid and vapour burn the skin. If taken by mouth there would be severe internal irritation and damage.

First aid *Vapour inhaled:* standard treatment (p. 136).
Affected eyes: standard treatment (p. 136).
Skin contact: standard treatment (p. 136).
If swallowed: standard treatment (p. 137).

Spillage disposal Instruct others to keep at a safe distance. Wear breathing apparatus and gloves. Spread soda ash liberally over the spillage and mop up cautiously with water — run this to waste, diluting greatly with running water.

Fluorotrichloromethane [75–69–4]

RL 1000 ppm (5600 mg m^{-3}).

Formaldehyde solution (formalin) [50–00–0]

Colourless, sometimes milky solution with pungent odour; bp 96 °C; miscible with water; the solution generally contains 37–41% formaldehyde and 11–14% methanol.

> TOXIC BY INHALATION, IN CONTACT WITH SKIN
> AND IF SWALLOWED
> MAY CAUSE SENSITISATION BY SKIN CONTACT

Avoid breathing vapour. Avoid contact with eyes and skin. CL 2 ppm (3 mg m^{-3}). Avoid formation of mixed vapours of CH_2O and HCl when the potent carcinogen bis(chloromethyl) ether is formed.

Toxic effects The vapour irritates all parts of the respiratory system. The liquid and vapour irritate the eyes severely. The liquid in contact with the skin has a hardening or tanning effect and causes irritation. Severe abdominal pains with nausea and vomiting and possibly loss of consciousness follow ingestion. **Chronic effects** High concentration of vapour inhaled for long periods can cause laryngitis, bronchitis or bronchial pneumonia; prolonged exposure may cause conjunctivitis; in contact with skin for long periods will cause cracking of skin and ulceration, particularly around fingernails. Nasal tumours have recently been reported.

Hazardous reactions	As an active reducing agent, it has been involved in various reactive incidents with oxidants like H_2O_2, $KMnO_4$, peroxyformic acid (B147, 1245).
First aid	*Vapour inhaled:* standard treatment (p. 136). *Affected eyes:* standard treatment (p. 136). *Skin contact:* standard treatment (p. 136). *If swallowed:* standard treatment (p. 137).
Fire hazard	Flash point (37%) 50 °C. *Extinguish fire with* water spray, dry powder, carbon dioxide or vaporising liquid.
Spillage disposal	Shut off all possible sources of ignition. Wear face-shield or goggles, and gloves. Mop up with plenty of water and run to waste, diluting greatly with running water. Ventilate area well to evaporate remaining liquid and dispel vapour.

Formamide   [75–12–7]

RL 20 ppm (30 mg m^{-3}).

Formdimethylamide
—*see* Dimethylformamide

Formic acid (methanoic acid)   [64–18–6]

Colourless liquid which fumes in the higher concentrations; 98/100% acid boils at 100.5 °C; miscible with water.

CAUSES SEVERE BURNS

Avoid breathing vapour. Avoid contact with eyes and skin. RL 5 ppm (9 mm m^{-3}).

Toxic effects	The vapour irritates all parts of the respiratory system. The vapour irritates the eyes and the liquid causes painful eye burns. The liquid burns the skin. If taken by mouth there is severe internal irritation and damage.
Hazardous reactions	Tightly sealed containers may burst from pressure of CO produced during slow decomposition in storage or on freezing at low temperatures; explosive decomposition on Ni (B148–149).
First aid	*Vapour inhaled:* standard treatment (p. 136). *Affected eyes:* standard treatment (p. 136). *Skin contact:* standard treatment (p. 136). *If swallowed:* standard treatment (p. 137).

Spillage disposal	Instruct others to keep at a safe distance. Wear breathing apparatus and gloves. Spread soda ash liberally over the spillage and mop up cautiously with water — run this to waste, diluting greatly with running water.

2-Furaldehyde (furfuraldehyde, furfural) [98–01–1]

Colourless to yellow liquid with distinctive odour; bp 162 °C; immiscible with water.

TOXIC BY INHALATION AND IF SWALLOWED

Avoid inhalation of vapour. Avoid contact with skin and eyes. RL (skin) 2 ppm (8 mg m^{-3}).

Toxic effects	The vapour irritates the eyes and respiratory system. The liquid irritates the eyes. Assumed to be poisonous if taken by mouth. **Chronic effects** Exposure over prolonged periods may result in nervous disturbance, eye inflammation and disturbance of vision.
Hazardous reaction	Reaction with NaHCO$_3$ in oil refining operations may give pyrophoric coke (B496–497).
First aid	*Vapour inhaled:* standard treatment (p. 136). *Affected eyes:* standard treatment (p. 136). *Skin contact:* standard treatment (p. 136). *If swallowed:* standard treatment (p. 137).
Spillage disposal	Instruct others to keep at a safe distance. Wear breathing apparatus and gloves. Apply nonflammable dispersing agent if available and work to an emulsion with brush and water — run this to waste, diluting greatly with running water. If dispersant not available, absorb on sand, shovel into bucket(s) and transport to safe, open area for burial. Site of spillage should be washed thoroughly with water and soap or detergent.

Furfuryl alcohol [98–00–0]

HARMFUL BY INHALATION, IN CONTACT WITH SKIN OR IF SWALLOWED

RL (skin) 5 ppm (20 mg m^{-3}).

Hazardous reactions	Reacts violently with formic acid; mixture with cyanoacetic acid exploded on heating; may react violently or explosively on contact with acids or acidic materials (B507).

Furoyl chloride [1300–32–9]

Hazardous reaction	Freshly distilled material polymerised explosively in storage (B492–493).

Gallium (metal) [7440–55–3]

Hazardous reactions Reacts violently with chlorine and bromine (B1090).

Gallium triperchlorate [17835–81–3]

Hazardous preparation Moist product may react violently with any organic material (B991).

Germane (germanium tetrahydride) [7782–65–2]

RL 0.2 ppm (0.6 mg m^{-3}).

Hazardous reactions Liable to ignite in air; contact with bromine may cause ignition at −112 °C (B1092).

Germanium [7440–56–4]

Hazardous reactions Powdered metal ignites in chlorine, lumps ignite on heating in chlorine or bromine; powdered metal reacts violently with nitric acid, mixtures with potassium chlorate and nitrate explode on heating (B1091).

Germanium monohydride [13572–99–1]

Hazardous reaction May decompose explosively in air (B1092).

Germanium tetrachloride [10038–98–9]

Hazardous reaction Reacts violently with water (B1003).

Germanium tetrahydride —*see* Germane

Glutaraldehyde (pentanedial) [111–30–8]

RL 0.2 ppm (0.7 mg m^{-3}).

Glutaryl diazide [64624–44–8]

Hazardous reaction   May explode on heating (B506).

Glycerol [56–81–5]

Hazardous reactions   Reacts violently or explosively with many oxidising agents (B382).

Glycidol —*see* 2,3-Epoxy-1-propanol

Glycol monoethyl ether —*see* 2-Ethoxyethanol

Glycolonitrile [107–16–4]

Storage hazard   Liable to polymerise explosively on storage (B246–247).

Glyoxal (ethanedial) [107–22–2]

IRRITATING TO EYES AND SKIN

Hazardous preparation   Oxidation of paraldehyde with nitric acid may become violent (B237).

Gold compounds Review of group (B1525).

Guanidinium nitrate [52470–25–4]

Hazardous preparation   Explosions have occurred in preparation from dicyanodiamide and ammonium nitrate (B178).

Hafnium [7440–58–6]

RL 0.5 mg m^{-3}.

Haloacetylene derivatives Review of class (B1526).

Haloalkanes Review of group (B1527).

Haloalkenes Review of group (B1529).

Haloarenemetal π-complexes Review of group (B1531).

Haloboranes Review of group (B1533).

Halocarbons Explanatory review (B1533).

Halogen azides Review of group (B1534).

N-Halogen compounds Review of class (B1535).

Halogen oxides Review of group (B1537).

N-Haloimides Review of group (B1537).

2-Halomethyl-furans or -thiophenes
Review of group (B1538).

Halophosphines Review of group (B1538).

Halosilanes Review of group (B1538).

Heavy metal derivatives Review of group (B1540).

Heptane (and heptane fraction from petroleum) [142–82–5]

Colourless liquid; bp 98 °C; insoluble in water.

HIGHLY FLAMMABLE
DANGER OF CUMULATIVE EFFECTS

Avoid breathing vapour. RL 400 ppm (1600 mg m^{-3}).

Toxic effects   May irritate the respiratory system as vapour which, in high concentrations, is narcotic.

First aid   *Vapour inhaled in high concentrations:* standard treatment (p. 136).
Affected eyes: standard treatment (p. 136).
If swallowed: standard treatment (p. 136).

Fire hazard   Flash point −4 °C; explosive limits 1.2–6.7%; ignition temp. 223 °C.
Extinguish fire with foam, dry powder, carbon dioxide or vaporising liquid.

Spillage disposal   Shut off all possible sources of ignition. Wear face-shield or goggles, and gloves. Apply non-flammable dispersing agent if available and work to an emulsion with brush and water —run this to waste, diluting greatly with running water. If dispersant not available, absorb on sand, shovel into bucket(s) and transport to safe, open area for atmospheric evaporation. Ventilate site of spillage well to evaporate remaining liquid and dispel vapour.

Heptan-2-one (methyl amyl ketone) [591–78–6]

FLAMMABLE
HARMFUL IF SWALLOWED

TLV 50 ppm (235 mg m^{-3}).

Heptan-3-one (ethyl butyl ketone) [106–35–4]

FLAMMABLE
HARMFUL BY INHALATION
IRRITATING TO EYES

RL 50 ppm (230 mg m^{-3}).

Hept-2-yn-1-ol [1002-36-4]

Hazardous preparation   Risk of explosion of distillation residue (B691).

Hexacarbonylchromium [13007-92-6]

Hazardous reaction   Explodes at 210 °C (B535).

Hexacarbonylmolybdenum [13939-06-5]

Storage hazard   Solutions liable to explode on long storage (B649).

Hexacarbonyltungsten [14040-11-0]

Hazardous preparation   Preparation from WCl_6, Al powder and CO in autoclave can be dangerous (B650).

Hexachlorobutadiene [76-68-3]

Suspected carcinogen. TLV 0.02 ppm (0.24 mg m^{-3}).

1,2,3,4,5,6-Hexachlorocyclohexane [608-73-1]

RL (γ-isomer) 0.5 mg m^{-3}.

Hazardous reaction   Reaction with dimethylformamide in presence of iron may become violent (B591).

Hexachlorocyclopentadiene [77-47-4]

TLV 0.01 ppm (0.1 mg m^{-3}).

Hexachloroethane [67-72-1]

RL (vapour, skin) 5 ppm (50 mg m^{-3}), respirable dust 5 mg m^{-3}.

Hexachloronaphthalene [1335–87–1]

TLV (skin) 0.2 mg m^{-3}.

Hexafluoroacetone [684–16–2]

TLV 0.1 ppm (0.7 mg m^{-3}).

Hexafluorochloronium salts Review of group (B1541).

Hexafluorosilicic acid and salts [16961–83–4]
(fluorosilicic acid and salts)

Fluorosilicic acid is a colourless fuming liquid miscible with water. The salts are generally colourless, crystalline and soluble in water.

                    CAUSES BURNS

Avoid inhaling vapour or dust. Prevent contact with eyes and skin.

Toxic effects	The vapour and dust irritate all parts of the respiratory system. The acid and salts burn the eyes severely. The acid burns the skin. If taken by mouth there would be severe internal irritation and damage.
First aid	*Vapour or dust inhaled:* standard treatment (p. 136). *Affected eyes:* standard treatment (p. 136). *Skin contact:* standard treatment (p. 136). *If swallowed:* standard treatment (p. 137).
Spillage disposal	Wear face-shield or goggles, and gloves. Mop up with plenty of water and run to waste, diluting greatly with running water.

Hexalithium disilicide [12136–61–7]

Hazardous reactions	Reacts violently with water; explodes with nitric acid; ignites on warming in fluorine; incandescent reactions with P, Se or Te (B1266).

Hexamethyldiplatinum [4711–74–4]

Hazardous reaction	Explodes on heating (B644).

Hexamethylphosphoramide [680–31–9]
(hexamethylphosphoric triamide)

Colourless liquid, slight basic odour; bp 234 °C; miscible with water.

> IRRITATING TO EYES AND SKIN
> POSSIBLE RISKS OF IRREVERSIBLE EFFECTS

Prevent inhalation of vapour. Prevent contact with skin and eyes. Potent experimental carcinogen, no TLV level set.

Toxic effects	No data is available on human toxicity, but animal experiments indicate acute toxic effects on lungs, kidneys, and the nervous and reproductive systems. **Chronic effects.** Animal experiments have led to tumours from inhalation of 0.4 ppm for long periods.
First aid	*Affected eyes:* standard treatment (p. 136). *Skin contact:* standard treatment (p. 136). *If swallowed:* standard treatment (p. 137).
Spillage disposal	Instruct others to keep at safe distance. Wear breathing apparatus and gloves. Mop up with plenty of water and run to waste, diluting greatly with running water.

Hexane (and hexane fraction from petroleum) [110–54–3]

Colourless liquid; bp about 69 °C; insoluble in water.

> HIGHLY FLAMMABLE
> HARMFUL BY INHALATION AND IN CONTACT WITH SKIN
> POSSIBLE RISK OF IRREVERSIBLE EFFECTS

Avoid breathing vapour. RL (n-) 100 ppm (360 mg m^{-3}). Other isomers RL 500 ppm (1800 mg m^{-3}).

Toxic effects	The vapour may irritate the respiratory system and, in high concentrations, have narcotic action. **Chronic effects** Loss of sensation in hands and feet.
First aid	*Vapour inhaled in high concentrations:* standard treatment (p. 136). *Affected eyes:* standard treatment (p. 136). *If swallowed:* standard treatment (p. 137).
Fire hazard	Flash point −23 °C; explosive limits 1.2–7.5%; ignition temp. 260 °C. *Extinguish fire with* foam, dry powder, carbon dioxide or vaporising liquid.
Spillage disposal	Shut off all possible sources of ignition. Wear face-shield or goggles, and gloves. Apply non-flammable dispersing agent if available and work to an emulsion with brush and water — run this to waste, diluting greatly with running water. If dispersant not available, absorb on sand, shovel into bucket(s) and transport to safe, open area for atmospheric evaporation. Ventilate site of spillage well to evaporate remaining liquid and dispel vapour.

Hexanoic acid (caproic acid) [142-62-1]

Colourless liquid with unpleasant, sweat-like odour, bp 205 °C; immiscible with water.

(CAUSES BURNS)

Avoid contact with eyes and skin.

Toxic effects	The liquid burns the skin and eyes. Corrosive if taken by mouth.
First aid	*Affected eyes:* standard treatment (p. 136).
	Skin contact: standard treatment (p. 136).
	If swallowed: standard treatment (p. 137).
Spillage disposal	Wear face-shield or goggles, and gloves. Spread soda ash liberally over the spillage and mop up cautiously with water — run this to waste, diluting greatly with running water.

Hexanolactam —*see* ε-Caprolactam

Hexan-2-one (butyl methyl ketone) [591-78-6]

Colourless liquid, sharp odour; bp 127 °C; slightly soluble in water.

FLAMMABLE
TOXIC BY INHALATION AND IN CONTACT WITH SKIN
POSSIBLE RISK OF IRREVERSIBLE EFFECTS

Avoid breathing vapour. Avoid contact with skin and eyes. RL (skin) 25 ppm (100 mg m^{-3}).

Toxic effects	The vapour irritates the eyes and respiratory system and has a depressant action on the nervous system. The liquid will irritate the eyes and may cause severe damage. If swallowed may cause gastric irritation. **Chronic effects.** Loss of sensation in hands and feet.
First aid	*Vapour inhaled:* standard treatment (p. 136).
	Affected eyes: standard treatment (p. 136).
	Skin contact: standard treatment (p. 136).
	If swallowed: standard treatment (p. 137).
Fire hazard	Flash point 23 °C; explosive limits 1.2–8%; ignition temperature 533 °C. *Extinguish fire with* foam, dry powder, carbon dioxide or vaporising liquid.
Spillage disposal	Shut off all possible sources of ignition. Wear breathing apparatus and gloves. Apply non-flammable dispersing agent if available and work to an emulsion with brush and water — run this to waste diluting greatly with running water. If no dispersant available, absorb on sand, shovel into bucket(s) and transport to safe open area for atmospheric evaporation. Ventilate site of spillage well to evaporate remaining liquid and dispel vapour.

Hex-3-enedinitrile (1,4-dicyano-2-butene) [1119–69–3]

Hazardous reaction	Accelerating polymerisation — decomposition due to overheating in vacuum evaporator (B592).

Hexone —*see* 4-Methylpentan-2-one

s-Hexyl acetate [142–92–7]

TLV 50 ppm (300 mg m^{-3}).

Hexylene glycol [107–41–5]

TLV 25 ppm (125 mg m^{-3}).

High-nitrogen compounds
Review of group (B1541–1542).

Hydrazinemetal nitrates Review of group (B1544).

Hydrazine salts

Colourless crystals; soluble in water.

> (CAUSE BURNS
> IRRITATING TO SKIN AND EYES)

Avoid contact with eyes and skin.

Toxic effects	The crystals and solutions irritate or burn the eyes severely. The crystals and solutions burn the skin. If taken by mouth there would be severe internal irritation and damage.
Hazardous reactions	See the hydrazinium salts below.

First aid	*Affected eyes:* standard treatment (p. 136).
	Skin contact: standard treatment (p. 136).
	If swallowed: standard treatment (p. 137).
Spillage disposal	Wear face-shield or goggles, and gloves. Mop up with plenty of water and run to waste, diluting greatly with running water.

Hydrazine solutions (hydrazine hydrate) [302–01–2]

Colourless liquid; bp 113 °C; miscible with water.

FLAMMABLE
VERY TOXIC IN CONTACT WITH SKIN AND IF SWALLOWED
CAUSES BURNS
POSSIBLE RISK OF IRREVERSIBLE EFFECTS

Avoid contact with eyes and skin. Suspected carcinogen. RL (skin) 0.1 ppm (0.1 mg m^{-3}).

Toxic effects	The liquid burns the eyes severely. The liquid burns the skin. If taken by mouth there is severe internal irritation and damage.
Hazardous reactions	Contact with many transition metals, their oxides (rust) or salts causes catalytic decomposition (violent if concentrated solutions) and possibly ignition of evolved hydrogen; reaction with oxidants is violent (B1190–1194).
First aid	*Affected eyes:* standard treatment (p. 136).
	Skin contact: standard treatment (p. 136).
	If swallowed: standard treatment (p. 137).
Spillage disposal	Wear face-shield or goggles, and gloves. Mop up with plenty of water and run to waste, diluting greatly with running water.

RSC *Lab. Haz. Data Sheet No. 14,* 1983 gives extended coverage.

Hydrazinium chlorate [66326–46–3]

Hazardous reaction	Explodes violently at its mp, 80 °C (B922).

Hydrazinium chlorite [66326–45–2]

Hazardous reaction	Spontaneously flammable when dry (B922).

Hydrazinium nitrate   [13464–97–6]

Hazardous   Explosive properties have been studied in detail (B1206).
reaction

Hydrazinium perchlorate   [13762–80–6]

Hazardous   Deflagration and thermal decomposition have been studied (B923).
reactions

Hydrazoic acid —*see* Hydrogen azide

Hydriodic acid   [10034–85–2]

Yellow to brown fuming liquid commonly available in concentrations of 55% and 66% HI; miscible with water.

CAUSES BURNS

Avoid breathing vapour. Prevent contact with eyes and skin.

Toxic effects   The vapour irritates the respiratory system. The vapour irritates and the liquid burns the eyes severely. The vapour and liquid burn the skin. If taken by mouth there would be severe internal irritation and damage.

Hazardous   Blockage of condenser in preparation from iodine and phosphorus caused
preparation   explosion (B919).

First aid   *Vapour inhaled:* standard treatment (p. 136).
Affected eyes: standard treatment (p. 136).
Skin contact: standard treatment (p. 136).
If swallowed: standard treatment (p. 137).

Spillage   Instruct others to keep at a safe distance. Wear breathing apparatus and
disposal   gloves. Spread soda ash liberally over the spillage and mop up cautiously with water — run this to waste, diluting greatly with running water.

Hydrobromic acid   [10035–10–6]

Colourless, fuming liquid with acrid smell commonly available in concentrations of 47%, 50% and 60%; miscible with water.

CAUSES BURNS
IRRITATING TO RESPIRATORY SYSTEM

Avoid breathing vapour. Prevent contact with eyes and skin. RL (as HBr) 3 ppm (10 mg m^{-3}).

357

Toxic effects   The vapour irritates all parts of the respiratory system. The vapour irritates, and the liquid burns, the eyes severely. The vapour and liquid irritate and may burn the skin. If taken by mouth there would be severe internal irritation and damage.

First aid   *Vapour inhaled:* standard treatment (p. 136).
Affected eyes: standard treatment (p. 136).
Skin contact: standard treatment (p. 136).
If swallowed: standard treatment (p. 137).

Spillage disposal   Instruct others to keep at a safe distance. Wear breathing apparatus and gloves. Spread soda ash liberally over the spillage and mop up cautiously with water — run this to waste, diluting greatly with running water.

Hydrochloric acid   [7647–01–0]

Colourless, fuming liquid with pungent smell, commonly available in concentrations of 32% and 36% HCl; miscible with water.

CAUSES BURNS
IRRITATING TO RESPIRATORY SYSTEM

Avoid breathing vapour. Prevent contact with eyes and skin. RL (as HCl) 5 ppm (7 mg m^{-3}).

Toxic effects   The vapour irritates all parts of the respiratory system. The vapour irritates, and the liquid burns, the eyes severely. The vapour and liquid irritate and may burn the skin. If taken internally there would be severe irritation and damage.

First aid   *Vapour inhaled:* standard treatment (p. 136).
Affected eyes: standard treatment (p. 136).
Skin contact: standard treatment (p. 136).
If swallowed: standard treatment (p. 137).

Spillage disposal   Instruct others to keep at a safe distance. Wear breathing apparatus and gloves. Spread soda ash liberally over the spillage and mop up cautiously with water — run this to waste, diluting greatly with running water.

Hydrofluoric acid   [7664–39–3]

Colourless, fuming liquid with pungent smell, commonly available in concentrations of 40%, 42% and 48% HF; miscible with water.

VERY TOXIC BY INHALATION, IN CONTACT WITH SKIN
AND IF SWALLOWED
CAUSES SEVERE BURNS

Prevent inhalation of fumes. Prevent contact with skin, eyes and clothing. RL (as HF) 3 ppm (2.5 mg m^{-3}).

Toxic effects The fume irritates severely all parts of the respiratory system. The fume or acid rapidly causes severe irritation of the eyes and eyelids; burns may develop. Skin burns do not usually cause pain until several hours have elapsed since contact. If taken by mouth there is immediate and severe internal irritation and damage.

First aid Special treatment (pp. 134 and 135).
Spillage Instruct others to keep at a safe distance. Wear breathing apparatus and
disposal gloves. Spread soda ash liberally over the spillage and mop up cautiously with water — run this to waste, diluting greatly with running water.

RSC *Lab. Haz. Data Sheet No. 19,* 1983 gives extended coverage.

Hydrogen [1333–74–0]

Colourless, odourless gas; bp −253 °C; sparingly soluble in water.

HIGHLY FLAMMABLE

Toxic effects Classed as a simple asphyxiant gas.

Hazardous Catalytic Pt and similar metals containing adsorbed O_2 will heat and cause
reactions ignition in contact with hydrogen; high-velocity jets may ignite sponta-neously particularly if rust is present; reaction hazards with such oxidants as chlorine dioxide, copper oxide, nitryl fluoride, NO, N_2O_4, PdO, O_2 are indicated — also those with BrF_3, Br_2, ClF_3, Cl_2, F_2; references given to literature on safe handling of liquid hydrogen (B1138–1143).

Fire hazard Explosive limits 4–75%; ignition temperature 585 °C. Extinguish small hydrogen fires with dry powder, carbon dioxide or vaporising liquid. Where the gas is supplied in a cylinder, turning off the valve will reduce fire.
Disposal Surplus gas or leaking cylinder can be vented slowly to air in a safe open space, or gas burnt off through a suitable burner.

Hydrogenated terphenyls [92–94–4]

TLV 0.5 ppm (5 mg m^{-3}).

Hydrogenation catalysts Review (B1544).

Hydrogen azide (hydrazoic acid) [7782–79–8]

Hazardous reactions	Safe in dilute solutions, extremely endothermic and violently explosive in concentrated or pure states, its properties are reviewed; readily forms explosive heavy metal azides (B1131–1132).

Hydrogen bromide [10035–10–6]

Colourless gas, fuming in moist air, with pungent acrid smell; dissolves readily in water forming hydrobromic acid.

CAUSES SEVERE BURNS
IRRITATING TO RESPIRATORY SYSTEM

Prevent inhalation of gas. Prevent contact with skin and eyes. RL 3 ppm (10 mg m^{-3}).

Toxic effects	The gas irritates the respiratory system severely; it irritates and may burn the skin and eyes.
Hazardous reactions	Ignites on contact with fluorine; explodes with ozone (B94).
First aid	*Gas inhaled:* standard treatment (p. 136). *Affected eyes:* standard treatment (p. 136). *Skin contact:* standard treatment (p. 136).
Disposal	Surplus gas or leaking cylinder can be vented slowly into a water-fed scrubbing tower or column in a fume cupboard, or into a fume cupboard served by such a tower.

Hydrogen chloride [7647–01–0]

Colourless gas, fuming in moist air, with pungent suffocating smell; bp −85 °C; dissolves readily in water forming hydrochloric acid.

CAUSES SEVERE BURNS
IRRITATING TO RESPIRATORY SYSTEM

Prevent inhalation of gas. Prevent contact with skin and eyes. RL 5 ppm (7 mg m^{-3}).

Toxic effects	The gas irritates the eyes and respiratory system severely; it irritates the skin and may cause severe burns to both eyes and skin.
Hazardous reactions	Violent reaction with aluminium; ignites on contact with fluorine; dangerous reactions with hexalithium disilicide, some acetylides, uranium dicarbide and tetraselenium tetranitride are recorded (B902–904).

| First aid | *Gas inhaled:* standard treatment (p. 136).
Affected eyes: standard treatment (p. 136).
Skin contact: standard treatment (p. 136). |
| Disposal | Surplus gas or leaking cylinder can be vented slowly into a water-fed scrubbing tower or column in a fume cupboard, or into a fume cupboard served by such a tower. |

Hydrogen cyanide (hydrocyanic acid gas)   [74–90–8]

Colourless liquid or gas with faint odour of bitter almonds; bp 26 °C; very soluble in water, the solution being only weakly acidic.

EXTREMELY FLAMMABLE
VERY TOXIC BY INHALATION, IN CONTACT WITH SKIN
AND IF SWALLOWED

Prevent inhalation of gas (some are unable to detect HCN by smell). Prevent contact with skin and eyes. RL (skin) 10 ppm (10 mg m^{-3}).

Toxic effects	Inhalation of high concentrations leads to shortness of breath, paralysis, unconsciousness, convulsions and death by respiratory failure. With lethal concentrations, death is extremely rapid although breathing may continue for some time. With low concentrations the effects are likely to be headache, vertigo, nausea and vomiting. Chronic exposure over long periods may induce fatigue and weakness. The average fatal dose is 55 mg which can also be assimilated by skin contact with the liquid. (*See* Ch 6, 7.)
Hazardous reaction	In absence of inhibitor, exothermic polymerisation occurs, at 184 °C this is explosive (B136).
First aid	Special treatment for hydrogen cyanide (p. 132).
Fire hazard	Flash point −18 °C; explosive limits 6–14%; ignition temp. 538 °C. Since the liquid is supplied in a cylinder, turning off the valve will reduce any fire involving it; if possible, cylinders should be removed quickly from an area in which a fire has developed. Breathing apparatus must be worn during these operations.
Disposal	Breathing apparatus must be worn. Surplus gas or a leaking cylinder can be vented slowly into a water-fed scrubbing tower or column in a fume cupboard.

Hydrogen fluoride   [7664–39–3]

Colourless, fuming gas or liquid; bp 19.5 °C; dissolves in water readily forming hydrofluoric acid.

VERY TOXIC BY INHALATION, IN CONTACT WITH SKIN,
AND IF SWALLOWED
CAUSES SEVERE BURNS

Prevent inhalation of gas. Prevent contact with skin and eyes. RL 3 ppm (2 mg m^{-3}).

Toxic effects   The gas irritates severely the eyes and respiratory system and may cause burns to the eyes. It irritates the skin and painful burns may develop after an interval. The liquid causes severe, painful burns on contact with all body tissues.

Hazardous reactions   Reference to handling precautions given; risk of violent reaction in fluorination of organic substances by passing HF into stirred suspension of mercury(II) oxide; arsenic trioxide and calcium oxide incandesce in contact with the liquid; violent evolution of SiF_4 when potassium tetrafluorosilicate comes in contact with the liquid (B1044–1046).

First aid   *Gas inhaled:* standard treatment (p. 136).
Affected eyes: special treatment as for hydrofluoric acid (p. 135).
Skin contact: special treatment as for hydrofluoric acid (p. 134).

Disposal   Surplus gas or leaking cylinder can be vented slowly into a water-fed scrubbing tower or column in a fume cupboard, or into a fume cupboard served by such a tower.

RSC *Lab. Haz. Data Sheet No. 19,* 1983 gives extended coverage.

Hydrogen iodide   [10034–85–2]

Colourless gas which dissolves readily in water to form hydriodic acid.

CAUSES SEVERE BURNS
IRRITATING TO RESPIRATORY SYSTEM

Avoid breathing gas. Avoid contact with skin and eyes.

Toxic effects   The gas irritates the respiratory system; it irritates and may burn the eyes and skin.

Hazardous reactions   Causes momentary ignition of magnesium metal; mixture of the anhydrous liquid and potassium metal explodes very violently; the gas is ignited by molten potassium chlorate (B1094).

First aid   *Gas inhaled:* standard treatment (p. 136).
Affected eyes: standard treatment (p. 136).
Skin contact: standard treatment (p. 136).

Disposal   Surplus gas or leaking cylinder can be vented slowly into a water-fed scrubbing tower or column in a fume cupboard, or into a fume cupboard served by such a tower.

Hydrogen peroxide   [7722–84–1]

Colourless liquid; miscible with water.

362

CONTACT WITH COMBUSTIBLE MATERIAL MAY CAUSE FIRE
CAUSES BURNS

Avoid contact with eyes and skin. RL 1 ppm (1.5 mg m^{-3}).

Toxic effects   Hydrogen peroxide, especially in higher concentrations, is irritant and caustic to the mucous membranes, eyes and skin. If swallowed, the sudden evolution of oxygen may cause injury by acute distension of the stomach and may cause nausea, vomiting and internal bleeding.

Hazardous reactions   Several reviews of the hazards of using strong hydrogen peroxide solutions are referred to; hazardous reactions ranging from ignition to explosion are recorded with acetic acid, acetone, alcohols, alcohols/sulphuric acid, charcoal, carboxylic acids, ketones/nitric acid, mercury(II) oxide/nitric acid, metals, metal oxides, nitric acid/thiourea, nitrogenous bases, phosphorus, P_2O_5, $SnCl_2$, vinyl acetate; violent reactions or explosions with reactions run in acetic acid (peroxyacetic acid formed), with acetaldehyde, acetic anhydride, benzenesulphonic anhydride, diethyl ether, ketene, H_2SO_4, tetrahydrothiophene, trioxane/lead or wood. Catalytic decomposition of concentrated peroxide by various metal catalysts may become explosive (B1149–1163).

First aid   *Affected eyes:* standard treatment (p. 136).
Skin contact: standard treatment (p. 136).
If swallowed: standard treatment (p. 137).
Spillage disposal   Wear face-shield or goggles, and gloves. Mop up with plenty of water and run to waste, diluting greatly with running water.

Hydrogen phosphide —*see* Phosphine

Hydrogen selenide [7783–07–5]

RL (as Se) 0.05 ppm (0.2 mg m^{-3}).

Hydrogen sulphide (sulphuretted hydrogen) [7783–06–4]

Colourless gas with an offensive odour; soluble in water.

EXTREMELY FLAMMABLE LIQUEFIED GAS
VERY TOXIC BY INHALATION

Avoid inhaling gas. RL 10 ppm (14 mg m^{-3}).

Toxic effects In high concentrations may cause immediate unconsciousness followed by respiratory paralysis; at lower concentrations causes irritation of all parts of the respiratory system and eyes, headache, dizziness and weakness, Irritates the eyes and may cause conjunctivitis. (*See* Ch 6 p 108.)

Hazardous reactions May ignite on contact with a wide range of metal oxides; finely divided tungsten glows red-hot in stream of the gas; dangerous reactions with a variety of oxidants including CrO_3, Cl_2O, F_2, HNO_3, Na_2O_2, PbO_2 have been recorded; exothermic reaction with soda lime, which may become explosive in presence of oxygen (B1172–174).

First aid *Gas inhaled:* standard treatment (p. 136).
Affected eyes: standard treatment (p. 136).

Fire hazard Explosive limits 4.3–46%; ignition temp. 260 °C. Since the gas is supplied in a cylinder, turning off the valve will reduce any fire involving it; if possible, cylinders should be removed quickly from an area in which a fire has developed.

Disposal Surplus gas or leaking cylinder can be vented slowly into a water-fed scrubbing tower or column in a fume cupboard, or into a fume cupboard served by such a tower.

RSC *Lab. Haz. Data Sheet No. 35,* 1985 gives extended coverage.

Hydrogen trisulphide [13845–23–3]

Hazardous reactions Some metal oxides cause violent decomposition and ignition; reactions with benzenediazonium chloride, NCl_3 and pentyl alcohol are explosive (B1174).

Hydroquinone (1,4-dihydroxybenzene) [123–31–9]

White crystals; mp 170 °C; soluble in water.

HARMFUL BY INHALATION AND IF SWALLOWED

Avoid inhalation of dust and contact with skin. RL 2 mg m^{-3}.

Toxic effects Ingestion has led to nausea, shortness of breath, delirium or collapse. Contact with skin may lead to dermatitis, and prolonged exposure of eyes to vapour may lead to corneal damage.

Hazardous reactions Decomposition in contact with NaOH; reaction with pressurised O_2 to form clathrate compound exploded (B597).

First aid	*Affected eyes:* standard treatment (p. 136).
	Skin contact: standard treatment (p. 136).
	If swallowed: standard treatment (p. 137).
Spillage disposal	Wear face shield or goggles, and gloves. Mop up with plenty of water and run to waste, diluting greatly with running water.

2-Hydroxybut-3-enonitrile [5809–59–6]

(1-cyano-2-propen-1-ol)

Hazardous reaction	Liable to polymerise explosively (B417).

2-Hydroxyethylamine —*see* 2-Aminoethanol

2-Hydroxyethyl methacrylate [868–77–9]

Pale yellow, mobile liquid; bp 84 °C at 7 mbar; miscible with water.

IRRITATING TO EYES AND SKIN

Prevent contact with skin, eyes and clothing.

Toxic effects	Splashes on skin and mucous surfaces can cause severe blistering.
First aid	*Affected eyes:* standard treatment (p. 136).
	Skin contact: standard treatment (p. 136).
	If swallowed: standard treatment (p. 137).
Spillage disposal	Wear face-shield or goggles, and gloves. Mop up with plenty of water and run to waste, diluting greatly with running water. Ventilate area of spillage until dry.

1-Hydroxyethyl peroxyacetate [7416–48–0]

Hazardous reaction	The low-melting solid is explosive (B453).

Hydroxylamine [7803–49–8]

Hazardous preparation and reactions	Risk of explosion in preparation from hydroxylamine hydrochloride and sodium hydroxide in methanol; ignites on contact with anhydrous copper(II) sulphate; dangerous reactions with Na, Ca and finely divided zinc; it reacts violently or explosively with a large number of oxidants (B1181–1182).

Hydroxylammonium salts

(hydroxylamine salts)

SO_4 [10039–54–0]
Cl [5470–11–1]

White crystals; soluble in water.

(CORROSIVE
IRRITATING TO SKIN AND EYES)

Avoid contact with eyes and skin.

Toxic effects	The salts and solutions irritate and burn the eyes severely. The salts and solutions burn the skin. If taken by mouth there is severe internal irritation and damage. **Chronic effects** Continued skin contact can cause dermatitis.
Hazardous reactions	Many salts decompose violently above 100 °C as solids or solutions (B1546).
First aid	*Affected eyes:* standard treatment (p. 136). *Skin contact:* standard treatment (p. 136). *If swallowed:* standard treatment (p. 137).
Spillage disposal	Wear face-shield or goggles, and gloves. Mop up with plenty of water and run to waste, diluting greatly with running water.

4-Hydroxy-4-methylpentan-2-one

[123–42–2]

(diacetone alcohol)

Colourless liquid with faint, pleasant odour; bp 168 °C; miscible with water.

HIGHLY FLAMMABLE
IRRITATING TO EYES

Avoid breathing vapour. Avoid contact with skin, eyes and clothing. RL 50 ppm (240 mg m^{-3}).

Toxic effects	The vapour irritates the eyes and respiratory system. The liquid irritates the eyes and mucous membranes and is absorbed by the skin, possibly with harmful effects. Taken by mouth, it has a narcotic effect; kidney and liver injury and anaemia have been produced in experimental animals.

First aid *Vapour inhaled:* standard treatment (p. 136).
Affected eyes: standard treatment (p. 136).
Skin contact: standard treatment (p. 136).
If swallowed: standard treatment (p. 137).

Spillage disposal Shut off all possible sources of ignition. Wear face-shield or goggles, and gloves. Mop up with plenty of water and run to waste, diluting greatly with running water. Ventilate area well to evaporate remaining liquid and dispel vapour.

2-Hydroxy-2-methylpropiononitrile [75–86–5]
(acetone cyanohydrin; 2-cyanopropan-2-ol)

Colourless liquid; bp 82 °C at 30 mbar; miscible with water.

VERY TOXIC BY INHALATION, IN CONTACT WITH SKIN
AND IF SWALLOWED

Prevent inhalation of vapour. Prevent contact with skin and eyes.

Toxic effects High vapour concentrations, if inhaled, rapidly cause giddiness, headache, unconsciousness, convulsions; in severe cases, breathing may cease due to hydrogen cyanide poisoning. It must be assumed that skin absorption and taking by mouth will have similar effects.

Hazardous reaction H_2SO_4 may dehydrate it to methacrylonitrile and catalyse exothermic polymerisation (B441).

First aid Special treatment for hydrogen cyanide (p. 132).
Spillage disposal Wear breathing apparatus and gloves. Instruct others to keep at a safe distance. Bleaching powder should be scattered liberally over the spillage, or an excess of sodium hypochlorite solution added. The treated spillage should then be mopped up into a bucket and allowed to stand for 24 hours before running to waste, diluting greatly with running water.

2-Hydroxypropyl acrylate [999–61–1]

RL (skin) 0.5 ppm (3 mg m^{-3}).

3-Hydroxypropyne —*see* Prop-2-yn-1-ol

3-Hydroxytriazenes Review of group (B1546).

Hypochlorous acid [7790–92–3]

Hazardous reactions   Contact with alcohols forms unstable alkyl hypochlorites; explodes violently on contact with ammonia gas or acetic anhydride; ignites on contact with arsenic (B904).

Hypohalites Review of class (B1546–1547).

Hyponitrous acid [14448–38–5]

Hazardous reaction   Extremely explosive solid (B1147).

3,3'-Iminodipropiononitrile [111–94–4]

Storage hazard   Bottles in store for 18 months exploded (B609).

Indium compounds

The chloride, nitrate and sulphate of indium are soluble in water. The antimonide, arsenide, selenide and telluride are insoluble compounds used in semiconductor research and attention to toxicity limits has arisen from the risk of inhaling this type of dust during working.

(TOXIC BY INHALATION AND IF SWALLOWED)

RL (as In) 0.1 mg m^{-3}.

Toxic effects   Evidence on human toxicity is scarce, but solutions of indium have been shown to be poisonous when injected into animals: the blood, heart, liver, and kidneys being injured.

First aid   *If soluble salts swallowed:* standard treatment (p. 137).
Spillage disposal   Soluble salts should be mopped up with plenty of water and the solution run to waste, diluting greatly with running water.

Iodic acid [7782–68–5]

Colourless crystals or powder; soluble in water.

(CONTACT WITH COMBUSTIBLE MATERIAL MAY CAUSE FIRE
TOXIC BY INHALATION
CAUSES BURNS)

Avoid breathing dust. Avoid contact with eyes and skin. Do not mix with combustible material.

Toxic effects The dust irritates all parts of the respiratory system. The acid and solutions burn the eyes and skin. If taken by mouth is assumed to cause severe internal irritation and damage.

Hazardous reactions Reacts with boron below 40 °C and incandesces; deflagrates with charcoal, phosphorus and sulphur on heating (B1094).

First aid *Dust inhaled:* standard treatment (p. 136).
Affected eyes: standard treatment (p. 136).
Skin contact: standard treatment (p. 136).
If swallowed: standard treatment (p. 137).

Spillage disposal Shovel into bucket of water and run solution or suspension to waste diluting greatly with running water. Site of spillage should be washed thoroughly to remove all oxidant, which is liable to render any organic matter (particularly wood, paper and textiles) with which it comes into contact, dangerously combustible when dry. Clothing wetted with the solution should be washed thoroughly.

Iodine [7753–56–2]

Bluish-black crystalline scales with a characteristic odour; almost insoluble in water.

HARMFUL BY INHALATION AND IN CONTACT WITH SKIN

Avoid breathing vapour. Avoid contact with eyes and skin. RL 0.1 ppm (1 mg m^{-3}).

Toxic effects The vapour irritates all parts of the respiratory system. The vapour and solid irritate the eyes. The solid burns the skin. If taken by mouth there is severe internal irritation and damage.

Hazardous reactions Forms highly explosive addition compounds with ammonia; reaction with ethanol and phosphorus considered dangerous as school experiment; incandescent reaction with BrF_3; violent reaction with BrF_5; ignites with ClF_3; ignites with F_2; several metal acetylides and carbides react very exothermally with iodine; mixture with potassium explodes weakly on impact; moisture ignites mixtures with Al, Mg, Ti and Zn powders, and Sb may ignite dry; explosion with tetraamminecopper(II) sulphate/ethanol (B1227–1230).

First aid *Vapour inhaled:* standard treatment (p. 136).
Affected eyes: standard treatment (p. 136).
Skin contact: standard treatment (p. 136).
If swallowed: standard treatment (p. 137).

Spillage disposal If the spillage is large and in a confined area, breathing apparatus should be worn. Large quantities are best disposed of by sweeping up with sand and burying in open waste land. Small quantities can be dealt with by dissolving in sodium thiosulphate or sodium metabisulphite solution and running the resulting solution to waste, diluting greatly with running water. Iodine stains on flooring can be cleared by mopping with thiosulphate or metabisulphite solution.

Iodine azide [14696–82–3]

Hazardous reaction Shock- and friction-sensitive explosive (B1226).

Iodine bromide [7789–33–5]

Hazardous reactions K, Na, Sn and P react violently or explosively (B96).

Iodine chloride [7790–99–0]

Reddish-brown liquid with pungent odour; soluble in water.

> (HARMFUL VAPOUR
> CAUSES SEVERE BURNS)

Avoid breathing vapour. Prevent contact with skin and eyes.

Toxic effects The vapour irritates all parts of the respiratory system. The vapour and liquid burn the eyes severely. The liquid burns the skin. Assumed to cause severe internal irritation and damage if taken by mouth.

Hazardous reactions Mixtures with sodium explode on impact, with potassium on contact. Al foil may ignite on long contact (B923).

First aid *Vapour inhaled:* standard treatment (p. 136).
Affected eyes: standard treatment (p. 136).
Skin contact: standard treatment (p. 136).
If swallowed: standard treatment (p. 137).

Spillage disposal Wear face-shield or goggles, and gloves; spread soda ash liberally over the spillage and mop up cautiously with plenty of water — run to waste, diluting greatly with running water.

Iodine compounds Review of Group (B1552).

Iodine heptafluoride [16921-96-3]

Hazardous reactions Activated carbon ignites in gas; mixtures with methane ignite, those with hydrogen explode on heating or sparking; violent reactions with Ba, K and Na on contact but Al, Mg and Sn require heating to react; benzene, light petroleum, ethanol and ether ignite on contact; vigorous reactions with acetic acid, acetone or ethyl acetate (B1080).

Iodine isocyanate [3607-48-5]

Storage hazard Solutions gradually deposit touch-sensitive explosive (B180).

Iodine(V) oxide (iodine pentaoxide) [12029-98-0]

Hazardous reactions Reacts violently with BrF_5; reacts explosively with warm Al, carbon, sulphur, resin, sugar or powdered easily oxidiable elements (B1230).

Iodine pentafluoride [7783-66-6]

Hazardous reactions Reacts violently with benzene above 50 °C; reactions with water and dimethyl sulphoxide are violent; incandescence may occur with B, Si, red P, S, Sb, Bi, Mo, W; K or molten Na explode on contact; calcium carbide or potassium hydride incandesce on contact. Violent reactions with organic compounds or KOH; explodes with Me_3SiNEt_2 at −80 °C (B1074–1075).

Iodine(III) perchlorate [38005-31-1]

Hazardous reaction Exploded on laser irradiation at low temperature (B993).

Iodine triacetate [6540-76-7]

Hazardous reaction Explodes at 140 °C (B608).

Iodine trichloride [865-44-1]

Fuming, orange-red crystalline masses with pungent odour; soluble in water.

(HARMFUL VAPOUR
CAUSES SEVERE BURNS)

Avoid breathing vapour. Prevent contact with eyes and skin.

Toxic effects	The vapour irritates all parts of the respiratory system. The vapour and solid irritate and burn the eyes severely. The solid burns the skin. Assumed to cause severe internal irritation and damage if taken by mouth.
Hazardous reaction	Ignites phosphorus (B982).
First aid	*If inhaled:* standard treatment (p. 136). *Affected eyes:* standard treatment (p. 136). *Skin contact:* standard treatment (p. 136). *If swallowed:* standard treatment (p. 137).
Spillage disposal	Wear face-shield or goggles, and gloves; spread soda ash liberally over the spillage and mop up cautiously with plenty of water — run to waste, diluting greatly with running water.

Iodoacetic acid [64–69–7]

Colourless crystals; mp 82 °C; soluble in water.

VERY TOXIC BY INHALATION, IN CONTACT WITH SKIN
AND IF SWALLOWED
CAUSES SEVERE BURNS

Prevent contact with skin, eyes and clothing.

Toxic effects	Causes severe irritation and burns to all tissues with which it comes in contact.
First aid	*Affected eyes:* standard treatment (p. 136). *Skin contact:* standard treatment (p. 136). *If swallowed:* standard treatment (p. 137).
Spillage disposal	Wear face-shield or goggles, and gloves. Spread soda ash liberally over the spillage and mop up cautiously with plenty of water — run to waste, diluting greatly with running water.

1-Iodobuta-1,3-diyne [6088–91–1]

Hazardous reaction	Crude material exploded during vacuum distillation; the pure compound exploded on scratching under illumination (B401).

Iododimethylarsine [676–75–5]

Hazardous reaction   Ignites when heated in air (B283).

Iodoethane (ethyl iodide) [75–03–6]

Colourless to brown liquid, sensitive to air and light; bp 72 °C; 4g dissolve in 100g water at 20 °C.

HARMFUL BY INHALATION

Avoid contact with skin, eyes and clothing.

Toxic effects   These are not well documented though it is considered to be moderately toxic by inhalation and narcotic in high concentrations.

Hazardous reactions   Reaction of ethanol, phosphorus and iodine considered too dangerous for preparation in school work; explosion with $AgClO_2$ (B277).

First aid   *Vapour inhaled:* standard treatment (p. 136).
Affected eyes: standard treatment (p. 136).
Skin contact: standard treatment (p. 136).
If swallowed: standard treatment (p. 137).

Spillage disposal   Wear face-shield or goggles, and gloves. Apply dispersing agent if available and work to an emulsion with brush and water — run this to waste, diluting greatly with running water. If dispersant not available, absorb on sand, shovel into bucket(s) and transport to safe, open area for atmospheric evaporation. Site of spillage should be washed thoroughly with water and soap or detergent.

Iodoform [75–47–8]

RL 0.6 ppm (10 mg m^{-3}).

Iodomethane (methyl iodide) [74–88–4]

Heavy, colourless to yellow-brown liquid with a sweetish smell, bp 43 °C; sparingly soluble in water.

TOXIC BY INHALATION, IN CONTACT WITH SKIN
AND IF SWALLOWED
CAUSES BURNS

Avoid breathing vapour. Prevent contact with skin and eyes. Suspected carcinogen. RL (skin) 5 ppm (28 mg m^{-3}).

Toxic effects Inhalation of vapour may cause dizziness, drowsiness, mental confusion, muscular twitching and delirium. The vapour and liquid irritate the eyes and distort the vision. The liquid irritates the skin and may cause blistering. The liquid must be assumed to be irritant and poisonous if taken by mouth.

Hazardous reaction Vigorously reactive with sodium dispersed in toluene (B156–157).

First aid *Vapour inhaled:* standard treatment (p. 136).
Affected eyes: standard treatment (p. 136).
Skin contact: standard treatment (p. 136).
If swallowed: standard treatment (p. 137).

Spillage disposal Instruct others to keep at a safe distance. Wear breathing apparatus and gloves. Apply dispersing agent if available and work to an emulsion with brush and water — run this to waste, diluting greatly with running water. If dispersant not available, absorb on sand, shovel into bucket(s) and transport to safe, open area for atmospheric evaporation. Site of spillage should be washed thoroughly with water and soap or detergent.

3-Iodo-1-phenylpropyne [73513–15–2]
(1-iodo-3-phenyl-2-propyne)

Hazardous reaction Exploded on distillation at about 180 °C (B740).

Iodosylbenzene [536–80–1]

Hazardous reaction Explodes at 210 °C (B578).

Iodylbenzene [696–33–1]

EXPLOSIVE WHEN DRY

Hazardous reaction Explodes at 230 °C; extreme care required in heating, compressing or grinding iodyl compounds (B578).

Iodylbenzene perchlorate [101672–14–4]

Hazardous reaction Small sample exploded violently while still damp (B580).

Ion exchange resins

Hazardous reactions   Expansion on moistening may fracture container; violent oxidation by dichromates or nitric acid with various metal ions (B1553–1554).

Iron(III) chloride anhydrous   [7705–08–0]
(ferric chloride anhydrous)

Black-brown crystals and aggregated masses; violently decomposed by water with the formation of hydrogen chloride.

CAUSES BURNS

Avoid contact with eyes and skin.

Toxic effects   Inhalation of fine crystals produces irritation or burns of the mucous membranes. Will cause painful eye burns. When moisture is present on skin, heat is produced on contact resulting in thermal and acid burns. If taken by mouth the immediate local reaction would cause severe burns.

First aid   *Dust inhaled:* standard treatment (p. 136).
Affected eyes: standard treatment (p. 136).
Skin contact: standard treatment (p. 136).
If swallowed: standard treatment (p. 137).

Spillage disposal   Wear face-shield or goggles, and gloves. Mop up with plenty of water and run to waste, diluting greatly with running water.

Iron(II) maleate   [101672–15–5]

Hazardous reaction   The finely divided compound (a product of phthalic anhydride manufacture) is rapidly oxidised in air above 150 °C and has been involved in plant fires (B795).

Iron(III) oxide   [1309–37–1]

RL (fume as Fe) 5 mg m^{-3}.

Iron pentacarbonyl   [13463–40–6]

RL (as Fe) 0.01 ppm (0.08 mg m^{-3}).

Iron(II) perchlorate [13933–94–3]

Hazardous preparation Violent explosion when mixture of iron (II) sulphate and perchloric acid was being heated (B966).

Iron salts

RL (soluble, as Fe) 1 mg m^{-3}.

Isobutane (2-methylpropane) [75–28–5]

Colourless gas; bp −12 °C; somewhat soluble in water.

EXTREMELY FLAMMABLE LIQUEFIED GAS.

Avoid breathing vapour.

Toxic effects The gas has an anaesthetic effect but is not toxic.

First aid *Gas inhaled in quantity:* standard treatment (p. 136).

Fire hazard Explosive limits 1.8–8.4%; ignition temp. 462 °C. Since the gas is supplied in a cylinder, turning off the valve will reduce any fire involving it; if possible cylinders should be removed quickly from an area in which a fire has developed.

Disposal Surplus gas or leaking cylinder can be vented slowly to air in a safe, open area or gas burnt off through a suitable burner.

Isobutene —*see* 2-Methylpropene

Isobutyl acetate [110–19–0]

HIGHLY FLAMMABLE

RL 150 ppm (700 mg m^{-3}).

Isobutyl alcohol [78–83–1]
(isobutanol; 2-methylpropan-1-ol)

Colourless liquid with characteristic odour; bp 106 °C; sparingly soluble in water (1 part dissolves in about 20 parts water at 25 °C).

FLAMMABLE
HARMFUL BY INHALATION

Avoid breathing vapour. Avoid contact with eyes. RL 50 ppm (150 mg m^{-3}).

Toxic effects   The vapour irritates the respiratory system and, in high concentrations, is narcotic. The liquid irritates the eyes and is harmful if taken internally.

First aid   *Vapour inhaled in high concentrations:* standard treatment (p. 136).
Affected eyes: standard treatment (p. 136).
If swallowed: standard treatment (p. 137).

Fire hazard   Flash point 28 °C; explosive limits at 100 °C 1.7–10.9%; ignition temp. 427 °C *Extinguish fire with* water spray, dry powder, carbon dioxide or vaporising liquid.

Spillage disposal   Shut off all possible sources of ignition. Wear face-shield or goggles, and gloves. Apply non-flammable dispersing agent if available and work to an emulsion with brush and water — run this to waste, diluting greatly with running water. If dispersant not available, absorb on sand, shovel into bucket(s) and transport to safe, open area for atmospheric evaporation. Ventilate site of spillage well to evaporate remaining liquid and dispel vapour.

Isobutylene —*see* 2-Methylpropene

Isobutyl methyl ketone
—*see* 4-Methylpentan-2-one

Isobutyric acid   [79–31–2]

Colourless, oily liquid with rancid odour; bp 154 °C; 1 part is soluble in 5 parts of water at 20 °C

HARMFUL IN CONTACT WITH SKIN AND IF SWALLOWED

Avoid contact with skin, eyes and clothing.

Toxic effects   The liquid burns the eyes and irritates or burns the skin. The liquid is irritant or corrosive if taken by mouth.

First aid   *Affected eyes:* standard treatment (p. 136).
Skin contact: standard treatment (p. 136).
If swallowed: standard treatment (p. 137).

Spillage disposal	Wear face-shield or goggles, and gloves. Spread soda ash liberally over the spillage and mop up with plenty of water — run to waste, diluting greatly with running water.

Isobutyric anhydride —*see* Butyric and Isobutyric anhydrides

Isocyanatomethane (methyl isocyanate)   [624–83–9]

Colourless lachrymatory liquid; bp 44 °C; sparingly soluble in water.

EXTREMELY FLAMMABLE
TOXIC BY INHALATION, IN CONTACT WITH SKIN
AND IF SWALLOWED
IRRITATING TO EYES, RESPIRATORY SYSTEM AND SKIN

Prevent inhalation of vapour. Prevent contact with skin, eyes and clothing. CL (skin) 0.02 mg m^{-3}.

Toxic effects	The vapour irritates the respiratory system and eyes severely, causing lachrymation. With high levels of vapour severe and extensive damage is caused, some still evident over a year after the Bhopal disaster. The liquid irritates the eyes and skin severely and is assumed to be highly irritant and damaging if taken internally.
Hazardous reactions	The underlying cause of the exothermic reaction in the Bhopal storage tank which led to the vaporisation and escape of several thousand kg of the highly toxic material appears to have been presence of a substantial amount of water, possibly accompanied by catalytically active impurities (B247).
First aid	*Vapour inhaled:* standard treatment (p. 136). *Affected eyes:* standard treatment (p. 136). *Skin contact:* standard treatment (p. 136). *If swallowed:* standard treatment (p. 137).
Fire hazard	Flash point below −7 °C. *Extinguish fire with* dry powder, carbon dioxide or vaporising liquid.
Spillage disposal	Shut off all possible sources of ignition. Instruct others to keep at a safe distance. Wear breathing apparatus and gloves. Apply non-flammable dispersing agent and work to an emulsion with brush and water — run this to waste, diluting greatly with running water. If dispersant not available, absorb on sand, shovel into bucket(s) and transport to safe open place for atmospheric evaporation or burial. Site of spillage should be washed thoroughly with water and soap or detergent.

Isophorone (3,5,5-trimethylcyclohex-2-en-1-one)   [78–59–1]

Colourless liquid; bp 215 °C; slightly soluble in water.

IRRITATING TO EYES, RESPIRATORY SYSTEM AND SKIN

Avoid breathing vapour. Avoid contact with skin, eyes and clothing. RL 5 ppm (25 mg m^{-3}).

Toxic effects	The vapour irritates the eyes and respiratory system. The liquid irritates the eyes and may cause corneal damage; it may also irritate the skin because of its degreasing action. It is assumed to be toxic if taken by mouth.
First aid	*Vapour inhaled:* standard treatment (p. 136). *Affected eyes:* standard treatment (p. 136). *Skin contact:* standard treatment (p. 136). *If swallowed:* standard treatment (p. 137).
Spillage disposal	Instruct others to keep at a safe distance. Wear breathing apparatus and gloves. Apply dispersing agent if available and work to an emulsion with brush and water — run this to waste, diluting greatly with running water. If dispersant not available, absorb on sand, shovel into bucket(s) and transport to safe, open area for burial. Site of spillage should be washed thoroughly wih water and soap or detergent.

Isophorone diisocyanate   [4098–71–9]

CL (skin) 0.02 mg m^{-3} (as −NCO).

Isoprene (2-methylbuta-1,3-diene)   [78–79–5]

Colourless volatile liquid; bp 34 °C; insoluble in water.

EXTREMELY FLAMMABLE

Avoid breathing vapour.

Toxic effects	The vapour is irritating to the respiratory system and is narcotic in high concentrations. The liquid irritates the skin and eyes, and is assumed to be irritant and harmful if taken internally.
Hazardous reactions	Absorbs oxygen from air forming explosive polymeric peroxide; explosion occurred during ozonisation in heptane (B509–510).
First aid	*Vapour inhaled:* standard treatment (p. 136). *Affected eyes:* standard treatment (p. 136). *Skin contact:* standard treatment (p. 136). *If swallowed:* standard treatment (p. 137).

Fire hazard   Flash point −53 °C; ignition temp. 220 °C. *Extinguish fire with* foam, dry powder, carbon dioxide or vaporising liquid.

Spillage disposal   Shut off all possible sources of ignition. Instruct others to keep at a safe distance. Wear breathing apparatus and gloves. Apply non-flammable dispersing agent if available and work to an emulsion with brush and water — run this to waste, diluting greatly with running water. If dispersant not available, absorb on sand, shovel into bucket(s) and transport to safe, open area for atmospheric evaporation. Site of spillage should be washed thoroughly with water and soap or detergent.

Isopropyl acetate   [108–21–4]

Colourless liquid; bp 93 °C; 1 part is soluble in 23 parts water at 27 °C.

HIGHY FLAMMABLE

Avoid breathing vapour. Avoid contact with eyes. RL 250 ppm (950 mg m^{-3}).

Toxic effects   The vapour may irritate the eyes and respiratory system and is narcotic in high concentrations. The liquid irritates the eyes and will be irritant and narcotic if taken by mouth.

First aid   *Vapour inhaled in high concentrations:* standard treatment (p. 136).
Affected eyes: standard treatment (p. 136).
If swallowed: standard treatment (p. 137).

Fire hazard   Flash point 4 °C; explosive limits 1.8–7.8%; ignition temp. 460 °C. *Extinguish fire with* dry powder, carbon dioxide or vaporising liquid.

Spillage disposal   Shut off all possible sources of ignition. Instruct others to keep at a safe distance. Wear breathing apparatus and gloves. Apply non-flammable dispersing agent and work to an emulsion with brush and water — run this to waste, diluting greatly with running water. If dispersant not available, absorb on sand, shovel into bucket and transport to safe, open area for atmospheric evaporation. Site of spillage should be washed thoroughly with water and soap or detergent.

Isopropyl alcohol —*see* Propan-2-ol

N-Isopropylaniline   [643–28–7]

TLV (skin) 2 ppm (10 mg m^{-3}).

Isopropylamine [75–31–0]

EXTREMELY FLAMMABLE
IRRITATING TO EYES, RESPIRATORY SYSTEM AND SKIN

TLV 5 ppm (12 mg m^{-3}).

Isopropylbenzene —*see* Cumene

Isopropyl ether —*see* Di-isopropyl ether

Isopropyl glycidyl ether (IGE)
—*see* 2,3-Epoxypropyl isopropyl ether

Isopropyl hydroperoxide [3031–75–2]

Hazardous reaction   Explodes just above bp, 107–109 °C (B381).

Isopropyl hypochlorite [53578–07–7]

Hazardous reaction   Has extremely low stability — explosions have occurred during preparation (B372–373).

Isopropyl nitrate [1712–64–7]

Storage and handling   Technical Bulletin available for this rocket propellant (B376).

Ketene [463–51–43]

RL 0.5 ppm (0.9 mg m^{-3}).

Ketone peroxides Review of group (B1555).

Lactic acid [598–82–3]

Colourless or slightly yellow, syrupy, hygroscopic liquid; mp 16·8 °C when anhydrous; miscible with water.

(IRRITATING TO EYES AND SKIN)

Avoid contact with skin, eyes and clothing.

Toxic effects	Irritates and may burn the eyes and skin. Irritant if taken by mouth.
First aid	*Affected eyes:* standard treatment (p. 136). *Skin contact:* standard treatment (p. 136). *If swallowed:* standard treatment (p. 137).
Spillage disposal	Wear face-shield or goggles, and gloves. Spread soda ash liberally over the spillage and mop up cautiously with plenty of water — run to waste, diluting greatly with running water.

Lanthanum hydride [13864–01–2]

Hazardous reaction	Ignites in air (B1177).

Lauroyl peroxide —*see* Dodecanoyl peroxide

Lead [7439–92–1]

(Fumes and dusts) CL (as Pb) 0.15 mg m^{-3}.

Lead arsenate [10102–48–4]

CL (as Pb) 0.15 mg m^{-3}.

Lead(II) azide [13424–46–9]

Hazardous reaction	A detonator that has been studied in detail; in prolonged contact with copper or zinc may form extremely sensitive azides of these metals (B1311).

Lead(IV) azide [73513–16–3]

Hazardous reaction   Liable to spontaneous decomposition which is sometimes explosive (B1313–1314).

Lead bromate [34018–28–5]

Hazardous reaction   An explosive salt (B106).

Lead chromate [7758–97–6]

Suspected carcinogen RL (as Cr) 0.05 mg m^{-3}: CL (as Pb) 0.15 mg m^{-3}.

Hazardous reactions   As an oxidant it has been involved in fires and explosions when mixed with organic pigments (B1027–1028).

Lead dichlorite [13453–57–1]

Hazardous reactions   Explodes on heating above 100 °C or on rubbing with antimony sulphide or fine sulphur (B981).

Lead diperchlorate [13637–76–8]

Hazardous reaction   A saturated solution of the anhydrous salt in dry methanol exploded violently (B983).

Lead dithiocyanate [592–87–0]

Hazardous reaction   Explosive (B315).

Lead imide [12397–26–1]

Hazardous reactions   Explodes on heating or in contact with water or dilute acids (B1130).

Lead salts

White or coloured crystals or powders.

HARMFUL BY INHALATION AND IF SWALLOWED
DANGER OF CUMULATIVE EFFECTS

Avoid breathing dust. CL (dust as Pb) (0.15 mg m^{-3}).

Toxic effects   Dust inhaled or material swallowed may cause severe internal injury with vomiting, diarrhoea and collapse. **Chronic effects** Loss of appetite, pallor, anaemia, constipation, colic, blue line on gums. (*See* Ch 6, 7.)

First aid   *If swallowed:* standard treatment (p. 137).
Spillage   Small quantities of soluble lead salts can be dissolved in water and the
disposal   solution run to waste, diluting greatly with running water. Insoluble compounds can be mixed with an excess of sand and disposed of as normal refuse.

Lead tetrachloride   [13463–30–4]

Hazardous   Liable to explode with potassium or on heating above 100° C; explodes on
reactions   warming with dilute sulphuric acid (B1005).

Linseed oil   [8001–26–1]

Hazardous   Cloths used to apply this to benches were dropped in waste bin —
reaction   laboratory destroyed by fire some hours later (B1158).

Liquid air

CONTACT WITH COMBUSTIBLE MATERIAL MAY CAUSE FIRE
CAUSES BURNS (Cold burns)

Hazardous   Review (B1559–1560).
reactions

Lithium (metal)   [7439–93–2]

Silver-white metal, becoming yellowish on exposure to moist air; mp 180 °C; reacts with water.

REACTS VIOLENTLY WITH WATER
CONTACT WITH WATER LIBERATES HIGHLY FLAMMABLE GASES
CAUSES BURNS

Avoid contact with skin, eyes and clothing.

Toxic effects   Lithium reacts with moisture forming lithium hydroxide which irritates the skin and mucous surfaces and may cause burns.

Hazardous reactions   Finely divided metal may ignite in air; will burn in nitrogen or carbon dioxide and is difficult to extinguish once alight; reacts violently with BrF_5 and explosively with diazomethane; mixtures with several halocarbons will explode on impact; formation of lithium amalgam may be explosive; ignites on contact with nitric acid; reacts violently or explosively with sulphur; heated metal may explode in humid air; powdered Li exploded with bromobenzene; incandesces in hot ethylene (B1259–1264).

First aid   *Affected eyes:* standard treatment (p. 136).
Skin contact: standard treatment (p. 136).
If swallowed: standard treatment (p. 137).

Fire hazard   Extinguish fire with special extinguishant; powdered graphite, LiCl, KCl or zirconium silicate may also be used.

Disposal   May be allowed to react in large excess of cold water and solution run to waste.

Lithium aluminium hydride
—*see* Aluminium lithium hydride

Lithium azide   [19597–69–4]

Hazardous reactions   Moist or dry salt decomposes explosively at 115–298 °C depending upon rate of heating (B1265).

Lithium benzenehexoxide ('carbonyllithium')   [101672–16–6]

Hazardous reaction   Explodes on contact with water (B648).

Lithium borohydride   [16949–15–8]
(lithium tetrahydroborate)

Hazardous reaction   Contact with limited amounts of water may cause ignition (B63).

Lithium hydride   [60380–67–8]

RL 0.025 mg m^{-3}.

Hazardous reactions	Ignites in warm air; mixtures with liquid oxygen are detonatable explosives (B1098).

Lithium hydroxide [1310–66–3]

White crystals; soluble in water.

(CAUSES BURNS)

Avoid contact with eyes and skin.

Toxic effects	The solid and solution irritate the eyes severely and burn the skin. It is assumed that there is internal irritation and damage if taken by mouth.
First aid	*Affected eyes:* standard treatment (p. 136). *Skin contact:* standard treatment (p. 136). *If swallowed:* standard treatment (p. 137).
Spillage disposal	Wear face-shield or goggles, and gloves. Mop up with plenty of water and run to waste, diluting greatly with running water.

Lithium tetrahydroborate
—*see* Lithium borohydride

Magnesium (powder or turnings) [7439–95–4]

HIGHLY FLAMMABLE
CONTACT WITH WATER LIBERATES HIGHLY FLAMMABLE GASES

Hazardous reactions	As fine powder dispersed in air it is a serious explosion hazard; reaction with beryllium fluoride is violent; explosive acetylide may be formed from traces of acetylene in ethylene oxide if magnesium is contained in fittings used in ethylene oxide service; the powdered metal reacts and may explode on contact with chloromethane, chloroform or carbon tetrachloride and mixtures with carbon tetrachloride or trichloroethylene will flash on heavy impact; mixtures with PTFE used as igniters; ignites, if moist, in fluorine or chlorine; may ignite, if finely divided, on heating in iodine vapour; reacts vigorously with certain cyanides; violently reduces some metal oxides; reaction with metal oxosalts may be explosive; reaction with methanol may become vigorous; heating with K_2CO_3, moist SiO_2, S, Te and numerous oxidants can be hazardous; mixtures of Mg dust with MeOH or H_2O are detonatable (B1267–1271).

Magnesium oxide [1309–48–4]

(Fume) RL (as Mg) 10 mg m^{-3}.

Magnesium perchlorate [10034–81–8]

Hazardous reactions Explosions have followed the use of the anhydrous salt for drying organic solvents and solutions of organic compounds in such solvents; an explosion occurred in a drying tube containing the desiccant between cotton wool wads, which had been used for drying O_2/N_2O_4 mixture; may explode on contact with trimethyl phosphite; explosions recorded with various materials attributed to traces of $HClO_4$ in the salt; explosion with Ar must also have involved an impurity (B971–974).

Maleic anhydride [108–31–6]

White crystalline powder or lumps; mp 53 °C; dissolves in water forming maleic acid.

HARMFUL IF SWALLOWED
IRRITATING TO EYES, RESPIRATORY SYSTEM AND SKIN
MAY CAUSE SENSITISATION BY INHALATION

Avoid breathing dust. Avoid contact with skin and eyes. RL 0.25 ppm (1 mg m^{-3}).

Toxic effects The dust and vapour irritate the eyes, skin and respiratory system; prolonged contact with tissues may result in burns. Assumed to be irritant and harmful if taken by mouth.

Hazardous reactions Decomposes exothermally in presence of alkali or alkaline earth metal or ammonium ions, dimethylamine, triethylamine, pyridine or quinoline at temperatures above 150 °C; sodium ions and pyridine are particularly active (B405).

First aid *Affected eyes:* standard treatment (p. 136).
Skin contact: standard treatment (p. 136).
If swallowed: standard treatment (p. 137).

Spillage disposal Sweep up and dissolve in water — run solution to waste, diluting greatly with running water.

Maleic anhydride ozonide [101672–17–7]

Hazardous reaction Explodes on warming to −40 °C (B406).

Malononitrile (dicyanomethane, methylene cyanide)   [109–77–3]

Colourless crystals, mp 30.5 °C. Polymerises violently on heating to 120 °C and on contact with alkaline materials. Soluble in water.

TOXIC BY INHALATION, IN CONTACT WITH SKIN
AND IF SWALLOWED

Prevent contact with skin and eyes.

Toxic effects   It is said that the toxicity of malononitrile is of the same order as that of hydrogen cyanide though no cases of poisoning of humans have been traced. It is therefore considered wisest to treat the compound with the respect given to the alkali cyanides.

Hazardous reactions   May polymerise violently on heating at 130 °C or in contact with strong bases at lower temperatures (B333–334).

First aid   Special treatment for hydrogen cyanide (p. 132).
Spillage disposal   Wear goggles or face-shield, and gloves. Sweep up, or absorb on sand, and place in a large volume of water to which is then added an excess of sodium hypochlorite solution. Allow to stand for 24 hours with occasional stirring and then run to waste, diluting greatly with running water. The site of the spillage should be mopped with sodium hypochlorite solution and rinsed with water.

Manganese and compounds   [7439–96–5]

RL (as Mn) 5 mg m^{-3}.

Manganese (fume)   [7439–96–5]

RL (as Mn) 1 mg m^{-3}.

Manganese dioxide (manganese(IV) oxide)   [1313–13–9]

Black powder; mp 535 °C (dec); insoluble in water.

HARMFUL BY INHALATION AND IF SWALLOWED

Avoid contact with eyes. RL (as Mn) 5 mg m^{-3}.

Toxic effects Inhalation of dust may lead to increased incidence of respiratory infections; assumed to be harmful if swallowed. **Chronic effects** Continued inhalation of dust may lead to excessive tiredness and effects on the central nervous system.

Hazardous reactions As an oxidant it reacts violently with Al powder, CaH_2, rubidium acetylide, H_2S or potassium azide. It also catalytically decomposes, usually violently or explosively, other oxidants such as concentrated hydrogen peroxide, peroxomonosulphuric acid or $KClO_3$ (B1275–1276).

First aid *Affected eyes:* standard treatment (p. 136).
If swallowed: standard treatment (p. 137).

Spillage disposal Wear face-shield or goggles, and gloves. Cover with sand, sweep up and transport to safe open area for burial. Site of spillage should be washed down thoroughly to remove traces of the oxidant.

Manganese diperchlorate [13770–16–6]

Hazardous reaction Explodes at 195 °C (B974)

Manganese tetraoxide [1317–35–7]

RL 1 mg m^{-3}.

Manganese trifluoride [7783–53–1]

Hazardous reaction Glass is attacked violently by heating in contact with it, SiF_4 being evolved (B1068).

Mercaptoacetic acid (thioglycolic acid) [68–11–1]

Colourless liquid with unpleasant smell; bp 123 °C at 39 mbar; miscible with water. It is readily decomposed by mineral acids with liberation of poisonous hydrogen sulphide.

CAUSES BURNS
TOXIC BY INHALATION, IN CONTACT WITH SKIN AND IF SWALLOWED

Avoid contact with skin and eyes. RL 1 ppm (5 mg m^{-3}).

Toxic effects The liquid irritates the eyes and skin and may cause burns. It must be assumed to be irritant and poisonous if taken by mouth.

First aid *Affected eyes:* standard treatment (p. 136).
Skin contact: standard treatment (p. 136).
If swallowed: standard treatment (p. 137).

Spillage disposal Wear face-shield or goggles, and gloves. Spread soda ash liberally over the spillage and mop up cautiously with plenty of water — run to waste diluting greatly with running water.

Mercury (quicksilver) [7439-97-6]

Heavy silvery liquid; insoluble in water.

TOXIC BY INHALATION
DANGER OF CUMULATIVE EFFECTS

Avoid breathing vapour. Avoid contact with eyes and skin. CL (skin) 0.05 mg m^{-3}.

Toxic effects High concentrations of vapour may cause metallic taste, nausea, abdominal pain, vomiting, diarrhoea and headache. **Chronic effects** Continued exposure to small concentrations of vapour may result in severe nervous disturbance, including tremor of the hands, insomnia, loss of memory, irritability and depression; other possible effects are loosening of teeth and excessive salivation. Continued skin contact with mercury may cause dermatitis and the above effects may be caused by absorption through the skin or following ingestion. Kidney damage may ensue. (*See* Ch 6, 7.)

Hazardous reactions Prolonged contact between mercury and ammonia may result in formation of explosive solid; ease with which it forms amalgams with laboratory and electrical contact metals can cause severe corrosion problems; reacts violently with dry bromine; chlorine dioxide explodes when shaken with mercury (B1217-1218).

First aid *Vapour inhaled:* standard treatment (p. 136).
Skin contact: standard treatment (p. 136).
If swallowed: standard treatment (p. 137).

Spillage disposal Because of the high toxicity of mercury vapour it is important to clean up mercury as thoroughly as possible, especially in confined areas. A small aspirator with a capillary tube and connected to a pump can be used for sucking up droplets. Mercury spilt into floor cracks can be made non-volatile by putting zinc dust down the cracks to form the amalgam. Smooth surfaces may be decontaminated by scattering and sweeping up a mixture of equal weights of zinc dust and sawdust, which should then be buried at an isolated site. (*See* handling note, p. 34)

Mercury(I) azide [38232-63-2]

Hazardous reaction Explodes in air on heating to over 270 °C (B1223).

Mercury(I) chlorite [101672–18–8]

Hazardous reaction   Explodes spontaneously when dry (B970).

Mercury compounds

Mercury compounds vary widely in appearance, solubility in water and toxicity. Mercury(II) compounds are generally more toxic than mercury(I) compounds. Some organic mercurial compounds are liquids with an extremely poisonous vapour, others are solids, the toxicity of which is not known with certainty. Thus phenylmercuric acetate appears relatively non-toxic but the alkyl mercurials are highly poisonous. Their toxicity by skin absorption is uncertain, but some, *e.g.* ethyl mercury phosphate and mercury fulminate, can cause dermatitis. The effects vary greatly according to the nature of the organic mercurial. Some compounds cause kidney damage while others can cause irreversible damage to the central nervous system.

> VERY TOXIC BY INHALATION, IN CONTACT WITH SKIN
> OR IF SWALLOWED
> DANGER OF CUMULATIVE EFFECTS

Avoid breathing dusts. Avoid contact with skin and eyes. RL (skin, as Hg) alkyl compounds 0.01 mg m^{-3}; all forms except alkyl, 0.05 mg m^{-3}.

Toxic effects   Inhalation of dust may cause nausea, abdomial pain, vomiting, diarrhoea and headache. Skin absorption may give rise to similar effects and the eyes may be damaged by direct contact with some salts. Abdominal pain, nausea, vomiting, diarrhoea and shock follow ingestion of the soluble mercuric salts. Kidney damage may ensue. **Chronic effects** The intake of small amounts of mercury compounds by inhalation, skin absorption or ingestion over a long period may cause nervous disturbance, including tremor of the hands, insomnia, loss of memory, irritability and depression; other possible effects are loosening of the teeth and excessive salivation. (*See* numerous references to the toxicity of mercury compounds in Ch 6.)

Hazardous reactions   Many Hg compounds show explosive instability (B1561–1562).

First aid   *Dusts inhaled:* standard treatment (p. 136).
Affected eyes: standard treatment (p. 136).
Skin contact: standard treatment (p. 136).
If swallowed: standard treatment (p. 137).

Spillage disposal   Small quantities of soluble mercury compounds can be swept up, dissolved in water or acid and run to waste at very high dilution. Recovery may be warranted in the case of large amounts but the disposal of these must be considered carefully in the light of local conditions and regulations. If burial is carried out in an isolated area the solid compound should first be diluted 10–20 times by weight with sand.

Mercury(II) cyanide [592–04–1]

Hazardous reactions

A friction- and impact-sensitive explosive which can detonate liquid HCN; reacts explosively with magnesium; explodes on heating with sodium nitrite (B309).

Mercury dichlorite [73513–17–4]

Hazardous reaction

Explodes spontaneously when dry (B969).

Mercury(II) fulminate [628–86–4]

TOXIC BY INHALATION, IN CONTACT WITH SKIN
AND IF SWALLOWED
DANGER OF CUMULATIVE EFFECTS

Hazardous reactions

A denotator initiated when dry by flame, heat, impact, friction or intense radiation; contact with sulphuric acid causes explosion (B309–310).

Mercury(II) nitrate [10045–94–0]

Hazardous reactions

Contact with acetylene in solution gives explosive mercury acetylide; with the aqueous solution ethanol gives mercury(II) fulminate; aqueous solution reacts with phosphine to give explosive complex; risk of violent reactions with petroleum hydrocarbons; violent reaction with phosphinic acid (B1219–1220).

Mercury(II) oxalate [3444–13–1]

Hazardous reactions

When dry, it explodes on percussion, grinding or heating to 105 °C; storage is inadvisable (B310).

Mercury(II) perbenzoate [18918–17–7]

Hazardous reaction

Explodes if heated above 110 °C (B827).

Mercury(II) perchlorate [7616–83–3]

Hazardous reactions

Aqueous solution and some solvated complexes showed explosive tendencies (B969–970).

Mesityl oxide —*see* 4-Methylpent-3-en-2-one

Mesoxalonitrile   [1115–12–4]

**Hazardous   Reacts explosively with water (B393).
reaction**

Metal abietates —*see* Abietates, metal

Metal acetylides Review of group (B1563).

Metal alkoxides Review of group (B1565).

Metal azides and azide halides Reviews of classes (B1566).

Metal cyanides and cyano complexes Review (B1570).

N-Metal derivatives Review of group (B1571).

Metal dusts Review of class (B1574).

Metal fulminates Review of group (B1574).

Metal hydrides Review of group (B1579).

Metal hypochlorites Review of group (B1580).

Metal perchlorates Review of group (B1588).

Metal pyruvate phenylhydrazones Review of group (B1591).

Methacrylic acid [79–41–4]

Colourless solid or liquid with unpleasant, acrid smell; mp 16 °C; bp 158 °C; soluble in water.

CAUSES BURNS

Avoid breathing vapour. Avoid contact with skin, eyes and clothing. RL 20 ppm (70 mg m^{-3}).

Toxic effects	The vapour irritates the eyes and respiratory system. The liquid irritates the eyes and skin and is assumed to be very irritant and harmful if taken internally.
First aid	*Vapour inhaled:* standard treatment (p. 136). *Affected eyes:* standard treatment (p. 136). *Skin contact:* standard treatment (p. 136). *If swallowed:* standard treatment (p. 137).
Spillage disposal	Wear goggles and gloves. Spread soda ash liberally over the spillage and mop up cautiously with plenty of water — run to waste, diluting greatly with running water.

Methacrylonitrile [126–98–7]

HIGHLY FLAMMABLE
TOXIC BY INHALATION, IN CONTACT WITH SKIN
AND IF SWALLOWED
MAY CAUSE SENSITISATION BY SKIN CONTACT

TLV (skin) 1 ppm (3 mg m^{-3}).

Methane [74–82–8]

Colourless gas, sparingly soluble in water.

EXTREMELY FLAMMABLE

Toxic effects	Methane is non-toxic but can have narcotic effects in high concentrations in the absence of oxygen.
First aid	*Vapour inhaled in high concentrations:* standard treatment (p. 136).
Fire hazard	Explosive limits 5–15%; ignition temp. 537 °C. Since the gas is supplied in cylinders, turning off the valve will reduce any fire involving it; if possible, cylinders should be removed quickly from an area in which a fire has developed.

Disposal   Surplus gas or leaking cylinder can be vented slowly to air in a safe, open area or gas burnt off through a suitable burner.

Methanesulphinyl chloride   [676–85–7]

Hazardous reaction   An unrefrigerated ampoule burst after extended storage (B155).

Methanethiol (methyl mercaptan)   [74–93–1]

Colourless gas with extremely disagreeable smell; bp 6 °C; sparingly soluble in water.

EXTREMELY FLAMMABLE LIQUEFIED GAS
HARMFUL BY INHALATION

Avoid breathing gas. Avoid contact with eyes. RL 0.5 ppm (1 mg m^{-3}).

Toxic effects   The gas is nauseous and may be narcotic in high concentrations.

Hazardous reaction   Reaction with HgO is rather violent (B173).

First aid   *Vapour inhaled:* standard treatment (p. 136).
Fire hazard   Flash point below −18 °C; explosive limits 3.9–21.8%. Since the gas is supplied in a cylinder, turning off the valve will reduce any fire in which it is involved; if possible, cylinders should be removed quickly from an area in which a fire has developed.
Disposal   Surplus gas or gas from a leaking cylinder should be burnt through a suitable gas burner in a fume cupboard.

Methanol (methyl alcohol)   [67–56–1]

Colourless, volatile liquid, bp 65 °C; miscible with water.

HIGHLY FLAMMABLE
TOXIC BY INHALATION AND IF SWALLOWED

Avoid breathing vapour. Avoid contact with skin and eyes. RL (skin) 200 ppm (260 mg m^{-3}).

Toxic effects   Inhalation of high concentrations of vapour may cause dizziness, stupor, cramps and digestive disturbance. Lower concentrations may cause headache, nausea, vomiting and irritation of the mucous membranes. The vapour and liquid are very dangerous to the eyes, the effects sometimes being delayed for many hours. Ingestion damages the central nervous system, particularly the optic nerve (causing temporary or permanent blindness), and injures the kidneys, liver, heart and other organs; apart from the effects

395

referred to above, unconsciousness may develop after some hours and this may be followed by death. **Chronic effects** Continued exposure to low concentrations of vapour may cause many of the above effects, continued skin contact may cause dermatitis. (Ch 6 p 104.)

Hazardous reactions

Violent explosion occurred when sodium was added to methanol/chloroform mixture; reaction with magnesium can be very vigorous; reaction with bromine can be violent, with sodium hypochlorite explosive; it is ignited by CrO_3; reactions with nitric acid or hydrogen peroxide may become explosive; violent reaction or ignition with alkylaluminium solutions, diethylzinc, P_2O_3, 2,4,6-trichloro-s-triazine (B170–172).

First aid

Vapour inhaled: standard treatment (p. 136).
Affected eyes: standard treatment (p. 136).
Skin contact: standard treatment (p. 136).
If swallowed: standard treatment (p. 137).

Fire hazard

Flash point 10 °C; explosive limits 7.3–36.5%; ignition temp. 464 °C. *Extinguish fire with* water spray, dry powder, carbon dioxide or vaporising liquid.

Spillage disposal

Shut off all possible sources of ignition. Instruct others to keep at a safe distance. Wear breathing apparatus and gloves. Mop up with plenty of water and run to waste, diluting greatly with running water. Ventilate area well to evaporate remaining liquid and dispel vapour.

RSC *Lab. Haz. Data Sheet No. 25,* 1984 gives extended coverage.

p-Methoxybenzoyl chloride [100–07–2]

Storage hazard

Bottles of this exploded on storage at room temperature (B705).

2-Methoxyethanol [109–86–4]

(ethylene glycol monomethyl ether; methyl cellosolve)

Colourless volatile liquid with pleasant smell; bp 125 °C; miscible with water; liable to form explosive peroxides on exposure to light and air which should be decomposed before distilling the ether to small volume.

FLAMMABLE
HARMFUL BY INHALATION, IN CONTACT WITH SKIN
AND IF SWALLOWED
IRRITATING TO RESPIRATORY SYSTEM

Avoid breathing the vapour. Avoid contact with skin and eyes. CL (skin) 5 ppm (16 mg m^{-3}).

Toxic effects

The vapour irritates the respiratory system. The vapour and liquid irritate the eyes. Assumed to be poisonous if taken by mouth. **Chronic effects** Anaemia, blood abnormalities and symptoms of central nervous system damage have resulted from prolonged exposure.

First aid	*Vapour inhaled:* standard treatment (p. 136). *Affected eyes:* standard treatment (p. 136). *If swallowed:* standard treatment (p. 137).
Fire hazard	Flash point 46 °C; explosive limits 2.5–14%; ignition temp. 288 °C. *Extinguish fire with* water spray, dry powder, carbon dioxide or vaporising liquid.
Spillage disposal	Shut off all possible sources of ignition. Instruct others to keep at a safe distance. Wear breathing apparatus and gloves. Mop up with plenty of water and run to waste, diluting greatly with running water. Ventilate area well to evaporate remaining liquid and dispel vapour.

2-Methoxyethyl acetate [110–49–6]

(ethylene glycol monomethyl ether acetate; methyl cellosolve acetate)

Colourless liquid; bp 143 °C; miscible with water; liable to form explosive peroxides on exposure to light and air which must be decomposed before the ether ester is distilled to small volume.

FLAMMABLE
HARMFUL BY INHALATION AND CONTACT WITH SKIN

Avoid contact with skin, eyes and clothing. CL (skin) 5 ppm (24 mg m^{-3}).

Toxic effects	The liquid irritates the eyes and may irritate the skin; it may cause headache, dizziness, fatigue, nausea, vomiting and more serious disorders if taken by mouth or absorbed extensively through the skin.
First aid	*Affected eyes:* standard treatment (p. 136). *Skin contact:* standard treatment (p. 136). *If swallowed:* standard treatment (p. 137).
Fire hazard	Flash point 54 °C. *Extinguish fire with* dry powder, carbon dioxide or vaporising liquid.
Spillage disposal	Shut off all possible sources of ignition. Wear face-shield or goggles, and gloves. Mop up with plenty of water and run to waste, diluting greatly with running water. Ventilate area of spillage well to evaporate remaining liquid and dispel vapour.

3-Methoxypropyne [627–41–8]

Hazardous reaction	Explodes on distillation at 61 °C (B428).

Methyl acetate [79–20–9]

Colourless, volatile liquid with pleasant odour; bp 58 °C; miscible with water.

HIGHLY FLAMMABLE

Avoid breathing vapour. Avoid contact with skin and eyes. RL 200 ppm (610 mg m^{-3}).

Toxic effects   The vapour irritates the eyes and respiratory system. High concentrations may cause dizziness and palpitations. Assumed to be poisonous if taken by mouth.

First aid   *Vapour inhaled:* standard treatment (p. 136).
Affected eyes: standard treatment (p. 136).
Skin contact: standard treatment (p. 136).
If swallowed: standard treatment (p. 137).

Fire hazard   Flash point −9 °C; explosive limits 3.1–16%; ignition temp. 502 °C. *Extinguish fire with* water spray, foam, dry powder, carbon dioxide or vaporising liquid.

Spillage disposal   Shut off all possible sources of ignition. Wear face-shield or goggles, and gloves. Mop up with plenty of water and run to waste, diluting greatly with running water. Ventilate area of spillage well to evaporate remaining liquid and dispel vapour.

Methylacetylene —*see* Propyne

Methyl acrylate (methyl propenoate)   [96–33–3]

Colourless liquid with acrid odour; bp 80 °C; immiscible with water.

HIGHLY FLAMMABLE
HARMFUL BY INHALATION AND IF SWALLOWED
IRRITATING TO EYES, RESPIRATORY SYSTEM AND SKIN

Avoid breathing vapour. Avoid contact with skin and eyes. RL (skin) 10 ppm (35 mg m^{-3}).

Toxic effects   The vapour irritates the eyes and respiratory system; high concentrations may cause lethargy and lead to convulsions. The liquid irritates the skin and eyes. Assumed to be poisonous if taken by mouth.

Hazardous reaction   Peroxidises and may polymerise violently. Store inhibited but with *access* of air (B429–430).

First aid   *Vapour inhaled:* standard treatment (p. 136).
Affected eyes: standard treatment (p. 136).
Skin contact: standard treatment (p. 136).
If swallowed: standard treatment (p. 137).

Fire hazard   Flash point −2.8 °C; explosive limits 2.8–25%. *Extinguish fire with* water spray, foam, dry powder, carbon dioxide or vaporising liquid.

Spillage disposal   Shut off all possible sources of ignition. Instruct others to keep at a safe distance. Wear breathing apparatus and gloves. Apply non-flammable dispersing agent if available and work to an emulsion with brush and water — run this to waste, diluting greatly with running water. If dispersant not available, absorb on sand, shovel into bucket(s) and transport to safe, open area for atmospheric evaporation or burial. Site of spillage should be washed thoroughly with water and soap or detergent.

Methylal —*see* Dimethoxymethane

Methyl alcohol —*see* Methanol

Methylamine [74–89–5]

Colourless gas with pungent, fishy smell, bp −6.3 °C; very soluble in water (see below for solutions).

> EXTREMELY FLAMMABLE LIQUEFIED GAS
> IRRITATING TO EYES AND RESPIRATORY SYSTEM

Avoid breathing gas. Avoid contact with skin and eyes. RL 10 ppm (12 mg m^{-3}).

Toxic effects	The gas irritates the skin, eyes and respiratory system and sustained contact may cause burns.
Hazardous reaction	Addition to nitromethane renders it susceptible to initiation by a detonator (B160).
First aid	*Gas inhaled:* standard treatment (p. 136). *Affected eyes:* standard treatment (p. 136). *Skin contact:* standard treatment (p. 136).
Fire hazard	Flash point 0 °C; explosive limits 4.9–20.7%; ignition temp. 430 °C. Since the gas is supplied in a cylinder, turning off the valve will reduce any fire involving it; if possible, cylinders should be removed quickly from an area in which fire has developed.
Disposal	Surplus gas or leaking cylinder can be vented slowly into a water-fed scrubbing tower or column in a fume cupboard, or into a fume cupboard served by such a tower.

Methylamine solutions [74–89–5]

Methylamine is commonly available in either aqueous or ethanolic solution in a concentration of 25–33%; these solutions are colourless and are miscible with water; they have a fishy smell.

> HIGHLY FLAMMABLE
> IRRITATING TO EYES AND RESPIRATORY SYSTEM

Avoid breathing vapour. Avoid contact with skin and eyes. RL (as CH_3NH_2) 10 ppm (12 mg m^{-3}).

Toxic effects	The vapour irritates the eyes and respiratory system. The solutions irritate the eyes and skin. The solution will cause irritation and damage if taken internally.

First aid
    Vapour inhaled: standard treatment (p. 136).
    Affected eyes: standard treatment (p. 136).
    Skin contact: standard treatment (p. 136).
    If swallowed: standard treatment (p. 137).

Fire hazard
    Flash point depends upon nature and strength of solution. *Extinguish fire with* water spray, foam, dry powder, carbon dioxide or vaporising liquid.

Spillage disposal
    Shut off all possible sources of ignition. Instruct others to keep at a safe distance. Wear breathing apparatus and gloves. Mop up with plenty of water and run to waste diluting greatly with running water. Ventilate area well to evaporate remaining liquid and dispel vapour.

Methyl n-amyl ketone —*see* Heptan-2-one

N-Methylaniline [100–61–8]

Colourless to brown liquid; bp 196 °C; insoluble in water.

TOXIC BY INHALATION, IN CONTACT WITH SKIN
AND IF SWALLOWED
DANGER OF CUMULATIVE EFFECTS

Avoid breathing vapour. Avoid contact with skin and eyes. TLV (skin) (0.5) ppm (2 mg m^{-3}).

Toxic effects
    Excessive breathing of the vapour or absorption of the liquid through the skin can cause headache, drowsiness, cyanosis, mental confusion and, in severe cases, convulsions. The liquid is dangerous to the eyes, and the above effects can also be experienced if it is swallowed. **Chronic effects** Continued exposure to the vapour, or slight skin exposure to the liquid over a period may affect the nervous system and the blood, causing fatigue, loss of appetite, headache and dizziness.

First aid
    Vapour inhaled: standard treatment (p. 136).
    Affected eyes: standard treatment (p. 136).
    Skin contact: standard treatment (p. 136).
    If swallowed: standard treatment (p. 137).

Spillage disposal
    Wear face-shield or goggles, and gloves. Mix with sand and shovel mixture into glass, enamel or polythene vessel for dispersion in an excess of dilute hydrochloric acid (1 volume of concentrated acid diluted with 2 volumes of water). Allow to stand, with occasional stirring, for 24 hours and then run acid extract to waste, diluting greatly with running water and washing the sand. The sand can be disposed of as normal refuse.

Methyl azide   [624–90–8]

Hazardous reactions	Stable at room temperature but may detonate on rapid heating; presence of mercury in the azide markedly reduces stability towards shock or electric discharge; a mixture with methanol and dimethyl malonate exploded while being sealed into a Carius tube (B164).

Methyl benzenediazoate   [66127–76–3]

Hazardous reaction	Explodes on heating or after about an hour's storage in a sealed tube at ambient temperature (B684).

Methyl bromide —*see* Bromomethane

2-Methylbuta-1,3-diene —*see* Isoprene

2-Methylbutan-2-ol   [75–85–4]
(t-amyl alcohol; t-pentyl alcohol)

Colourless, volatile liquid with characteristic odour and burning taste; bp 102 °C.

HIGHLY FLAMMABLE
HARMFUL BY INHALATION

Avoid breathing vapour. Avoid contact with skin and eyes.

Toxic effects	Vapour may irritate the eyes and respiratory system. Liquid irritates the eyes severely and may irritate skin. If swallowed may cause headache, vertigo, nausea, vomiting, excitement and delirium followed by coma.
First aid	*Vapour inhaled:* standard treatment (p. 136). *Affected eyes:* standard treatment (p. 136). *Skin contact:* standard treatment (p. 136). *If swallowed:* standard treatment (p. 137).
Fire hazard	Flash point 19 °C; explosive limits 1.2–9%; *Extinguish fire with* water spray, dry powder, carbon dioxide or vaporising liquids.
Spillage disposal	Shut off all possible sources of ignition. Wear face-shield and gloves. Apply non-flammable dispersing agent if available and work to an emulsion with brush and water — run this to waste diluting greatly with running water. If dispersant not available, absorb on sand, shovel into bucket(s) and transport to safe open area for atmospheric evaporation or burial. Ventilate well to evaporate remaining liquid and dispel vapour.

Methyl butyl ketone —*see* Hexan-2-one

Methyl carbonate —*see* Dimethyl carbonate

Methyl cellosolve —*see* 2-Methoxyethanol

Methyl cellosolve acetate —*see* 2-Methoxyethyl acetate

Methyl chloride —*see* Chloromethane

Methylchloroform —*see* 1,1,1-Trichloroethane

Methyl chloroformate [79–22–1]

Colourless liquid, bp 71 °C; immiscible with water.

HIGHLY FLAMMABLE
TOXIC BY INHALATION
IRRITATING TO RESPIRATORY SYSTEM, EYES AND SKIN

Avoid breathing vapour. Avoid contact with skin and eyes.

Toxic effects	The vapour irritates the respiratory system. The vapour irritates and the liquid burns the eyes. The liquid can irritate and burn the skin. Assumed to be irritant and poisonous if taken by mouth.
First aid	*Vapour inhaled:* standard treatment (p. 136). *Affected eyes:* standard treatment (p. 136). *Skin contact:* standard treatment (p. 136). *If swallowed:* standard treatment (p. 137).
Fire hazard	Flash point 12 °C; ignition temp. 504 °C. *Extinguish fire with* water spray, foam, dry powder, carbon dioxide or vaporising liquid.
Spillage disposal	Shut off all possible sources of ignition. Instruct others to keep at a safe distance. Wear breathing apparatus and gloves. Apply non-flammable dispersing agent and work to an emulsion with brush and water — run this to waste, diluting greatly with running water. If dispersant not available absorb on sand, shovel into bucket and transport to safe, open area for atmospheric evaporation. Site of spillage should be washed thoroughly with water and soap or detergent.

Methyl cyanide —*see* Acetonitrile

Methyl 2-cyanoacrylate   [137–05–3]

RL 2 ppm (8 mg m^{-3}).

Methylcyclohexane   [108–87–2]

Colourless liquid; bp 100 °C; insoluble in water.

HIGHLY FLAMMABLE

Avoid breathing vapour. Avoid contact with skin and eyes. RL 400 ppm (1600 mg m^{-3}).

Toxic effects	These are not well documented but animal experiments suggest that it is more toxic than cyclohexane. In high concentrations the vapour causes narcosis and anaesthesia.
First aid	*Vapour inhaled in high concentrations:* standard treatment (p. 136). *Affected eyes:* standard treatment (p. 136). *Skin contact:* standard treatment (p. 136). *If swallowed:* standard treatment (p. 137).
Fire hazard	Flash point −4 °C; ignition temp. 285 °C. *Extinguish fire with* foam, dry powder, carbon dioxide or vaporising liquid.
Spillage disposal	Shut off all possible sources of ignition. Wear face-shield or goggles, and gloves. Apply non-flammable dispersing agent if available and work to an emulsion with brush and water — run this to waste, diluting greatly with running water. If dispersant not available, absorb on sand, shovel into bucket(s) and transport to safe, open area for atmospheric evaporation. Ventilate site of spillage well to evaporate remaining liquid and dispel vapour.

Methylcyclohexanol (mixed isomers)   [25639–42–3]

Colourless, viscous liquid with odour similar to that of menthol; bp of mixed isomers 155–180 °C; slightly soluble in water.

HARMFUL BY INHALATION

Avoid breathing vapour. Avoid contact with skin, eyes and clothing. RL 50 ppm (235 mg m^{-3}).

Toxic effects	The vapour irritates the eyes and respiratory system. The liquid irritates the eyes and may irritate the skin. Assumed to be irritant and harmful if taken by mouth.

Hazardous chemicals

| First aid | *Vapour inhaled:* standard treatment (p. 136).
Affected eyes: standard treatment (p. 136).
Skin contact: standard treatment (p. 136).
If swallowed: standard treatment (p. 137). |

First aid *Vapour inhaled:* standard treatment (p. 136).
Affected eyes: standard treatment (p. 136).
Skin contact: standard treatment (p. 136).
If swallowed: standard treatment (p. 137).

Spillage disposal Wear face-shield or goggles, and gloves. Apply dispersing agent if available and work to an emulsion with brush and water — run this to waste, diluting greatly with running water. If dispersant not available, absorb on sand, shovel into bucket(s) and transport to safe, open area for burial. Site of spillage should be washed thoroughly with water and soap or detergent.

2-Methylcyclohexanone [583–60–8]
(and isomers)

Colourless to pale yellow liquid with smell like that of acetone; bp of mixed isomers 160–170 °C; insoluble in water.

FLAMMABLE
HARMFUL BY INHALATION

Avoid breathing vapour. Avoid contact with skin, eyes and clothing. RL (skin) 50 ppm (230 mg m^{-3}).

Toxic effects The vapour irritates the eyes and respiratory system. The liquid irritates the eyes and prolonged contact with the skin may result in kidney and liver damage. Assumed to cause irritation and damage if taken internally.

Hazardous reactions Oxidation of the 4-isomer by addition to nitric acid at about 75 °C caused a detonation; reaction with mixtures of hydrogen peroxide and nitric acid caused the 3-isomer to form an oily explosive peroxide (B691).

First aid *Vapour inhaled:* standard treatment (p. 136).
Affected eyes: standard treatment (p. 136).
Skin contact: standard treatment (p. 136).
If swallowed: standard treatment (p. 137).

Fire hazard Flash point 48 °C. *Extinguish fire with* foam, dry powder, carbon dioxide or vaporising liquid.

Spillage disposal Wear face-shield or goggles, and gloves. Apply dispersing agent if available and work to an emulsion with brush and water — run this to waste, diluting greatly with running water. If dispersant not available, absorb on sand, shovel into bucket(s) and transport to safe, open area for burial. Site of spillage should be washed thoroughly with water and soap or detergent.

Methylcyclopentadienylmanganese tricarbonyl

RL (skin, as Mn) 0.2 mg m^{-3}. [12108–13–3]

.

3-Methyldiazirine [765–31–1]

Hazardous reaction The gas explodes on heating (B260).

Methylenebis(2-chloroaniline) [101–14–4]

Suspected carcinogen, CL (skin) 0.005 mg m^{-3}.

Methylenebis(4-cyclohexyl isocyanate) [28615–81–4]

CL (as –NCO) 0.02 mg m^{-3}.

Methylenebis(4-phenyl isocyanate) [101–68–8]
(MDI)

HARMFUL BY INHALATION

CL (as –NCO) 0.02 mg m^{-3}.

Methyl diazoacetate [6832–16–2]

Hazardous reaction Explodes with violence when heated (B347).

Methylene chloride —*see* Dichloromethane

Methylene chlorobromide —*see* Bromochloromethane

Methylene cyanide —*see* Malononitrile

4-Methyleneoxetan-2-one —*see* Diketene

Methyl ether —*see* Dimethyl ether

Methyl ethyl ketone —*see* Butanone

Methyl ethyl ketone peroxide —*see* Butanone peroxide

Methyl fluorosulphate   [421–20–5]
(methyl fluorosulphonate)

Liquid, ethereal odour; bp 92 °C; decomposed by water.

(VERY TOXIC BY INHALATION, IN CONTACT WITH SKIN
AND IF SWALLOWED
DANGER OF VERY SERIOUS IRREVERSIBLE EFFECTS)

Prevent absolutely inhalation of vapour, or skin or eye contact with vapour or liquid. Use only in fume cupboard of proven effectiveness. Potent experimental mutagen.

Toxic effects   Vapour irritates eyes severely and may lead to temporary corneal damage. Inhalation of vapour in small amounts has led to a temporary cough, succeeded after six hours by fatal lung inflammation.

First aid   *Vapour inhaled:* Application of dexamethason isonicotinate ("Auxiloson") has been recommended; keep this available.
Affected eyes: standard treatment (p. 136).
Skin contact: standard treatment (p. 136).
If swallowed: standard treatment (p. 137).

Spillage disposal   Evacuate the area, wear breathing apparatus and full protective clothing. Spread soda-ash liberally over the spillage and leave to stand with occasional stirring for a few minutes. Shovel up, transfer to fume cupboard and add to excess 10% sodium hydroxide solution and stir well. Wash down spillage site well with water, running to waste and diluting greatly with water. Run alkaline solution to waste in fume cupboard, diluting greatly with water.

Methyl formate (methyl methanoate)   [107–31–3]

Colourless liquid with pleasant odour; bp 32 °C; moderately soluble in water.

EXTREMELY FLAMMABLE

Avoid breathing vapour. Avoid contact with skin, eyes and clothing. RL 100 ppm (250 mg m^{-3}).

Toxic effects	The vapour irritates the eyes and respiratory system; in severe cases there may be retching, narcosis and pulmonary irritation that can result in death. Assumed to cause severe irritation and damage if taken internally.
First aid	*Vapour inhaled:* standard treatment (p. 136). *Affected eyes:* standard treatment (p. 136). *Skin contact:* standard treatment (p. 136). *If swallowed:* standard treatment (p. 137).
Fire hazard	Flash point $-19\,°C$; explosive limits 5.9–20%; ignition temp. $456\,°C$. *Extinguish fire with* dry powder, carbon dioxide or vaporising liquid.
Spillage disposal	Shut off all possible sources of ignition. Instruct others to keep at a safe distance. Wear breathing apparatus and gloves. Mop up with plenty of water and run to waste, diluting greatly with running water. Ventilate area well to evaporate remaining liquid and dispel vapour.

5-Methyl-3-heptanone (ethyl amyl ketone) [41–85–5]

FLAMMABLE
IRRITATING TO EYES AND RESPIRATORY SYSTEM

RL 25 ppm (130 mg m^{-3}).

5-Methylhexan-2-one (methyl isoamyl ketone) [110–12–3]

FLAMMABLE

RL 100 (50) ppm (475 mg m^{-3}).

Methylhydrazine [60–34–4]

Suspected carcinogen, TLV (skin) 0.2 ppm (0.35 mg m^{-3}).

Hazardous reactions	May ignite in air when extended (on fibre, or as film). A powerful reducing agent and fuel, hypergolic with many oxidants (B176).

Methyl hydroperoxide [3031–73–0]

Hazardous reactions	Violently explosive, shock-sensitive, especially on warming (B172).

Methyl hypochlorite [593–78–2]

Hazardous reaction   Superheated vapour explodes as does liquid on ignition (B154–155).

Methyl iodide —*see* Iodomethane

Methyl isoamyl ketone
—*see* 5-Methylhexan-2-one

Methyl isobutyl carbinol
—*see* 4-Methylpentan-2-ol

Methyl isobutyl ketone
—*see* 4-Methylpentan-2-one

Methyl isocyanate —*see* Isocyanatomethane

Methyl isocyanide [593–75–9]

Hazardous reaction   Exploded when heated in a sealed ampoule (B246).

Methyllithium [917–54–4]

Hazardous reaction   Ignites and burns in air (B158).

Methyl mercaptan —*see* Methanethiol

Methyl methacrylate [80–62–6]

Colourless liquid; bp 101 °C; almost insoluble in water. This monomer normally contains a small quantity of stabiliser.

HIGHLY FLAMMABLE
IRRITATING TO EYES, RESPIRATORY SYSTEM AND SKIN
MAY CAUSE SENSITISATION BY SKIN CONTACT

Avoid breathing vapour. Avoid contact with skin, eyes and clothing. RL 100 ppm (410 mg m⁻³).

Toxic effects	The vapour may irritate the eyes and respiratory system. The liquid will irritate the eyes and alimentary system if taken by mouth.
Hazardous reactions	Exposure to air at room temperature for two months generated an explosive ester/oxygen interpolymer; ignition occurred when dibenzoyl peroxide was added to a small amount of the ester. Store inhibited monomer with *slight access* of air (B513–514).
First aid	*Vapour inhaled:* standard treatment (p. 136). *Affected eyes:* standard treatment (p. 136). *Skin contact:* standard treatment (p. 136). *If swallowed:* standard treatment (p. 137).
Fire hazard	Flash point 10 °C; explosive limits 2.1–12.5%; ignition temp. 421 °C. *Extinguish fire with* water spray, foam, dry powder, carbon dioxide or vaporising liquid.
Spillage disposal	Shut off all possible sources of ignition. Wear face-shield or goggles, and gloves. Apply non-flammable dispersing agent if available and work to an emulsion with brush and water — run this to waste, diluting greatly with running water. If dispersant not available, absorb on sand, shovel into bucket(s) and transport to safe, open area for atmospheric evaporation. Ventilate site of spillage well to evaporate remaining liquid and dispel vapour.

Methyl nitrate   [598–59–3]

Hazardous reaction	Explodes at 65 °C and has high shock-sensitivity (B163–164).

Methyl 2-nitrobenzenediazoate   [62375–91–1]

Hazardous reactions	Explodes on heating or on disturbing after 24 hours in a sealed tube at ambient temperature (B680).

1-Methyl-3-nitro-1-nitrosoguanidine   [70–25–7]

Hazardous reactions	A former diazomethane precursor, this compound detonates on high impact; sample exploded when heated in sealed capillary tube (B281).

2-Methyl-2-nitropropane [594–70–7]

(t-nitrobutane)

Hazardous reaction   Sample exploded during distillation (B460).

N-Methyl-*N*-nitrosotoluene-4-sulphonamide

[80–11–5]

Yellow crystals, insoluble in water; commonly used as a source of the highly reactive, toxic and explosive gas diazomethane which is generated when it is treated with alkalies; it may itself explode when heated to above 45 °C.

(HEATING MAY CAUSE AN EXPLOSION)

Avoid contact with skin and eyes.

Toxic effects   No evidence has been found that this reagent is irritant or otherwise toxic, but its close association with the methylating technique involving highly toxic diazomethane suggests that protection against skin and eye contact should be used when it is being handled.

First aid   *Affected eyes:* standard treatment (p. 136).
Skin contact: standard treatment (p. 136).
If swallowed: standard treatment (p. 137).

Spillage disposal   Wear face-shield or goggles, and gloves. Mix with sand and transport to safe, open area for burial. Site of spillage should be washed thoroughly with water and soap or detergent.

N-Methyl-*N*-nitrosourea [684–93–5]

Suspected carcinogen.

Hazardous reaction   Material stored at 20 °C exploded after 6 months (B281).

2-Methyloxiran —*see* Propylene oxide

4-Methylpentan-2-ol [105–30–6]

(methyl isobutyl carbinol)

FLAMMABLE
IRRITATING TO RESPIRATORY SYSTEM

RL (skin) 25 ppm (100 mg m^{-3}).

4-Methylpentan-2-one   [108–10–1]
(isobutyl methyl ketone, hexone)

Colourless liquid with faint camphor-like smell; bp 126 °C; slightly soluble in water.

> HIGHLY FLAMMABLE

Avoid breathing vapour. Avoid contact with eyes. RL (skin) 100 (50) ppm (410 mg m^{-3}).

Toxic effects	The vapour is somewhat irritating to the eyes and respiratory system and narcotic in high concentrations. The liquid irritates the eyes and will cause irritation and damage if taken internally.
Hazardous reaction	Unusual peroxide explosion (B625).
First aid	*Vapour inhaled:* standard treatment (p. 136). *Affected eyes:* standard treatment (p. 136). *If swallowed:* standard treatment (p. 137).
Fire hazard	Flash point 17 °C; explosive limits 1.2–8%; ignition temp. 460 °C. *Extinguish fire with* water spray, dry powder, carbon dioxide or vaporising liquid.
Spillage disposal	Shut off all possible sources of ignition. Wear face-shield or goggles, and gloves. Apply non-flammable dispersing agent if available and work to an emulsion with brush and water — run this to waste, diluting greatly with running water. If dispersant not available, absorb on sand, shovel into bucket(s) and transport to safe, open area for atmospheric evaporation or burial. Ventilate site of spillage well to evaporate remaining liquid and dispel vapour.

4-Methylpent-3-en-2-one   [141–79–7]
(mesityl oxide)

Colourless, oily liquid with smell somewhat like honey; bp 130 °C; sparingly soluble in water.

> FLAMMABLE
> HARMFUL BY INHALATION, IN CONTACT WITH SKIN
> AND IF SWALLOWED

Avoid breathing vapour. Avoid contact with skin, eyes and clothing. RL 15 ppm (60 mg m^{-3}).

Toxic effects	The vapour irritates the eyes and respiratory system. The liquid is highly irritating to the eyes and skin and will cause internal irritation and damage if taken by mouth.
First aid	*Vapour inhaled:* standard treatment (p. 136). *Affected eyes:* standard treatment (p. 136). *Skin contact:* standard treatment (p. 136). *If swallowed:* standard treatment (p. 137).

411

Fire hazard   Flash point 31 °C; ignition temp. 344 °C. *Extinguish fire with* foam, dry powder, carbon dioxide or vaporising liquid.

Spillage disposal   Shut off all possible sources of ignition. Instruct others to keep at a safe distance. Wear breathing apparatus and gloves. Apply non-flammable dispersing agent if available and work to an emulsion with brush and water — run this to waste, diluting greatly with running water. If dispersant not available, absorb on sand, shovel into bucket(s) and transport to safe, open area for burial. Ventilate site of spillage well to evaporate remaining liquid and dispel vapour.

Methyl perchlorate   [17043–56–0]

Hazardous reaction   Explosive (B155).

Methylphosphine   [593–54–4]

Hazardous reaction   Readily ignites in air (B175).

Methylpotassium   [17814–73–2]

Hazardous reaction   Dry material is highly pyrophoric (B157–158).

2-Methylpropane —*see* Isobutane

2-Methylpropan-1-ol —*see* Isobutyl alcohol

2-Methylpropan-2-ol (t-butyl alcohol)   [75–65–0]

Colourless crystalline solid or liquid with camphor-like odour; mp 25 °C; bp 83 °C; miscible with water.

HIGHLY FLAMMABLE
HARMFUL BY INHALATION

Avoid breathing vapour. Avoid contact with skin and eyes. RL 100 ppm (300 mg m^{-3}).

Toxic effects   Vapour may irritate eyes and respiratory system. The liquid irritates the eyes and may irritate the skin causing dermatitis. If taken by mouth may cause headache, dizziness, drowsiness and narcosis.

First aid	*Vapour inhaled:* standard treatment (p. 136).
	Affected eyes: standard treatment (p. 136).
	Skin contact: standard treatment (p. 136).
	If swallowed: standard treatment (p. 137).
Fire hazard	Flash point 10 °C; explosive limits 2.4–8%; ignition temp. 478 °C. *Extinguish fire with* dry powder, carbon dioxide or vaporising liquid.
Spillage disposal	Shut off all possible sources of ignition. Wear face-shield or goggles, and gloves. Mop up with plenty of water and run to waste, diluting greatly with running water. Ventilate area of spillage well to evaporate remaining liquid and dispel vapour.

2-Methylpropene (isobutylene; isobutene)　　　[115–11–7]

Colourless gas with smell like that of coal gas; bp −7 °C; practically insoluble in water.

(EXTREMELY FLAMMABLE LIQUEFIED GAS)

Avoid breathing gas.

Toxic effects	The gas has an anaesthetic effect but is not toxic.
First aid	*Gas inhaled in quantity:* standard treatment (p. 136).
Fire hazard	Flash point below −7 °C; explosive limits 1.8–8.8%; ignition temp. 465 °C. Since the gas is supplied in a cylinder, turning off the valve will reduce any fire involving it; if possible, cylinders should be removed quickly from an area in which a fire has developed.
Disposal	Surplus gas or leaking cylinder can be vented slowly to air in a safe, open area or gas burnt off through a suitable burner.

2-Methylpropenoic acid (methacrylic acid)　　　[79–41–4]

Colourless crystals or liquid with acrid odour; mp 16 °C; bp 163 °C; soluble in hot water. This monomer usually contains low levels of stabiliser(s).

CAUSES BURNS

Avoiding breathing vapour. Prevent contact with skin and eyes. RL 20 ppm (70 mg m^{-3}).

Toxic effects	Irritates the eyes and skin (direct contact with the liquid severely so), though less than acrylic acid. Assumed to be severely irritant if taken by mouth.
Hazardous reaction	Uninhibited material stored under cold conditions was brought into a warm room to thaw and subsequently polymerised exothermically (B429).

413

First aid	*Vapour inhaled:* standard treatment (p. 136).
	Affected eyes: standard treatment (p. 136).
	Skin contact: standard treatment (p. 136).
	If swallowed: standard treatment (p. 137).
Fire hazard	Flash point 77 °C. *Extinguish fire with* water spray, dry powder, carbon dioxide or vapourising liquid.
Spillage disposal	Wear face shield or goggles, and gloves. Mop up with plenty of water and run to waste, diluting greatly with running water. Ventilate area of spillage well to evaporate remaining traces and dispel vapour.

Methyl propyl ketone —*see* Pentan-2-one

Methylpyridines   2- [109–06–8] 3- [108–99–6]
4- [108–89–4]

(picolines)

2-, 3- and 4-picolines are colourless liquids, boiling at 129 °C, 144 °C and 143 °C respectively; the 2- and 4-isomers have unpleasant smells. All are very soluble in water.

(FLAMMABLE
IRRITATING TO EYES AND RESPIRATORY SYSTEM)

Avoid breathing vapour. Avoid contact with skin and eyes.

Toxic effects	The vapours irritate the respiratory tract to some extent. The liquids irritate the eyes and may be assumed to cause irritation and damage if taken internally.
First aid	*Vapour inhaled:* standard treatment (p. 136).
	Affected eyes: standard treatment (p. 136).
	Skin contact: standard treatment (p. 136).
	If swallowed: standard treatment (p. 137).
Fire hazard	Flash points 28 °C, 40 °C and 57 °C respectively; 2-picoline has an ignition temperature of 538 °C. *Extinguish fire with* water spray, dry powder, carbon dioxide or vaporising liquid.
Spillage disposal	Shut off all possible sources of ignition. Wear face-shield or goggles, and gloves. Mop up with plenty of water and run to waste, diluting greatly with running water. Ventilate area well to evaporate remaining liquid and dispel vapour.

Methyl silicate —*see* Tetramethyl silicate

Methylsodium   [18356–02–0]

Hazardous reaction	Ignites immediately in air (B165).

α-Methylstyrene (2-phenylpropene)   [98–83–9]

Colourless liquid; bp 167 °C; insoluble in water.

HARMFUL BY INHALATION

Avoid breathing vapour. Avoid contact with skin and eyes. RL 100 ppm (480 mg m^{-3}).

Toxic effects	The vapour irritates the eyes and respiratory system. The liquid irritates the eyes and may cause conjunctivitis. The liquid irritates the skin and may cause dermatitis. It is assumed to be irritant and harmful if taken by mouth.
First aid	*Vapour inhaled:* standard treatment (p. 136). *Affected eyes:* standard treatment (p. 136). *Skin contact:* standard treatment (p. 136). *If swallowed:* standard treatment (p. 137).
Fire hazard	Flash point 54 °C; explosive limits 1.9–6.1%; ignition temp. 494 °C. *Extinguish fire with* foam, dry powder, carbon dioxide or vaporising liquid.
Spillage disposal	Shut off all possible sources of ignition. Wear face-shield or goggles, and gloves. Apply non-flammable dispersing agent if available and work to an emulsion with brush and water — run this to waste, diluting greatly with running water. If dispersant not available, absorb on sand, shovel into bucket(s) and transport to safe, open area for burial. Ventilate site of spillage well to evaporate remaining liquid and dispel vapour.

2-, 3-, and 4-Methylstyrenes   [25013–15–4]
(vinyltoluenes, mixed isomers)

RL 100 ppm (480 mg m^{-3}).

Methyl sulphate —*see* Dimethyl sulphate

Methyltrichlorosilane   [75–79–6]
[trichloro(methyl)silane]

Colourless to pale yellow, volatile liquid with pungent smell, which fumes strongly in moist air; bp 65.5 °C; it reacts vigorously with water forming hydrochloric acid and polymeric gels.

HIGHLY FLAMMABLE
REACTS VIOLENTLY WITH WATER
IRRITATING TO EYES, RESPIRATORY SYSTEM AND SKIN

Prevent inhalation of vapour. Prevent contact with skin, eyes and clothing.

Toxic effects	The vapour irritates the eyes and respiratory system severely. The liquid burns the skin and eyes and will cause severe internal damage if taken by mouth.
First aid	*Vapour inhaled:* standard treatment (p. 136). *Affected eyes:* standard treatment (p. 136). *Skin contact:* standard treatment (p. 136). *If swallowed:* standard treatment (p. 137).
Fire hazard	Flash point 8.3 °C; explosive limits 8.5% to over 17%; ignition temp. 580 °C. *Extinguish fire with* dry powder, carbon dioxide or vaporising liquid.
Spillage disposal	Shut off all possible sources of ignition. Instruct others to keep at safe distance. Wear breathing apparatus and gloves. Absorb on sand, shovel into bucket(s), transport to safe, open area and tip into large volume of water; leave to decompose before decanting the water to waste, diluting greatly with running water. Site of spillage should be ventilated after washing thoroughly with water and soap or detergent.

Methyl vinyl ether (vinyl methyl ether)   [107–25–5]

Colourless liquid or gas with sweetish odour; bp 8 °C; slightly soluble in water.

EXTREMELY FLAMMABLE LIQUEFIED GAS

Avoid breathing gas.

Toxic effects	These have not been fully investigated. It has narcotic properties.
Hazardous reactions	Forms peroxides; acids hydrolyse the latter to acetaldehyde and cause rapid polymerisation (B366).
First aid	*Vapour inhaled:* standard treatment (p. 136).
Fire hazard	Flash point −51 °C. Since the liquid is supplied in a cylinder, turning off the valve will reduce any fire involving it; if possible, cylinders should be removed from an area in which a fire has developed.
Disposal	Surplus gas or leaking cylinder can be vented slowly to air in a safe, open area or gas burnt off through a suitable burner.

Molecular sieves

Hazardous reactions	Catalysed ignition of ethylene; fire when heated to regenerate, explosions with benzyl bromide, nitromethane (B1595).

Molybdenum [7439-98-77]

Soluble compounds, RL (as Mo) 5 ppm.
Insoluble compounds, RL (as Mo) 10 ppm.

Morpholine (diethylene oximide) [110-91-8]

Colourless, mobile liquid with amine-like odour; bp 128 °C; miscible with water.

> FLAMMABLE
> HARMFUL BY INHALATION, IN CONTACT WITH SKIN
> AND IF SWALLOWED
> CAUSES BURNS

Avoid breathing vapour. Avoid contact with skin, eyes and clothing. RL (skin) 20 ppm (70 mg m^{-3}).

Toxic effects	The vapour irritates the eyes and respiratory system. The liquid irritates the eyes and skin; it is also irritant when taken internally and may cause kidney and liver injury.
Hazardous reaction	Its addition to nitromethane makes it susceptible to initiation by a detonator (B160).
First aid	*Vapour inhaled:* standard treatment (p. 136).
	Affected eyes: standard treatment (p. 136).
	Skin contact: standard treatment (p. 136).
	If swallowed: standard treatment (p. 137).
Fire hazard	Flash point 38 °C (open cup); ignition temp. 310 °C. *Extinguish fire with* dry powder, carbon dioxide or vaporising liquid.
Spillage disposal	Shut off all possible sources of ignition. Wear face-shield or goggles, and gloves. Mop up with plenty of water and run to waste, diluting greatly with running water. Ventilate area of spillage well to evaporate remaining liquid and dispel vapour.

Naphthalene [91-20-3]

RL 10 ppm (50 mg m^{-3}).

Hazardous reaction	Reacts explosively with dinitrogen pentaoxide (B764).

Naphthalene-1- ion [15511-25-8]
and -2-diazonium salts ion [36097-38-8]

Hazardous reactions	They react with ammonium sulphide or hydrogen sulphide to form explosive compounds (B763).

1-Naphthylamine [134–32–7]
and salts (containing <1% of 2-isomer)

Colourless crystals when pure, darkening on exposure to light and air; the base, which has an unpleasant smell, melts at 50 °C and is insoluble in water; the hydrochloride is soluble in water. The use of 1-naphthylamine and its salts is controlled in the United Kingdom by The Carcinogenic Substances Regulations 1967 (see p. 148).

> HARMFUL BY INHALATION, IN CONTACT WITH SKIN
> AND IF SWALLOWED
> DANGER OF CUMULATIVE EFFECTS

Prevent inhalation of dust. Prevent contact with skin, eyes and clothing.

Toxic effects	The salts or their solutions irritate the eyes. **Chronic effects** Exposure to the dust or absorption through the skin may cause bladder tumours.
First aid	*Affected eyes:* standard treatment (p. 136). *Skin contact:* standard treatment (p. 136). *If swallowed:* standard treatment (p. 137).
Spillage disposal	Wear breathing apparatus and gloves. Mix with sand and shovel mixture into glass, enamel or polythene vessel for dispersion in an excess of dilute hydrochloric acid (1 volume of concentrated acid diluted with 2 volumes of water). Allow to stand, with occasional stirring, for 24 hours and then run acid extract to waste, diluting greatly with running water and washing the sand. The sand can be disposed of as normal refuse.

2-Naphthylamine and salts [91–59–8]
Human carcinogen.

The use of these compounds in the United Kingdom is now prohibited under The Carcinogenic Substances Regulations 1967 (see p. 148). Inhalation or absorption through the skin of the dust has been recognised as a cause of bladder tumours. It is not therefore considered appropriate to deal with their hazards more fully in this book.

Neopentane —*see* 2,2-Dimethylpropane

Nickel (metal) [7440–02–0]

RL 1 mg m^{-3}.

Nickel carbonyl —*see* Tetracarbonylnickel

Nickel salts   Cl [7791–20–0] NO₃ [13478–00–7]

Green crystals or powder; mostly soluble in water.

> (HARMFUL DUST
> IRRITATING TO SKIN AND EYES)

Avoid breathing dust. Avoid contact with eyes and skin. RL (metal and soluble compounds) 0.1 mg m^{-3} (as Ni).

Toxic effects   The salts and their solutions will irritate the eyes. Assumed to be poisonous if taken by mouth. **Chronic effects** Continued skin contact can cause dermatitis.

First aid   *Inhaled dust:* standard treatment (p. 136).
Affected eyes: standard treatment (p. 136).
Skin contact: standard treatment (p. 136).
If swallowed: standard treatment (p. 137).

Spillage disposal   Soluble nickel salts should be dissolved in water and the solution run to waste diluting greatly with running water. Insoluble compounds can be mixed with sand and put out as ordinary refuse.

Nitric acid   [7797–37–2]

Colourless or pale yellow fuming liquid; miscible with water.

> CONTACT WITH COMBUSTIBLE MATERIALS MAY CAUSE FIRE
> CAUSES SEVERE BURNS

Avoid breathing vapour. Prevent contact with eyes and skin. RL 2 ppm (5 mg m^{-3}).

Toxic effects   The vapour irritates all parts of the respiratory system. The vapour irritates and the liquid burns the eyes severely. The vapour and liquid burn the skin. If taken by mouth there is severe internal irritation and damage.

Hazardous reactions   The range of vigorous, violent and explosive reactions, in which the stronger forms of nitric acid participate, is very wide. Several pages in *Handbook of reactive chemical hazards* are devoted to them, the other reactive participants including acetic acid, acetic anhydride, acetone, acetonitrile, acrylonitrile, alcohols, ammonia, aromatic amines, BrF₅, butanethiol, cellulose, crotonaldehyde, copper nitride, cyclohexylamine, dichloromethane, diethyl ether, 1,1-dimethylhydrazine, divinyl ether, fluorine, hydrazine, hydrocarbons, hydrogen iodide, hydrogen peroxide, ion exchange resins, iron(II) oxide, lactic acid, metal acetylides, metals, metal salicylates, 4-methylcyclohexanone, nitroaromatics, nitrobenzene, nitromethane, non-metal hydrides, non-metals, organic matter, phenylacetylene, phosphine derivatives, phosphorus halides, phthalic anhydride/sulphuric acid, polyalkenes, sulphur dioxide, sulphur halides, thioaldehydes, thioketones, thiophene,

419

2,4,6-trimethyltrioxane. This powerful oxidant is the compound most frequently involved in hazardous reactions. Fuming nitric acid will attack unprotected plastic screw caps of storage bottles, and causes alcohols, amines, unsaturated hydrocarbons, phosphorus compounds or wood to ignite (B1100–1129).

First aid	*Vapour inhaled:* standard treatment (p. 136). *Affected eyes:* standard treatment (p. 136). *Skin contact:* standard treatment (p. 136). *If swallowed:* standard treatment (p. 137).
Spillage disposal	Instruct others to keep at a safe distance. Wear breathing apparatus and gloves. Spread soda ash liberally over the spillage and mop up cautiously with water — run this to waste, diluting greatly with running water.

RSC *Lab. Haz. Data Sheet No. 32,* 1985 gives extended coverage.

Nitric amide (nitramide, nitroamine)   [7782–94–7]

Hazardous reactions	Various preparations have been violent or explosive; drop of concentrated alkali added to solid causes a flame and explosive decomposition; it explodes on contact with concentrated sulphuric acid (B1147–1148).

Nitric oxide —*see* Nitrogen oxide

Nitrites of nitrogenous bases Review of group (B1599).

Nitro acyl halides Review of group (B1600).

Nitroalkanes Review of group (B1600).

Nitroalkenes Review of group (B1600).

Nitroanilines   *o*- [88–74–4] *m*- [99–09–2] *p*- [100–01–6]

The nitroanilines (*o*-, *m*- and *p*-) are yellow to orange-red crystals or powders; slightly soluble in water.

TOXIC BY INHALATION, IN CONTACT WITH SKIN
AND IF SWALLOWED
DANGER OF CUMULATIVE EFFECTS

Avoid breathing dusts. Prevent contact with skin and eyes. RL (skin) (*p*-isomer) 1 ppm (6 mg m^{-3}).

Toxic effects   Inhalation of dusts or excessive skin absorption of the solids may result in headache, flushing of the face, difficulty in breathing, nausea and vomiting; weakness, drowsiness, irritability and cyanosis may follow. Dermatitis may follow skin contact. The dusts will damage the eyes and effects similar to the above may be expected if the substances are taken by mouth.

Hazardous reactions   Thermal stability of the isomers is reduced by various impurities (B593–594).

First aid   *Dust inhaled:* standard treatment (p. 136).
Affected eyes: standard treatment (p. 136).
Skin contact: standard treatment (p. 136).
If swallowed: standard treatment (p. 137).

Spillage disposal   Wear face-shield or goggles, and gloves. Mix with sand and shovel mixture into glass, enamel or polythene vessel for dispersion in an excess of dilute hydrochloric acid (1 volume of concentrated acid diluted with 2 volumes of water). Allow to stand, with occasional stirring, for 24 hours and then run acid extract to waste, diluting greatly with running water and washing the sand. The sand can be disposed of as normal refuse.

Nitroaromatic—alkali hazards Review of topic (B1602).

Nitrobenzene [98–95–3]

Pale yellow or yellow liquid with odour of bitter almonds; bp 211 °C; immiscible with water.

VERY TOXIC BY INHALATION, IN CONTACT WITH SKIN
AND IF SWALLOWED
DANGER OF CUMULATIVE EFFECTS

Avoid breathing vapour. Prevent contact with skin and eyes. RL (skin) 1 ppm (5 mg m^{-3}).

Toxic effects   Inhalation of the vapour may cause a burning sensation in the chest, difficulty in breathing, cyanosis and, in severe cases, unconsciousness. The liquid injures the eyes and if it is absorbed excessively through the skin may give rise to the above symptoms. Drowsiness, vomiting, cyanosis and unconsciousness may follow ingestion.

Hazardous reactions	Addition of $AlCl_3$ to large volume of nitrobenzene containing 5% of phenol caused violent explosion; mixture with sodium chlorate is highly explosive; plant explosion occurred when it was being reacted with nitric acid and water; mixtures with dinitrogen tetraoxide once used as liquid explosives; explosion on heating with NaOH, KOH or $AlCl_3$; mixtures with K, HNO_3 or NH_4NO_3 are capable of detonation (B579–581).
First aid	*Vapour inhaled:* standard treatment (p. 136). *Affected eyes:* standard treatment (p. 136). *Skin contact:* standard treatment (p. 136). *If swallowed:* standard treatment (p. 137).
Spillage disposal	Instruct others to keep at a safe distance. Wear breathing apparatus and gloves. Apply dispersing agent if available and work to an emulsion with brush and water — run this to waste, diluting greatly with running water. If dispersant not available, absorb on sand, shovel into bucket(s) and transport to safe, open area for burial. Site of spillage should be washed thoroughly with water and soap or detergent.

m-Nitrobenzenediazonium perchlorate [22751–24–2]

Hazardous reaction Explosive, very sensitive to heat and shock (B560).

2-Nitrobenzonitrile [612–24–8]

Hazardous preparation Explosion occurred (B655).

2-Nitrobenzoyl chloride [610–14–0]

Hazardous preparation Explosion of distillation residue (B653).

4-Nitrobenzoyl chloride [122–04–3]

Yellow crystals with pungent odour; mp 75 °C; reacts with water forming benzoic and hydrochloric acids.

(CAUSES BURNS
IRRITATING TO EYES AND SKIN)

Avoid breathing vapour. Avoid contact with skin and eyes.

Toxic effects These result mainly from its reaction with moisture on the tissues to form hydrochloric acid, which is the primary irritant. Thus irritation or burns may be caused at the point of contact.

First aid *Vapour inhaled:* standard treatment (p. 136).
Affected eyes: standard treatment (p. 136).
Skin contact: standard treatment (p. 136).
If swallowed: standard treatment (p. 137).

Spillage disposal Wear face-shield or goggles, and gloves. Spread soda ash liberally over the spillage and mop up cautiously with plenty of water — run to waste, diluting greatly with running water.

3-Nitrobenzoyl nitrate [101672–19–9]

Hazardous reaction Explodes if heated rapidly (B655).

Nitrobenzyl compounds Review of group (B1605).

4-Nitrobiphenyl [92–93–3]

Human carcinogen, use prohibited in the United Kingdom under The Carcinogenic Substances Regulations 1967 (see p. 148). Inhalation or absorption through the skin of the dust has been recognised as a cause of bladder tumours. It is not therefore considered appropriate to deal with the hazards more fully in this book.

t-Nitrobutane —*see* 2-Methyl-2-nitropropane

Nitrocellulose —*see* Cellulose nitrate

o-, *m-*, *p-*Nitrochlorobenzene —*see* Chloronitrobenzenes

Nitroethane [79–24–3]

Colourless, oily liquid with pleasant odour; bp 114 °C; immiscible with water; may form explosive compounds with certain amines and alkalies.

FLAMMABLE
HARMFUL BY INHALATION AND IF SWALLOWED

Avoid breathing vapour. Avoid contact with skin and eyes. TLV 100 ppm (310 mg m^{-3}).

423

Toxic effects The vapour irritates the eyes and respiratory system. The liquid irritates the eyes. Absorption by skin contact or ingestion may give rise to liver and kidney damage.

First aid *Vapour inhaled:* standard treatment (p. 136).
Affected eyes: standard treatment (p. 136).
Skin contact: standard treatment (p. 136).
If swallowed: standard treatment (p. 137).

Fire hazard Flash point 28 °C; ignition temp. 415 °C. *Extinguish fire with* dry powder, carbon dioxide or vaporising liquid.

Spillage disposal Shut off all possible sources of ignition. Instruct others to keep at a safe distance. Wear breathing apparatus and gloves. Apply non-flammable dispersing agent and work to an emulsion with brush and water — run this to waste, diluting greatly with running water. If dispersant not available, absorb on sand, shovel into bucket and transport to safe, open area for atmospheric evaporation or burial. Site of spillage should be washed thoroughly with water and soap or detergent.

2-Nitroethanol [625–48–9]

Hazardous reaction Explosion towards end of vacuum distillation (B280).

Nitrogen dioxide —*see* Dinitrogen tetraoxide

Nitrogen oxide (nitric oxide) [10102–43–9]

Colourless gas which, on release to atmosphere, is rapidly oxidised to nitrogen dioxide, a red gas with a pungent odour; bp −152 °C; about 7 cm^3 dissolves in 100 g water at 0 °C.

(VERY TOXIC BY INHALATION)

Prevent inhalation of gas. TLV 25 ppm (30 mg m^{-3}).

Toxic effects The toxic effects must be assumed to be those of nitrogen dioxide which is rapidly formed when nitric oxide mixes with air; nitrogen dioxide is a particularly dangerous gas because of its insidious mode of attack — several hours may elapse before the person exposed to it develops lung irritation and great discomfort. If gassing has been extensive, pulmonary oedema (flooding of lungs) may develop and this can result, several days later, in death.

Hazardous reactions Highly endothermic and an active oxidant (53% O). The liquid is sensitive to detonation in the absence of fuel; when mixed with carbon disulphide an explosion occurred; it ignited hydrogen/oxygen mixtures; pyrophoric chromium incandesces in the gas while Ca, K or U need heating before ignition occurs; acts as initiator to explosion of NCl_3; reacts with boron at ambient temperature with brilliant flashes while charcoal and phosphorus burn more brilliantly than in air; carbon black ignites at 100 °C if potassium hydrogentartrate is present (B1285–1288).

First aid   *It is important to treat any case of considerable exposure as serious and to obtain medical attention even if symptoms of respiratory irritation have not shown themselves. If exposed to other than very low concentrations of the gas, the casualty should be made to rest and kept warm until medical attention is received.*

Disposal   Surplus gas or leaking cylinder can be vented slowly into a water-fed scrubbing tower or column in a fume cupboard, or into a fume cupboard served by such a tower.

Nitrogen trichloride [10025–85–1]

Hazardous reaction   Wide variety of solids, liquids and gases will initiate the violent and often explosive decomposition of NCl_3 (B993–994).

Nitrogen trifluoride [7783–54–2]

Colourless gas with pungent, 'mouldy' smell; bp −129 °C; insoluble in water. Shock exposure of the gas to heat, flame or electric spark, or active contact with organic material, may cause fire and possibly explosion.

(VERY TOXIC BY INHALATION
IRRITATING TO EYES, RESPIRATORY SYSTEM AND SKIN)

Prevent inhalation of gas. Prevent contact with skin, eyes and clothing. RL 10 ppm (29 mg m^{-3}).

Toxic effects   The gas irritates eyes, skin and respiratory system severely. Prolonged exposure to low concentrations may cause mottling of the teeth and skeletal change.

Hazardous reactions   Explosion when adsorbed on activated charcoal at −100 °C. Violent explosions when mixtures with NH_3, H_2, H_2S, CO, CH_4 or C_2H_4 are sparked (B1068–1069).

First aid   *Vapour inhaled:* standard treatment (p. 136).
Affected eyes: standard treatment (p. 136).
Skin contact: standard treatment (p. 136).

Disposal   If the cylinder develops a leak it should be vented slowly in a well-ventilated fume cupboard until discharged.

425

Nitroguanidine   [556–88–7]

Hazardous   Explosive though difficult to detonate (B170).
reaction

Nitroindane   4- [34701–14–9] 5- [7436–07–9]

Hazardous   Crude mixture of 4- and 5-isomers obtained by nitration of indane is
reaction   explosive in final stages of preparation (B744).

Nitromethane   [75–52–5]

Colourless, oily liquid; bp 101 °C; slightly soluble in water; may form shock-sensitive, explosive compounds with certain amines and alkalies.

HEATING MAY CAUSE AN EXPLOSION
FLAMMABLE
HARMFUL IF SWALLOWED

Avoid breathing vapour. RL 100 ppm (250 mg m^{-3}).

Toxic effects   These arise out of ingestion, and inhalation of the vapour. The vapour irritates the respiratory system while there will be irritation and damage internally if the liquid is taken by mouth.

Hazardous   May explode by detonation, heat or shock; addition of bases or acids
reactions   renders it susceptible to initiation by detonator; risk of explosion in preparation of 2-nitroethanol after reaction with formaldehyde; risk of explosion on heating with hydrocarbons; explosions have occurred with lithium perchlorate, molecular sieve and sodium hydride; mixture with nitric acid is extremely explosive; solutions in acetone, $CHCl_3$ or $CHBr_3$ are detonatable (B159–163).

First aid   *Vapour inhaled:* standard treatment (p. 136).
Affected eyes: standard treatment (p. 136).
Skin contact: standard treatment (p. 136).
If swallowed: standard treatment (p. 137).

Fire hazard   Flash point 35 °C; ignition temp. 418 °C. *Extinguish fire with* dry powder, carbon dioxide or vaporising liquid. Shock or heat may cause nitromethane to explode.

Spillage   Shut off all possible sources of ignition. Instruct others to keep at a safe
disposal   distance. Wear breathing apparatus and gloves. Apply non-flammable dispersing agent and work to an emulsion with brush and water — run this to waste, diluting greatly with running water. If dispersant not available, absorb on sand, shovel into buckets and transport to safe, open area for atmospheric evaporation or burial. Site of spillage should be washed thoroughly with water and soap or detergent.

N-Nitromethylamine [598–57–2]

Hazardous reaction	Decomposed explosively by concentrated sulphuric acid (B169).

3-Nitroperchlorylbenzene [20731–44–6]

Hazardous reactions	Explosive; shock-sensitive (B557).

Nitrophenols   *o*- [88–75–5] *m*- [554–84–7] *p*- [100–02–7]

o-Nitrophenol is more volatile than the *m*- and *p*- compounds, and has a phenolic (carbolic) odour; the nitrophenols are pale yellow or yellow crystals or powders; sparingly soluble in water.

> (*p*-) HARMFUL BY INHALATION, IN CONTACT WITH SKIN
> AND IF SWALLOWED
> DANGER OF CUMULATIVE EFFECTS

Avoid breathing vapour or dust. Prevent contact with skin and eyes.

Toxic effects	Excessive intake by inhalation of the dust, absorption by the skin or ingestion may cause irritation, headache, drowsiness and cyanosis. Effects may be cumulative. Assumed to irritate and injure the eyes.
Hazardous reaction	Violent reaction with KOH, and reaction product from *o*-nitrophenol and chlorosulphuric acid decomposed violently. *p*-Nitrophenol reacts vigorously with diethyl phosphate (B582).
First aid	*Vapour or dust inhaled:* standard treatment (p. 136). *Affected eyes:* standard treatment (p. 136). *Skin contact:* standard treatment (p. 136). But also see note on phenol (p. 135). *If swallowed:* standard treatment (p. 137).
Spillage disposal	Wear face-shield or goggles, and gloves. Mix with sand and transport to a safe, open area for burial. Site of spillage should be washed thoroughly with water and soap or detergent.

2-Nitrophenylacetyl chloride [22751–23–1]

Hazardous preparation	Distillation residue liable to explode (B702).

p-Nitrophenylhydrazine   [100–16–3]

Orange-yellow powder; sparingly soluble in water.

(HARMFUL IN CONTACT WITH SKIN)

Prevent contact with skin and eyes.

Toxic effects	These have not been recorded but it must be assumed that intake, whether by inhalation of the dust, absorption through the skin or ingestion, will result in irritation or poisoning. Irritation of the eyes by the dust must also be assumed.
First aid	*Dust inhaled:* standard treatment (p. 136).
	Affected eyes: standard treatment (p. 136).
	Skin contact: standard treatment (p. 136).
	If swallowed: standard treatment (p. 137).
Spillage disposal	Wear face-shield or goggles, and gloves. Mix with sand and shovel mixture into glass, enamel or polythene vessel for dispersion in an excess of dilute hydrochloric acid (1 volume of concentrated acid diluted with 2 volumes of water). Allow to stand, with occasional stirring, for 24 hours and then run acid extract to waste, diluting greatly with running water and washing the sand. The sand can be disposed of as normal refuse.

3-Nitrophthalic acid   [603–11–2]

Hazardous preparation	Eruptive decomposition in nitration of phthalic anhydride (B701).

1-Nitropropane   [108–03–2]

TLV 25 ppm (90 mg m^{-3}).

2-Nitropropane   [79–46–9]

Suspected carcinogen RL 10 ppm (36 mg m^{-3}).

RSC *Lab. Haz. Data Sheet No. 10,* 1983 gives extended coverage.

Nitroso compounds Review of group (B1608).

N-Nitrosodimethylamine [62–75–9]

Suspected carcinogen, no TLV set.

Nitrosodimethylaniline
—*see N,N*-Dimethyl-*p*-nitrosoaniline

1-Nitroso-2-naphthol [131–91–9]

Yellow-brown powder; insoluble in water.

(IRRITATING TO EYES, RESPIRATORY SYSTEM AND SKIN)

Avoid breathing dust. Avoid contact with skin and eyes.

Toxic effects The dust irritates the respiratory system. The dust irritates the eyes and skin and may cause dermatitis. Assumed to be irritant and poisonous if taken by mouth.

Hazardous reaction Becomes unstable on heating and may ignite spontaneously (B763).

First aid *Dust inhaled:* standard treatment (p. 136).
Affected eyes: standard treatment (p. 136).
Skin contact: standard treatment (p. 136).
If swallowed: standard treatment (p. 137).

Spillage disposal Wear face-shield or goggles, and gloves. Mix with sand and transport to a safe, open area for burial. Site of spillage should be washed thoroughly with water and soap or detergent.

Nitrosophenols *o*- [13168–78–0] *p*- [637–62–7]

Hazardous reactions The *o*-isomer explodes on heating or in contact with concentrated acids. Barrels of the *p*-isomer heated spontaneously and caused a fire: this tendency increases after compaction. These materials are now supplied wet with 10% water (B581–582).

3-Nitrosotriazenes Review of group (B1609).

Nitrosyl chloride [2696–92–6]

Reddish brown gas with irritant, penetrating smell; bp −5.8 °C; decomposes on contact with water or moisture.

(TOXIC BY INHALATION
IRRITATING TO EYES, RESPIRATORY SYSTEM AND SKIN)

Prevent inhalation of gas. Prevent contact with skin.

Toxic effects   These have not been fully investigated but may be expected to be intermediate between those of chlorine and nitrogen oxides. The gas is intensely irritating but provides good warning of its dangers because of its penetrating odour.

Hazardous reaction   Cold sealed tube containing NOCl, Pt wire and traces of acetone exploded on warming up (B934).

First aid   *Vapour inhaled:* standard treatment (p. 136).
Affected eyes: standard treatment (p. 136).
Skin contact: standard treatment (p. 136).

Disposal   Surplus gas or leaking cylinder can be vented slowly into a water-fed scrubbing tower or column in a fume cupboard, or into a fume cupboard served by such a tower.

Nitrosyl fluoride   [7789-25-5]

Hazardous reactions   Reaction of mixture with unspecified haloalkene in pressure vessel at $-78\,°C$ caused it to rupture when moved; incandescent reactions with Sb, Bi, As, B, red P and Si; explodes on mixing with oxygen difluoride (B1047).

Nitrosyl perchlorate   [15605-28-4]

Hazardous reactions   Explodes on contact with pinene; it ignites acetone and ethanol and then explodes; with ether there is gassing followed by explosion; small amounts of primary aromatic amines are ignited on contact while explosions result with larger amounts; urea ignites on stirring with it; reaction mixture with phenyl isocyanate and pentaammineazidocobalt(III) perchlorate exploded when stirring stopped (B935-936).

Nitrosylsulphuric acid   [7782-78-7]

Hazardous reactions   In preparation from SO_2 and nitric acid, absence of N_2O_4 may result in explosion; explosion occurred during plant-scale diazotisation of a dinitroaniline hydrochloride; explosion during plant-scale diazotisation of 6-chloro-2,4-dinitroaniline (B1130).

Nitrosyl tribromide [13444–89–8]

Hazardous reaction Powdered sodium antimonide ignites when dropped into the vapour (B109).

5-Nitrotetrazole [55011–46–6]

Hazardous reaction An evaporated solution of the sodium salt exploded after 2 weeks (B138).

Nitrotoluenes *o*- [88–72–2] *m*- [99–08–1] *p*- [99–99–0]

o-Nitrotoluene is a yellow liquid (bp 220 °C) *m*-nitrotoluene a yellow or brown-yellow liquid or crystalline solid (mp 15 °C) and *p*-nitrotoluene is a pale yellow or yellow crystalline solid (mp 52 °C). All are insoluble in water.

TOXIC BY INHALATION, IN CONTACT WITH SKIN
AND IF SWALLOWED
DANGER OF CUMULATIVE EFFECTS

Avoid breathing vapour or dust. Avoid contact with skin and eyes. RL (skin) 5 ppm (30 mg m^{-3}).

Toxic effects Inhalation of vapour or dust or skin absorption may cause difficulty in breathing, cyanosis and, in severe cases, unconsciousness. Assumed to injure the eyes and cause the above effects if taken by mouth.

Hazardous reactions Distillation residues exploded; mixtures of *p*-isomer with oleum decomposed at 52 °C (B675).

First aid *Vapour or dust inhaled:* standard treatment (p. 136).
Affected eyes: standard treatment (p. 136).
Skin contact: standard treatment (p. 136).
If swallowed: standard treatment (p. 137).

Spillage disposal Wear face-shield or goggles, and gloves. Mix with sand and transport to a safe, open area for burial. Site of spillage should be washed thoroughly with water and soap or detergent.

Nitrourea [556–89–8]

Hazardous reaction Unstable explosive (B164).

Nitrous acid [7782–77–6]

Hazardous reaction Causes phosphorus trichloride to explode (B1099–1100).

431

Nitrous fumes —*see* Dinitrogen tetraoxide

Nitrous oxide —*see* Dinitrogen oxide

Nitryl chloride [13444–90–1]

Hazardous reactions Reacts violently with ammonia or sulphur trioxide even at −75 °C; it attacks organic matter rapidly, sometimes explosively (B934–935).

Nitryl fluoride [10022–50–1]

Hazardous reactions Reaction with various metals causes incandescence or glowing to occur, and H_2 explodes at 200 °C (B1047–1048).

Nitryl hypochlorite ('chlorine nitrate') [14545–72–3]

Hazardous reactions Reacts explosively with alcohols, ethers and most organic materials. Explosive reactions with metal halides if above −40 °C (B935).

Nitryl hypofluorite ('fluorine nitrate') [7789–26–6]

Hazardous reactions Dangerously explosive as solid, liquid or gas. Ignition in gas phase with NH_3, N_2O or H_2S; it may explode on contact with alcohol, ether, aniline or grease (B1048).

Nonacarbonyldiiron [20982–74–5]

Hazardous reaction Commercial material (flash point 35 °C) has autoignition temperature of 93 °C. (B738).

Nonane [111–84–2]

TLV 200 ppm (1050 mg m^{-3}).

Octane [111–65–9]

RL 300 ppm (1450 mg m^{-3}).

Oleum (fuming sulphuric acid, Nordhausen acid) 30% SO$_3$ [8014–95–7]

Colourless to yellow viscous fuming liquid; reacts violently with water.

REACTS VIOLENTLY WITH WATER
CAUSES SEVERE BURNS
IRRITATING TO RESPIRATORY SYSTEM

Avoid inhaling vapour. Prevent contact with eyes and skin.

Toxic effects The vapour irritates all parts of the respiratory system. The vapour irritates and the liquid burns the eyes severely. The vapour and liquid burn the skin. If swallowed there would be most severe internal irritation and damage.

Hazardous reactions These are covered under sulphur trioxide (B1365–1366).

First aid *Vapour inhaled:* standard treatment (p. 136).
Affected eyes: standard treatment (p. 136).
Skin contact: standard treatment (p. 136).
If swallowed: standard treatment (p. 137).

Spillage disposal Wear face-shield or goggles, and gloves. Spread soda ash liberally over the spillage and mop up cautiously with plenty of water — run to waste, diluting greatly with running water.

Organic peroxides Review of group (B1623).

Organolithium reagents Review of group (B1624).

Organosilyl perchlorates Review of group (B1627).

Orthophosphoric acid —*see* Phosphoric acid

Osmic acid (osmium tetroxide) [20816–12–0]

Colourless or pale yellow crystals with pungent odour; mp 40 °C; soluble in water.

VERY TOXIC BY INHALATION, IN CONTACT WITH SKIN
AND IF SWALLOWED
CAUSES BURNS

Prevent inhalation of vapour. Prevent contact with eyes and skin. RL (as Os) 0.0002 ppm (0.002 mg m^{-3}).

Toxic effects	The vapour irritates all parts of the respiratory system. The vapour, solid and solution irritate and burn the eyes severely. The acid and its solution burn the skin. If taken by mouth there would be severe internal irritation and damage. **Chronic effects** Continued exposure to vapour causes disturbances of the vision; continued skin contact results in dermatitis and ulceration.
Hazardous reactions	Amorphous form, prepared by low temperature dehydration, is pyrophoric; explosion with 1-methylimidazole (B1369).
First aid	*Vapour inhaled:* standard treatment (p. 136). *Affected eyes:* standard treatment (p. 136). *Skin contact:* standard treatment (p. 136). *If swallowed:* standard treatment (p. 137).
Spillage disposal	Wear face-shield or goggles, and gloves. Mop up with plenty of water and run to waste, diluting greatly with running water.

Oxalates   NH$_4$ [6009–70–7] Na [62–76–0] K [6487–48–5]

Most oxalates are colourless; ammonium, sodium and potassium oxalates are soluble in water.

HARMFUL IN CONTACT WITH SKIN AND IF SWALLOWED

Avoid contact with eyes and skin.

Toxic effects	If swallowed there would be severe internal pain followed by collapse.
First aid	*Affected eyes:* standard treatment (p. 136). *Skin contact:* standard treatment (p. 136). *If swallowed:* standard treatment (p. 137).
Spillage disposal	Wear face-shield or goggles, and gloves. Mop up with plenty of water and run to waste, diluting greatly with running water.

Oxalic acid [144–62–7]

Colourless crystals; soluble in water.

HARMFUL IN CONTACT WITH SKIN AND IF SWALLOWED

Avoid contact with eyes and skin. RL 1 mg m^{-3}.

Toxic effects	The dust irritates the respiratory system. The dust and solutions irritate the eyes. If swallowed there would be severe internal pain followed by collapse.
Hazardous reaction	Dry mixture of oxalic acid and sodium chlorite exploded on addition of water (B237).
First aid	*Dust inhaled:* standard treatment (p. 136). *Affected eyes:* standard treatment (p. 136). *Skin contact:* standard treatment (p. 136). *If swallowed:* standard treatment (p. 137).
Spillage disposal	Wear face-shield or goggles, and gloves. Mop up with plenty of water and run to waste, diluting greatly with running water.

Oxalyl chloride [79-37-8]

Colourless fuming liquid with acrid odour; bp 64 °C; reacts vigorously with water forming hydrochloric and oxalic acids.

(TOXIC BY INHALATION
CAUSES BURNS)

Avoid breathing vapour. Prevent contact with skin and eyes.

Toxic effects	The vapour irritates the respiratory system severely. The vapour and liquid irritate the eyes. The liquid irritates the skin and may cause burns; must be assumed to be very irritant and poisonous if taken by mouth.
Hazardous reactions	Mixture with K or with dimethyl sulphoxide is explosive (B207).
First aid	*Vapour inhaled:* standard treatment (p. 136). *Affected eyes:* standard treatment (p. 136). *Skin contact:* standard treatment (p. 136). *If swallowed:* standard treatment (p. 137).
Spillage disposal	Wear face-shield or goggles, and gloves. Spread soda ash liberally over the spillage and mop up cautiously with plenty of water — run to waste, diluting greatly with running water.

Oximes Review of group (B1633).

435

Oxodisilane [22755-00-6]

Hazardous reaction   Ignites in air (B1202).

2-Oxohexamethyleneimine
—*see* ε-Caprolactam

Oxosilane [22755-01-7]

Hazardous reaction   Ignites in air (B1149).

Oxygen [7782-44-7]

Handbook of reactive chemical hazards (pp. 1105–1116) covers many incidents with liquid and gaseous oxygen. Leakage of the latter into unventilated spaces to cause oxygen enrichment by only 3–4% above the normal atmospheric level of 21% dramatically increases fire risks.

Oxygen difluoride [7783-41-7]

TLV 0.05 ppm (0.1 mg m^{-3}).

Oxygen fluorides Review of group (B1637).

1-Oxyperoxy compounds Review of group (B1638).

Ozone (trioxygen) [10028-15-6]

Colourless gas or dark blue liquid (bp −111 °C) with characteristic odour.

(TOXIC BY INHALATION
IRRITATING TO RESPIRATORY SYSTEM)

Avoid inhalation. RL 0.1 ppm (0.2 mg m^{-3}).

Toxic effects   The gas irritates the upper respiratory system strongly and may cause headache. High concentrations have caused death by lung congestion in animals.

Hazardous reactions   Solid and liquid ozone are highly explosive; reacts with alkenes to form peroxides which are often explosive; gelatinous explosive ozonides formed with benzene, aniline and other aromatic compounds; reactions with bromine, N_2O_4 and HBr are usually explosive; C_2H_4 at low pressure explodes at $-150\,°C$; isopropylidene compounds give explosive acetone peroxide; N_2 or NO are oxidised explosively (B1121–1125).

First aid   *Gas inhaled:* standard treatment (p. 136).

Ozonides Review of group (B1639).

Paraffin wax (fume)   [8002–74–2]

RL 2 mg m^{-3}.

Paraformaldehyde   [9002–81–7]

White crystalline powder; depolymerised on heating to form gaseous formaldehyde; slowly soluble in cold water, more readily in hot.

(HARMFUL IF SWALLOWED)

Avoid breathing dust. Avoid contact with skin and eyes.

Toxic effects   The dust irritates the eyes, skin and respiratory system and is irritant and damaging if taken internally.

First aid   *Dust inhaled:* standard treatment (p. 136).
Affected eyes: standard treatment (p. 136).
If swallowed: standard treatment (p. 137).

Spillage disposal   Wear face-shield or goggles, and gloves. Mop up with plenty of water and run to waste, diluting greatly with running water.

Paraldehyde (2,4,6-trimethyl-1,3,5-trioxane)   [123–63–7]

Colourless liquid with characteristic, aromatic odour; bp 128 °C; 1 part dissolves in about 8 parts of water at 25 °C.

HIGHLY FLAMMABLE

Avoid breathing vapour. Avoid contact with skin, eyes and clothing.

Toxic effects	The vapour has narcotic effects, and large doses taken internally cause prolonged unconsciousness, respiratory difficulty and pulmonary oedema.
Hazardous reaction	Reaction with nitric acid to form glyoxal is liable to become violent (B628).
First aid	*Vapour inhaled:* standard treatment (p. 136). *Affected eyes:* standard treatment (p. 136). *Skin contact;* standard treatment (p. 136). *If swallowed:* standard treatment (p. 137).
Fire hazard	Flash point 36 °C; ignition temp. 238 °C. *Extinguish fire with* water spray, dry powder, carbon dioxide or vaporising liquid.
Spillage disposal	Shut off all possible sources of ignition. Instruct others to keep at a safe distance. Wear breathing apparatus and gloves. Apply non-flammable dispersing agent and work to an emulsion with brush and water — run this to waste, diluting greatly with running water. If dispersant not available, absorb on sand, shovel into bucket(s) and transport to safe, open area for burial. Site of spillage should be washed thoroughly with water and soap or detergent.

Pentaboranes   (9) [19624–22–7] (11) [18433–84–6]

TLV 0.005 ppm (0.01 mg m^{-3}).

Hazardous reactions	B_5H_9 ignites spontaneously if impure; forms shock-sensitive solution in solvents containing carbonyl, ether or ester functional groups and/or halogen substituents. B_5H_{11} ignites in air (B75–76).

Pentachloroethane   [76–01–7]

Colourless, heavy liquid of relatively low volatility with chloroform-like odour; bp 162 °C; immiscible with water.

VERY TOXIC BY INHALATION AND IN CONTACT WITH SKIN

Avoid breathing vapour. Avoid contact with skin, eyes and clothing.

Toxic effects	The vapour irritates the eyes, nose and lungs and may cause drowsiness, giddiness, headache, and in high concentrations, unconsciousness. Assumed to be poisonous if taken by mouth. **Chronic effects** Continuous breathing of low concentrations of vapour over a long period may cause jaundice by action on the liver. It may also affect the nervous system and the blood.
Hazardous reaction	Mixture with potassium may explode after short delay (B1236).

First aid   *Vapour inhaled:* standard treatment (p. 136).
                      Affected eyes: standard treatment (p. 136).
                      Skin contact: standard treatment (p. 136).
                      If swallowed: standard treatment (p. 137).

Spillage   Instruct others to keep at a safe distance. Wear breathing apparatus and
disposal   gloves. Apply dispersing agent if available and work to an emulsion with
                      brush and water — run this to waste, diluting greatly with running water. If
                      dispersant not available, absorb on sand, shovel into bucket(s) and
                      transport to safe, open area for burial. Site of spillage should be washed
                      thoroughly with water and soap or detergent.

Pentachlorophenol and   [87–86–5]
sodium pentachlorophenate   [131–52–2]

Pentachlorophenol is a colourless to yellow crystalline solid with a phenolic (carbolic) odour;
mp 190 °C; insoluble in water. The sodium salt is a buff powder or flaked solid, soluble in water.

> TOXIC BY INHALATION, IN CONTACT WITH SKIN
> AND IF SWALLOWED

Avoid breathing dust. Prevent contact with skin and eyes. Pentachlorophenol RL (skin) 0.5 mg
m^{-3}.

Toxic effects   The dusts irritate the nose and eyes. Absorption through the skin or
                      ingestion may cause accelerated breathing, feverishness and muscular
                      weakness; in severe cases, convulsions and unconsciousness may follow.
                      Skin contact may cause dermatitis.

First aid   *Dust inhaled:* standard treatment (p. 136).
                      Affected eyes: standard treatment (p. 136).
                      Skin contact: standard treatment (p. 136).
                      If swallowed: standard treatment (p. 137).

Spillage   Wear face-shield or goggles, and gloves. Mix with sand and transport to a
disposal   safe, open area for burial. Site of spillage should be washed thoroughly with
                      water and soap or detergent.

Penta-1,3-diyne   [4911–55–1]

Hazardous   Explodes on distillation (B494).
reaction

Pentafluoroguanidine   [10051–06–6]

Hazardous   Extremely explosive (B129).
reaction

Pentanal —*see* Valeraldehyde

Pentane [109–66–0]

Colourless liquid; bp 30 °C; almost insoluble in water.

HIGHLY FLAMMABLE

Avoid breathing vapour. RL 600 ppm (1800 mg m^{-3}).

Toxic effects	The vapour is narcotic in high concentrations.
First aid	*Vapour inhaled in high concentrations:* standard treatment (p. 136). *Affected eyes:* standard treatment (p. 136). *If swallowed:* standard treatment (p. 137).
Fire hazard	Flash point −49 °C; explosive limits 1.4–8%; ignition temp. 309 °C. *Extinguish fire with* foam, dry powder, carbon dioxide or vaporising liquid.
Spillage disposal	Shut off all possible sources of ignition. Wear face-shield or goggles, and gloves. Apply non-flammable dispersing agent if available and work to an emulsion with brush and water — run this to waste, diluting greatly with running water. If dispersant not available, absorb on sand, shovel into bucket(s) and transport to safe, open area for atmospheric evaporation. Ventilate site of spillage well to evaporate remaining liquid and dispel vapour.

Pentane-2,4-dione (acetylacetone) [123–54–6]

Colourless or slightly yellow liquid with pleasant smell; bp 140 °C, 1 part dissolves in about 8 parts of water at 25 °C.

FLAMMABLE
HARMFUL IF SWALLOWED

Avoid breathing vapour. Avoid contact with skin and eyes.

Toxic effects	The vapour may irritate the respiratory system. The liquid irritates the eyes. The liquid may irritate the skin. If swallowed, may cause internal irritation and more severe damage.
First aid	*Vapour inhaled:* standard treatment (p. 136). *Affected eyes:* standard treatment (p. 136). *Skin contact:* standard treatment (p. 136). *If swallowed:* standard treatment (p. 137).
Fire hazard	Flash point 34 °C. *Extinguish fire with* water spray, dry powder, carbon dioxide or vaporising liquids.

Spillage
disposal

Shut off all possible sources of ignition. Wear face-shield and gloves. Apply non-flammable dispersing agent and work to an emulsion with brush and water — run this to waste diluting greatly with running water. If dispersant not available, absorb on sand, shovel into bucket(s) and transport to safe open area for burial. Ventilate area well to evaporate remaining liquid and dispel vapour.

Pentan-1-ol and   [71–41–0]
pentan-2-ol (n-and s-amyl alcohols)   [6032–29–7]

Colourless liquids; bp 138 °C and 119 °C respectively; both are slightly soluble in water.

> FLAMMABLE
> HARMFUL BY INHALATION

Avoid breathing vapour. Avoid contact with skin and eyes.

Toxic effects

The vapours may irritate the eyes and respiratory system. The alcohols are not readily absorbed by the skin but, as liquids, will irritate the eyes. They are both harmful when taken by mouth and may cause giddiness, headache, coughing, deafness and delirium.

First aid

Vapour inhaled in high concentrations: standard treatment (p. 136).
Affected eyes: standard treatment (p. 136).
If swallowed: standard treatment (p. 137).

Fire hazard

Pentan-1-ol: flash point 33 °C; explosive limits 1–10%; ignition temp. 300 °C. Pentan-2-ol: flash point 40 °C. *Extinguish fire with* water spray, foam, dry powder, carbon dioxide or vaporising liquid.

Spillage
disposal

Shut off all possible sources of ignition. Wear face-shield or goggles, and gloves. Apply non-flammable dispersing agent if available and work to an emulsion with brush and water — run this to waste, diluting greatly with running water. If dispersant not available, absorb on sand, shovel into bucket(s) and transport to safe, open area for atmospheric evaporation or burial. Ventilate site of spillage well to evaporate remaining liquid and dispel vapour.

Pentan-2-one (methyl propyl ketone)   [107–87–9]

Colourless liquid with acetone/ether-like odour; bp 102 °C; about 5.5 g dissolves in 100 g water at 25 °C.

> (HIGHLY FLAMMABLE)

Avoid breathing vapour. Avoid contact with skin and eyes. RL 200 ppm (700 mg m^{-3}).

Toxic effects   The vapour may irritate the eyes and respiratory system; it is narcotic in high concentrations. The liquid will irritate the eyes and is assumed to be irritant and narcotic if taken internally.

Hazardous reaction   May react explosively with BrF_3 (B522).

First aid   *Vapour inhaled:* standard treatment (p. 136).
Affected eyes: standard treatment (p. 136).
Skin contact: standard treatment (p. 136).
If swallowed: standard treatment (p. 137).

Fire hazard   Flash point 7 °C; explosive limits 1.5–8%; ignition temp. 505 °C. *Extinguish fire with* water spray, dry powder, carbon dioxide or vaporising liquid.

Spillage disposal   Shut off all possible sources of ignition. Wear face-shield or goggles, and gloves. Apply non-flammable dispersing agent if available and work to an emulsion with brush and water — run this to waste, diluting greatly with running water. If dispersant not available, absorb on sand, shovel into bucket(s) and transport to safe, open area for atmospheric evaporation. Ventilate site of spillage well to evaporate remaining liquid and dispel vapour.

Pentan-3-one (diethyl ketone)   [96–22–0]

HIGHLY FLAMMABLE

RL 200 ppm (700 mg m^{-3}).

Hazardous reaction   Gives shock- and heat-sensitive oily peroxide with hydrogen peroxide and nitric acid (B522).

Pent-2-en-4-yn-3-ol   [101672–20–2]

Hazardous reaction   Distillation residue exploded at over 90 °C (B507).

Pentyl acetate (n-amyl acetate   [628–63–7]
and commercial amyl acetate)   [123–92–2]

(Latter consisting mainly of 3-methylbutyl acetate). Colourless liquids with pear-like odour; bp 148 °C and 128–132 °C respectively; slightly soluble in water.

FLAMMABLE

Avoid breathing vapour. Avoid contact with eyes. RL 100 ppm (530 mg m^{-3}).

Toxic effects   High concentrations of the vapour irritate the eyes and may cause headache and fatigue. Must be considered harmful if taken by mouth.

First aid   *Vapour inhaled in high concentrations:* standard treatment (p. 136).
Affected eyes: standard treatment (p. 136).
If swallowed: standard treatment (p. 137).

Fire hazard   Flash points 25 °C and 23 °C respectively; explosive limits 1–7.5% for both; ignition temps. 379 °C and 380 °C. *Extinguish fire with* dry powder, carbon dioxide or vaporising liquid.

Spillage disposal   Shut off all possible sources of ignition. Wear face-shield or goggles, and gloves. Apply non-flammable dispersing agent if available and work to an emulsion with brush and water — run this to waste, diluting greatly with running water. If dispersant not available, absorb on sand, shovel into bucket(s) and transport to safe, open area for atmospheric evaporation or burial. Ventilate site of spillage well to evaporate remaining liquid and dispel vapour.

Perchlorates   NH$_4$ [7790–98–9] Mg [10034–81–8]
Na [7791–07–3] K [7778–74–7]

The perchlorates of ammonium, magnesium, sodium and potassium are colourless crystalline solids, all except that of potassium being readily soluble in water. Ammonium perchlorate is explosive when dry.

EXPLOSIVE WHEN MIXED WITH COMBUSTIBLE MATERIAL
HARMFUL IF SWALLOWED

Avoid contact with combustible materials. Avoid contact with skin, eyes and clothing.

Toxic effects   The dust and strong solutions will irritate the skin, eyes and respiratory system. They are also irritant and harmful if taken internally.

Hazardous reactions   Mixtures of inorganic perchlorates with combustible materials are readily ignited; mixtures with finely divided combustible materials frequently react explosively. Organic perchlorates are self-contained explosives. Specific accidents with perchlorates of Ag, Be, Ga, K, Li, Mg, Mn, Na, Ni, Pb, Ti, U, V and Zr are recorded in *Handbook of reactive chemical hazards* (*see* Metal Oxohalogenates, p. 1586) and some of these appear in this text.

First aid   *Affected eyes:* standard treatment (p. 136).
Skin contact: standard treatment (p. 136).
If swallowed: standard treatment (p. 137).

Spillage disposal   Shovel into bucket of water and run solution to waste, diluting greatly with running water. Site of spillage should be washed thoroughly to remove all oxidant, which is liable to render any organic matter (particularly paper, textiles and wood) with which it comes into contact dangerously combustible when dry. Clothing wetted with the solution should be washed thoroughly.

443

Perchloric acid [7601–90–3]

Colourless liquid; miscible with water; commonly sold commercially as 60–62% or 73% constant boiling acid.

HEATING MAY CAUSE AN EXPLOSION
CONTACT WITH COMBUSTIBLE MATERIAL MAY CAUSE FIRE
CAUSES SEVERE BURNS

Prevent contact with eyes and skin.

Toxic effects   Burns eyes and skin severely. Assumed to cause severe internal irritation and damage if taken by mouth.

Hazardous reactions   Everyone using perchloric acid or perchlorates should become familiar with the chapter 'Handling perchloric acid and perchlorates' by Everett and Graf in *Handbook of laboratory safety* edited by N. V. Steere (The Chemical Rubber Co., Cleveland). There is a very long history of accidents with perchloric acid in particular and a study of the summaries contained on pp 906–917 of *Handbook of reactive chemical hazards* shows that the partners of perchloric acid in explosions and violent reactions include acetic acid, acetic anhydride, acetonitrile, alcohols, aniline/formaldehyde, antimony(III) compounds, bismuth and alloys, 2-butoxyethanol, carbon, cellulose and derivatives, dehydrating agents, diethyl ether, dimethyl ether, glycerol/lead oxide cement, glycols and their ethers, hydrogen, iodides, iron(II) sulphate, ketones, 2-methylpropene, nitric acid/organic matter, oleic acid, phenols, phosphine, pyridine, sodium phosphinate, organic sulphoxides, trichloroethylene, trizinc diphosphide (B906–917).

The distillation of 70–72% commercial perchloric acid under vacuum concentrates the acid to 75% and over, and these stronger acids are liable to explode on heating; such distillation, when justified, should be carried out at atmospheric pressure so that the constant-boiling (and much safer) acid is obtained. The use of wooden bench tops for analytical and other work with perchloric acid and perchlorates is very undesirable. Such benches, used for these purposes regularly and subjected to occasional perchloric acid spills, have been known to explode on percussion.

First aid   *Affected eyes:* standard treatment (p. 136).
Skin contact: standard treatment (p. 136).
If swallowed: standard treatment (p. 137).

Spillage disposal   Wear face-shield or goggles, and gloves. Spread soda ash liberally over the spillage and mop up cautiously with plenty of water — run to waste, diluting greatly with running water.

Perchloroethylene —*see* Tetrachloroethylene

Perchloromethyl mercaptan
—*see* Trichloromethanesulphenyl chloride

Perchlorylbenzene [5390–07–8]

Hazardous reaction After an interval, a mixture with aluminium chloride suddenly exploded (B575).

Perchloryl compounds Review of group (B1642).

Perchloryl fluoride [7616–94–6]

RL 3 ppm (14 mg m^{-3}).

Hazardous reactions Reference to booklet on safe handling given; at temperatures over 100 °C, calcium acetylide, potassium cyanide, potassium thiocyanate may react explosively in the gas; may react with carbon powder, foamed elastomers, lampblack, sawdust very violently even at −78 °C; reactions with hydrocarbons, H_2S, NO, SCl_2 or vinylidene chloride may be explosive above 100 °C, or if mixtures are ignited; explosive compounds may be formed by reaction with nitrogenous bases; explosive reactions with amines, esters, lithiated compounds and laboratory materials (B892–897).

1-Perchlorylpiperidine [768–34–3]

Hazardous reactions Has exploded on storage, on heating or in contact with piperidine (B520).

Periodic acid [13444–71–8]

Hazardous reactions 1.5M solutions in dimethyl sulphoxide explode after short interval; evaporation with Et_4NOH led to explosion (B1094–1095).

Permanganic acid [13465–41–3]

Hazardous reactions Likely to explode at room temperature; solution of acid produced by action of H_2SO_4 on permanganates will explode on contact with benzene, CS_2, ether and other organic substances (B1099).

Peroxides and peroxides in solvents

Reviews of groups (B1643).

445

Peroxidisable compounds

A particularly useful review which has been extended to include 80 especially susceptible compounds (B1644–1647).

Peroxoacids and salts Reviews of group (B1647).

Peroxodisulphuric acid [13445–49–3]

Hazardous reactions   An extreme oxidant, which may cause aniline, benzene, ethanol, ether, nitrobenzene or phenol to explode on contact (B1171).

Peroxomonophosphoric acid [13598–52–5]

Hazardous reactions   80% solution may ignite organic matter (B1185).

Peroxomonosulphuric acid [7722–86–3]

Hazardous reactions   Explosions have occurred alone, or with acetone, alcohols (secondary and tertiary), aromatic compounds, catalysts; it rapidly carbonises wool or cellulose while cotton is ignited (B1170–1171).

Peroxonitric acid [26604–66–0]

Hazardous reaction   Decomposes explosively at −30 °C (B1130).

Peroxyacetic acid [79–21–0]

Hazardous reaction   Explodes violently at 110 °C; Cl ions catalyse decomposition; ether solvents are incompatible (B271–272).

Peroxy acids Review of group (B1648).

Peroxybenzoic acid   [93–59–4]

Hazardous   Explodes weakly on heating (B670).
reaction

Peroxycarbonate esters Review of group (B1649).

Peroxyesters Review of group (B1650).

Peroxyhexanoic acid   [5106–46–7]

Hazardous   Explodes and ignites on rapid heating (B628).
reaction

Peroxytrifluoroacetic acid
—*see* Trifluoroperoxyacetic acid

Petroleum spirit (petroleum ether)   [8032–32–4]

Petroleum spirits are supplied in a variety of fractions boiling between 30 °C and 160 °C; all are colourless liquids whose smell varies with the volatilities of the fractions.

EXTREMELY FLAMMABLE or
HIGHLY FLAMMABLE

Avoid breathing vapour. Avoid contact with skin and eyes. TLV will lie between 50 and 600 ppm depending on composition.

Toxic effects   Inhalation of high concentrations of the vapour, particularly of the lower boiling fractions, can cause intoxication, headache, nausea and coma. The liquids irritate the eyes and skin contact results in defatting of the area of contact, increasing the risk of dermatitis from other agents. If taken by mouth they may cause burning sensation, vomiting, diarrhoea and drowsiness.

First aid   *Vapour inhaled in high concentrations:* standard treatment (p. 136).
Affected eyes: standard treatment (p. 136).
Skin contact: standard treatment (p. 136).
If swallowed: standard treatment (p. 137).

Fire hazard   Flash point (lower fractions) below −17 °C; explosive limits approx. 1–6%; ignition temperatures range from about 290 °C. *Extinguish fire with* foam, dry powder, carbon dioxide or vaporising liquid.

Spillage
disposal

Shut off all possible sources of ignition. Wear face-shield or goggles and gloves. Apply non-flammable dispersing agent if available and work to an emulsion with brush and water — run this to waste, diluting greatly with running water. If dispersant not available, absorb on sand, shovel into bucket(s) and transport to safe, open area for atmospheric evaporation. Ventilate site of spillage well to evaporate remaining liquid and dispel vapour.

Phenacyl bromide (ω-bromoacetophenone) [70–11–1]

Colourless to greenish crystals; mp 50 °C; insoluble in water.

and

Phenacyl chloride (ω-chloroacetophenone) [532–57–4]

Colourless to yellow crystals or powder; mp 54 °C; insoluble in water.

(IRRITATING TO SKIN, EYES AND RESPIRATORY SYSTEM)

Avoid breathing vapour. Avoid contact with skin and eyes. RL 0.05 ppm (0.3 mg m^{-3}).

Toxic effects

The vapour and dust irritate the respiratory system and skin; they irritate the eyes severely. Assumed to be very irritant if taken by mouth.

First aid

Vapour inhaled: standard treatment (p. 136).
Affected eyes: standard treatment (p. 136).
Skin contact: standard treatment (p. 136).
If swallowed: standard treatment (p. 137).

Spillage
disposal

Wear face-shield or goggles, and gloves. Mix with sand and transport to a safe, open area for burial. Site of spillage should be washed thoroughly with water and soap or detergent.

Phenetidines —*see* Ethoxyanilines

Phenol (carbolic acid) [108–95–2]

Colourless to pink crystalline substance with distinctive odour; mp 43 °C; somewhat soluble in water (1 g dissolves in about 15 cm^3 water at 25 °C).

TOXIC IN CONTACT WITH SKIN AND IF SWALLOWED
CAUSES BURNS

Avoid breathing vapour. Prevent contact with skin and eyes. RL (skin) 5 ppm (19 mg m^{-3}).

Toxic effects The vapour irritates the respiratory system and eyes. Skin contact causes softening and whitening followed by the development of painful burns; its rapid absorption through the skin may cause headache, dizziness, rapid and difficult breathing, weakness and collapse. If taken by mouth it causes severe burns, abdominal pain, nausea, vomiting and internal damage. **Chronic effects** The inhalation of vapour over a long period may cause digestive disturbances, nervous disorders, skin eruptions and damage to the liver and kidneys; dermatitis may result from prolonged contact with weak solutions. (*See* notes in Ch 6 p. 107 and Ch 7 p. 120.)

Hazardous reactions Violent or explosive reactions with $AlCl_3$, formaldehyde, peroxo- mono- and di-sulphuric acids, $NaNO_2$, $NaNO_3$/trifluoroacetic acid (B596).

First aid *Vapour inhaled:* standard treatment (p. 136).
Affected eyes: standard treatment (p. 136).
Skin contact: standard treatment (p. 136). But see also note on phenol (p. 135).
If swallowed: standard treatment (p. 137).

Spillage disposal Wear face-shield or goggles, and gloves. Mix with sand and transport to a safe, open area for burial. Site of spillage should be washed thoroughly with water and soap or detergent.

Phenol-disulphonic and -sulphonic acids
—*see* **Sulphonic acids**

Phenothiazine (thiodiphenylamine)   [92–84–2]

TLV 5 mg m^{-3}.

Phenoxyacetylene   [4279–76–9]

Hazardous reaction Samples rapidly heated in sealed tubes to about 100 °C exploded (B703).

Phenylacetonitrile (benzyl cyanide)   [140–29–4]

Colourless to yellow liquid; bp 234 °C; insoluble in water.

(HARMFUL BY INHALATION, IN CONTACT WITH SKIN AND IF SWALLOWED
IRRITATING TO EYES)

Avoid breathing vapour. Avoid contact with skin and eyes.

Toxic effects	Inhalation of the vapour may cause pallor, faintness, headache and possibly vomiting. The liquid irritates the eyes and may irritate the skin. If taken by mouth, internal irritation and poisoning must be assumed.
Hazardous reaction	When sodium hypochlorite solution was used to destroy acidified residues, a violent explosion occurred (B706).
First aid	*Vapour inhaled:* standard treatment (p. 136). *Affected eyes:* standard treatment (p. 136). *Skin contact:* standard treatment (p. 136). *If swallowed:* standard treatment (p. 137).
Spillage disposal	Wear face-shield or goggles, and gloves. Apply dispersing agent if available and work to an emulsion with brush and water — run this to waste, diluting greatly with running water. If dispersant not available, absorb on sand, shovel into bucket(s) and transport to safe, open area for burial. Site of spillage should be washed thoroughly with water and soap or detergent.

α-Phenylazo hydroperoxides Review of group (B1651).

Phenylchlorodiazirine [4460–46–2]

Hazardous reaction	Is about three times as shock-sensitive as glyceryl nitrate (B657).

Phenylenediamines *o*- [95–54–5] *m*- [108–45–2]
p- [106–50–3]

p-Phenylenediamine — pale mauve to mauve crystals or powder. *m*-Phenylenediamine — colourless to brown or black crystals or lumps. The *m*-isomer is moderately soluble in water, the *p*-isomer only sparingly so.

TOXIC BY INHALATION, IN CONTACT WITH SKIN
AND IF SWALLOWED
MAY CAUSE SENSITISATION BY SKIN CONTACT

Avoid contact with skin and eyes and inhalation of dust. RL (skin) (*p*-) 0.1 mg m^{-3}.

Toxic effects	Eye-irritation and injury follow contact. May irritate the skin causing dermatitis; the skin becomes blackened. Assumed to be poisonous if taken by mouth. Inhalation of dust may cause bronchial asthma.
First aid	*Affected eyes:* standard treatment (p. 136). *Skin contact:* standard treatment (p. 136). *If swallowed:* standard treatment (p. 137).

Spillage disposal Wear face-shield or goggles, and gloves. Shovel into a glass, enamel or polythene vessel for dispersion in an excess of dilute hydrochloric acid (1 volume concentrated acid diluted with 2 volumes of water). Allow to stand, with occasional stirring, for 24 hours and then run the base hydrochloride solution to waste, diluting greatly with running water. Site of spillage should be washed with water and detergent.

Phenyl ether —*see* Diphenyl ether

Phenylethylene —*see* Styrene

Phenyl glycidyl ether (PGE) —*see* 2,3-Epoxypropyl phenyl ether

Phenylhydrazine and hydrochloride

[100–63–0]
[59–88–1]

Phenylhydrazine is a yellow to red-brown liquid or solid (mp 19 °C); insoluble in water. The hydrochloride consists of colourless to brown crystals or powder, soluble in water.

TOXIC BY INHALATION, IN CONTACT WITH SKIN
AND IF SWALLOWED
IRRITATING TO EYES

Avoid inhalation of vapour or dust, powerful allergen. Prevent contact with skin and eyes. RL (phenylhydrazine, skin) 5 ppm (20 mg m^{-3}).

Toxic effects Absorption resulting from inhalation, skin contact or ingestion may result in blood and liver damage, giving rise to nausea, vomiting and jaundice. Dermatitis or hypersensitisation may follow skin contact. Must be considered injurious to the eyes. (*See* note in Ch 6 p. 105.)

Hazardous reactions Reacts violently with PbO_2 or benzaldehyde phenylhydrazone oxide; forms explosive solid with ClO_3F (B605).

First aid *Vapour or dust inhaled:* standard treatment (p. 136).
Affected eyes: standard treatment (p. 136).
Skin contact: standard treatment (p. 136).
If swallowed: standard treatment (p. 137).

Spillage
disposal

Wear face-shield or goggles, and gloves. Mix the base with sand and shovel mixture into glass, enamel or polythene vessel for dispersion in an excess of dilute hydrochloric acid (1 volume of concentrated acid diluted with 2 volumes of water. Allow to stand, with occasional stirring, for 24 hours and then run acid extract to waste, diluting greatly with running water and washing the sand. The sand can be disposed of as normal refuse. A spillage of the hydrochloride can be mopped up with water and the solution run to waste, diluting greatly with running water.

N-Phenylhydroxylamine   [100–65–2]

Hazardous
reactions

Reduction of nitrobenzene with Zn gives pyrophoric residues, and the hydrochloride salt may decompose in storage (B602, 604).

Phenyl mercaptan —*see* Benzenethiol

Phenylmercury salts
—*see* Mercury compounds

N-Phenyl-2-naphthylamine   [135–88–6]
Suspected carcinogen.

Phenylphosphine   [638–21–1]
TLV 0.05 ppm (0.25 mg m^{-3}).

Phenylsilver   [5274–48–6]

Hazardous
reactions

Explodes on warming to room temperature or on light friction (B571).

5-Phenyltetrazole   [3999–10–8]

Hazardous
reaction

Explodes on attempted distillation (B668).

Phosgene (carbonyl chloride) and solutions [75–44–5]

The gas is colourless and has a musty smell. As a pale yellow liquid it boils at 7.6 °C. Solution of the gas in toluene (HIGHLY FLAMMABLE) is available commercially. Gas slightly soluble in water.

VERY TOXIC BY INHALATON

Do not breath gas. Avoid contact of solutions with skin and eyes. RL 0.1 ppm (0.4 mg m^{-3}).

Toxic effects The gas produces delayed secretion of fluid into the lung (pulmonary oedema) when inhaled and there may be delay of several hours before effects develop. These include breathlessness, cyanosis and the coughing up of frothy fluid. (*See* note in Ch 6 p. 103.)

Hazardous reaction Mixture with potassium is shock-sensitive (B120).

First aid *Vapour inhaled:* standard treatment (p. 136).
Affected eyes: standard treatment (p. 136).
Skin contact with solution: standard treatment (p. 136).
If solution swallowed: standard treatment (p. 137).

Spillage disposal (a) Solution in toluene: Shut off all possible sources of ignition. Instruct others to keep at a safe distance. Wear breathing apparatus and gloves. Apply non-flammable dispersing agent and work to an emulsion with brush and water — run this to waste, diluting greatly with running water. If dispersant not available, absorb on sand, shovel into bucket(s) and transport to safe, open area for burial. Site of spillage should be washed thoroughly with water and soap or detergent. Ventilate area of spillage well to dispel remaining vapour. (b) Gas: Surplus gas or leaking cylinder can be vented slowly into a water-fed scrubbing tower or column in a fume cupboard, or into a fume cupboard served by such a tower.

Phosphine (hydrogen phosphide) [3803–51–2]

Colourless gas with smell somewhat like that of rotting fish; bp −88 °C; slightly soluble in water. With a trace of P_2H_4 in it, phosphine (PH_3) is spontaneously flammable in air, burning with a luminous flame.

(EXTREMELY FLAMMABLE
VERY TOXIC BY INHALATION)

Prevent inhalation of gas. RL 0.3 ppm (0.4 mg m^{-3}).

Toxic effects The sequence of effects of inhalation have been stated to be pain in chest, sensation of coldness, weakness, vertigo, shortness of breath, lung damage, convulsions, coma and death.

Hazardous reactions	The impure gas ignites spontaneously in air, the liquified gas can be detonated; ignition occurs on contact with chlorine or bromine or their aqueous solutions; oxidised explosively by fuming nitric acid (B1185–1187).
First aid	*Gas inhaled:* standard treatment (p. 136).
Fire hazard	Impure phosphine must be regarded as spontaneously flammable. Since the gas is supplied in a cylinder, turning off the valve will reduce any fire involving it; if possible, cylinders should be removed quickly from an area in which a fire has developed.
Disposal	Surplus gas or leaking cylinder can be vented slowly to air in a safe, open area or gas burnt off through a suitable burner in a fume cupboard.

Phosphines Review of group (B1651).

Phosphinic acid   [6303–21–5]

Hazardous reaction	Redox reaction with mercury(II) oxide is explosive (B1183–1184).

Phosphonium perchlorate   [101672–21–3]

Hazardous reaction	An explosive, sensitive to moist air, friction or heat (B922).

Phosphoric acid (orthophosphoric acid)   [7664–38–2]

Colourless viscous liquid (88–93%) or moist white crystals (100%); miscible with water.

CAUSES BURNS

Avoid contact with eyes and skin. RL 1 mg m^{-3}.

Toxic effects	The liquid burns the eyes severely. The liquid burns the skin. If taken by mouth there would be severe internal irritation and damage.
First aid	*Affected eyes:* standard treatment (p. 136).
	Skin contact: standard treatment (p. 136).
	If swallowed: standard treatment (p. 137).
Spillage disposal	Wear face-shield or goggles, and gloves. Spread soda ash liberally over the spillage and mop up cautiously with plenty of water — run to waste, diluting greatly with running water.

Phosphoric oxide
—*see* **Phosphorus pentaoxide**

Phosphorus, red
and white (yellow)

[7723–14–0]

Red (amorphous) phosphorus sublimes at 416 °C; insoluble in water. White (yellow) phosphorus is a pale yellow, waxy, translucent solid; mp 44 °C; usually stored under water; if exposed to the air it rapidly ignites giving off fumes of the oxide.

RED	**WHITE (YELLOW)**
HIGHLY FLAMMABLE	SPONTANEOUSLY FLAMMABLE IN
EXPLOSIVE WHEN MIXED	AIR
WITH OXIDISING	VERY TOXIC BY INHALATION
SUBSTANCES	AND IF SWALLOWED
	CAUSES SEVERE BURNS
	Prevent contact with skin and eyes.
	RL 0.1 mg m^{-3}.

Toxic effects Red phosphorus is relatively harmless physiologically unless it contains the white allotrope. The vapour from ignited phosphorus irritates the nose, throat, lungs and eyes. Solid white phosphorus burns the skin and eyes and causes severe internal damage if taken by mouth. **Chronic effects** Continued absorption of small amounts can cause anaemia, intestinal weakness, pallor, bone and liver damage.

Hazardous reactions Reference to brochure on handling is given; contact of both allotropes with boiling alkalies evolves phosphine, which usually ignites in air; yellow phosphorus reacts vigorously with chlorosulphuric acid, accelerating to explosion — the same occurs at a higher temperature with the red allotrope; both forms ignite on contact with fluorine and chlorine, the red allotrope in liquid bromine; the white form explodes in liquid bromine or chlorine, and ignites in contact with bromine vapour or solid iodine; the white form reacts far more readily than the red with air or oxygen; both forms react violently with hydrogen peroxide if incompletely immersed; both forms are liable to explode when mixed with perchlorates, chlorates, bromates and subjected to friction, impact or heat; explodes with chromyl chloride when moist and on impact or grinding with ammonium, mercury(I) or silver nitrates or potassium permanganate; incandescent or violent reactions occur with some metals and oxides (B1376–1381).

First aid *Vapour of burning phosphorus inhaled:* standard treatment (p. 136).
Affected eyes: standard treatment (p. 136).
Skin contact: red P, standard treatment (p. 136); white P, swab with 3% CuSO$_4$ solution, then standard treatment (p. 136).
If swallowed: standard treatment (p. 137).

Fire hazard Water is the best medium for fighting a phosphorus fire caused by its spontaneous ignition.

Spillage disposal	*Red form:* Moisten with water, shovel into a bucket, transport to a safe, isolated area where the moisture can be allowed to dry off and the phosphorus burnt off. This applies to small spillages only. *White form:* Wear face-shield or goggles, and gloves; cover with wet sand, shovel into bucket and remove to safe, open, isolated area where the phosphorus can be burnt off under supervision after drying out. Small spillages of phosphorus can be burnt off in a fume cupboard.

Phosphorus esters Review of group (B1652).

Phosphorus(III) oxide   [10248-58-5]

Hazardous reactions	Reacts violently with disulphur dichloride, liquid bromine, PCl_5, sulphur, sulphuric acid, hot water; ignition likely with air and oxygen at elevated temperatures; violent reaction or ignition with ethanol, methanol, dimethyl-formamide, dimethyl sulphite or dimethyl sulphoxide (B1372-1373).

Phosphorus oxychloride
—*see* Phosphoryl chloride

Phosphorus pentachloride   [10026-13-8]

White to pale yellow, fuming, crystalline masses with pungent, unpleasant odour. Violently decomposed by water with formation of hydrochloric acid and phosphoric acid.

CAUSES BURNS
IRRITATING TO RESPIRATORY SYSTEM

Avoid breathing vapour and dust. Prevent contact with eyes and skin. RL 0.1 ppm (1 mg m^{-3}).

Toxic effects	Vapour and dust severely irritate the mucous membranes and all parts of the respiratory system. The vapour severely irritates and the solid burns the eyes. The vapour and solid burn the skin. If taken by mouth there would be severe internal irritation and damage. **Chronic effects** Continued exposure to low concentrations of vapour may cause damage to lungs.
Hazardous reactions	Reacts violently with P_2O_5, nitrobenzene and water, explosively with sodium and urea; ignites with Al powder and fluorine (B1008-1009).
First aid	*Vapour inhaled:* standard treatment (p. 136). *Affected eyes:* standard treatment (p. 136). *Skin contact:* standard treatment (p. 136). *If swallowed:* standard treatment (p. 137).

Spillage
disposal
Wear goggles and gloves. Mix with dry sand, shovel into an enamel or polythene bucket, transport to a safe, open area and add, a little at a time, to a large volume of water; after reaction is complete, run to waste, diluting greatly with running water. Dispose of sand as normal refuse. Wash spillage site with water.

Phosphorus pentafluoride   [7647–19–0]

Colourless gas, fuming strongly in moist air, with a pungent smell; bp −85 °C; it is hydrolysed by water with formation of hydrogen fluoride and, ultimately, phosphoric acid.

(VERY TOXIC BY INHALATION AND IN CONTACT WITH SKIN
CAUSES SEVERE BURNS)

Prevent inhalation of gas. Prevent contact with skin and eyes.

Toxic effects
Because of its ready hydrolysis by moisture or water to hydrogen fluoride its toxic effects are similar to the latter. It irritates severely the eyes and respiratory system and may cause burns to the eyes. It irritates the skin and, if exposure has been considerable, painful burns may develop after an interval.

First aid
Gas inhaled: standard treatment (p. 136).
Affected eyes: special treatment for hydrofluoric acid (p. 135).
Skin contact: special treatment for hydrofluoric acid (p. 134).

Disposal
Surplus gas or leaking cylinder can be vented slowly into a water-fed scrubbing tower or column in a fume cupboard, or into a fume cupboard served by such a tower.

Phosphorus pentaoxide   [1314–56–3]
(phosphorus(V) oxide; phosphoric oxide)

White crystalline deliquescent powder; reacts violently with water.

CAUSES SEVERE BURNS

Prevent inhalation of dust or contact with eyes and skin.

Toxic effects
The dust irritates all parts of the respiratory system, burns the eyes severely and burns the skin. If taken by mouth there would be severe internal irritation and damage.

Hazardous reactions	Reference given to brochure on safe handling; attempted dehydration of 95% formic acid caused rapid evolution of CO; reacts vigorously with HF below 20 °C; dry mixtures with Na_2O or CaO react violently if warmed or moistened; reaction with warm Na or K is incandescent; reaction explosive when heated with Ca; reacts violently with concentrated hydrogen peroxide; ignition occurs on contact with F_2O; the reaction with water is highly exothermic and enough to ignite combustible materials in contact; incandescence with barium sulphide; violent reaction with iodides; explosive with methyl hydroperoxide (B1374–1375).
First aid	*Dust inhaled:* standard treatment (p. 136). *Affected eyes:* standard treatment (p. 136). *Skin contact:* standard treatment (p. 136). *If swallowed:* standard treatment (p. 137).
Spillage disposal	Wear face-shield or goggles, and gloves. Mix with dry sand, shovel into an enamel or polythene bucket, transport to a safe, open area and add, a little at a time, to a large volume of water; after reaction is complete, run to waste, diluting greatly with running water. Dispose of sand as normal refuse. Wash site of spillage with water.

Phosphorus(III) and (V) sulphides

[1314–80–3]
[15857–57–5]

Yellow crystalline powders with a peculiar odour; react with water, forming phosphorus acids and hydrogen sulphide.

HIGHLY FLAMMABLE
HARMFUL BY INHALATION AND IF SWALLOWED
CONTACT WITH WATER LIBERATES TOXIC GAS

Avoid breathing dust. Prevent contact with eyes and skin. RL (P_4S_{10}) 1 mg m^{-3}.

Toxic effects	The dust irritates the mucous membranes and all parts of the respiratory system; it irritates the eyes and burns the skin. Assumed to cause severe internal irritation and damage if taken by mouth.
Hazardous reactions	P_4S_{10} ignites by friction, sparks or flames; it heats and may ignite with limited amounts of water (B1382).
First aid	*Dust inhaled:* standard treatment (p. 136). *Affected eyes:* standard treatment (p. 136). *Skin contact:* standard treatment (p. 136). *If swallowed:* standard treatment (p. 137).
Fire hazard	Ignition temp. (P_4S_3) 282 °C, (P_4S_{10}) 142 °C, the latter temperature being readily obtained by friction. *Extinguish fire with* dry powder, carbon dioxide or sand.
Spillage disposal	Wear face-shield or goggles, and gloves. Spread soda ash liberally over the spillage and mop up cautiously with plenty of water — run to waste, diluting greatly with running water.

Phosphorus tribromide [7789-60-8]

Colourless fuming liquid; bp 175 °C; reacts violently with water, forming hydrobromic acid and phosphorous acid.

REACTS VIOLENTLY WITH WATER
CAUSES BURNS
IRRITATING TO RESPIRATORY SYSTEM

Avoid breathing vapour. Prevent contact with eyes and skin.

Toxic effects Vapour severely irritates the mucous membranes and all parts of the respiratory system. The vapour irritates and the liquid burns the eyes. The vapour and liquid burn the skin. Assumed to cause severe internal irritation and damage if taken by mouth. **Chronic effects** Continued exposure to low concentrations of vapour may cause damage to lungs.

Hazardous reactions Reacts rapidly with warm water, violently with limited quantities; sodium floats on PBr_3 without reaction but the addition of a little water causes violent explosion; potassium ignites on contact with the liquid or vapour (B109–110).

First aid *Vapour inhaled:* standard treatment (p. 136).
Affected eyes: standard treatment (p. 136).
Skin contact: standard treatment (p. 136).
If swallowed: standard treatment (p. 137).

Spillage disposal Instruct others to keep at a safe distance. Wear breathing apparatus and gloves. Spread soda ash liberally over the spillage and mop up cautiously with water — run this to waste, diluting greatly with running water.

Phosphorus trichloride [7719-12-2]

Colourless fuming liquid; bp 75 °C; violently decomposed by water, forming hydrochloric acid and phosphorous acid.

CAUSES BURNS
IRRITATING TO RESPIRATORY SYSTEM

Avoid breathing vapour. Prevent contact with eyes and skin. RL 0.5 (0.2) ppm (3 mg m^{-3}).

Toxic effects Vapour severely irritates the mucous membranes and all parts of the respiratory system. The vapour and liquid burn the eyes and skin. Assumed to cause severe internal irritation and damage if taken by mouth. **Chronic effects** Continued exposure to low concentrations of vapour may cause lung damage.

Hazardous reactions	Residue in preparation of acetyl chloride from PCl_3 and acetic acid may decompose violently with evolution of flammable phosphine; reacts violently or explosively with dimethyl sulphoxide; potassium ignites in PCl_3, molten sodium explodes on contact; may explode on contact with chromyl chloride; ignites on contact with fluorine; explodes with nitric acid; reacts violently or explosively with sodium peroxide; reacts violently with water with liberation of some diphosphane which ignites; Al powder incandesces; violent reaction with CrF_5; explosion with tetravinyl lead (B998–1000).
First aid	*Vapour inhaled:* standard treatment (p. 136). *Affected eyes:* standard treatment (p. 136). *Skin contact:* standard treatment (p. 136). *If swallowed:* standard treatment (p. 137).
Spillage disposal	Instruct others to keep at a safe distance. Wear breathing apparatus and gloves. Mix with dry sand, shovel into enamel or polythene bucket(s) and add, a little at a time, to a large volume of water; when reaction is complete run solution to waste, diluting greatly with running water. Wash area of spillage thoroughly with water.

Phosphorus tricyanide [1116–01–4]

Hazardous reaction	Explosions occurred during vacuum sublimation (B393–394).

Phosphorus trifluoride [7783–55–3]

Hazardous reactions	Ignites on contact with fluorine; explosion has occurred at low temperatures with dioxygen difluoride (B1069–1070).

Phosphorus trisulphide
—*see* **Phosphorus(III) sulphide**

Phosphoryl chloride (phosphorus oxychloride) [10026–13–8]

Colourless fuming liquid; bp 107 °C; violently decomposed by water, forming hydrochloric acid and phosphoric acid.

CAUSES BURNS
IRRITATING TO RESPIRATORY SYSTEM

Avoid breathing vapour. Prevent contact with eyes and skin. RL 0.2 ppm (1.2 mg m^{-3}).

Toxic effects   The vapour severely irritates the mucous membranes and all parts of the respiratory system; there may be sudden or delayed pulmonary oedema. The vapour severely irritates the eyes and skin. The liquid burns the eyes and skin. Assumed to cause severe internal irritation and damage if taken by mouth.

Hazardous reactions   There is a considerable delay in its reaction with water which ultimately becomes violent — closed or only slightly vented vessels therefore represent a hazard; may react explosively with dimethyl sulphoxide or heated sodium; zinc dust ignites on contact with a little $POCl_3$ (B996–997).

First aid   *Vapour inhaled:* standard treatment (p. 136).
Affected eyes: standard treatment (p. 136).
Skin contact: standard treatment (p. 136).
If swallowed: standard treatment (p. 137).

Spillage disposal   Instruct others to keep at a safe distance. Wear breathing apparatus and gloves. Mix with dry sand, shovel into enamel or polythene bucket(s) and add, a little at a time, to a large volume of water; when reaction is complete run solution to waste, diuting greatly with running water. Wash area of spillage thoroughly with water.

Phthalic anhydride   [85–44–9]

White crystalline needles; mp 131 °C; dissolves in hot water forming phthalic acid.

IRRITATING TO EYES AND RESPIRATORY SYSTEM

Avoid contact with skin and eyes. RL 1 ppm (6 mg m^{-3}).

Toxic effects   The dust irritates the eyes, skin and respiratory system. It will cause internal irritation if taken by mouth.

Hazardous reactions   Mixture with anhydrous CuO exploded on heating; nitration with nitric acid/sulphuric acid presents dangers; mixture with sodium nitrite will explode on heating (B699).

First aid   *Affected eyes:* standard treatment (p. 136).
Skin contact: standard treatment (p. 136).
If swallowed: standard treatment (p. 136).

Spillage disposal   Clear up with dust-pan and brush or bucket and shovel, placing in large volume of water. The acid solution produced on standing may be run to waste, diluting greatly with running water.

m-Phthalodinitrile   [626–17–5]

TLV 5 mg m^{-3}.

461

Phthaloyl diazide [50906–29–1]

Hazardous reaction	Extremely explosive (B698–699).

Phthaloyl peroxide [4733–52–2]

Hazardous reactions	Detonatable by impact or by melting at 123 °C (B700).

Picolines —*see* Methylpyridines

Picrates Review of group (B1653)

Picric acid [88–89–1]

Picric acid (yellow crystals) should be kept moist with not less than half its own weight of water. It is commonly used in alcoholic solutions in the laboratory.

(RISK OF EXPLOSION BY SHOCK, FRICTION, FIRE OR OTHER
SOURCE OF IGNITION
FORMS VERY SENSITIVE EXPLOSIVE METALLIC COMPOUNDS
TOXIC BY INHALATION, IN CONTACT WITH SKIN
AND IF SWALLOWED)

Avoid contact with skin and eyes. RL (skin) 0.1 mg m^{-3}.

Toxic effects	Skin contact may result in dermatitis. Poisonous if taken by mouth. **Chronic effects** Absorption through the skin or inhalation of dust over a long period may result in skin eruptions, headache, nausea, vomiting or diarrhoea; the skin may become yellow.
Hazardous reactions	Explosive which is usually stored as water-wet paste; forms salts with many metals some of which (Pb, Hg, Cu or Zn) are sensitive to heat, friction or impact; contact of acid with concrete floors may form friction-sensitive calcium salt; moisture ignites dry mixtures with Al powder. Handling and disposal procedures reviewed (B552).
First aid	*Affected eyes:* standard treatment (p. 136). *Skin contact:* standard treatment (p. 136). *If swallowed:* standard treatment (p. 137).
Spillage disposal	Wear face-shield or goggles, and gloves. Moisten well with water and mix with sand. Transport to isolated area for burial. Site of spillage must be washed thoroughly with water and detergent.

RSC *Lab. Haz. Data Sheet No. 34,* 1985 gives extended coverage.

Picryl azide [1600–31–3]

Hazardous reaction   Explodes weakly on impact (B546).

Piperazine [110–85–0]
and piperazine hydrate [142–63–2]

Both take the form of colourless crystals; mp 106 °C and 44 °C respectively; both are very soluble in water, the solutions being strongly alkaline.

CAUSE BURNS

Avoid breathing vapour. Avoid contact with skin, eyes and clothing.

Toxic effects   The vapours irritate the eyes and respiratory system. The solutions irritate and may burn the eyes and skin. The solids and aqueous solutions will cause internal burning and damage if taken by mouth.

First aid   *Vapour inhaled:* standard treatment (p. 136).
Affected eyes: standard treatment (p. 136).
Skin contact: standard treatment (p. 136).
If swallowed: standard treatment (p. 137).

Spillage disposal   Wear face-shield or goggles, and gloves. Mop up with plenty of water and run to waste, diluting greatly with running water.

Piperidine (hexahydropyridine) [110–89–4]

Colourless liquid with amine-like odour; bp 106 °C; miscible with water.

HIGHLY FLAMMABLE
TOXIC BY INHALATION AND IN CONTACT WITH SKIN
CAUSES BURNS

Avoid breathing vapour. Avoid contact with skin, eyes and clothing.

Toxic effects   The vapour irritates the eyes and respiratory system. The liquid irritates the eyes and skin and is assumed to be irritant and harmful if taken internally.

First aid   *Vapour inhaled:* standard treatment (p. 136).
Affected eyes: standard treatment (p. 136).
Skin contact: standard treatment (p. 136).
If swallowed: standard treatment (p. 137).

Fire hazard   Flash point 16 °C. *Extinguish fire with* dry powder, carbon dioxide or vaporising liquid.

463

Spillage
disposal

Shut off all possible sources of ignition. Wear face-shield or goggles, and gloves. Mop up with plenty of water and run to waste, diluting greatly with running water. Ventilate area of spillage well to evaporate remaining liquid and dispel vapour.

Pivaloyl azide [4981–48–0]

Hazardous
reaction

Explodes on warming (B517).

Platinum (metal) [7440–06–4]

Hazardous
reactions

Used Pt catalysts present explosion risk; mixture of Pt and As in sealed tube at 270 °C exploded; addition of Pt-black catalyst to ethanol caused ignition; addition of Pt-black to concentrated hydrogen peroxide may cause explosion; reacts violently with Li at about 540 °C; platinum burns in phosphorus vapour below red-heat; finely divided Pt reacts incandescently when heated with selenium or tellurium; Pt sponge causes methyl hydroperoxide to explode (B1385–1387).

Platinum compounds Review of group (B1654).

RL (soluble salts as Pt) 0.002 mg m^{-3}.

Poly(dimercuryimmonium) salts Review of group (B1655).

Polymerisation incidents Review of topic (B1656)

Polynitroalkyl and polynitroaryl compounds Reviews of classes (B1658–1659).

Polyperoxides Review of group (B164).

Potassium (metal) [7440–09–7]

Soft silvery white masses normally coated with a grey oxide or hydroxide skin; mp 64 °C; reacts violently with water with the formation of potassium hydroxide and hydrogen gas, which will ignite. Explosions have occurred when old heavily-crusted potassium metal has been cut with a knife. Such old stock should be disposed of by dissolving, uncut, in propan-2-ol.

> REACTS VIOLENTLY WITH WATER
> CONTACT WITH WATER LIBERATES HIGHLY FLAMMABLE GASES
> CAUSES BURNS

Avoid contact with skin and eyes.

Toxic effects   The metal in contact with moisture on the skin can cause thermal and caustic burns. In the same way the potassium hydroxide resulting from the reaction of the metal with water can cause burning of the skin and eyes.

Hazardous reactions   Reference given to detailed procedures for safe handling; comparison of properties of K and Na reviewed showing former to be invariably more hazardous; K reacts readily with CO to form explosive carbonyl; K does not react with air or oxygen in complete absence of moisture, but in its presence oxidation becomes fast and melting and ignition take place; prolonged but restricted access to air results in the formation of coatings of yellow superoxide on top of the monoxide — percussion or dry cutting of the metal brings traces of residual oil into contact with the superoxide and a very violent explosion occurs [*see* potassium dioxide (p. 468)]; the disposal of scrap potassium is described, the use of ethanol or methanol being excluded because of their violent reactions with the metal; the metal reacts vigorously with various forms of carbon and, if air is present, ignition and explosion are likely; mixtures of the metal with a wide range of halocarbons are shock sensitive and may explode with violence on slight impact; K ignites in fluorine and dry chlorine; it explodes violently in liquid bromine and incandesces in the vapour; mixture of metal with anhydrous HI explodes violently; reaction with mercury to form amalgams is vigorous or violent; reacts violently or explosively with many metal halides or oxides; reactions with non-metal halides and oxides are usually explosive as are those with oxidants; the reaction with sulphuric acid is explosive; with water, the heat evolved is enough to ignite the hydrogen evolved — large pieces of the metal explode on the surface of water scattering burning particles over a wide area (B1233–1240).

First aid   *Affected eyes:* standard treatment (p. 136).
Skin contact: standard treatment (p. 136).
Mouth contact: standard treatment (p. 137).

Fire hazard   A fire resulting from the ignition of potassium metal is best extinguished by smothering it with dry soda ash.

Spillage disposal   Instruct others to keep at a safe distance. Wear face-shield or goggles, and gloves. Cover with dry soda ash, shovel into dry bucket, transport to safe, open area and add, a little at a time, to a large excess of dry propan-2-ol. Leave to stand for 24 hours and run solution to waste, diluting greatly with water.

465

Potassium amide [17242–52–3]

Hazardous reactions   More violently reactive than sodium amide; reaction with water is violent and ignition may occur; explodes when heated with potassium nitrite under vacuum (B1143–1144).

Potassium antimonate
—*see* Antimony compounds

Potassium arsenate and arsenite
—*see* Arsenic and compounds

Potassium azide [20762–60–1]

Hazardous reactions   Melts on heating then decomposes evolving nitrogen, the residue igniting with a feeble explosion; reacts violently with manganese dioxide on warming; explodes at 120 °C when heated in liquid sulphur dioxide (B1251–1252).

Potassium benzenehexoxide [3264–86–6]
('carbonylpotassium')

Hazardous reactions   Reacts violently with oxygen; explodes on heating in air or on contact with water (B648).

Potassium bichromate
—*see* Chromates and dichromates

Potassium bisulphate
—*see* Potassium hydrogen sulphate

Potassium bromate [7758–01–2]

Hazardous reactions   Sulphur, selenium and sulphur bromide all ignite with the salt; a heated mixture with aluminium and dinitrotoluene evolves huge volumes of gas rapidly (B96–97).

Potassium chlorate

[3811–04–9]

Colourless crystals or crystalline powder; mp 368 °C; soluble in water.

EXPLOSIVE WHEN MIXED WITH COMBUSTIBLE MATERIAL
HARMFUL BY INHALATION AND IF SWALLOWED

Avoid contact with skin, eyes and clothing.

Toxic effects   When taken internally may irritate the intestinal tract and kidneys.

Hazardous reactions   Fabric gloves (wrongly used) became impregnated with the chlorate during handling operations and were subsequently ignited by cigarette ash; molten $KClO_3$ ignites HI gas; impact- *etc* -sensitive explosives are formed by mixtures of the chlorate with tricopper diphosphide, trimercury tetraphosphide, finely divided Al, Cu, Mg or Zn, many metal sulphides, or thiocyanates, arsenic, carbon, phosphorus, sulphur or other readily oxidised materials; addition of $KClO_3$ in portions to H_2SO_4 below 60 °C and above 200 °C causes brisk effervescence — between these temperatures, explosions are caused by the ClO_2 produced; a mixture with sodium amide exploded; explosions with NH_4Cl, cyanoguanidine, As_2S_3, hydrocarbons, metals, reducing agents, steel wool, sugar or tannic acid (B924–930).

First aid   *Affected eyes:* standard treatment (p. 136).
If swallowed: standard treatment (p. 137).

Spillage disposal   Shovel into bucket of water and run solution to waste, diluting greatly with running water. Site of spillage and clothing should be washed thoroughly to remove all oxidant.

Potassium chromate
—*see* **Chromates and dichromates**

Potassium cyanide
—*see* **Cyanides** (water-soluble)

Potassium dichromate
—*see* **Chromates and dichromates**

467

Potassium dioxide (potassium superoxide)   [12030-88-5]

Hazardous reactions   Stable when pure, impact-sensitivity of old peroxidised potassium now known to be due to presence of oil or organic residues; reacts violently with diselenium dichloride; explosions may occur with hydrocarbons; oxidises arsenic, antimony, copper, tin or zinc with incandescence. May ignite filter paper, explosive oxidation of 2-aminophenol in THF (B1252-1254).

Potassium ferricyanide and ferrocyanide
—*see* **Potassium hexacyanoferrate**

Potassium fluoride —*see* Fluorides (water soluble)

Potassium fluorosilicate
—*see* **Hexafluorosilicic acid and salts**

Potassium hexacyanoferrate   3- [13746-66-2]
4- [13943-58-3]

Hazardous reactions   Contact of the hexacyanoferrate(4-) with ammonia may be explosive; mixtures of the (4-) salt with CrO_3 explode on heating above 196 °C or with $Cu(NO_3)_2$ at 220 °C; both of the salts mixed with sodium nitrite explode on heating (B537-538).

Potassium hydride   [7693-26-7]

Hazardous reactions   Ignites on contact with fluorine; reacts slowly with oxygen in dry air, more rapidly in moist, explosion with fluoroalkene at 12 °C (B1095-1096).

Potassium hydrogen fluoride
—*see* Fluorides (water soluble)

Potassium hydrogen sulphate   [7646-93-7]
(potassium bisulphate)

Colourless crystals or fused masses; soluble in water.

(CAUSES BURNS
IRRITATING TO EYES AND SKIN)

Avoid contact with skin and eyes.

Toxic effects   The solid and its solutions severely irritate and burn the eyes and skin. If taken by mouth there is severe internal irritation and damage.

First aid   *Affected eyes:* standard treatment (p. 136).
Skin contact: standard treatment (p. 136).
If swallowed: standard treatment (p. 137).

**Spillage
disposal**   Wear face-shield or goggles, and gloves. Spread soda ash liberally over the spillage and mop up cautiously with plenty of water — run to waste, diluting greatly with running water.

Potassium hydroxide (caustic potash) [1310–58–3]

Colourless sticks, flakes, powder or pellets soluble in water.

CAUSES SEVERE BURNS

Prevent contact with eyes and skin. RL 2 mg m^{-3}.

Toxic effects   The solid and its solutions severely irritate and burn the eyes and skin. If taken by mouth there would be severe internal irritation and damage.

**Hazardous
reactions**   Liquid or gaseous ClO_2 will explode on contact with solid KOH or its concentrated solution; germanium is oxidised by fused KOH with incandescence; as little as two flakes of moist KOH dropped on 2 kg potassium peroxodisulphate caused a vigorous self-sustained fire; may cause the explosion of peroxidised tetrahydrofuran and must not be used to dry this (B1097–1098).

First aid   *Affected eyes:* standard treatment (p. 136).
Skin contact: standard treatment (p. 136).
If swallowed: standard treatment (p. 137).

**Spillage
disposal**   Wear face-shield or goggles, and gloves. Shovel into large volume of water in an enamel or polythene vessel and stir to dissolve; run the solution to waste diluting greatly with running water.

Potassium methylamide [13427–02–6]

**Hazardous
reactions**   Extremely hygroscopic and pyrophoric; may explode on contact with air (B167).

469

Hazardous chemicals

Potassium nitrate [7757-79-1]

Colourless crystals, soluble in water.

(EXPLOSIVE WHEN MIXED WITH COMBUSTIBLE MATERIALS)

Keep out of contact with all combustible material.

Hazardous reactions Mixture of KNO_3 and calcium silicide is readily ignited, high temperature primer capable of initiating many high temperature reactions; mixtures with powdered Ti, Sb or Ge explode on heating; mixtures with antimony trisulphide, barium or calcium sulphides all explode on heating; mixtures with arsenic disulphide or molybdenum disulphide are detonatable; finely divided mixture with boron ignites and explodes on percussion; the oldest explosive, gunpowder, is a mixture of the nitrate with charcoal and sulphur; mixture with red phosphorus reacts vigorously on heating; mixtures of the nitrate with arsenic explode when ignited; in general, mixtures with combustible materials or organic impurities are readily ignited, if finely divided they can be explosive; CrN deflagrates, and sulphides react violently with the molten salt (B1249–1251).

Fire hazard Mixtures of potassium nitrate and combustible materials are readily ignited; mixtures with finely divided combustible materials can react explosively.

Spillage disposal Shovel into bucket of water and run solution to waste, diluting greatly with running water. Site of spillage should be washed thoroughly to remove all oxidant, which is liable to render any organic matter (particularly wood, paper and textiles) with which it comes into contact, dangerously combustible when dry. Clothing wetted with the solution should be washed thoroughly.

Potassium nitride [29285-24-3]

Hazardous reactions Usually ignites in air; on heating with phosphorus a highly flammable mixture is formed which evolves ammonia and phosphine with water; with sulphur a similar mixture is formed which evolves ammonia and hydrogen sulphide on contact with water (B1258).

Potassium nitrite [7758-09-0]

White or slightly yellow, deliquescent, crystalline solid; mp 441 °C; soluble in water.

CONTACT WITH COMBUSTIBLE MATERIAL MAY CAUSE FIRE
TOXIC IF SWALLOWED

Avoid contact with clothing and other absorbent fabrics.

Hazardous reactions   Addition of ammonium sulphate to the fused nitrite causes effervescence and ignition; addition of boron to fused nitrite causes violent decomposition; explosion occurs when mixture with potassium amide is heated under vacuum. Double salt with KCN explosive (B1248–1249).

Spillage disposal   Shovel into bucket of water and run solution to waste diluting greatly with running water. Site of spillage should be washed thoroughly to remove all oxidant. Clothing wetted by a solution of the nitrite should be washed thoroughly.

Potassium perchlorate —*see* Perchlorates

Potassium permanganate   [7722–64–7]

CONTACT WITH COMBUSTIBLE MATERIAL MAY CAUSE FIRE
HARMFUL IF SWALLOWED

Hazardous reactions   Risk of explosion with acetic acid or acetic anhydride; antimony ignites on grinding in a mortar with the solid permanganate, while arsenic explodes; contact of glycerol with solid $KMnO_4$ caused a vigorous fire; reaction with concentrated hydrochloric acid is occasionally explosive; hydrogen peroxide and permanganate (solid) react violently and cause a fire; mixtures with phosphorus or sulphur react explosively on grinding or heating respectively; explosion follows the addition of concentrated sulphuric acid to damp $KMnO_4$; violent reaction with t-butylamine in aqueous acetone; ignition on contact with dichloro(methyl)silane and various oxygenated organic compounds; violent exotherm with 60% HF solution; explosions with oil content of mineral fibre ('slag wool'). Explodes soon after mixing with 4 vols. DMF, and DMSO ignites (B1242–1243).

Potassium peroxodisulphate   [7727–21–1]

Hazardous reactions   A vigorous self-sustaining fire resulted when as little as two flakes of moist KOH was dropped on the surface of 2 kg of the dry salt. Dry salt gives off oxygen rapidly at 100 °C, but at only 50 °C when wet (B1257).

Potassium selenate and selenite
—*see* Selenium and compounds

Potassium tellurite
—*see* Tellurium and compounds

Propadiene (allene) [463–49–0]

Hazardous reaction	May decompose explosively under a pressure of 2 bar (2×10^5 Pa) (B344).

Propanal —*see* Propionaldehyde

Propane [74–98–6]

Colourless gas which burns with a smoky flame; bp −42 °C; 100 volumes of water dissolve 6.5 volumes of the gas at 17.8 °C and 1003 mbar.

EXTREMELY FLAMMABLE LIQUEFIED GAS

Avoid breathing gas.

Toxic effects The gas is anaesthetic in high concentrations but is not toxic.

First aid *Gas inhaled in high concentrations:* standard treatment (p. 136).

Fire hazard Flash point −104 °C; explosive limits 2.2–9.5%; ignition temp. 468 °C. Since the gas is supplied in a cylinder, turning off the valve will reduce any fire involving it; if possible, cylinders should be removed quickly from an area in which a fire has developed.

Disposal Surplus gas or leaking cylinder can be vented slowly to air in a safe, open area or gas burnt off through a suitable burner.

1,3-Propanesultone [1120–71–4]

Suspected carcinogen, no TLV set.

Propan-1-ol (propyl alcohol) [71–23–8]

Colourless liquid with alcoholic odour; bp 97 °C; miscible with water.

HIGHLY FLAMMABLE

Avoid breathing vapour. RL (skin) 200 ppm (500 mg m^{-3}).

Toxic effects The vapour may irritate the eyes and respiratory system and may be narcotic in high concentrations. The liquid irritates the eyes and is narcotic if taken internally.

472

First aid	*Vapour inhaled:* standard treatment (p. 136). *Affected eyes:* standard treatment (p. 136). *If swallowed:* standard treatment (p. 137).
Fire hazard	Flash point 25 °C; explosive limits 2.1–13.5%; ignition temp. 433 °C. *Extinguish fire with* water spray, dry powder, carbon dioxide or vaporising liquid.
Spillage disposal	Shut off all possible sources of ignition. Wear face-shield or goggles, and gloves. Mop up with plenty of water and run to waste, diluting greatly with running water. Ventilate area well to evaporate remaining liquid and dispel vapour.

Propan-2-ol (isopropyl alcohol, 'IPA') [67–63–0]

Colourless liquid with somewhat alcoholic odour; bp 82 °C; miscible with water.

HIGHLY FLAMMABLE

Avoid breathing vapour. RL (skin) 400 ppm (980 mg m^{-3}).

Toxic effects	Inhalation of the vapour in high concentrations and ingestion of the liquid may result in headache, dizziness, mental depression, nausea, vomiting, narcosis, anaesthesia and coma; the fatal dose is about 100 cm^3. The liquid may damage the eyes severely.
Hazardous reactions	In the preparation of aluminium isopropoxide, the dissolution of Al in propan-2-ol becomes vigorously exothermic but its onset is frequently delayed — only small amounts of Al should be present until the reaction starts; in recovering propan-2-ol from the reduction of crotonaldehyde by aluminium isopropoxide, a violent explosion occurred which was attributed to either peroxidised by-product diisopropyl ether or peroxidised crotonaldehyde; an explosion occurred during the thawing of a frozen mixture of the alcohol with trinitromethane; ignition occurs on grinding with CrO$_3$; it forms explosive mixtures with hydrogen peroxide; a bottle of the alcohol, exposed to sunlight for a long period, became 0.36M in peroxide and potentially explosive: traces of 2-butanone promote this (B379–380).
First aid	*Vapour inhaled in high concentrations:* standard treatment (p. 136). *Affected eyes:* standard treatment (p. 136). *If swallowed:* standard treatment (p. 137).
Fire hazard	Flash point 12 °C; explosive limits 2.3–12.7%; ignition temp. 399 °C. *Extinguish fire with* water spray, dry powder, carbon dioxide or vaporising liquid.
Spillage disposal	Shut off all possible sources of ignition. Wear face-shield or goggles, and gloves. Mop up with plenty of water and run to waste, diluting greatly with running water. Ventilate area of spillage well to evaporate remaining liquid and dispel vapour.

473

Propargyl alcohol —*see* 3-Hydroxypropyne

Propargyl bromide —*see* 3-Bromopropyne

Propene (propylene) [115–07–1]

Colourless gas which burns with a sooty flame; bp −48 °C; slightly soluble in water.

EXTREMELY FLAMMABLE LIQUEFIED GAS

Avoid breathing gas.

Toxic effects	The gas is a simple asphyxiant which is anaesthetic if inhaled in high concentrations; it has no significant toxic properties.
Hazardous reaction	Mixture with $LiNO_3$ and SO_2 in pressure bottle at 20 °C polymerised explosively (B361).
First aid	*Vapour inhaled in high concentrations:* standard treatment (p. 136).
Fire hazard	Flash point −108 °C; explosive limits 2–11.1%; ignition temp. 460 °C. Since the gas is supplied in a cylinder, turning off the valve will reduce any fire involving it; if possible, cylinders should be removed quickly from an area in which a fire has developed.
Spillage disposal	Surplus gas or leaking cylinder can be vented slowly to air in a safe, open area or gas burnt off through a suitable burner.

Propene ozonide [38787–96–1]
(3-methyl-1,2,4-trioxolane)

Hazardous reaction	Liable to explode at ambient temperature (B370).

Prop-2-en-1-ol —*see* Allyl alcohol

β-Propiolactone (oxetan-2-one) [57–57–8]

Colourless liquid; bp 155 °C with decomposition; slowly hydrolysed in water to hydracrylic acid.

(TOXIC IN CONTACT WITH SKIN
IRRITATING TO EYES, RESPIRATORY SYSTEM AND SKIN)

Prevent contact with skin, eyes and clothing. Prevent inhalation of vapour. Suspected carcinogen. TLV 0.5 ppm (1.5 mg m^{-3}).

Toxic effects	It is highly irritant to the skin and has produced skin cancer in experimental animals. It is assumed to be similarly irritant and dangerous if taken by mouth. It must be regarded as a potential human carcinogen.
First aid	*Affected eyes:* standard treatment (p. 136). *Skin contact:* standard treatment (p. 136). *If swallowed:* standard treatment (p. 137).
Spillage disposal	Wear breathing apparatus, and gloves. Apply dispersing agent if available and work to an emulsion with brush and water — run this to waste, diluting greatly with running water. If dispersant not available, absorb on sand, shovel into bucket(s) and transport to safe, open area for burial. Site of spillage should be washed thoroughly with water and soap or detergent.

Propiolaldehyde [624–67–9]

Hazardous reaction	Undergoes almost explosive polymerisation in the presence of alkalies or pyridine (B335).

Propionaldehyde (propanal) [123–38–6]

Colourless liquid with suffocating odour; bp 49 °C; 1 part dissolves in about 5 parts water at 20 °C. It may form explosive peroxides.

HIGHLY FLAMMABLE
IRRITATING TO EYES, RESPIRATORY SYSTEM AND SKIN

Avoid breathing vapour.

Toxic effects	The vapour irritates the respiratory system. The liquid irritates the eyes and is assumed to be irritant and damaging if taken internally.
First aid	*Vapour inhaled:* standard treatment (p. 136). *Affected eyes:* standard treatment (p. 136). *If swallowed:* standard treatment (p. 137).
Fire hazard	Flash point −9 °C; explosive limits 2.9–17%; ignition temp. 207 °C. *Extinguish fire with* water spray, dry powder, carbon dioxide or vaporising liquid.
Spillage disposal	Shut off all possible sources of ignition. Instruct others to keep at a safe distance. Wear breathing apparatus and gloves. Apply non-flammable dispersing agent and work to an emulsion with brush and water — run this to waste, diluting greatly with running water. If dispersant not available, absorb on sand, shovel into bucket(s) and transport to safe, open area for atmospheric evaporation. Site of spillage should be washed thoroughly with water and soap or detergent.

Propionic acid [79-04-4]

Colourless, oily liquid with rancid odour; bp 141 °C; miscible with water.

CAUSES BURNS

Avoid contact with skin and eyes. RL 10 ppm (30 mg m^{-3}).

Toxic effects	The vapour irritates the eyes and respiratory system. The liquid burns the skin and eyes. The acid is corrosive if taken by mouth.
First aid	*Affected eyes:* standard treatment (p. 136). *Skin contact:* standard treatment (p. 136). *If swallowed:* standard treatment (p. 137).
Fire hazard	Flash point 54 °C; ignition temp. 513 °C. *Extinguish fire with* water spray, dry powder, carbon dioxide or vaporising liquid.
Spillage disposal	Shut off all possible sources of ignition; wear face-shield or goggles, and gloves. Spread soda ash liberally over the spillage and mop up cautiously with plenty of water — run to waste, diluting greatly with running water.

Propiononitrile (ethyl cyanide) [107-12-0]

Colourless liquid, ethereal odour; bp 97 °C: miscible with water.

(HIGHLY FLAMMABLE
VERY TOXIC BY INHALATION, IN CONTACT WITH SKIN
AND IF SWALLOWED)

Prevent inhalation of vapour. Prevent contact with skin and eyes.

Toxic effects	Inhalation of vapour may cause headache, dizziness, rapid breathing, nausea, and in severe cases unconsciousness, convulsions and death. Evidence is lacking on the effects of skin absorption and ingestion, but these may be similar to those from inhalation.
First aid	Special treatment as for hydrogen cyanide (p. 132).
Fire hazard	Flash point 16 °C. *Extinguish fire with* water spray, foam, dry powder, carbon dioxide or vaporising liquids.
Spillage disposal	Shut off all possible sources of ignition. Instruct others to keep at safe distance. Wear breathing apparatus and gloves. Mop up with plenty of water — run to waste, diluting greatly with running water. Ventilate area well to evaporate remaining liquid and dispel vapour.

Propionyl chloride [79-03-8]

Colourless liquid with acrid odour; bp 80 °C; reacts with water forming propionic and hydrochloric acids.

HIGHLY FLAMMABLE
REACTS VIOLENTLY WITH WATER
CAUSES BURNS

Avoid breathing vapour. Prevent contact with skin and eyes.

Toxic effects	The vapour irritates the eyes and respiratory system. The liquid burns the skin and eyes. Assumed to cause severe internal irritation and damage if taken by mouth.
Hazardous reaction	A mixture with diisopropyl ether evolved gas and burst the container (B355).
First aid	*Vapour inhaled:* standard treatment (p. 136). *Affected eyes:* standard treatment (p. 136). *Skin contact:* standard treatment (p. 136). *If swallowed:* standard treatment (p. 137).
Fire hazard	Flash point 12 °C. *Extinguish fire with* dry powder, carbon dioxide or vaporising liquid.
Spillage disposal	Shut off all possible sources of ignition. Instruct others to keep at a safe distance. Wear breathing apparatus and gloves. Absorb on sand, shovel into bucket(s), transport to safe, open area and tip into large volume of water; leave to decompose before decanting the water to waste, diluting greatly with running water. Site of spillage should be ventilated after washing thoroughly with water and soap or detergent.

Propyl acetate   [109-60-4]

HIGHLY FLAMMABLE

RL 200 ppm (840 mg m^{-3}).

isoPropyl acetate *—see* Isopropyl acetate

Propyl alcohol *—see* Propan-1-ol

isoPropyl alcohol *—see* Propan-2-ol

Propylamines n- [107–10–8] iso- [75–31–0]

The primary propylamines (n- and iso-) are colourless, volatile liquids with an ammoniacal odour; bp 49 °C and 32 °C respectively; miscible with water.

(iso-) EXTREMELY FLAMMABLE
IRRITATING TO EYES, RESPIRATORY SYSTEM AND SKIN

Avoid breathing vapour. Avoid contact with skin and eyes. TLV (iso-) 5 ppm (12 mg m^{-3}).

Toxic effects The vapour irritates the respiratory system. The vapour and liquid irritate and may burn the eyes. The liquid may cause skin burns. Assumed to be very irritant and poisonous if taken by mouth.

First aid *Vapour inhaled:* standard treatment (p. 136).
Affected eyes: standard treatment (p. 136).
Skin contact: standard treatment (p. 136).
If swallowed: standard treatment (p. 137).

Fire hazard Flash point (n- and iso-) −37 °C; explosive limits (n- and iso-) about 2–10%; ignition temp. 318 °C (n-), 420 °C (iso-). *Extinguish fire with* water spray, dry powder, carbon dioxide or vaporising liquid.

Spillage disposal Shut off all possible sources of ignition. Instruct others to keep at a safe distance. Wear breathing apparatus and gloves. Mop up with plenty of water and run to waste, diluting greatly with running water. Ventilate area well to evaporate remaining liquid and dispel vapour.

Propyl chlorides —*see* Chloropropanes

Propyl cyanide —*see* Butyronitrile

3-Propyldiazirine [70348–66–2]

Hazardous reaction Exploded on attempted distillation at about 75 °C (B445).

Propylene dichloride —*see* 1,2-Dichloropropane

Propylene glycol dinitrate [6423–43–4]

RL (skin) 0.2 ppm (1.4 mg m^{-3}).

Propylene glycol monomethyl ether   [107–98–2]

RL 100 ppm (360 mg m^{-3}).

Propylene imine (2-methylaziridine)   [75–55–8]

Suspected carcinogen TLV (skin) 2 ppm (5 mg m^{-3}).

Propylene oxide   [75–56–9]
(1,2-epoxypropane; 2-methyloxiran)

Colourless, volatile liquid with ethereal odour; bp 35 °C; somewhat soluble in water.

EXTREMELY FLAMMABLE
HARMFUL BY INHALATION, IN CONTACT WITH SKIN
AND IF SWALLOWED

Avoid breathing vapour. Avoid contact with skin and eyes. RL 100 (20) ppm (240 mg m^{-3}).

Toxic effects   The vapour irritates the eyes and respiratory system. The liquid irritates the eyes and has a delayed blistering action upon the skin in which it is rapidly absorbed. Assumed to be irritant and poisonous when taken by mouth.

Hazardous reactions   Use as a biological sterilant is hazardous because of ready formation of explosive mixtures with air; drum of crude oxide and sodium hydroxide catalyst exploded and ignited. Handling procedures detailed (B367–368).

First aid   *Vapour inhaled:* standard treatment (p. 136).
Affected eyes: standard treatment (p. 136).
Skin contact: standard treatment (p. 136).
If swallowed: standard treatment (p. 137).

Fire hazard   Flash point −37 °C; explosive limits 2.1–28.5%. *Extinguish fire with* dry powder, carbon dioxide or vaporising liquid.

Spillage disposal   Shut off all possible sources of ignition. Instruct others to keep at a safe distance. Wear breathing apparatus and gloves. Organise ventilation of the area to dispel vapour completely.

Propyl nitrate   [627–13–4]

TLV 25 ppm (105 mg m^{-3}).

Hazardous reaction   Shock-sensitive liquid (B377).

Propyne (allylene; methylacetylene) [74–99–7]

Colourless gas; bp −23 °C; slightly soluble in water.

EXTREMELY FLAMMABLE

Avoid breathing gas. TLV 1000 ppm (1650 mg m^{-3}).

Toxic effects   The gas is anaesthetic in high concentrations but is not toxic.

First aid   *Gas inhaled in high concentrations:* standard treatment (p. 136).
Fire hazard   Explosive limits about 2.4–11.7%. Since the gas is supplied in a cylinder, turning off the valve will reduce any fire involving it; if possible, cylinders should be removed quickly from an area in which a fire has developed.
Disposal   Surplus gas or leaking cylinder can be vented slowly to air in a safe, open area or burnt off through a suitable burner.

Prop-2-yn-1-ol (propargyl alcohol, 3-hydroxypropyne) [107–19–7]

RL (skin) 1 ppm (2 mg m^{-3}).

Hazardous reactions   If dried with alkali or P_2O_5 before distillation, the residue is liable to explode; violent exothermic eruption may occur in preparation of hydroxyacetone from mercury(II) sulphate, sulphuric acid and the alcohol (B349–350).

Prop-2-yn-1-thiol [27846–30–6]

Hazardous reaction   Polymerised explosively when distilled at atmospheric pressure (B351–352).

Prop-2-ynyl vinyl sulphide [21916–66–5]

Hazardous reaction   Decomposed explosively above 85 °C (B508).

Pyridine [110–86–1]

Colourless liquid with a sharp penetrating odour; bp 115 °C; miscible with water.

HIGHLY FLAMMABLE
HARMFUL BY INHALATION, IN CONTACT WITH SKIN
AND IF SWALLOWED

Avoid breathing vapour. Avoid contact with skin and eyes. RL 5 ppm (15 mg m^{-3}).

Toxic effects The vapour irritates the respiratory system and may cause headache, nausea, giddiness and vomiting. The vapour and liquid irritate the eyes and may cause conjunctivitis. The liquid may irritate the skin causing dermatitis. Affects the central nervous system if taken by mouth and large doses act as a heart poison.

Hazardous reactions Maleic anhydride decomposes exothermally in the presence of pyridine; the solid obtained by reaction with BrF_3 ignites when dry; the pyridine complex with CrO_3 is unstable; it incandesces on contact with fluorine; reacts violently with N_2O_4 (B500).

First aid *Vapour inhaled:* standard treatment (p. 136).
Affected eyes: standard treatment (p. 136).
Skin contact: standard treatment (p. 136).
If swallowed: standard treatment (p. 137).

Fire hazard Flash point 20 °C; explosive limits 1.8–12.4%; ignition temp. 482 °C. *Extinguish fire with* water spray, foam, dry powder, carbon dioxide or vaporising liquid.

Spillage disposal Shut off all possible sources of ignition. Instruct others to keep at a safe distance. Wear breathing apparatus and gloves. Mop up with plenty of water and run to waste, diluting greatly with running water. Ventilate area well to evaporate remaining liquid and dispel vapour.

RSC *Lab. Haz. Data Sheet No. 8,* 1983 gives extended coverage.

Pyridinium dichromate [20039–37–6]

Hazardous preparation Preparation from pyridine and chromium trioxide in water is explosion-prone (B773).

Pyridinium nitrate [543–53–3]

Hazardous reaction Explodes on heating (B506).

Pyridinium perchlorate [15598–34–2]

Hazardous reactions Can be detonated by impact and has occasionally exploded when disturbed (B504).

Pyrocatechol —*see* Catechol

Pyrophoric metals and alloys, catalysts and compounds Reviews (B1670–1673).

Pyrrolidine [123-75-1]

Colourless mobile liquid; bp 89 °C; miscible in water.

> (HIGHLY FLAMMABLE
> IRRITATING TO EYES AND SKIN)

Avoid breathing vapour. Avoid contact with skin, eyes and clothing.

Toxic effects The vapour irritates the respiratory system; the liquid irritates the skin and mucous surfaces.

First aid *Vapour inhaled:* standard treatment (p. 136).
Affected eyes: standard treatment (p. 136).
Skin contact: standard treatment (p. 136).
If swallowed: standard treatment (p. 137).

Fire hazard Flash point 3 °C; *Extinguish fire with* water spray, foam, dry powder or vaporising liquid.

Spillage disposal Shut off all possible sources of ignition. Instruct others to keep at a safe distance. Wear breathing apparatus and gloves. Mop up with plenty of water and run to waste diluting greatly with running water. Ventilate area well to evaporate remaining liquid and dispel vapour.

Quinoline [91-22-5]

Colourless to yellow, hygroscopic liquid; bp 238 °C; sparingly soluble in cold water.

> (HARMFUL IF SWALLOWED
> IRRITATING TO EYES AND SKIN)

Avoid breathing vapour. Avoid contact with skin, eyes and clothing.

Toxic effects Irritates skin and mucous surfaces.

First aid *Vapour inhaled:* standard treatment (p. 136).
Affected eyes: standard treatment (p. 136).
Skin contact: standard treatment (p. 136).

Spillage disposal Wear face-shield or goggles, and gloves. Mop up with water and detergent and run to waste, diluting greatly with running water. Ventilate area well to evaporate remaining liquid.

Quinone —see *p*-Benzoquinone

Redox compounds and reactions Review of group (B1676–1677).

Resorcinol (1,3-dihydroxybenzene)   [108–46–3]

Colourless to pink crystals; mp 110 °C; soluble in water.

> HARMFUL IF SWALLOWED
> IRRITATING TO EYES AND SKIN

Avoid contact with skin and eyes. RL 10 ppm (45 mg m^{-3}).

Toxic effects	Absorption by the skin may result in itching and dermatitis; in severe cases, restlessness, cyanosis and convulsions may follow. The solid irritates the eyes. It may cause dizziness, drowsiness and tremors if taken by mouth.
First aid	*Affected eyes:* standard treatment (p. 136). *Skin contact:* standard treatment (p. 136). *If swallowed:* standard treatment (p. 137).
Spillage disposal	Wear face-shield or goggles, and gloves. Mix with sand and transport to a safe, open area for burial. Site of spillage should be washed thoroughly with water and soap or detergent.

Rhodium   [7440–16–6]

TLV (metal fume and dusts) 0.1 mg m^{-3} (soluble salts, as Rh) 0.001 mg m^{-3}.

Rubidium (metal)   [7440–17–7]

Hazardous properties	Ignites on exposure to air or dry oxygen; ignites on contact with fluorine or chlorine or the vapours of bromine or iodine; reaction with mercury may be violent; hydrogen is evolved vigorously on contact with cold water and ignites (B1387–1388).

Rubidium hydride   [13446–75–8]

Hazardous reactions	Reacts violently with water; its reaction with acetylene is vigorous at -60 °C (B1137–1138).

Sebacoyl dichloride (decanedioyl dichloride) [111–19–3]

Hazardous reaction At the end of vacuum distillation, the residue frequently decomposes spontaneously, producing voluminous black foam (B778).

Seleninyl bromide [7789–51–7]

Hazardous reactions The liquid bromide reacts explosively with sodium and potassium, and ignites zinc dust; red phosphorus ignites and the white allotrope explodes in contact with the liquid bromide (B105–106).

Seleninyl chloride [7791–23–3]

Hazardous reactions Potassium and phosphorus (white) explode on contact with the liquid, powdered antimony ignites (B979).

Selenium and compounds Se [7782–49–2]

Selenium is a steel grey or purplish powder, also fabricated into pellets, sticks or plates; insoluble in water. Selenium dioxide, selenous acid and the alkali-metal selenites and selenates are colourless powders or crystals, soluble in water; selenium chloride (reddish yellow), selenyl chloride (colourless or yellow) (bp 176 °C); and selenic acid (colourless) are liquids, whereas selenium tetrachloride is a cream-coloured crystalline solid. Highly poisonous hydrogen selenide (offensive smell) is generated when an acid solution of a selenium compound is reduced by metals such as tin and zinc. Selenic acid reacts vigorously with water.

TOXIC BY INHALATION AND IF SWALLOWED
DANGER OF CUMULATIVE EFFECTS

Avoid breathing dust. Avoid contact with skin and eyes. RL (selenium compounds as Se) 0.2 mg m^{-3}.

Toxic effects Hydrogen selenide irritates the nose and eyes and causes inflammation of the lungs and disturbance of the digestive and nervous systems; it imparts a garlic-like odour to the breath. The chloride and solutions of the acids and salts may burn the skin — severe pain may be experienced under the finger-nails by skin absorption at the finger-tips. Selenium dioxide dust is particularly penetrating and irritates the respiratory system, eyes and skin. Assumed to be irritant and poisonous if taken by mouth. **Chronic effects** Inhalation of selenium dust over a prolonged period may cause fatigue, loss of appetite, digestive disturbance and bronchitis; dermatitis may result from prolonged exposure of skin to small amounts of selenium and its compounds. (*See* note in Ch 6 p. 104).

Hazardous reactions   Nickel, sodium and potassium interact with selenium with incandescence; the particle size of cadmium and selenium must be below a critical size to prevent explosions when making cadmium selenide — this also applies to zinc and selenium; the oxidation of recovered selenium with nitric acid is made vigorous by the presence of organic impurities; selenium may react explosively with BrF_5, ClF_3 or Na_2O_2; it ignites on contact with fluorine; explosive products formed with metal amides; exothermic reaction with hot Sn; violent reactions with Ag_2O, BaO_2, $KBrO_3$ or CrO_3 (B1396–1398).

First aid   *Dust or vapour inhaled:* standard treatment (p. 136).
Affected eyes: standard treatment (p. 136).
Skin contact: standard treatment (p. 136).
If swallowed: standard treatment (p. 137).

Spillage disposal   Selenium powder may be mixed with sand and treated as normal refuse, as may the disulphide. Soluble selenites and selenates can be dissolved in water and run to waste, diluting greatly with running water. Soda ash should be applied liberally to spillages of selenium dioxide, selenic and selenous acids and selenyl and selenium chlorides which may then be mopped up cautiously with plenty of water, this being run to waste, diluting greatly with running water

Selenium hexafluoride [7783–79–1]

RL (as Se) 0.05 ppm (0.02 mg m^{-3}).

Selenium tetrafluoride [13465–66–2]

Hazardous reaction   Reacts violently with water (B1073).

Silane (silicon tetrahydride) [7803–62–5]

RL 0.5 (5) ppm (0.7 mg m^{-3}).

Hazardous reactions   The very pure material ignites in air; it burns in contact with bromine, chlorine or some covalent chlorides (B1203–1204).

Silanes Review (B1684).

Silicon monohydride [13774–94–2]

Hazardous reaction   The polymeric hydride reacts violently with aqueous alkali, evolving hydrogen (B1138).

Silicon tetraazide [27890–58–0]

Hazardous reaction Spontaneously explosive at times (B1314).

Silicon tetrachloride (silicon chloride) [10026–04–7]

Colourless fuming liquid; bp 59 °C; reacts violently with water, forming hydrochloric acid and silica.

REACTS VIOLENTLY WITH WATER
IRRITATING TO EYES, RESPIRATORY SYSTEM AND SKIN

Avoid breathing vapour. Prevent contact with eyes and skin.

Toxic effects The vapour severely irritates all parts of the respiratory system. The vapour severely irritates the eyes and the liquid burns the eyes and skin. If taken by mouth there would be severe internal irritation and damage.

First aid *Vapour inhaled:* standard treatment (p. 136).
Affected eyes: standard treatment (p. 136).
Skin contact: standard treatment (p. 136).
If swallowed: standard treatment (p. 137).

Spillage disposal Instruct others to keep at a safe distance. Wear breathing apparatus and gloves. Spread soda ash liberally over the spillage and mop up cautiously with water — run to waste, diluting greatly with running water.

Silicon tetrahydride—*see* Silane

Silver (metal) [7740–22–4]

RL 0.1 mg m^{-3}.

Hazardous reactions Silver powder reacts vigorously with several halides and halogen compounds, catalytically decomposes several oxidants, and forms unstable N–Ag compounds (B3–4).

Silver acetylide [7659–31–6]

Hazardous reaction Touch-sensitive explosive (B196–197).

Silver amide   [65235–79–2]

Hazardous reaction   Very explosive when dry (B10).

Silver chlorate   [7783–92–8]

Hazardous reactions   Explosively unstable and a powerful oxidant (B7).

Silver chlorite   [7783–91–7]

Hazardous reactions   Reacted explosively with iodomethane or iodoethane with or without solvent dilution; the salt itself is impact-sensitive, cannot be ground and explodes at 105 °C; it explodes in contact with hydrochloric acid or on rubbing with sulphur (B6–7).

Silver compounds   Review of group (B1686–1688).

RL (soluble compounds, as Ag) 0.01 mg m^{-3}.

Silver fluoride   [7775–41–9]

Hazardous reactions   Reactions with calcium hydride (on grinding mixture) and titanium (at 320 °C) are incandescent; silicon reacts violently, boron explosively when ground with AgF (B9).

Silver fulminate   [5610–59–3]

Hazardous reactions   A powerful detonator, it explodes violently in contact with hydrogen sulphide (B114).

Silvering solutions   Review (B1688).

Silver nitrate   (lunar caustic)   [7761–88–8]

White crystals; soluble in water.

CAUSES BURNS

Avoid contact with eyes and skin. RL (soluble silver compounds as Ag) 0.01 mg m^{-3}.

Toxic effects   The solid and its solutions severely irritate the eyes and can cause skin burns. If taken by mouth silver nitrate can cause internal damage due to absorption in the blood followed by deposition of silver in various tissues of the body.

Hazardous reactions   Reacts with acetylene in the presence of ammonia to form silver acetylide, a powerful detonator; finely divided mixture of arsenic with excess nitrate ignited when shaken on to paper; reacts violently with chlorosulphuric acid forming nitrosulphuric acid; crystals damp with ethanol exploded when touched with spatula; intimate mixture with dry magnesium powder may ignite explosively on contact with drop of water; mixtures with phosphorus and sulphur explode under hammer blow; mixture with charcoal ignites on percussion; explosive precipitates readily formed with ammonia and alkali; such solutions must not be stored (B11–14).

First aid   *Affected eyes:* standard treatment (p. 136).
Skin contact: standard treatment (p. 136).
If swallowed: standard treatment (p. 137).

Spillage disposal   Wear face-shield or goggles, and gloves. Mop up with plenty of water and run to waste, diluting greatly with running water.

Silver oxalate [533–51–7]

Hazardous reaction   Liable to explode when heated above 140 °C (B198).

Silver(I) oxide [20667–12–3]

Hazardous reactions   Slowly forms explosive silver nitride with ammonia or hydrazine; oxidises carbon monoxide exothermically and may ignite hydrogen sulphide; Se, S, P or metal sulphides are ignited on grinding with the oxide (B17–18).

Silver perchlorate [7783–93–9]

White deliquescent crystals; decomposes at 486 °C; freely soluble in water.

(EXPLOSIVE WHEN MIXED WITH COMBUSTIBLE MATERIAL
IRRITATING TO SKIN AND EYES)

Avoid contact with skin, eyes and clothing.

Toxic effects   It irritates the skin and mucous surfaces and would be harmful if taken by mouth.

Hazardous reaction	Forms explosive solvates with several organic solvents; that with ether exploded violently on crushing in a mortar; explodes with ethylene diamine (B7–9).
First aid	*Affected eyes:* standard treatment (p. 136). *Skin contact:* standard treatment (p. 136). *If swallowed:* standard treatment (p. 137).
Spillage disposal	Wear face-shield or goggles, and gloves. Mop up with plenty of water and run to waste, diluting greatly with running water. Large amounts may be worth recovering.

Soda asbestos

Grey or brown granules; largely soluble in water. The fibre content is not asbestos.

(CAUSES BURNS
IRRITATING TO EYES, RESPIRATORY SYSTEM AND SKIN)

Avoid breathing dust. Prevent contact with eyes and skin.

Toxic effects	The dust severely irritates the nose and mouth. The solid and dust burn the eyes and skin. If swallowed there would be severe internal irritation and damage.
First aid	*Dust inhaled:* standard treatment (p. 136). *Affected eyes:* standard treatment (p. 136). *Skin contact:* standard treatment (p. 136). *If swallowed:* standard treatment (p. 137).
Spillage disposal	Wear face-shield or goggles, and gloves. Shovel into enamel or polythene bucket, add, a little at a time with stirring, to a large volume of water. After reaction is complete, run solution to waste, diluting greatly with running water. Wash down site of spillage thoroughly with water and dispose of solid residue.

Soda-lime (mixed calcium/sodium hydroxide granules)   [8006–28–8]

Hazardous reactions	Fires have been caused in waste bins into which soda-lime that has absorbed hydrogen sulphide has been thrown — considerable heat develops when this spent material is exposed to moisture and air (B179).

Sodamide —*see* Sodium amide

Sodium and sodium amalgam   metal [7740-23-5]

Sodium — soft, silvery white sticks, pellets, wire or granules, normally coated with a grey oxide or hydroxide skin; reacts vigorously with water with formation of sodium hydroxide and hydrogen gas which may ignite. Sodium amalgam — silvery or grey spongy masses; reacts with water forming sodium hydroxide and hydrogen gas.

> REACTS VIOLENTLY WITH WATER
> CONTACT WITH WATER LIBERATES HIGHLY FLAMMABLE GASES
> CAUSES BURNS

Prevent contact with skin and eyes.

Toxic effects   The metal or amalgam in contact with moisture on the skin can cause thermal and caustic burns. In the same way the sodium hydroxide resulting from the reaction of the metal with water can cause burning of the skin and eyes.

Hazardous reactions   References given to special papers, booklets, *etc.* on safe handling of sodium on laboratory and large scale — also the disposal of unwanted residues; anhydrous HCl, HF or sulphuric acids react slowly with sodium while the aqueous solution reacts explosively; dispersions of Na in volatile solvents become pyrophoric when solvent evaporates — serum cap closures on bottles are safest; the metal reacts with different degrees of violence with chloroform/methanol, diazomethane, *N,N*-dimethylformamide, fluorinated compounds, halocarbons, halogens, interhalogens, mercury, metal halides and oxides, non-metal halides and oxides, non-metals, oxidants, water; violent or explosive reactions with bromobenzene, SnO_2, $POCl_3$, sulphides; mixtures with SO_2F_2, $SiCl_4$, CS_2, alkyl oxalates or aromatic nitro compounds are shock-sensitive (B1315–1323).

First aid   *Affected eyes:* standard treatment (p. 136).
Skin contact: standard treatment (p. 136).
Mouth contact: standard treatment (p. 137).

Spillage disposal   (a) Sodium metal; instruct others to keep at a safe distance. Wear face-shield or goggles, and gloves. Cover with dry soda ash, shovel into dry bucket, transport to safe, open area and add, a little at a time, to a large excess of dry propan-2-ol. Leave to stand for 24 hours and run to waste, diluting greatly with running water.
(b) Sodium amalgam: cover with large volume of water in suitable vessel and allow to stand until there is no further reaction; the mercury may then be separated and recovered, and the sodium hydroxide solution run to waste, diluting greatly with running water.

Sodium acetylide   [2281-62-1]

Hazardous reactions   Decomposes at 150 °C evolving gas which ignites in air; vigorous or violent reactions with oxidants (B319–320).

Sodium amide (sodamide)   [7782–92–5]

Colourless or greyish white lumps or powder smelling of ammonia. Reacts violently with water with the formation of sodium hydroxide and ammonia.

(REACTS VIOLENTLY WITH WATER
CAUSES BURNS
IRRITATING TO EYES, RESPIRATORY SYSTEM AND SKIN)

Avoid breathing dust. Prevent contact with eyes and skin.

Toxic effects   The dust severely irritates the mouth and nose. The solid and dust severely irritate or burn the eyes. The solid in contact with moisture on the skin can cause thermal and caustic burns. If taken by mouth there would be severe internal irritation and damage.

Hazardous reactions   May ignite or explode on heating or grinding in air; explodes with potassium chlorate or sodium nitrite; fresh material behaves like sodium on contact with water but old material may give delayed explosion; reference to disposal given (B1145–1146).

First aid   *Dust inhaled:* standard treatment (p. 136).
Affected eyes: standard treatment (p. 136).
Skin contact: standard treatment (p. 136).
If swallowed: standard treatment (p. 137).

Spillage disposal   Wear face-shield or goggles, and gloves. Mix with dry sand, shovel into an enamel or polythene bucket, transport to safe, open area, and add, a little at a time with stirring, to a large volume of water; after reaction is complete, run solution to waste, diluting greatly with running water. Dispose of sand as normal refuse. Wash down site of spillage thoroughly with water.

Sodium arsenate and arsenite
—*see* Arsenic and compounds

Sodium azide   [26628–22–8]

Colourless crystalline powder; soluble in water.

VERY TOXIC IF SWALLOWED
CONTACT WITH ACIDS LIBERATES VERY TOXIC GAS

Prevent contact with skin and eyes. Avoid breathing dust. RL 0.1 ppm (0.3 mg m^{-3}).

Toxic effects   The dust and solution irritate the eyes. The solid and solution irritate the skin and can cause blistering. Assumed to be very irritant and poisonous if taken by mouth.

491

Hazardous reactions	Decomposes somewhat explosively above its mp, particularly if heated rapidly; liable to explode with bromine, carbon disulphide or chromyl chloride; when water is added to the strongly heated azide there is a violent reaction; heavy metals form explosive deposits (B1304–1306).
First aid	*Affected eyes:* standard treatment (p. 136). *Skin contact:* standard treatment (p. 136). *If swallowed:* standard treatment (p. 137).
Spillage disposal	Wear face-shield or goggles, and gloves. Mop up with plenty of water and run to waste, diluting greatly with running water.

Sodium azidosulphate   [67880–15–3]

Hazardous reaction	Weak explosive of variable sensitivity (B1306).

Sodium benzenehexoxide ('carbonylsodium')   [101672–22–4]

Hazardous reactions	Explodes on heating in air or on contact with water (B650).

Sodium bisulphate
—*see* Sodium hydrogen sulphate

Sodium borohydride (sodium tetrahydroborate)   [16940–66–2]

White to pale grey microcrystalline powder or lumps; decomposed by water with evolution of hydrogen.

(HARMFUL IF SWALLOWED
IRRITATING TO SKIN, EYES AND RESPIRATORY SYSTEM)

Avoid breathing dust. Avoid contact with skin, eyes and clothing.

Toxic effects	Dust irritates respiratory system. Irritating to skin and mucous surfaces.
Hazardous reaction	A large volume of the alkaline solution spontaneously heated and decomposed, liberating hydrogen; anhydrous acids form diborane violently; ruthenium salts give explosive precipitates; explosion with $AlCl_3$ in bis-2-methoxyethyl ether; hot solutions in DMF decompose violently (B63–65).

First aid   *Affected eyes:* standard treatment (p. 136).
           Skin contact: standard treatment (p. 136).
           If swallowed: standard treatment (p. 137).

Spillage   Wear face-shield or goggles, and gloves. Mop up with plenty of water and
disposal   run to waste, diluting greatly with running water.

RSC *Lab. Haz. Data Sheet No. 39,* 1985, gives extended coverage.

Sodium chlorate   [775–09–9]

Colourless crystals, soluble in water.

            EXPLOSIVE WHEN MIXED WITH COMBUSTIBLE MATERIAL
            HARMFUL BY INHALATION AND IF SWALLOWED

Avoid contact with combustible materials and acids. Avoid contact with skin, eyes and clothing.

Toxic effects   The dust or strong solutions may irritate the eyes and skin. The solid or
            solutions are damaging if taken internally, symptoms of poisoning being
            nausea, vomiting and abdominal pain; kidney damage may follow.

Hazardous   Mixtures with ammonium salts, powdered metals, phosphorus, silicon,
reactions   sulphur or sulphides are readily ignited and potentially explosive; mixture
            with nitrobenzene is powerful explosive; mixtures with fibrous or absorbent
            organic materials (charcoal, flour, shellac, sawdust, sugar) are hazardous
            and can be caused to explode by static, friction or shock; addition of
            concentrated sulphuric acid to solid chlorate causes explosion of ClO_2
            generated; violent or explosive reactions with Al/rubber, $(NH_4)S_2O_3$, 1,3-
            bis(trichloromethyl)benzene, As_2O_3, grease, leather shoes, Os. The salt
            alone may also explode under fire conditions (B942–945).

First aid   *Affected eyes:* standard treatment (p. 136).
           Skin contact: standard treatment (p. 136).
           If swallowed: standard treatment (p. 137).

Fire hazard   Mixtures of sodium chlorate and combustible materials are readily ignited;
            mixtures with finely divided combustible materials can react explosively.
            Extinguish fire with water spray.

Spillage   Shovel into bucket of water and run solution or suspension to waste,
disposal   diluting greatly with running water. Site of spillage should be washed
            thoroughly to remove all oxidant, which is liable to render any organic
            matter (particularly wood, paper and textiles) with which it comes into
            contact, dangerously combustible when dry. Clothing wetted with the
            solution should be washed thoroughly.

Sodium chlorite   [7758–19–2]

Hazardous   Explodes on impact; intimate mixtures with finely divided or fibrous organic
reactions   matter may be very sensitive to heat, impact or friction (B940–941).

Sodium *N*-chloro-*p*-toluenesulphonamide [127–65–1]
(Chloramine-T)

White or pale cream crystals; decomposes on heating or exposure to atmosphere; soluble in water.

IRRITATING TO EYES, RESPIRATORY SYSTEM AND SKIN

Prevent inhalation of dust and contact with skin or eyes.

Toxic effects Inhalation of dust may lead to asthma. Ingestion of the oxidant would be assumed to cause internal irritation and damage

First aid *Affected eyes:* standard treatment (p. 136).
If swallowed: standard treatment (p. 137).

Spillage disposal Wear face shield or goggles, and gloves. Shovel up into polythene or enamel bucket of water, stir thoroughly to dissolve, then run solution of the oxidant to waste, diluting greatly with running water. Site of spillage should be washed thoroughly to remove last traces of the oxidant.

Sodium chromate and dichromate
—*see* **Chromates and dichromates**

Sodium cyanide
—*see* **Cyanides** (water soluble)

Sodium dichromate
—*see* **Chromates and dichromates**

Sodium dihydrogenphosphide [24167–76–8]

Hazardous reaction Ignites in air (B1149).

Sodium disulphite (sodium metabisulphite) [7681–57–4]

RL 5 mg m^{-3}.

Sodium dithionite (sodium hydrosulphite)   [7775–14–6]

White or off-white crystalline powder; very soluble in water, but addition of small amount of water to the salt causes hazardous reaction.

> MAY CAUSE FIRE
> HARMFUL IF SWALLOWED
> CONTACT WITH ACIDS LIBERATES TOXIC GAS

Avoid contact with moist flammable materials including clothing.

Hazardous reaction	Addition of 10% of water caused heating and spontaneous ignition (B1329).
Fire hazard	Fires resulting from contact with moist combustible materials are best extinguished by means of water spray.
Spillage disposal	Shovel into bucket of water and run solution or suspension to waste, diluting greatly with running water. Site of spillage should be washed thoroughly to remove all traces.

Sodium ethoxide (sodium ethylate)   [141–52–6]

and sodium methoxide (sodium methylate)   [124–41–4]

Although these compounds are sometimes used industrially as solids, they are more frequently prepared in the laboratory by reacting sodium metal with ethanol and methanol. The hazards of these solutions are dealt with here.

> (HIGHLY FLAMMABLE
> CAUSE BURNS
> IRRITATING TO EYES AND SKIN)

Prevent contact with skin and eyes.

Toxic effects	The solutions are extremely irritant to the eyes, causing burns. The solutions irritate the skin and may produce burns. They will cause severe internal irritation and damage if taken by mouth.
Hazardous reaction	Mixture of solid NaOMe, MeOH and $CHCl_3$ boiled violently, then exploded (B165–166).
First aid	*Affected eyes:* standard treatment (p. 136). *Skin contact:* standard treatment (p. 136). *If swallowed:* standard treatment (p. 137).
Spillage disposal	Shut off all possible sources of ignition. Wear face-shield or goggles, and gloves. Mop up with plenty of water and run to waste, diluting greatly with running water. Ventilate area of spillage well to evaporate remaining liquid and dispel vapour.

Sodium ethoxyacetylide [73506–39–5]

Hazardous reaction	Extremely pyrophoric; it may explode after prolonged contact with air (B418).

Sodium fluoride
—*see* Fluorides (water-soluble)

Sodium fluoroacetate [62–74–8]

RL (skin) 0.05 mg m^{-3}.

Sodium fluoroborate
—*see* Tetrafluoroboric acid and salts

Sodium fluorosilicate
—*see* Hexafluorosilicic acid and salts

Sodium hydrazide [13598–47–5]

Hazardous reactions	Heating to 100 °C or contact with traces of air, alcohol or water cause violent explosion (B1183).

Sodium hydride [7646–69–7]

CONTACT WITH WATER LIBERATES HIGHLY FLAMMABLE GAS

Hazardous reactions	Addition to small amount of water causes explosion; reacts vigorously with acetylene in the presence of moisture even at −60 °C; the finely divided dry powder ignites in dry air; may react explosively with dimethyl sulphoxide; interaction with chlorine or fluorine is incandescent; reacts vigorously with sulphur, explosively with sulphur dioxide; oil-coated powder available and safer to use; exothermic reaction with DMF sets in at 40°C (B1132–1134).

Sodium hydrogen difluoride
—*see* **Fluorides** (water-soluble)

Sodium hydrogen sulphate (sodium bisulphate)   [7681–38–1]

Colourless crystals or fused masses; soluble in water.

> (CAUSES BURNS
> IRRITATING TO SKIN AND EYES)

Avoid contact with skin and eyes.

Toxic effects	The solid and its strong solutions in water cause severe burns of the eyes and skin. Causes severe internal irritation and damage if taken by mouth.
First aid	*Affected eyes:* standard treatment (p. 136). *Skin contact:* standard treatment (p. 136). *If swallowed:* standard treatment (p. 137).
Spillage disposal	Wear face-shield or goggles, and gloves. Spread soda ash liberally over the spillage and mop up cautiously with plenty of water — run to waste, diluting greatly with running water.

Sodium hydrosulphite
—*see* **Sodium dithionite**

Sodium hydroxide (caustic soda)   [1310–73–2]

Colourless sticks, flakes, powder or pellets; soluble in water.

> CAUSES SEVERE BURNS

Prevent contact with eyes and skin. RL 2 mg m^{-3}.

Toxic effects	The solid and its strong solutions cause severe burns of the eyes and skin. If taken by mouth there would be severe internal irritation and damage. Solutions as weak as 2.5M can damage eyes severely.
Hazardous reactions	Very exothermic reaction with limited amounts of water; reacts vigorously with chloroform/methanol; explosion results when it is heated with zirconium; accidental contamination of metal scoop with flake NaOH, prior to its use with Zn dust, caused the latter to ignite (B1135–1137).

497

First aid	*Affected eyes:* standard treatment (p. 136).
	Skin contact: standard treatment (p. 136).
	If swallowed: standard treatment (p. 137).
Spillage disposal	Wear face-shield or goggles, and gloves. Shovel into an enamel or polythene bucket, and add, a little at a time with stirring, to a large volume of water. After solution is complete run this to waste, diluting greatly with running water. Wash down site of spillage thoroughly with water.

RSC *Lab. Haz. Data Sheet No. 36,* 1985, gives extended coverage.

Sodium hypochlorite solution [7681–52–9]
(containing over 5% active chlorine)

Colourless solution smelling of chlorine.

> CONTACT WITH ACIDS LIBERATES TOXIC GAS
> IRRITATING TO EYES AND SKIN

Avoid contact with skin, eyes and clothing.

Toxic effects	Extremely irritating to the eyes and gives rise to burns. Bleaches and may burn the skin. Will cause internal irritation and damage if ingested.
Hazardous reactions	Reacts with nitrogen compounds to form unstable or explosive *N*-chloro compounds; explosions with methanol and with benzyl cyanide have been reported; violent reaction with hot formic acid (B938–940).
First aid	*Affected eyes:* standard treatment (p. 136).
	Skin contact: standard treatment (p. 136).
	If swallowed: standard treatment (p. 137).
Spillage disposal	Wear face-shield or goggles, and gloves. Mop up with plenty of water and run to waste, diluting greatly with running water.

Sodium hypophosphite
—*see* **Sodium phosphinate**

Sodium iodate [7681–55–2]

White crystals which decompose on heating; 1 part dissolves in about 11 parts of cold water.

> CONTACT WITH COMBUSTIBLE MATERIAL MAY CAUSE FIRE

Avoid contact with combustible materials.

Fire hazard   Mixture of sodium iodate and combustible materials are readily ignited; mixtures with finely divided combustible materials can react explosively. *Extinguish fire with* water spray.

Spillage disposal   Shovel into bucket of water and run solution to waste, diluting greatly with running water. Site of spillage should be washed thoroughly to remove all oxidant, which is liable to render any organic matter (particularly wood, paper and textiles) with which it comes into contact, dangerously combustible when dry. Clothing wetted with the solution should be washed thoroughly.

Sodium metabisulphite
—*see* **Sodium disulphite**

Sodium metaperiodate
—*see* **Sodium periodate**

Sodium methoxide —*see* **Sodium ethoxide**

Sodium nitrate   [7631-99-4]

Colourless, deliquescent crystals; soluble in water.

(EXPLOSIVE WHEN MIXED WITH COMBUSTIBLE MATERIALS)

Avoid contact with combustible materials.

Toxic effects   Ingestion may cause gastro-enteritis, abdominal pains, vomiting, muscular weakness, irregular pulse, convulsions and collapse; 15–30 g in one dose may be fatal.

Hazardous reactions   Reacts explosively with Al, Al_2O_3, Sb, barium thiocyanate, sodium phosphinate, sodium thiosulphate; reacts violently with acetic anhydride; may ignite fibrous organic material, charcoal, *etc* on heating; an explosive compound is formed by interaction with sodium; Al powder and moisture react at 70 °C; Mg powder ignites in the molten salt; amidosulphates or cyanides react violently on heating (B1282–1285).

First aid   *If swallowed:* standard treatment (p. 136).
Fire hazard   Mixtures of sodium nitrate and combustible materials are readily ignited; mixtures with finely divided, combustible materials can react explosively. *Extinguish fire with* water spray.

Spillage **disposal**	Shovel into bucket of water and run solution or suspension to waste, diluting greatly with running water. Site of spillage should be washed thoroughly to remove all oxidant, which is liable to render any organic matter (particularly wood, paper and textiles) with which it comes into contact, dangerously combustible when dry. Clothing wetted with the solution should be washed thoroughly.

Sodium nitrite [7632–00–0]

White or slightly yellow granules, rods or powder; hygroscopic and soluble in water. Poisonous nitrous fumes produced by the action of acids; used as food additive.

> CONTACT WITH COMBUSTIBLE MATERIAL MAY CAUSE FIRE
> TOXIC IF SWALLOWED

Avoid contact with combustible materials or acids.

Hazardous **reactions**	Explosions are likely to occur on heating mixtures of the nitrite with ammonium salts, metal cyanides, phthalic acid or anhydride, sodium amide, sodium thiocyanate or thiosulphate; wood impregnated with solutions of nitrite over a long period may be accidentally ignited and burn fiercely; violent reaction on heating with amidosulphates; explosion with phenol; explosive products formed with dienes (B1060–1062).
Fire hazard	Mixtures of sodium nitrite and combustible materials are readily ignited; mixtures with finely divided combustible materials can react explosively. *Extinguish fire with* water spray.
Spillage **disposal**	Shovel into bucket of water and run solution to waste, diluting greatly with running water. Site of spillage should be washed thoroughly to remove all oxidant, which is liable to render any organic matter (particularly wood, paper and textiles) with which it comes into contact, dangerously combustible when dry. Clothing wetted with the solution should be washed thoroughly.

Sodium oxalate —*see* Oxalates

Sodium pentachlorophenate
—*see* Pentachlorophenol

Sodium perchlorate —*see* Perchlorates

Sodium periodate (sodium metaperiodate) [7790–28–5]

White crystals, soluble in cold water.

(CONTACT WITH COMBUSTIBLE MATERIALS MAY CAUSE FIRE)

Avoid contact with combustible materials. Avoid contact with skin, eyes and clothing.

Toxic effects	No documentary evidence on the toxicity of this compound has been traced. However, as potassium metaperiodate is rated 'high' for its irritant characteristics and toxicity by ingestion, it must clearly be treated with respect.
First aid	*Affected eyes:* standard treatment (p. 136). *Skin contact:* standard treatment (p. 136). *If swallowed:* standard treatment (p. 137).
Fire hazard	Mixtures with combustible materials are readily ignited; mixtures with finely divided combustible materials can react explosively. *Extinguish fire with water spray.*
Spillage disposal	Shovel into bucket of water and run solution to waste, diluting greatly with running water. Site of spillage should be washed thoroughly to remove all oxidant, which is liable to render any organic matter (particularly wood, paper and textiles) with which it comes into contact, dangerously combustible when dry. Clothing wetted with the solution should be washed thoroughly.

Sodium peroxide [1313–60–6]

Pale yellow powder; reacts violently with water forming sodium hydroxide and oxygen.

CONTACT WITH COMBUSTIBLE MATERIALS MAY CAUSE FIRE
CAUSES SEVERE BURNS

Do not breathe dust. Prevent contact with skin and eyes.

Toxic effects	The dust irritates the respiratory system and eyes. Because of the vigour of its reaction with water it may cause both thermal and caustic burns on moist skin. Would cause severe internal irritation and damage if taken by mouth.
Hazardous reactions	Admixture with acetic acid causes explosion; mixture with ammonium peroxodisulphate explodes on heating above 75 °C, grinding in mortar or exposure to drops of water or carbon dioxide; mixtures with calcium acetylide explode; contact with fibrous materials often causes ignition; mixture with hydroxy compounds usually results in fires or explosions; reacts violently or explosively with some metals, non-metals and non-metal halides; reacts vigorously or explosively with water depending on relative quantities; presence of moisture assists ignition of many finely-divided or fibrous solids in contact with Na_2O_2. Use of too little peroxide and too much dextrose sample ruptured a bomb calorimeter (B1324–1328).

First aid	*Dust inhaled:* standard treatment (p. 136).
	Affected eyes: standard treatment (p. 136).
	If swallowed: standard treatment (p. 137).
Fire hazard	If in contact, or mixed with organic or other oxidisable substances, ignition or explosion may take place.
Spillage disposal	Wear face-shield or goggles, and gloves. Mix with dry sand, shovel into an enamel bucket, transport to a safe, open area and add, a little at a time, to a large volume of water; after reaction is complete, run solution to waste, diluting greatly with running water. Dispose of sand as normal refuse.

Sodium peroxyacetate [64057–57–4]

Hazardous reaction	Dry salt exploded at room temperature (B252).

Sodium phenylacetylide [1004–22–4]

Hazardous reaction	Ether-moist powder ignites in air (B701).

Sodium phosphide [12058–85–4]

Hazardous reaction	Decomposed by water or moist air, evolving phosphine which often ignites (B1331).

Sodium phosphinate (sodium hypophosphite) [7681–53–0]

Hazardous reactions	Evaporation of aqueous solution by heating may cause explosion, phosphine being evolved; reaction of this powerful reducing agent with oxidants ($NaClO_3$, $NaNO_3$, I_2) is often violent or explosive (B1148–1149).

Sodium pyrosulphate [13870–29–6]

This colourless crystalline salt of variable composition, produced by dehydrating sodium hydrogen sulphate, is soluble in water giving an acidic solution.

(CAUSES BURNS)

Avoid contact with skin, eyes and clothing.

Toxic effects	The solid and its strong solutions in water cause burns to the skin and eyes. Causes severe internal irritation and damage if taken by mouth.

First aid	*Affected eyes:* standard treatment (p. 136).
	Skin contact: standard treatment (p. 136).
	If swallowed: standard treatment (p. 137).
Spillage disposal	Wear face-shield or goggles, and gloves. Spread soda ash liberally over the spillage and mop up cautiously with plenty of water—run to waste, diluting greatly with running water.

Sodium selenate and selenite
—*see* Selenium and compounds

Sodium silicide [12164–12–4]

Hazardous reaction	Ignites in air (B1324).

Sodium sulphide [1313–82–2]

The hydrated salt consists of colourless crystalline masses; the fused salt forms brownish lumps or powder; soluble in water. With acids, poisonous hydrogen sulphide gas is evolved.

> CONTACT WITH ACIDS LIBERATES TOXIC GAS
> CAUSES BURNS

Prevent contact with skin and eyes.

Toxic effects	The solid or solution cause severe burns of the eyes. The solid or solution may irritate and burn the skin. Irritant and poisonous if taken by mouth.
Hazardous reactions	After exposure to moisture and air, small lumps of fused sodium sulphide are liable to spontaneous heating; mixtures with finely divided carbon react exothermally (B1330).
First aid	*Affected eyes:* standard treatment (p. 136).
	Skin contact: standard treatment (p. 136).
	If swallowed: standard treatment (p. 137).
Spillage disposal	Wear face-shield or goggles and gloves. Mop up with plenty of water and run to waste, diluting greatly with running water.

Sodium tellurate and tellurite
—*see* Tellurium and compounds

Sodium tetrahydroborate
—*see* **Sodium borohydride**

Sodium tetrahydroaluminate
—*see* **Aluminium sodium hydride**

Sodium vanadate
—*see* **Vanadium compounds**

Stannic chloride —*see* Tin(IV) chloride

Stannous chloride —*see* Tin(II) choride

Stibine (antimony trihydride)   [7803–52–3]

RL 0.1 ppm (0.5 mg m^{-3}).

Hazardous reactions   During evaporation of the liquid at −17 °C, a weakly explosive decomposition may occur; a heated mixture with ammonia explodes; explodes with chlorine, nitric acid or ozone (B1187).

Strontium acetylide   [12071–29–3]

Hazardous reactions   Incandesces with chlorine, bromine and iodine in the temperature range 174–197 °C (B321).

Strontium nitrate   [10042–76–9]

White granules or crystalline powder; 1 part dissolves in about 3 parts of water at 20 °C.

(CONTACT WITH COMBUSTIBLE MATERIALS MAY CAUSE FIRE)

Avoid contact with all combustible materials.

Fire hazard   Mixtures of strontium nitrate and combustible materials are readily ignited; mixtures with finely divided combustible materials can react explosively. *Extinguish fire with* water spray.

Spillage disposal Shovel into bucket of water and run solution to waste, diluting greatly with running water. Site of spillage should be washed thoroughly to remove all oxidant, which is liable to render any organic matter (particularly wood, paper and textiles) with which it comes into contact, dangerously combustible when dry. Clothing wetted with the solution should be washed thoroughly.

Styrene (phenylethylene) [100–42–5]

Colourless liquid with penetrating disagreeable odour; bp 146 °C; immiscible with water.

FLAMMABLE
IRRITATING TO EYES AND RESPIRATORY SYSTEM

Avoid breathing vapour. Avoid contact with skin and eyes. CL 100 ppm (420 mg m^{-3}).

Toxic effects The vapour irritates the eyes and respiratory system. The liquid irritates the eyes and is reported to cause severe eye injuries. Assumed to be poisonous if taken by mouth.

Hazardous reactions Autocatalytic exothermic polymerisation becomes self-sustaining above 65 °C; on exposure to oxygen at 40–60 °C an interpolymeric peroxide was formed which exploded on gentle heating (B707–709).

First aid *Vapour inhaled:* standard treatment (p. 136).
Affected eyes: standard treatment (p. 136).
Skin contact: standard treatment (p. 136).
If swallowed: standard treatment (p. 137).

Fire hazard Flash point 31 °C; explosive limits 1.1–6.1%; ignition temp. 490 °C. *Extinguish fire with* foam, dry powder, carbon dioxide or vaporising liquid.

Spillage disposal Shut off all possible sources of ignition. Wear face-shield or goggles, and gloves. Apply non-flammable dispersing agent if available and work to an emulsion with brush and water—run this to waste, diluting greatly with running water. If dispersant not available, absorb on sand, shovel into bucket(s) and transport to safe, open area for burial. Ventilate site of spillage well to evaporate remaining liquid and dispel vapour.

RSC *Lab. Haz. Data Sheet No. 22,* 1984, gives extended coverage.

Succinoyl diazide [40428–75–9]

Hazardous reaction Exploded violently during isolation (B412).

Sulphamic acid —*see* Amidosulphuric acid

Sulphinyl azides Review of group (B1694).

Sulphonic acids   tosic acid [104-15-4]

The simpler sulphonic acids, such as benzenesulphonic, benzenedisulphonic, phenolsulphonic, phenoldisulphonic, and cresolsulphonic acids are generally supplied as solutions in water or sulphuric acid. Toluene-*p*-sulphonic (tosic) acid is a colourless, hygroscopic solid.

> (tosic acid) CAUSES BURNS
> IRRITATING TO EYES, RESPIRATORY SYSTEM AND SKIN

Avoid contact with skin and eyes.

Toxic effects   The solutions irritate the skin and eyes and may cause burns. Will cause internal irritation and damage if taken by mouth.

First aid   *Affected eyes:* standard treatment (p. 136).
                       Skin contact: standard treatment (p. 136).
                       If swallowed: standard treatment (p. 137).

Spillage   Wear face-shield or goggles, and gloves. Spread soda ash liberally over the
disposal   spillage and mop up cautiously with plenty of water—run to waste, diluting greatly with running water.

Sulphur   [7704-34-9]

Hazardous   Evaporation of an ethereal extract of sulphur exploded violently; ignites in
reactions   fluorine and in chlorine dioxide gas (explosion possible); reacts with varying degrees of vigour with interhalogens, metal acetylides, metal oxides, metals, some non-metals, many oxidants and sodium hydride; dust very easily ignited by static sparks; violent or explosive reactions with Al/Cu powders, Gd, Zn, K, Ca or P (B1390-1393).

Sulphur esters

Review of unstable esters of various sulphur-derived acids (B1695).

Sulphur dichloride (sulphur chloride)   [10025-67-9]

Red-brown, fuming liquid; bp 59 °C; decomposed by water with the liberation of sulphur dioxide and hydrogen chloride.

> REACTS VIOLENTLY WITH WATER
> CAUSES BURNS
> IRRITATING TO RESPIRATORY SYSTEM

Avoid breathing vapour. Prevent contact with skin and eyes. Do not put water into container.

Toxic effects   The vapour irritates the respiratory system. The vapour and liquid irritate the eyes severely and the liquid may cause burns. The liquid irritates the skin and may cause burns; its ingestion would result in severe internal irritation and damage.

Hazardous reactions   Reacts vigorously with acetone, violently or explosively with dimethyl sulphoxide; mixture with sodium is impact-sensitive, as is also the case with potassium; reacts violently with nitric acid and explosively with N_2O_5; exothermic reaction with water or steam (B985).

First aid   *Vapour inhaled:* standard treatment (p. 136).
Affected eyes: standard treatment (p. 136).
Skin contact: standard treatment (p. 136).
If swallowed: standard treatment (p. 137).

Spillage disposal   Instruct others to keep at a safe distance. Wear breathing apparatus and gloves. Spread soda ash liberally over the spillage and mop up cautiously with water—run to waste, diluting greatly with running water.

Sulphur dioxide   [7446-09-5]

Colourless gas with a distinctive odour; supplied in liquefied form in canisters or cylinders; somewhat soluble in water.

TOXIC BY INHALATION
IRRITATING TO EYES AND RESPIRATORY SYSTEM

Avoid inhalation of gas. RL 2 ppm (5 mg m^{-3}).

Toxic effects   The vapour irritates the respiratory system and may cause bronchitis and asphyxia. High concentrations of vapour irritate the eyes and may cause conjunctivitis.

Hazardous reactions   Sulphur dioxide reacts violently with BrF_5, explodes with fluorine or chlorine trifluoride; incandescence and ignition occur with some metal acetylides and metal oxides; some metals react with sulphur dioxide with varying degress of vigour; solutions of SO_2 in ethanol or ether explode on contact with potassium chlorate; reacts explosively with sodium hydride; BaO_2, Cs_2O, FeO, SnO_2 or PbO_2 ignite and incandesce in hot SO_2 (B1355–1356).

First aid   *Vapour inhaled:* standard treatment (p. 136).
Affected eyes: standard treatment (p. 136).

Spillage disposal   Surplus gas or leaking cylinder can be vented slowly into a water-fed scrubbing tower or column in a fume cupboard, or into a fume cupboard served by such a tower.

Sulphur hexafluoride [2551-62-4]

RL 1000 ppm (6000 mg m^{-3}).

Sulphuric acid [7664-93-9]

Concentrated sulphuric acid is a colourless, viscous liquid reacting vigorously with water.

CAUSES SEVERE BURNS

Prevent contact with skin and eyes. Do not put water into container. RL 1 mg m^{-3}.

Toxic effects	The concentrated acid burns the eyes and skin severely. The dilute acid irritates the eyes and may cause burns; it will irritate the skin and may give rise to dermatitis. The concentrated acid, if taken by mouth, will cause severe internal irritation and damage.
Hazardous reactions	Many substances and classes of substance react with concentrated sulphuric acid (a powerful oxidising desiccant) with varying degrees of violence. These include acetone/nitric acid, acetonitrile/SO$_3$, acrylonitrile, alkyl nitrates, BrF$_5$, copper, 2-cyanopropan-2-ol, cyclopentadiene, metal acetylides or carbides, metal chlorates and perchlorates, nitramide, nitric acid/organic matter, nitric acid/toluene, nitrobenzene, nitromethane, *p*-nitrotoluene, permanganates, phosphorus, phosphorus trioxide, potassium, sodium, sodium carbonate. The dilution of sulphuric acid with water is vigorously exothermic and must be effected by adding acid to water to avoid local boiling; violent or explosive reactions with acetaldehyde, benzyl alcohol at 180 °C, 1-chloro-2,3-epoxypropane, *p*-chloronitrobenzene/SO$_3$, 2-cyano-4-nitro-benzenediazonium hydrogen sulphate, 1,3-diazido-benzene, *p*-dimethylaminobenzaldehyde, hexalithium disilicide, H$_2$O$_2$, NaBH$_4$, 1,2,4,5-tetrazine or ZnI$_2$. In many of these reactions, the exothermic dehydrating action of H$_2$SO$_4$ augments its oxidising action (B1163–1170).
First aid	*Affected eyes:* standard treatment (p. 136). *Skin contact:* standard treatment (p. 136). *If swallowed:* standard treatment (p. 137).
Spillage disposal	Wear face-shield or goggles, and gloves. Spread soda ash liberally over the spillage and mop up cautiously with plenty of water — run to waste, diluting greatly with running water.

RSC *Lab. Haz. Data Sheet No. 33,* 1985, gives extended coverage.

Sulphur pentafluoride [5714-22-7]

RL 0.025 ppm (0.25 mg m^{-3}).

508

Sulphur tetrafluoride [7783-60-0]

Colourless gas with smell resembling that of sulphur dioxide; bp −40 °C; reacts violently with water forming hydrogen fluoride and sulphur dioxide; attacks glass.

(REACTS VIOLENTLY WITH WATER
VERY TOXIC BY INHALATION
CAUSES SEVERE BURNS)

Prevent inhalation of gas. Prevent contact with skin and eyes. RL 0.1 ppm (0.4 mg m^{-3}).

Toxic effects	The gas reacts with moisture on the body tissues forming highly toxic and corrosive hydrogen fluoride. It is thus extremely irritant to the eyes, skin and respiratory system, and will cause severe burns if exposure is considerable.
First aid	*Gas inhaled:* special treatment for hydrofluoric acid (p. 136).
	Affected eyes: special treatment for hydrofluoric acid (p. 136).
	Skin contact: special treatment for hydrofluoric acid (p. 136).
Spillage disposal	Surplus gas or leaking cylinder can be vented slowly into a water-fed scrubbing tower or column in a fume cupboard, or into a fume cupboard served by such a tower.

Sulphur trioxide [7446-11-9]

Hazardous reactions	Dissolution of SO_3 in dimethyl sulphoxide is very exothermic; the 1:1 addition complex with dioxan or dimethylformamide sometimes decomposes violently on storing at ambient temperature; reacts violently with diphenylmercury; reacts with barium and lead oxides with incandescence; reaction with water is vigorously exothermic, sometimes explosive; reaction with deficiency of tetrafluoroethylene is explosive (B1365–1366).

Sulphuryl chloride [7791-25-5]

Colourless or yellow, fuming liquid with a pungent odour; bp 69 °C; decomposed by water with formation of hydrochloric and sulphuric acids.

REACTS VIOLENTLY WITH WATER
CAUSES BURNS
IRRITATING TO RESPIRATORY SYSTEM

Avoid breathing vapour. Prevent contact with skin and eyes. Do not put water into container.

Toxic effects	The vapour irritates the respiratory system. The vapour and liquid irritate the eyes and the liquid will cause burns. The liquid irritates the skin and causes burns; its ingestion would result in severe internal irritation and damage.

Hazardous reactions	Reactions with alkalies may be violently explosive; solution of sulphuryl chloride in ether decomposed vigorously; reacts violently or explosively with dimethyl sulphoxide or lead dioxide; reacts vigorously with red phosphorus on warming, explosively with N_2O_5 (B979–980).
First aid	*Vapour inhaled:* standard treatment (p. 136). *Affected eyes:* standard treatment (p. 136). *Skin contact:* standard treatment (p. 136). *If swallowed:* standard treatment (p. 137).
Spillage disposal	Instruct others to keep at a safe distance. Wear breathing apparatus and gloves. Spread soda ash liberally over the spillage and mop up cautiously with water — run to waste, diluting with running water.

Sulphuryl diazide [72250–07–8]

Hazardous reaction	Explodes violently when heated and often spontaneously at ambient temperatures (B1310–1311).

Sulphuryl fluoride [2699–79–8]

RL 5 ppm (20 mg m^{-3}).

Tantalum [7440–25–7]

RL 5 mg m^{-3}.

Tellurium and compounds Te [13494–80–9]

Metallic ingots breaking with a white lustrous fracture, or a greyish-black powder. Telluric acid and the alkali-metal tellurites and tellurates are colourless powders or crystals, soluble in water; tellurium oxide (TeO_2) is also colourless, but is insoluble in water. Tellurium tetrachloride is a cream-coloured solid. Poisonous hydrogen telluride (offensive smell) is generated when an acid solution of tellurium compound is reduced by metals such as tin and zinc.

> (HARMFUL BY INHALATION
> HARMFUL IF SWALLOWED)

Avoid breathing dust. Avoid contact with skin and eyes. RL (as Te) 0.1 mg m^{-3}.

Toxic effects   Inhalation of dust or tellurium fume gives rise to a dry mouth, metallic taste, drowsiness, loss of appetite, excessive salivation, nausea, vomiting and a foul garlic-like odour of the breath. Skin absorption and ingestion also result in foul breath and it can be assumed that effects similar to those of inhalation will be experienced. Ingestion of soluble tellurium salts may, in addition, lead to cyanosis and unconsciousness. **Chronic effects** These are similar to the above, but arise from prolonged absorption of very small amounts of tellurium compounds.

First aid   *Inhalation of dust or fume:* standard treatment (p. 136).
Affected eyes: standard treatment (p. 136).
Skin contact: standard treatment (p. 136).
If swallowed: standard treatment (p. 137).

Spillage disposal   Because of its low toxicity, tellurium powder can be mixed with a large quantity of sand and dealt with as normal refuse. The soluble tellurates and tellurites and telluric acid should be dissolved in a large volume of water and run to waste, diluting greatly with running water.

Tellurium hexafluoride   [7783-80-4]

RL (as Te) 0.02 ppm (0.2 mg m^{-3}).

Tellurium tetrachloride   [10026-07-0]

Hazardous reaction   Interaction with liquid ammonia at $-15\,°C$ forms an explosive nitride (B1006).

Terphenyls   [92-94-4]

RL 0.5 ppm (5 mg m^{-3}).

Tetraaluminium tricarbide   [1299-86-1]

Hazardous reactions   Incandescent reaction with lead peroxide or potassium permanganate (B323).

Tetraborane   [18283-93-7]

Hazardous reactions   Ignites in air or oxygen and explodes with concentrated nitric acid (B74).

1,1,2,2-Tetrabromoethane [79–27–6]

(acetylene tetrabromide)

Very dense yellowish liquid with chloroform-like odour; mp 0 °C; bp 151 °C at 72 mbar; immiscible with water.

VERY TOXIC BY INHALATION
IRRITATING TO EYES

Avoid breathing vapour. Avoid contact with skin and eyes. TLV 1 ppm (15 mg m^{-3}).

Toxic effects	Records of these have not been traced though it is suggested that it has narcotic properties and may cause damage to the liver. It is probably similar in action to tetrachloroethane though not so toxic.
First aid	*Vapour inhaled:* standard treatment (p. 136). *Affected eyes:* standard treatment (p. 136). *If swallowed:* standard treatment (p. 137).
Spillage disposal	Instruct others to keep at a safe distance. Wear breathing apparatus and gloves. Apply dispersing agent if available and work to an emulsion with brush and water—run this to waste, diluting greatly with running water. If dispersant not available, absorb on sand, shovel into bucket(s) and transport to safe, open area for burial. Site of spillage should be washed thoroughly with water and soap or detergent.

Tetracarbonylnickel (nickel carbonyl) [13463–39–3]

Colourless, mobile liquid normally supplied in cylinders; bp 43 °C; practically insoluble in water.

HIGHLY FLAMMABLE
VERY TOXIC BY INHALATION
POSSIBLE RISK OF IRREVERSIBLE EFFECTS

Avoid breathing vapour. Avoid contact with skin, eyes and clothing. RL 0.05 ppm (0.35 mg m^{-3}).

Toxic effects	The vapour is exceedingly poisonous when inhaled. With low concentrations the initial symptoms are giddiness and slight headache. Heavier exposure causes nausea, tightness of the chest, weakness of limbs, perspiring, coughing, vomiting, cold and clammy skin and shortness of breath. The toxicity is believed to be derived from both the nickel and carbon monoxide liberated in the lungs. Cases of lung cancer have occurred as a result of prolonged exposure to vapour.
Hazardous reactions	Reacts explosively with liquid bromine, but smoothly in the gaseous state; a mixture with mercury will explode on vigorous shaking; on exposure to air, the carbonyl produces a deposit which becomes peroxidised and may ignite; violent reaction with N_2O_4 (B490).

First aid	*Vapour inhaled:* standard treatment (p. 136). *Affected eyes:* standard treatment (p. 136). *Skin contact:* standard treatment (p. 136). *If swallowed:* standard treatment (p. 137).
Fire hazard	The liquid is highly flammable and is liable to explode if heated to 60 °C or above. *Extinguish fire with* dry powder, carbon dioxide or vaporising liquid.
Spillage disposal	Shut off all possible sources of ignition. Instruct others to keep at a safe distance. Wear breathing apparatus and gloves. Apply dispersing agent if available and work to an emulsion with brush and water—run this to waste, diluting greatly with running water. If dispersant not available, absorb on sand, shovel into bucket(s) and transport to safe, open area for burial. Site of spillage should be washed thoroughly with water and soap or detergent.

1,2,4,5-Tetrachlorobenzene [95-94-3]

Hazardous reaction	Explosions have occurred in the commercial production of 2,4,5-trichloro-phenol by alkaline hydrolysis of tetrachlorobenzene, leading to formation of the extremely toxic tetrachlorodibenzodioxin (TCDD) as in the Seveso incident, which is discussed in detail (B541–544).

Tetrachlorodifluoroethanes 1,1,1,2- [76-11-9]
TLV 500 ppm (4170 mg m^{-3}). 1,1,2,2- [76-12-0]

1,1,2,2-Tetrachloroethane [79-34-5]
(acetylene tetrachloride)

Colourless, dense liquid with chloroform-like odour; bp 146 °C; immiscible with water.

(VERY TOXIC BY INHALATION AND IN CONTACT WITH SKIN)

Avoid breathing vapour. Avoid contact with eyes. TLV (skin) 5 ppm (35 mg m^{-3}).

Toxic effects	The vapour irritates the eyes, nose and lungs and may cause drowsiness, giddiness and headache and, in high concentrations, unconsciousness. Assumed to be poisonous if taken by mouth. **Chronic effects** Continuous breathing of low concentrations of vapour over a period may cause jaundice by action on the liver; it may also affect the nervous system and blood. (*See* note in Ch 6 p. 105.)
Hazardous reactions	When heated with solid NaOH, chloro- or dichloro- acetylene, which ignite in air, are formed; forms impact-sensitive mixtures with K and Na (B233).
First aid	*Vapour inhaled:* standard treatment (p. 136). *Affected eyes:* standard treatment (p. 136). *If swallowed:* standard treatment (p. 137).

Spillage disposal	Instruct others to keep at a safe distance. Wear breathing apparatus and gloves. Apply dispersing agent if available and work to an emulsion with brush and water — run this to waste, diluting greatly with running water. If dispersant not available, absorb on sand, shovel into bucket(s) and transport to safe, open area for burial. Site of spillage should be washed thoroughly with water and soap or detergent.

Tetrachloroethylene (perchloroethylene) [127–18–4]

Colourless liquid with faint ethereal odour; bp 121 °C; immiscible with water.

HARMFUL BY INHALATION AND IF SWALLOWED

Avoid breathing vapour. Avoid contact with skin and eyes. RL (skin) 100 ppm (670 mg m^{-3}).

Toxic effects	The inhalation of the vapour may cause dizziness, nausea and vomiting — in high concentrations, stupor. The liquid irritates the eyes and has a degreasing action on the skin. Assumed to cause symptoms similar to those of inhalation if taken by mouth.
Hazardous reactions	Explosive mixtures are formed with N_2O_4, Ba, Li; impure material containing trichloroethylene, when treated with solid NaOH and subsequently distilled, may have a volatile, explosive fore-run (B208).
First aid	*Vapour inhaled:* standard treatment (p. 136). *Affected eyes:* standard treatment (p. 136). *Skin contact:* standard treatment (p. 136). *If swallowed:* standard treatment (p. 137).
Spillage disposal	Wear face-shield or goggles, and gloves. Apply dispersing agent if available and work to an emulsion with brush and water — run this to waste, diluting greatly with running water. If dispersant not available, absorb on sand, shovel into bucket(s) and transport to safe, open area for burial. Site of spillage should be washed thoroughly with water and soap or detergent.

RSC *Lab. Haz. Data Sheet No. 6,* 1982, gives extended coverage.

Tetrachloromethane —*see* Carbon tetrachloride

Tetrachloronaphthalene [1335–88–2]

RL (all isomers) 2 mg m^{-3}.

Tetraethylenepentamine [112–57–2]

Pale yellow-brown, viscous, hygroscopic liquid; bp 340 °C; miscible with water.

HARMFUL IN CONTACT WITH SKIN AND IF SWALLOWED
CAUSES BURNS
MAY CAUSE SENSITISATION BY SKIN CONTACT

Avoid contact with skin, eyes and clothing.

Toxic effects   The liquid irritates the eyes and skin and may cause burns. It is irritating and
harmful if taken internally.

First aid   *Affected eyes:* standard treatment (p. 136).
Skin contact: standard treatment (p. 136).
If swallowed: standard treatment (p. 137).

Spillage   Wear face-shield or goggles, and gloves. Mop up with plenty of water and
disposal   run to waste, diluting greatly with running water.

Tetraethyllead   [78-00-2]

CL (skin, as Pb) 0.10 mg m^{-3}.

Tetraethyl silicate   [78-10-4]

Colourless liquid; bp 166 °C; practically insoluble in water, by which it is slowly hydrolysed.

FLAMMABLE
HARMFUL BY INHALATION
IRRITATING TO EYES AND RESPIRATORY SYSTEM

Avoid breathing vapour. Avoid contact with skin, eyes and clothing. RL 10 ppm (85 mg m^{-3}).

Toxic effects   The vapour irritates the eyes and respiratory system and is narcotic in high
concentrations. The liquid irritates the eyes and may irritate the skin; it is
irritant and damaging if taken by mouth.

First aid   *Vapour inhaled:* standard treatment (p. 136).
Affected eyes: standard treatment (p. 136).
Skin contact: standard treatment (p. 136).
If swallowed: standard treatment (p. 137).

Fire hazard   Flash point 52 °C. *Extinguish fire with* foam, dry powder, carbon dioxide or
vaporising liquid.

Spillage   Shut off all possible sources of ignition. Wear face-shield or goggles, and
disposal   gloves. Apply non-flammable dispersing agent if available and work to an
emulsion with brush and water — run this to waste, diluting greatly with
running water. If dispersant not available absorb on sand, shovel into
bucket(s) and transport to safe, open area for burial. Ventilate site of spillage
well to evaporate remaining liquid and dispel vapour.

Tetraethynyltin [16413–88–0]

Hazardous   Explodes on rapid heating (B700).
reaction

Tetrafluoroboric acid and salts   acid [16872–11–0]
(fluoroboric acid and salts)

Fluoroboric acid is a colourless fuming liquid miscible with water. The salts are generally colourless, crystalline and soluble in water.

CAUSES BURNS

Avoid inhaling vapour or dust. Prevent contact with eyes and skin.

Toxic effects   The vapour and dust irritate all parts of the respiratory system. The acid and salts burn the eyes severely. The acid burns the skin. If taken by mouth there would be severe internal irritation and damage.

First aid   *Vapour or dust inhaled:* standard treatment (p. 136).
Affected eyes: standard treatment (p. 136).
Skin contact: standard treatment (p. 136).
If swallowed: standard treatment (p. 137).

Spillage   Wear face-shield or goggles, and gloves. Mop up with plenty of water and
disposal   run to waste, diluting greatly with running water.

Tetrafluoroethylene [116–14–3]

Hazardous   The monomer explodes spontaneously at pressures above 27 bar (2.7×10^5
reactions   Pa); the inhibited monomer will explode if ignited; the liquid monomer, exposed to air, will form an explosive peroxidic polymer; explodes with iodine pentafluoride; forms explosive, polymeric peroxide with oxygen; mixture with hexafluoropropene formed explosive peroxide; explosive reactions with chloroperoxytrifluoromethane, difluoromethylene dihypofluorite, F_2O_2, B_3F_5 (B212–213).

Tetrafluorohydrazine [10036–47–2]

Hazardous   Gas above $-73\,°C$ which explodes on contact with air or combustible
reactions   vapours or on irradiation; mixtures with hydrocarbons are potentially highly explosive (B1071).

Tetrahydrofuran   [109–99–9]

Colourless, volatile liquid with ethereal odour; bp 66 °C; liable to form explosive peroxides on exposure to light and air which should be decomposed before distilling to small volume.

> HIGHLY FLAMMABLE
> MAY FORM EXPLOSIVE PEROXIDES
> IRRITATING TO EYES AND RESPIRATORY SYSTEM

Avoid breathing vapour. Avoid contact with eyes. RL 200 ppm (590 mg m^{-3}).

Toxic effects	The vapour irritates the eyes and respiratory system; high concentrations have a narcotic effect. It is believed that liver damage may result from skin absorption or ingestion.
Hazardous reactions	Readily forms peroxides by autoxidation and reference is given to brochure on handling. Peroxidised material should not be dried with NaOH or KOH as explosions may occur; explosion in reflux with CaH$_2$; vigorous reaction with Br$_2$ (B448–449).
First aid	*Vapour inhaled:* standard treatment (p. 136). *Affected eyes:* standard treatment (p. 136). *Skin contact:* standard treatment (p. 136). *If swallowed:* standard treatment (p. 137).
Fire hazard	Flash point −17 °C; explosive limits 2.3–11.8%; ignition temp. 321 °C. *Extinguish fire with* water spray, foam, dry powder, carbon dioxide or vaporising liquid.
Spillage disposal	Shut off all possible sources of ignition. Instruct others to keep at a safe distance. Wear breathing apparatus and gloves. Mop up with plenty of water and run to waste, diluting greatly with running water. Ventilate area well to evaporate remaining liquid and dispel vapour.

RSC *Lab. Haz. Data Sheet No. 12,* 1983, gives extended coverage.

Tetrahydropyranyl ethers Review of group (B1696).

Tetrahydrothiophene   [110–01–0]
(thiophane; tetramethylene sulphide)

Colourless liquid with smell like that of coal gas for which it is an established odorant; bp 121 °C; immiscible with water.

> (HIGHLY FLAMMABLE)

Avoid breathing vapour. Avoid contact with skin, eyes and clothing.

517

Toxic effects The evidence available is derived from animal experiments which showed that it was not irritant to the eyes and skin of rabbits and caused no permanent damage to the eyes. Oral injection produced acute toxicity 'no worse than that of benzene'. It is thought wisest to handle with caution and avoid breathing the vapour, which advertises its presence unmistakably.

First aid *Vapour inhaled:* standard treatment (p. 136).
Affected eyes: standard treatment (p. 136).
Skin contact: standard treatment (p. 136).
If swallowed: standard treatment (p. 137).

Fire hazard Flash point 18 °C; *Extinguish fire with* foam, dry powder, carbon dioxide or vaporising liquid.

Spillage disposal Shut off all possible sources of ignition. Instruct others to keep at a safe distance. Wear breathing apparatus and gloves. Apply non-flammable dispersing agent and work to an emulsion with brush and water — run this to waste, diluting greatly with running water. If dispersant not available, absorb on sand, shovel into bucket(s) and transport to safe, open area for burial. Site of spillage should be washed thoroughly with water and soap or detergent.

Tetramethylammonium chlorite [67922–18–3]

Hazardous reaction The dry solid explodes on impact (B478).

Tetramethyllead [75–74–1]

Hazardous reaction Liable to explode above 90 °C (B483).

Tetramethyl silicate (methyl silicate) [681–84–5]

(VERY IRRITATING TO EYES
DANGER OF VERY SERIOUS IRREVERSIBLE EFFECTS)

RL 1 ppm (6 mg m^{-3}).

Tetramethylurea [632–22–4]

Colourless liquid, slight mint odour; bp 176 °C; miscible with water.

(IRRITATING BY INHALATION, IN CONTACT WITH SKIN
OR IF SWALLOWED
POSSIBLE RISKS OF IRREVERSIBLE EFFECTS)

Avoid inhalation of vapour. Avoid contact of liquid with skin or eyes. Prevent contact during pregnancy; experimental teratogen.

Toxic effects   Contact of liquid with eyes may lead to conjunctivitis or temporary opacity of the cornea. Prolonged inhalation or skin contact may lead to systemic injury.

First aid   *Affected eyes:* standard treatment (p. 136).
Skin contact: standard treatment (p. 136).
If swallowed: standard treatment (p. 137).

Spillage disposal   Instruct others to keep at safe distance. Wear breathing apparatus and gloves. Mop up with plenty of water — run to waste, diluting greatly with running water.

Tetraselenium tetranitride   [12033–88–6]

Hazardous reactions   Dry material explodes on compression or on heating at 130–230 °C; contact with bromine, chlorine or fuming hydrochloric acid also causes explosion (B1309).

Tetrasodium pyrophosphate
—*see* Sodium pyrophosphate

Tetrasilane   [7783–29–1]

Hazardous reactions   Ignites and explodes in air or oxygen; reacts vigorously with carbon tetrachloride (B1214–1215).

Tetrasulphur tetranitride   [28950–34–7]

Hazardous reactions   Liable to explosive decomposition on friction, shock or heating above 100 °C; forms explosive complexes with metal halides (B1308–1309).

Tetrazoles Review of group (B1697).

Thallium and salts   Tl [7440–28–0]

Thallium–soft, silvery sticks often stored under water. The carbonate, nitrate and sulphate are colourless, crystalline solids, soluble in water; the oxide is black and insoluble in water.

VERY TOXIC BY INHALATION AND IF SWALLOWED
DANGER OF CUMULATIVE EFFECTS

Avoid breathing dust. Avoid contact with skin and eyes. RL (skin, soluble salts as Tl) 0.1 mg m^{-3}.

Toxic effects   The dusts irritate the nose and eyes and may cause nausea and abdominal pain by absorption. The metal on contact with moist skin produces a white film of the hydroxide. Skin absorption of the soluble salts and ingestion may cause nausea, vomiting, abdominal pains, weakness of the legs and mental confusion. **Chronic effects** Exposure over a long period to small amounts of the dust or solutions may result in loss of appetite, falling out of hair, pain or weakness of limbs, insomnia and mental disturbance. (*See* note in Ch 6 p. 104).

First aid   *Dust inhaled:* standard treatment (p. 136).
Affected eyes: standard treatment (p. 136).
Skin contact: standard treatment (p. 136).
If swallowed: standard treatment (p. 137).

Spillage disposal   Small quantities of soluble thallium salts should be converted to the insoluble sulphide by treatment in solution with a slight excess of sodium sulphide and filtration. Insoluble thallium can be thoroughly mixed with a large amount of sand and the mixture buried in a safe, open area.

Thallium azide   [13847–66–0]

Hazardous reaction   Relatively stable, it can be exploded on heavy impact or by heating at 350–400 °C (B1307).

Thiocarbonyl tetrachloride
—*see* Trichloromethanesulphenyl chloride

Thiocyanogen   [505–14–6]

Hazardous reaction   Polymerises explosively above its mp −2 °C (B315).

Thiodiphenylamine —*see* Phenothiazine

Thioglycollic acid —*see* Mercaptoacetic acid

Thionyl chloride (sulphinyl chloride) [7719-09-7]

Pale yellow or yellow fuming liquid with an odour like sulphur dioxide; bp 79 °C; it reacts with water to form hydrochloric acid and sulphur dioxide.

REACTS VIOLENTLY WITH WATER
CAUSES BURNS
IRRITATING TO RESPIRATORY SYSTEM

Avoid breathing vapour. Prevent contact with skin and eyes. Do not put water into container.

Hazardous reactions	Addition of concentrated ammonia may cause an explosion; reacts violently or explosively with dimethyl sulphoxide; explosive reactions with Cl_2O_6, bis(dimethylamino) sulphoxide, 1,2,3-cyclohexanetrione trioxime; reaction with H_2O generates 990 volumes of mixed corrosive gases. If the chloride is dissolved in toluene, hydrolysis is very slow (B977–979).
First aid	*Vapour inhaled:* standard treatment (p. 136). *Affected eyes:* standard treatment (p. 136). *Skin contact:* standard treatment (p. 136). *If swallowed:* standard treatment (p. 137).
Spillage disposal	Instruct others to keep at a safe distance. Wear breathing apparatus and gloves. Spread soda ash liberally over spillage and mop up cautiously with water — run to waste, diluting greatly with running water.

RSC *Lab. Haz. Data Sheet No. 26,* 1984, gives extended coverage.

Thiophane —*see* Tetrahydrothiophene

Thiophene [110-02-1]

Clear, colourless liquid; bp 84 °C; insoluble in water.

(HIGHLY FLAMMABLE)

Avoid breathing vapour. Avoid contact with skin, eyes and clothing.

Toxic effects	Animal experiments indicate moderate toxicity.
Hazardous reaction	Reacts very violently with fuming nitric acid (B413).

First aid	*Affected eyes:* standard treatment (p. 136).
	Skin contact: standard treatment (p. 136).
	If swallowed: standard treatment (p. 137).
Fire hazard	Flash point −6 °C. *Extinguish fire with* foam, dry powder, carbon dioxide or vaporising liquid.
Spillage disposal	Shut off all possible sources of ignition. Instruct others to keep at a safe distance. Wear breathing apparatus and gloves. Apply non-flammable dispersing agent and work to an emulsion with brush and water — run this to waste, diluting greatly with running water. If dispersant not available, absorb on sand, shovel into bucket and transport to safe, open area for burial or slow evaporation.

Thiophenol —*see* Benzenethiol

Thiophosphoryl fluoride [2404–52–6]

Hazardous reactions	Ignites or explodes on contact with air; heated sodium ignites in the gas (B1070).

Thorium (metal) [7440–29–1]

Hazardous reactions	Finely divided metal is pyrophoric in air; incandesces in chlorine, bromine and iodine vapour; reactions with P and S are also incandescent (B1402–1403).

Thorium hydride [15457–87–1]

Hazardous reactions	Explodes on heating in air, powdered material readily ignites in air on handling (B1204).

Tin, and inorganic compounds of Sn [7440–31–5]
(except SnH_4, SnO_2)

RL (as Sn) 2 mg m^{-3}.

Tin, organic compounds of

Certain alkyltin compounds — notably di-n-butyltin diacetate, dilaurate, maleate and oxide — are used quite extensively as stabilisers for pvc resins, as catalysts and biocides. These are harmful by skin absorption and some, as dust, irritate the respiratory system. RL (skin, as Sn) 0.1 mg m^{-3}.

First aid	*Affected eyes:* standard treatment (p. 136).
	Skin contact: standard treatment (p. 136).
	If swallowed: standard treatment (p. 137).
Spillage disposal	Wear face-shield or goggles, and gloves. Mix with sand and transport to safe, open area for burial. Site of spillage should be washed thoroughly with water and soap or detergent.

Tin(II) chloride (stannous chloride) [7772–99–8]

Hazardous reaction	Reaction with hydrogen peroxide is strongly exothermic, even in solution (B986).

Tin(IV) choride [7646–78–8]
(stannic chloride; tin tetrachloride)

Colourless fuming liquid; bp 114 °C; reacts with water forming hydrochloric acid.

(CAUSES BURNS
IRRITATING TO RESPIRATORY SYSTEM)

Avoid contact with skin and eyes. Do not put water into container. RL (inorganic tin compounds, as Sn) 2 mg m^{-3}.

Toxic effects	The vapour irritates the respiratory system. The vapour irritates the eyes. The liquid irritates the eyes and skin and may cause burns. Will result in internal irritation and damage if taken by mouth.
Hazardous reaction	Traces may catalyse delayed decomposition of alkyl nitrates or violent polymerisation of ethylene oxide (B851).
First aid	*Vapour inhaled:* standard treatment (p. 136).
	Affected eyes: standard treatment (p. 136).
	Skin contact: standard treatment (p. 136).
	If swallowed: standard treatment (p. 137).
Spillage disposal ·	Instruct others to keep at a safe distance. Wear breathing apparatus and gloves. Spread soda ash liberally over the spillage and mop up cautiously with water — run to waste, diluting greatly with running water.

Titanium [7440–32–6]

Hazardous reactions	The finely divided metal is pyrophoric and once burning is difficult to extinguish as it burns in both CO_2 and N_2. Violent reactions with halogens, halocarbons, metal salts and oxidants (B1403–1405).

523

Titanium(II) chloride [10049-06-6]

Hazardous reaction Ignites readily in air, particularly if moist (B986–987).

Titanium(III) chloride [7705-07-9]

Hazardous reactions Reacts vigorously with air or water, pyrophoric if finely divided (B1002).

Titanium(IV) chloride [7550-45-0]
(titanium tetrachloride, titanic chloride)

Colourless, fuming liquid with a pungent odour; bp 136 °C; reacts with water with liberation of hydrogen chloride.

> REACTS VIOLENTLY WITH WATER
> CAUSES BURNS
> IRRITATING TO EYES AND RESPIRATORY SYSTEM

Avoid breathing vapour. Prevent contact with skin and eyes. Do not put water into container.

Toxic effects The vapour irritates the respiratory system. The vapour and liquid irritate the eyes and may cause burns. The liquid irritates the skin. Assumed to cause severe internal irritation and damage if taken by mouth.

First aid *Vapour inhaled:* standard treatment (p. 136).
Affected eyes: standard treatment (p. 136).
Skin contact: standard treatment (p. 136).
If swallowed: standard treatment (p. 137).

Spillage disposal Instruct others to keep at a safe distance. Wear breathing apparatus and gloves. Spread soda ash liberally over the spillage and mop up cautiously with water — run to waste, diluting greatly with running water.

o-Tolidine (2,2'-dimethylbenzidine) [119-93-7]
and *o*-tolidine dihydrochloride [612-82-8]

Colourless to grey or brown powder; slightly soluble in water. The use of *o*-tolidine and its salts is controlled in the United Kingdom by The Carcinogenic Substances Regulations 1967 (see p. 148).

> HARMFUL BY INHALATION, IN CONTACT WITH SKIN
> AND IF SWALLOWED

Prevent inhalation of dust. Prevent contact with skin and eyes.

Toxic effects The dihydrochloride and its solutions irritate the skin and eyes. **Chronic effects** There is evidence that *o*-tolidine, through continued absorption, can cause cancer of the bladder.

First aid *Affected eyes:* standard treatment (p. 136).
Skin contact: standard treatment (p. 136).
If swallowed: standard treatment (p. 137).

Spillage disposal Wear breathing apparatus and gloves. Mix spillage with moist sand and shovel mixtures into glass, enamel or polythene vessels for dispersion in an excess of dilute hydrochloric acid (1 volume of concentrated acid diluted with 2 volumes of water). Allow to stand, with occasional stirring, for 24 hours and then run extract to waste, diluting greatly with running water and washing the sand. The residual sand can be treated as normal refuse. The site of the spillage should be washed with water and soap or detergent.

Toluene (toluol) [108–88–3]

Colourless liquid with characteristic odour; bp 111 °C; immiscible with water.

HIGHLY FLAMMABLE
HARMFUL BY INHALATION

Avoid breathing vapour. Avoid contact with skin and eyes. RL 100 ppm (375 mg m^{-3}).

Toxic effects Inhalation of the vapour may cause dizziness, headache, nausea and mental confusion. The vapour and liquid irritate the eyes and mucous membranes. Absorption through the skin and ingestion would cause poisoning. **Chronic effects** If the toluene contains more than traces of benzene as an impurity, breathing of vapour over long periods may cause blood disease. Prolonged skin contact may cause dermatitis.

Hazardous reactions Reacts violently with BrF_3 at −80 °C; inadequate control in nitration of toluene with mixed acids may lead to runaway or explosive reaction; violent or explosive reactions with a range of oxidants, including BrF_3, 1,3-dichloro-5,5-dimethyl-2,4-imidazolidindione, N_2O_4, HNO_3, tetranitromethane (extremely violent explosion), UF_6 (B682).

First aid *Vapour inhaled:* standard treatment (p. 136).
Affected eyes: standard treatment (p. 136).
Skin contact: standard treatment (p. 136).
If swallowed: standard treatment (p. 137).

Fire hazard Flash point 4.4 °C; explosive limits 1.4–6.7%; ignition temp. 536 °C. *Extinguish fire with* foam, dry powder, carbon dioxide or vaporising liquid.

Spillage disposal Shut off all possible sources of ignition. Wear face-shield or goggles, and gloves. Apply non-flammable dispersing agent if available and work to an emulsion with brush and water — run this to waste, diluting greatly with running water. If dispersant not available, absorb on sand, shovel into

525

bucket(s) and transport to safe, open area for atmospheric evaporation. Ventilate site of spillage well to evaporate remaining liquid and dispel vapour.

RSC *Lab. Haz. Data Sheet No. 15,* 1983, gives extended coverage.

Toluene-2-diazonium perchlorate [69597–04–6]

Hazardous Explosive when wet (B672).
reaction

Toluene-2-, -3- and -4-diazonium salts

Covers hazardous reactions with ammonium sulphide, hydrogen sulphide or potassium iodide (B679–680).

Toluene di-isocyanate
—*see* 2,4-Di-isocyanatotoluene

Toluene-4-sulphonic acid
—*see* Sulphonic acids

Toluene-4-sulphonyl azide [941–55–9]

Hazardous Distillation residue containing this will explode if temperature exceeds
reaction 120 °C (B680).

Toluidines *o-* [95–53–4] *m-* [108–44–1] *p-* [106–49–0]

o-Toluidine and *m*-toluidine are red to dark brown liquids; *p*-toluidine consists of pale brown crystals. Insoluble in water.

TOXIC BY INHALATION, IN CONTACT WITH SKIN
AND IF SWALLOWED
DANGER OF CUMULATIVE EFFECTS

Avoid breathing vapour. Avoid contact with skin and eyes. RL (skin) (*o-*) 2 ppm (9 mg m^{-3}).

Toxic effects　　Excessive breathing of the vapour or absorption through the skin may cause headache, drowsiness, cyanosis, mental confusion and, in severe cases, convulsions. They are dangerous to the eyes and the above effects may also be experienced if they are taken by mouth.

Hazardous　　　Ignite in contact with fuming nitric acid (B1106).
reaction

First aid　　　*Vapour inhaled:* standard treatment (p. 136).
Affected eyes: standard treatment (p. 136).
Skin contact: standard treatment (p. 136).
If swallowed: standard treatment (p. 137).

Spillage　　　Wear face-shield or goggles, and gloves. Mix with sand and shovel mixture
disposal　　　into glass, enamel or polythene vessel for dispersion in an excess of dilute hydrochloric acid (1 volume of concentrated acid diluted with 2 volumes of water). Allow to stand, with occasional stirring, for 24 hours and then run acid extract to waste, diluting greatly with running water and washing the sand. The sand can be disposed of as normal refuse.

Tolylcoppers　　　*o-* [20854–03–9] *m-* [20854–05–1]
p- [5588–74–9]

Hazardous　　　*o-*, *m-* and *p-*isomers usually explode strongly on exposure to oxygen at
reaction　　　0 °C, or weakly above 100 °C in vacuo (B673).

Trialkyl-aluminiums and -bismuths Reviews (B1702–1703).

Triallyl cyanurate (2,4,6-trisallyloxy-*s*-triazine)　　　[1025–15–6]

Colourless liquid or solid; mp 27 °C; bp 162 °C at 2mmHg; hydrolysed by water forming allyl alcohol which is irritant.

(HARMFUL IF SWALLOWED)

Avoid contact with skin, eyes and clothing.

Toxic effects　　Arise because of the allyl alcohol formed on hydrolysis.

First aid　　　*Affected eyes:* standard treatment (p. 136).
Skin contact: standard treatment (p. 136).
If swallowed: standard treatment (p. 137).

Spillage　　　Wear face-shield or goggles, and gloves. Clear up with dust-pan and brush.
disposal　　　May be disposed of, after mixing with sand, as normal refuse or flushed away to waste with water.

Triallyl phosphate [1624-19-4]

Hazardous　　Distillation residue exploded (B753).
reaction

Triazenes Review of group (B1703).

2,4,6-Triazido-1,3,5-triazine [5637-83-2]

Hazardous　　Explodes on impact, shock or rapid heating to 170–180 °C (B394–395).
reaction

Tribenzylarsine [5888-61-9]

Hazardous　　Oxidises slowly at first in air but becomes violent (B857).
reaction

Tribromomethane —*see* Bromoform

Tribromosilane [7789-57-3]

Hazardous　　Usually ignites when poured in air (B108).
reaction

Tributylamine —*see* Butylamines

Tributylbismuth [3692-81-7]

Hazardous　　Explodes in oxygen and ignites in air (B816).
reactions

Tributylborane n- [122-56-8] iso- [1116-39-8]

Hazardous　　Mixture of n- and iso- isomers ignited on exposure to air (B815).
reaction

Tributyl phosphate [126-73-8]

HARMFUL IF SWALLOWED

RL 5 mg m^{-3}.

Tributylphosphine [998-40-3]

Colourless liquid with garlic-like odour; bp 240 °C; practically insoluble in water; usually packed under nitrogen to avoid oxidation.

(FLAMMABLE
HARMFUL BY INHALATION)

Avoid breathing vapour. Avoid contact with skin, eyes and clothing.

Toxic effects   Irritates respiratory system and skin; very irritant to eyes; moderately toxic by ingestion.

First aid   *Vapour inhaled:* standard treatment (p. 136).
Affected eyes: standard treatment (p. 136).
Skin contact: standard treatment (p. 136).
If swallowed: standard treatment (p. 137).

Fire hazard   Flash point 40 °C; ignition temperature 200 °C. *Extinguish fire with* foam, dry powder, carbon dioxide or vaporising liquid.

Spillage disposal   Shut off all possible sources of ignition. Wear face-shield or goggles and gloves. Apply non-flammable dispersing agent if available and work to an emulsion with brush and water — run this to waste, diluting greatly with running water. If dispersant not available, absorb on sand, shovel into bucket(s) and transport to safe open area for atmospheric evaporation. Ventilate site of spillage well to evaporate remaining liquid and dispel vapour.

Tricadmium dinitride [12380-95-9]

Hazardous reactions   Explodes on heating and on contact with water (B887–888).

Tricalcium dinitride [12013-82-0]

Hazardous reactions   Spontaneously flammable in air; incandesces in chlorine or bromine vapour (B884).

Tricalcium diphosphide [1305-99-3]

Hazardous reaction Liberates phosphine with water — this ignites spontaneously (B885).

Trichloroacetaldehyde —*see* Chloral

Trichloroacetic acid [76-03-9]

Colourless, hygroscopic crystals; mp 58 °C; soluble in water.

CAUSES SEVERE BURNS

Prevent contact with skin and eyes. RL 1 ppm (5 mg m^{-3}).

Toxic effects Severely irritates the eyes and skin, producing blisters after a latent period. Assumed to cause severe irritation and damage if taken by mouth.

Hazardous reaction Violent reaction with Cu and DMSO (B292–293).

First aid *Affected eyes:* standard treatment (p. 136).
Skin contact: standard treatment (p. 136).
If swallowed: standard treatment (p. 137).

Spillage disposal Wear face-shield or goggles, and gloves. Spread soda ash liberally over the spillage and mop up cautiously with plenty of water — run to waste, diluting greatly with running water.

Trichloroacetyl chloride [76-02-8]

Colourless liquid with acrid, pungent odour; bp 118 °C; reacts with water forming trichloroacetic and hydrochloric acids.

(CAUSES SEVERE BURNS
IRRITATING TO EYES AND RESPIRATORY SYSTEM)

Prevent inhalation of vapour. Prevent contact with skin and eyes.

Toxic effects The vapour irritates the eyes and respiratory system. The liquid burns the eyes and skin. Assumed to cause severe internal irritation and damage if taken by mouth.

First aid *Vapour inhaled:* standard treatment (p. 136).
Affected eyes: standard treatment (p. 136).
Skin contact: standard treatment (p. 136).
If swallowed: standard treatment (p. 137).

Spillage disposal	Instruct others to keep at a safe distance. Wear breathing apparatus and gloves. Spread soda ash liberally over the spillage and mop up cautiously with water — run to waste, diluting greatly with running water.

1,2,4-Trichlorobenzene [120–82–1]

Colourless liquid or solid; mp 17 °C; insoluble in water.

(IRRITATING TO EYES, RESPIRATORY SYSTEM AND SKIN)

Avoid breathing vapour. Avoid contact with skin, eyes and clothing. RL 5 ppm (40 mg m^{-3}).

Toxic effects	These are not well documented, but it has been ascribed moderate acute and chronic systemic toxicity when it is ingested or the vapour inhaled.
First aid	*Vapour inhaled:* standard treatment (p. 136). *Affected eyes:* standard treatment (p. 136). *Skin contact:* standard treatment (p. 136). *If swallowed:* standard treatment (p. 137).
Spillage disposal	Wear face-shield or goggles, and gloves. Mix with sand and transport to a safe, open area for burial. The site of the spillage should be washed thoroughly with water and soap or detergent.

1,1,1-Trichloroethane (methylchloroform) [71–55–6]

Colourless liquid; bp 74 °C; insoluble in water.

HARMFUL BY INHALATION AND IF SWALLOWED

Avoid breathing vapour. Avoid contact with skin, eyes and clothing. CL 350 ppm (1900 mg m^{-3}).

Toxic effects	The vapour may irritate the eyes and respiratory system; it is narcotic in high concentrations. The liquid may irritate the skin; it irritates the eyes without causing serious damage and must be assumed to be harmful if taken by mouth.
Hazardous reactions	Mixture with potassium may explode on light impact; violent decomposition, with evolution of HCl, may occur when it comes into contact with aluminium, magnesium, or their alloys (B241).
First aid	*Vapour inhaled:* standard treatment (p. 136). *Affected eyes:* standard treatment (p. 136). *Skin contact:* standard treatment (p. 136). *If swallowed:* standard treatment (p. 137).

531

Spillage
disposal
Wear face-shield or goggles, and gloves. Apply dispersing agent if available and work to an emulsion with brush and water — run this to waste, diluting greatly with running water. If dispersant not available, absorb on sand, shovel into bucket(s) and transport to safe, open area for atmospheric evaporation. Site of spillage should be washed thoroughly with water and soap or detergent.

RSC *Lab. Haz. Data Sheet No. 1,* 1982, gives extended coverage.

1,1,2-Trichloroethane [79-00-5]

HARMFUL BY INHALATION, SKIN CONTACT
AND IF SWALLOWED.

RL (skin) 10 ppm (45 mg m^{-3}).

Trichloroethylene [79-01-6]

Colourless liquid with sweetish chloroform-like odour; bp 87 °C; immiscible with water.

HARMFUL BY INHALATION AND IF SWALLOWED

Avoid breathing vapour. Avoid contact with skin and eyes. CL 100 ppm (535 mg m^{-3}).

Toxic effects
Inhalation of the vapour may cause headache, dizziness and nausea — with high concentrations, unconsciousness. The vapour and liquid irritate the eyes. Ingestion produces similar effects to inhalation of the vapour.

Hazardous
reactions
Decomposes with strong alkalies with evolution of spontaneously flammable dichloroacetylene; reacts violently with many metals, *e.g.* Al, Ba, Be, Li, Mg, Ti; reacts violently with anhydrous perchloric acid, explosively with N$_2$O$_4$ and with liquid oxygen (when initiated). Corrosion products from hydrolysis during large scale distillation led to exothermic decomposition (B218–220).

First aid
Vapour inhaled: standard treatment (p. 136).
Affected eyes: standard treatment (p. 136).
If swallowed: standard treatment (p. 137).

Spillage
disposal
Wear face-shield or goggles, and gloves. Apply dispersing agent if available and work to an emulsion with brush and water — run this to waste, diluting greatly with running water. If dispersant not available, absorb on sand, shovel into bucket(s) and transport to safe, open area for atmospheric evaporation. Site of spillage should be washed thoroughly with water and soap or detergent.

RSC *Lab. Haz. Data Sheet No. 13,* 1983, gives extended coverage.

Trichloromethanesulphenyl chloride [594-42-3]
(perchloromethyl mercaptan; thiocarbonyl tetrachloride)

Oily, yellow liquid; bp 149 °C with slight decomposition; insoluble in water — hydrolysis is slow.

(VERY TOXIC IF SWALLOWED
IRRITATING TO EYES, RESPIRATORY SYSTEM AND SKIN)

Toxic effects The vapour irritates the eyes and respiratory system. The liquid irritates the skin and eyes and is highly toxic if taken by mouth. TLV 0.1 ppm (0.8 mg m^{-3}).

First aid *Vapour inhaled:* standard treatment (p. 136).
Affected eyes: standard treatment (p. 136).
Skin contact: standard treatment (p. 136).
If swallowed: standard treatment (p. 137).

Spillage disposal Instruct others to keep at a safe distance. Wear breathing apparatus and gloves. Apply dispersing agent if available and work to an emulsion with brush and water — run this to waste, diluting greatly with running water. If dispersant not available, absorb on sand, shovel into bucket(s) and transport to safe, open area for burial. Wash site of spillage thoroughly with water and soap or detergent.

Trichloro(methyl)silane
—*see* **Methyltrichlorosilane**

Trichloronaphthalene [1321-65-9]

TLV 5 mg m^{-3}.

Trichloronitromethane (chloropicrin) [76-06-2]

Colourless liquid with intense penetrating odour; bp 112 °C; almost insoluble in water.

VERY TOXIC BY INHALATION, SKIN CONTACT
AND IF SWALLOWED
IRRITATING TO EYES, RESPIRATORY SYSTEM AND SKIN

Do not breath vapour. Prevent contact with skin and eyes. RL 0.1 ppm (0.7 mg m^{-3}).

Toxic effects The vapour irritates the respiratory system leading in severe cases to bronchitis and recurrent asthmatic attacks through lung damage; it causes nausea and vomiting. The vapour irritates the eyes severely, causing intense lachrymation. The vapour and liquid irritate the skin. If swallowed, the liquid causes vomiting and diarrhoea.

Hazardous reactions Above a critical volume, bulk containers can be shocked into detonation; reacts violently with aniline at 145 °C, and with alcoholic sodium hydroxide; mixture with 3-bromopropyne is shock- and heat-sensitive explosive (B121).

First aid *Vapour inhaled:* standard treatment (p. 136).
Affected eyes: standard treatment (p. 136).
Skin contact: standard treatment (p. 136).
If swallowed: standard treatment (p. 137).

Spillage disposal Instruct others to keep at a safe distance. Wear breathing apparatus and gloves. Apply dispersing agent if available and work to an emulsion with brush and water — run this to waste diluting greatly with running water. If dispersant not available, absorb on sand, shovel into bucket(s) and transport to safe open area for burial. Site of spillage should be washed thoroughly with water and soap or detergent.

2,4,5-and 2,4,6-Trichlorophenols

2,4,5-[95–95–4]
2,4,6-[88–06–2]

Colourless crystals or grey flakes with strong, phenolic odours; mps 67 and 69 °C; practically insoluble in water.

HARMFUL IF SWALLOWED
IRRITATING TO EYES AND SKIN

Avoid breathing dusts. Avoid contact with skin, eyes and clothing.

Toxic effects Inhalation, ingestion or skin absorption of the dust or solid may result in lung, liver or kidney damage; symptoms of poisoning are an increase followed by a decrease in respiratory rate and urinary output, fever, increased bowel action, weakness of movement, collapse and convulsions. Skin contact may cause dermatitis.

First aid *Affected eyes:* standard treatment (p. 136).
Skin contact: standard treatment (p. 136). But see note on phenol p. 135.
If swallowed: standard treatment (p. 137).

Spillage disposal Wear face-shield or goggles, and gloves. Mix with sand and transport to safe, open area for burial. Site of spillage should be washed thoroughly with water and soap or detergent.

1,2,3-Trichloropropane

[96–18–4]

RL 50 ppm (300 mg m^{-3}).

α,α,α-Trichlorotoluene
—*see* Benzylidyne chloride

2,4,6-Trichloro-*s*-triazine [108–77–0]
(cyanuric chloride; cyanuryl chloride)

Colourless crystals with pungent odour; hydrolyses in presence of water, forming hydrochloric acid.

IRRITATING TO EYES, RESPIRATORY SYSTEM AND SKIN
Avoid breathing dust. Avoid contact with skin and eyes.

Toxic effects	The dust irritates the respiratory system, eyes, and skin. Assumed to be irritating and damaging to the alimentary system if taken by mouth.
Hazardous reactions	Violent or explosive reactions with dimethylformamide, dimethyl sulphoxide, 2-ethoxyethanol, methanol or water (B324–326).
First aid	*Dust inhaled:* standard treatment (p. 136). *Affected eyes:* standard treatment (p. 136). *Skin contact:* standard treatment (p. 136). *If swallowed:* standard treatment (p. 137).
Spillage disposal	Wear face-shield or goggles, and gloves. Mop up with plenty of water and run to waste, diluting greatly with running water.

1,1,2-Trichloro-1,2,2-trifluoroethane [76–13–1]

RL 1000 ppm (7600 mg m^{-3}).

Tricopper diphosphide [12134–35–9]

Hazardous reactions	Powder burns in chlorine; mixtures with potassium chlorate explode on impact, and with potassium nitrate on heating; CuP and CuP$_2$ behave similarly (B1043).

Tricresyl phosphate —*see* Tritolyl phosphate

Tricyclo[3.1.0.0^{2,6}]hex-3-ene
—*see* Benzvalene

Triethylaluminium [97-93-8]

REACTS VIOLENTLY WITH WATER
SPONTANEOUSLY FLAMMABLE IN AIR
CAUSES BURNS

Hazardous reactions Ignites in air; reacts explosively with methanol, ethanol, propan-2-ol, vigorously with 2-methylpropan-2-ol; a mixture with *N,N*-dimethylformamide exploded when heated; complex with CCl_4 exploded (B634-635).

Triethylamine [121-44-8]

Colourless liquid with stong ammoniacal odour; bp 89 °C; miscible with water.

HIGHLY FLAMMABLE
IRRITATING TO EYES, RESPIRATORY SYSTEM AND SKIN

Avoid breathing vapour. Avoid contact with skin and eyes. RL 10 ppm (40 mg m^{-3}).

Toxic effects The vapour irritates the eyes and respiratory system. The liquid irritates the eyes. Assumed to be irritant and poisonous if taken by mouth.

Hazardous reaction Complex with N_2O_4, containing excess of latter, exploded below 0 °C when free of solvent (B638).

First aid *Vapour inhaled:* standard treatment (p. 136).
Affected eyes: standard treatment (p. 136).
Skin contact: standard treatment (p. 136).
If swallowed: standard treatment (p. 137).

Fire hazard Flash point −7 °C; explosive limits 1.2–8.0%. *Extinguish fire with* water spray, dry powder, carbon dioxide or vaporising liquid.

Spillage disposal Shut off all possible sources of ignition. Instruct others to keep at a safe distance. Wear breathing apparatus and gloves. Mop up with plenty of water and run to waste, diluting greatly with running water. Ventilate area well to evaporate remaining liquid and dispel vapour.

RSC *Lab. Haz. Data Sheet No. 27*, 1984, gives extended coverage.

Triethylantimony [617-85-6]

Hazardous reaction Inflames in air (B639).

Triethylarsine [617-75-4]

Hazardous reaction Inflames in air (B635).

Triethylbismuth [617–77–6]

Hazardous reactions   Ignites in air, and explodes at about 150 °C (B636)

Triethylborane [97–94–9]

Hazardous reaction   Ignites in air (B636).

Triethylenetetramine [112–24–3]

Colourless to yellowish liquid; bp 278 °C; miscible with water.

> HARMFUL IN CONTACT WITH SKIN
> CAUSES BURNS
> MAY CAUSE SENSITISATION BY SKIN CONTACT

Avoid breathing vapour. Avoid contact with skin and eyes.

Toxic effects   The vapour irritates the eyes and respiratory system. The liquid burns the skin and eyes. Assumed to be irritant and poisonous if taken by mouth.

First aid   *Vapour inhaled:* standard treatment (p. 136).
Affected skin: standard treatment (p. 136).
Skin contact: standard treatment (p. 136).
If swallowed: standard treatment (p. 137).

Spillage disposal   Wear face-shield or goggles, and gloves. Mop up with plenty of water and run to waste, diluting greatly with running water.

Triethylgallium [1115–99–7]

Hazardous reaction   Ignites in air (B637).

Triethylphosphine [554–70–1]

Hazardous reaction   Explosive product by reaction of oxygen at low temperature (B639).

Triethynylaluminium [61204–16–8]

Hazardous reactions	Residue from sublimation of dioxan complex is explosive; trimethylamine complex may also explode on sublimation (B547).

Triethynylantimony [687–81–0]

Hazardous reaction	Explodes on strong friction (B554).

Triethynylarsine [687–78–5]

Hazardous reaction	Explodes on strong friction (B547).

1,3,5-Triethynylbenzene [17814–74–3]

Hazardous reactions	Exploded on rapid heating and compression (B794).

Triethynylphosphine [687–80–9]

Hazardous reactions	Explodes on strong friction and may explode spontaneously on standing (B553).

Trifluoroacetic acid [76–05–1]
and anhydride [407–25–0]

Colourless liquids with pungent odour; bp 72 °C and 40 °C respectively; the acid is miscible with water, and the anhydride reacts with water forming the acid.

HARMFUL BY INHALATION
CAUSES SEVERE BURNS

Avoid breathing vapour. Prevent contact with skin and eyes.

Toxic effects	The vapours irritate the eyes and respiratory system. The liquids burn the eyes and quickly penetrate the skin to cause deep-seated burns. Assumed to cause severe burning and damage if taken by mouth. There is no reported toxic effect due to the presence of the fluorine as in the case of highly toxic fluoroacetic acid.

Hazardous reactions	The acid reacts violently with LiAlH$_4$, and the anhydride explosively with dimethyl sulphoxide (B222, 398).
First aid	*Vapour inhaled:* standard treatment (p. 136).
	Affected eyes: standard treatment (p. 136).
	Skin contact: standard treatment (p. 136).
	If swallowed: standard treatment (p. 137).
Spillage disposal	Instruct others to keep at a safe distance. Wear breathing apparatus and gloves. Spread soda ash liberally over the spillage and mop up cautiously with water — run to waste, diluting greatly with running water.

Trifluoromethanesulphonic acid [1493-13-6]

Hazardous reactions	As the strongest acid known, it can exert powerful catalytic effects, *e.g.*, on Friedel-Craft reactions (B134-135).

Trifluoroperoxyacetic acid [359-48-8]

(peroxytrifluoroacetic acid)

Hazardous reactions	An extremely powerful oxidising agent that must be used with great care (B222).

3,3,3-Trifluoropropyne [661-54-1]

Hazardous reaction	Liable to explode (B331).

Tri-isobutylaluminium [100-99-2]

Hazardous reaction	Powerful reductant supplied in hydrocarbon solvent; undiluted material ignites in air (B815).

Tri-isopropylphosphine [6476-36-4]

Hazardous reactions	Reacts rather vigorously with most peroxides, ozonides, *N*-oxides, and chloroform (B758).

Trilead dinitride [58572-21-7]

Hazardous reaction	Decomposes explosively during vacuum degassing (B1303).

Trimagnesium diphosphide [12057-74-8]

Hazardous reactions	Ignites on heating in chlorine, or in bromine or iodine vapours; reaction with nitric acid causes incandescence; with water, phosphine is evolved and may ignite (B1273).

Trimercury dinitride [12136-15-1]

Hazardous reactions	Explodes on friction, impact or in contact with sulphuric acid (B1224).

Trimercury tetraphosphide [12397-29-4]

Hazardous reactions	Ignites in chlorine or when warmed in air; mixture with potassium chlorate explodes on impact (B1017).

Trimethylaluminium [75-24-1]

REACTS VIOLENTLY WITH WATER
SPONTANEOUSLY FLAMMABLE IN AIR
CAUSES BURNS

Hazardous reaction	Extremely pyrophoric (B383–384).

Trimethylamine and solutions [75-50-3]

Colourless gas with fishy odour that clings to clothes; bp 3 °C; available in liquefied form in cylinders and also in aqueous and ethanolic solutions.

EXTREMELY FLAMMABLE LIQUEFIED GAS
IRRITATING TO EYES AND RESPIRATORY SYSTEM

Avoid breathing vapour. Avoid contact with skin and eyes. RL 10 ppm (24 mg m^{-3}).

Toxic effects	The vapour irritates the mucous membranes and respiratory system; in high concentrations it may affect the nervous system. The vapour and solutions irritate the eyes. The solutions may irritate the skin. Assumed to be poisonous if taken by mouth.
First aid	*Vapour inhaled:* standard treatment (p. 136). *Affected eyes:* standard treatment (p. 136). *Skin contact:* standard treatment (p. 136). *If swallowed:* standard treatment (p. 137).

540

Fire hazard (a) Gas: explosive limits 2.0–11.6% ignition temp. 190 °C. Since the gas is supplied in a cylinder, turning off the valve will reduce any fire involving it; if possible, cylinders should be removed quickly from an area in which fire has developed. (b) Solutions in water and ethanol: *extinguish fire with* spray, foam, dry powder, carbon dioxide or vaporising liquid.

Disposal Surplus gas or leaking cylinder can be vented slowly into a water-fed scrubbing tower or column in a fume cupboard, or into a cupboard served by such a tower. Solutions may be run to waste, diluting greatly with running water.

Trimethylamine oxide [1184–78–7]

Hazardous preparation The oxide exploded during concentration (B388).

Trimethylarsine [593–88–4]

Hazardous reactions Inflames in air; reaction with halogens is violent (B384).

Trimethylbenzenes 1,2,3- [526–73–8] 1,2,4- [95–63–6]
1,3,5- [108–67–8]

RL (all isomers) 25 ppm (125 mg m^{-3}).

Trimethylbismuth [593–91–9]

Hazardous reaction Ignites in air (B385).

Trimethylborane [593–90–8]

Hazardous reaction Ignites in air (B384).

Trimethylchlorosilane [chloro(trimethyl)silane] [75–77–4]

Colourless, volatile, fuming liquid with pungent odour; bp 57 °C; reacts violently with water.

> HIGHLY FLAMMABLE
> CAUSES BURNS
> IRRITATING TO EYES AND RESPIRATORY SYSTEM

Prevent inhalation of vapour. Prevent contact with skin, eyes and clothing.

Toxic effects	The vapour and fumes are strongly irritant to the eyes, skin and respiratory system. The liquid burns the skin and eyes and will cause severe damage if taken internally.
First aid	*Vapour inhaled:* standard treatment (p. 136). *Affected eyes:* standard treatment (p. 136). *Skin contact:* standard treatment (p. 136). *If swallowed:* standard treatment (p. 137).
Fire hazard	Flash point −18 °C. *Extinguish fire with* dry sand, dry powder or carbon dioxide.
Spillage disposal	Instruct others to keep at a safe distance. Wear breathing apparatus and gloves. Spread soda ash liberally over the spillage and mop up cautiously with water — run to waste, diluting greatly with running water.

Trimethylgallium [1445-79-0]

Hazardous reactions	Ignites in air and reacts violently with water (B386).

2,4,4-Trimethylpentene (di-isobutylene) [107-39-1]

Colourless liquid; bp 102 °C; insoluble in water.

HIGHLY FLAMMABLE

Avoid breathing vapour. Avoid contact with skin and eyes.

Toxic effects	The vapour is slightly irritant at low concentrations, more so and narcotic at high concentrations. The liquid irritates the eyes and may irritate the skin.
First aid	*Vapour inhaled:* standard treatment (p. 136). *Affected eyes:* standard treatment (p. 136). *Skin contact:* standard treatment (p. 136). *If swallowed:* standard treatment (p. 137).
Fire hazard	Flash point below 2 °C. *Extinguish fire with* foam, dry powder, carbon dioxide or vaporising liquids.
Spillage disposal	Shut off all possible sources of ignition. Wear face-shield or goggles, and gloves. Apply non-flammable dispersing agent if available and work to an emulsion with brush and water — run this to waste, diluting greatly with running water. If dispersant not available, absorb on sand, shovel into bucket(s) and transport to safe open area for atmospheric evaporation. Ventilate site of spillage well to evaporate remaining liquid and dispel vapour.

Trimethyl phosphate [512-56-1]

Hazardous reaction   Distillation residue exploded (B389).

Trimethylphosphine [594-09-2]

Hazardous reaction   May ignite in air (B370).

Trimethyl phosphite [121-45-9]

TLV 2 ppm (10 mg m^{-3}).

Trimethylthallium [3003-15-4]

Hazardous reactions   Liable to explode above 90 °C; ignites in air (B390).

2,4,6-Trimethyl-1,3,5-trioxane
—*see* Paraldehyde

Trinitroacetonitrile [630-72-8]

Hazardous reaction   Explodes if heated quickly to 220 °C (B317).

2,2,2-Trinitroethanol [918-54-7]

Hazardous reaction   Shock-sensitive explosive which has exploded during distillation (B251).

Trinitromethane (nitroform) [517-25-9]

Hazardous reactions   Exploded during distillation; exploded in mixture with an impure ketone; frozen mixtures of trinitromethane and propan-2-ol exploded during thawing (B137-138).

2,4,6–Trinitrotoluene (TNT)   [118–96–7]

(RISK OF EXPLOSION BY SHOCK, FRICTION, FIRE
OR OTHER SOURCES OF IGNITION
TOXIC BY INHALATION, IN CONTACT WITH SKIN
AND IF SWALLOWED
DANGER OF CUMULATIVE EFFECTS)

Material now supplied wetted with not less than 30 wt% water. Risk phrases applicable to dry material. RL (skin) 0.5 mg m^{-3}.

Hazardous reactions The explosion temperature of TNT was reduced by addition of 1% red lead, sodium carbonate or potassium hydroxide (B661–662).

Triphenylaluminium   [841–76–9]

Hazardous reaction Evolves heat and sparks on contact with water (B846).

Triphenyl phosphate   [115–86–6]

RL 3 mg m^{-3}.

Triphenyltin hydroperoxide   [4150–34–9]

Hazardous reaction Explodes reproducibly at 75 °C (B847).

Trisilane   [7783–26–8]

Hazardous reaction Ignites or explodes in air or oxygen (B1213–1214).

Trisilver nitride   [20737–02–4]

Hazardous reaction Sensitive explosive when dry (B19).

Trisilylamine   [13862–16–3]

Hazardous reaction Ignites in air (B1214).

544

Trithorium tetranitride   [12033–90–8]

Hazardous reaction   Burns incandescently in air, vividly in oxygen (B1309).

Tritolyl phosphate (tricresyl phosphate)   [78–30–8]

Pale brown, almost colourless, liquid; bp 410 °C; immiscible with water.

> TOXIC BY INHALATION, IN CONTACT WITH SKIN
> AND IF SWALLOWED
> DANGER OF VERY SERIOUS IRREVERSIBLE EFFECTS

Avoid contact with skin and eyes. RL (*o*-isomer) 0.1 mg m^{-3}.

Toxic effects   When absorbed through the skin or ingested, tritolyl phosphate may cause serious damage to the nervous and digestive systems. Poisoning may show itself in degrees of muscular pain and paralysis.

First aid   *Affected eyes:* standard treatment (p. 136).
Skin contact: standard treatment (p. 136).
If swallowed: standard treatment (p. 137).

Spillage disposal   Wear face-shield or goggles, and gloves. Apply dispersing agent if available and work to an emulsion with brush and water — run this to waste, diluting greatly with running water. If dispersant not available, absorb on sand, shovel into bucket(s) and transport to safe, open area for burial. Site of spillage should be washed thoroughly with water and soap or detergent.

Trivinylbismuth   [65313–35–1]

Hazardous reaction   Ignites in air (B608).

Tungsten and compounds   [7440–33–7]

RL (as W) soluble 1 mg m^{-3}, insoluble 5 mg m^{-3}.

Uranium (metal)   [7440–61–1]

Hazardous reactions   Storage of foil in closed containers in presence of air and moisture may produce a pyrophoric surface; the metal incandesces in ammonia at dull red heat; ignites in fluorine at ambient temperature, in chlorine at 150–180 °C, in bromine vapour at 210–240 °C and in iodine vapour at 260 °C; incandesces in sulphur vapour and with selenium; ignites in CO_2 at 750–800 °C; explosion when CCl_4 extinguisher used on U fire. (B1406–1407).

Uranium compounds

The commonest uranium compounds encountered in the laboratory are the acetate, nitrate and double zinc and magnesium acetates, all of which are yellow crystalline salts soluble in water. Uranium hexafluoride is a colourless or pale yellow crystalline solid which sublimes readily at about 56 °C. Users in the United Kingdom who stock appreciable quantities of uranium compounds are advised to ascertain their responsibilities under the 1985 legislation for unsealed sources (see Chapter 9).

> VERY TOXIC BY INHALATION AND IF SWALLOWED
> DANGER OF CUMULATIVE EFFECTS

Avoid breathing dust or vapour. RL (as U) 0.2 mg m^{-3}. (*See* Ch 6 p. 108.)

Toxic effects	The dust may irritate the lungs and cause retention of uranium in the body with subsequent damage to the kidneys. The vapour of the hexachloride irritates the respiratory system and may injure the kidneys. Assumed to cause internal damage if taken by mouth.
First aid	*Dust or vapour inhaled:* standard treatment (p. 136). *Affected eyes:* standard treatment (p. 136). *If swallowed:* standard treatment (p. 137).
Spillage disposal	Small quantities of soluble uranium salts can be dissolved in a large volume of water and run to waste, diluting greatly with running water. Small amounts of insoluble compounds can be mixed with a large excess of sand and buried in a safe, open area.

Uranium dicarbide   [12071–33–9]

Hazardous reactions	Ignites on grinding in a mortar or on heating in air to 400 °C; incandescent reactions with fluorine, chlorine and bromine; violent reaction with hot water (B322).

Uranium hexafluoride   [7783–81–5]

Hazardous reactions	Reacts very vigorously with benzene, toluene or xylene, violently with water or ethanol (B1078–1079).

Uranium(III) hydride   [13598–56–6]

Hazardous reaction	Dry powdered hydride ignites in air (B1188).

Uranium(IV) hydride [51680–55–8]

Hazardous reaction Finely divided hydride ignites on contact with oxygen (B1204–1205).

Uranyl diperchlorate [13093–00–0]

Hazardous reaction Attempted recrystallisation from ethanol caused explosion (B984).

Valeraldehyde (pentanal) [110–62–3]

TLV 50 ppm (175 mg m^{-3}).

Vanadium compounds oxide [1314–62–1]

Vanadium pentoxide is a red-brown to dark brown powder. The other compounds most commonly encountered are the sodium and ammonium vanadates, which are colourless, crystalline, and soluble in water.

HARMFUL BY INHALATION

Avoid breathing dust. TLV pentoxide dust 0.5 (as V) mg m^{-3}, pentoxide fume 0.05 mg m^{-3}.

Toxic effects The dust or fume of vanadium pentoxide causes irritation of the respiratory system, chest constriction, coughing, and the tongue assumes a blackish-green colour. The dust or fume irritates the eyes and may cause conjunctivitis. If taken by mouth vanadium compounds cause vomiting, excessive salivation and diarrhoea; large doses may damage the nervous system, causing drowsiness, convulsions and unconsciousness.

First aid *Dust or fume inhaled:* standard treatment (p. 136).
Affected eyes: standard treatment (p. 136).
If swallowed: standard treatment (p. 137).

Spillage disposal Soluble vanadium compounds can be dissolved in water and run to waste, diluting greatly with running water. Insoluble compounds can be mixed with moist sand and buried in a safe, open area.

Vanadium trichloride [7718–98–1]

Hazardous reaction Reaction with Grignard reagents is almost explosively violent under some conditions (B1002).

547

Vanadyl triperchlorate [67632–69–3]

Hazardous reactions   Explodes above 80 °C, and ignites many organic solvents on contact (B998).

Vinyl acetate [108–05–4]

Colourless liquid; bp 73 °C; 1 g dissolves in 50 cm^3 water at 20 °C.

HIGHLY FLAMMABLE

Avoid breathing vapour. Avoid contact with skin and eyes. RL 10 ppm (30 mg m^{-3}).

Toxic effects   The vapour may be narcotic when inhaled in high concentrations. The liquid irritates the eyes and may irritate the skin by its defatting action; it is assumed to be harmful if taken by mouth.

Hazardous reactions   Polymerisation may accelerate to dangerous extent; the vapour reacts vigorously in contact with silica gel or alumina; unstabilised polymer exposed to oxygen at 50 °C gave interpolymeric peroxide which was explosive (B430–431).

First aid   *Vapour inhaled in high concentrations:* standard treatment (p. 136).
Affected eyes: standard treatment (p. 136).
Skin contact: standard treatment (p. 136).
If swallowed: standard treatment (p. 137).

Fire hazard   Flash point −8 °C; explosive limits 2.6–13.4%; ignition temp 427 °C. *Extinguish fire with* foam, dry powder, carbon dioxide or vaporising liquid.

Spillage disposal   Shut off all possible sources of ignition. Instruct others to keep at a safe distance. Wear breathing apparatus and gloves. Apply non-flammable dispersing agent if available and work to an emulsion with brush and water — run this to waste, diluting greatly with running water. If dispersant not available, absorb on sand, shovel into bucket(s) and transport to safe, open area for atmospheric evaporation. Site of spillage should be washed thoroughly with water and soap or detergent.

Vinyl acetate ozonide [101672–23–5]

Hazardous reaction   Explosive when dry (B437).

Vinylbenzene —*see* Styrene

Vinyl bromide (bromoethylene) [593–60–2]

Colourless liquid or gas; bp 16 °C; insoluble in water.

EXTREMELY FLAMMABLE LIQUEFIED GAS

Prevent inhalation of vapour. Prevent contact with skin and eyes. Suspected carcinogen. RL 5 ppm (20 mg m^{-3}).

Toxic effects	Inhalation of vapour in high concentrations may produce dizziness and narcosis. The liquid irritates the eyes and may irritate the skin by its defatting action; it is assumed to be harmful if taken by mouth. In view of the recent observation that vinyl chloride (*see* below) can cause a cancer of the liver, it must be assumed that vinyl bromide is likely to behave in a similar manner.
First aid	*Vapour inhaled in high concentrations:* standard treatment (p. 136). *Affected eyes:* standard treatment (p. 136). *Skin contact:* standard treatment (p. 136). *If swallowed:* standard treatment (p. 137).
Fire hazard	Flash point below −8 °C. *Extinguish fire with* water spray, foam, dry powder, carbon dioxide or vaporising liquid.
Spillage disposal	Shut off all possible sources of ignition. Instruct others to keep at a safe distance. Wear breathing apparatus and gloves. Apply non-flammable dispersing agent if available and work to an emulsion with brush and water — run this to waste, diluting greatly with running water. If dispersant not available, absorb on sand, shovel into bucket(s) and transport to safe, open area for atmospheric evaporation. Site of spillage should be ventilated thoroughly.

Vinyl chloride (chloroethylene) [75–01–4]

Colourless gas with pleasant, sweet odour; bp −14 °C; slightly soluble in water.

EXTREMELY FLAMMABLE LIQUEFIED GAS
DANGER OF VERY SERIOUS IRREVERSIBLE EFFECTS

Prevent inhalation of gas. Prevent contact with liquid. Human carcinogen. CL 3 ppm (annual figure for manufacturing operations). For other exposures use CL 7 ppm.

Toxic effects	Inhalation of vapour in high concentrations produces dizziness and narcosis. The liquid may irritate and burn the skin, the latter due to its freezing action. **Chronic effects.** Exposure to the vapour may lead to loss of feeling in hands and feet. More seriously, it is now proven that exposure to working atmospheres of vinyl chloride may lead to a rare liver cancer, up to 20 years after initial exposure.
Hazardous reaction	Formation of unstable polyperoxide may occur (B238–239).

First aid	*Vapour inhaled in high concentrations:* standard treatment (p. 136).
	Skin contact with liquid: standard treatment (p. 136).
Fire hazard	Flash point −78 °C; explosive limits 4–22%; ignition temp. 472 °C. Since the gas is supplied in a cylinder, turning off the valve will reduce any fire involving it; if possible, cylinders should be removed quickly from an area in which a fire has developed.
Spillage disposal	Surplus gas or leaking cylinder can be vented slowly to air in a safe, open area or gas burnt off through a suitable burner in a fume cupboard.

Vinylcyclohexene dioxide —*see* 1-Epoxyethyl-3,4-epoxycyclohexane

Vinylethylene —*see* Buta-1,3-diene

Vinyl fluoride (fluoroethylene) [75–02–5]

Hazardous reaction	Explosive when mixed with air (explosive limits 2.6–22%).

Vinylidene chloride (1,1-dichloroethylene) [75–35–4]

Colourless, volatile liquid with chloroform-like odour; bp 32 °C; almost insoluble in water.

EXTREMELY FLAMMABLE
HARMFUL BY INHALATION
POSSIBLE RISK OF IRREVERSIBLE EFFECTS

Avoid breathing vapour. Avoid contact with skin and eyes. RL 10 ppm (40 mg m^{-3}).

Toxic effects	Inhalation of vapour may cause drowsiness and anaesthesia; maximum safe working concentration is now 10 ppm. The frequent inhalation of small quantities can result in chronic effects which take the form of liver and kidney damage. The liquid irritates the skin and eyes and must be assumed to be poisonous if taken by mouth.
Hazardous reactions	Rapidly absorbs oxygen from air forming explosive peroxide; reaction products formed with ozone are particularly dangerous (B231–232).
First aid	*Vapour inhaled:* standard treatment (p. 136).
	Affected eyes: standard treatment (p. 136).
	Skin contact: standard treatment (p. 136).
	If swallowed: standard treatment (p. 137).

Fire hazard   Flash point −15 °C (open cup); explosive limits 5.6–11.4%; ignition temp. 458 °C. *Extinguish fire with* water spray, foam, dry powder, carbon dioxide or vaporising liquid.

Spillage   Shut off all possible sources of ignition. Instruct others to keep at a safe
disposal   distance. Wear breathing apparatus and gloves. Apply non-flammable dispersing agent if available and work to an emulsion with brush and water — run this to waste, diluting greatly with running water. If dispersant not available, absorb on sand, shovel into bucket(s) and transport to safe, open area for atmospheric evaporation. Site of spillage should be ventilated thoroughly.

Vinyllithium   [917-57-7]

Hazardous   Violently pyrophoric when freshly prepared (B244).
reaction

2-Vinylpyridine   [100-69-6]

Colourless liquid rapidly darkening to red-brown due to polymerisation; bp 158 °C; t-butylcatechol is commonly added to minimise polymer formation; 2.7 g dissolves in 100 g water at 20 °C.

(FLAMMABLE
HARMFUL BY INHALATION AND SKIN CONTACT)

Avoid breathing vapour. Avoid contact with skin, eyes and clothing.

Toxic effects   These are not well documented, but animal experiments suggest high toxicity. The vapour irritates the skin, eyes and respiratory system. The liquid irritates the skin and eyes and must be assumed to be irritant and injurious if taken by mouth.

Hazardous   Polymerisation is sometimes spontaneous and may become violent (B
reaction   674).

First aid   *Vapour inhaled:* standard treatment (p. 136).
                     Affected eyes: standard treatment (p. 136).
                     Skin contact: standard treatment (p. 136).
                     If swallowed: standard treatment (p. 137).

Fire hazard   Flash point 42 °C. *Extinguish fire with* foam, dry powder, carbon dioxide or vaporising liquid.

Spillage   Shut off all possible sources of ignition. Instruct others to keep at a safe
disposal   distance. Wear breathing apparatus and gloves. Apply non-flammable dispersing agent if available and work to an emulsion with brush and water — run this to waste, diluting greatly with running water. If dispersant not available, absorb on sand, shovel into bucket(s) and transport to safe, open area for burial. Site of spillage should be washed thoroughly with water and soap or detergent.

Vinyltoluene —*see* 2-Methylstyrene

Xenon compounds Review (B1708-1709).

Xylenes (xylols) *o*- [95–47–6] *m*- [108–38–3] *p*- [106–42–3]

Colourless liquids; bps 144 °C (*o*-), 139 °C (*m*-) and 138 °C (*p*-); immiscible with water.

FLAMMABLE
HARMFUL BY INHALATION

Avoid breathing vapour. Avoid contact with skin and eyes. RL (all isomers) 100 ppm (435 mg m^{-3}).

Toxic effects	Inhalation of the vapour may cause dizziness, headache, nausea and mental confusion. The vapour and liquid irritate the eyes and mucous membranes. Absorption through the skin and ingestion would cause poisoning. **Chronic effects** If the xylene contains benzene as an impurity, repeated breathing of vapour over long periods may cause blood disease. Prolonged skin contact may cause dermatitis.
Hazardous reactions	Aerobic and nitric acid oxidations of *p*-xylene to terephthalic acid both carry special hazards (B713–714).
First aid	*Vapour inhaled:* standard treatment (p. 136). *Affected eyes:* standard treatment (p. 136). *Skin contact:* standard treatment (p. 136). *If swallowed:* standard treatment (p. 137).
Fire hazard	Flash points 17 °C (*o*-) and 25 °C (*m*- and *p*-); explosive limits approximately 1–7%; ignition temperatures 464 °C (*o*-), 528 °C (*m*-) and 529 °C (*p*-). *Extinguish fire with* foam, dry powder, carbon dioxide or vaporising liquid.
Spillage disposal	Shut off all possible sources of ignition. Wear face-shield or goggles, and gloves. Apply non-flammable dispersing agent if available and work to an emulsion with brush and water — run this to waste, diluting greatly with running water. If dispersant not available, absorb on sand, shovel into bucket(s) and transport to safe, open area for atmospheric evaporation or burial. Ventilate site of spillage well to evaporate remaining liquid and dispel vapour.

Xylenols   *mixo*- [1300–71–6]

With the exception of 2,4-xylenol, which is often encountered as a yellow-brown liquid, the commoner xylenols are colourless crystalline solids, slightly soluble in water.

TOXIC IN CONTACT WITH SKIN AND IF SWALLOWED
CAUSES BURNS

Avoid breathing vapour. Avoid contact with skin and eyes.

Toxic effects   The vapour of heated xylenols is irritant to the respiratory system. They irritate or burn the eyes and skin severely. Considerable absorption through the skin or ingestion may cause headache, dizziness, nausea, vomiting, stomach pain, exhaustion and coma. **Chronic effects** Repeated inhalation or absorption of small amounts may result in damage to the liver or kidneys.

First aid   *Vapour inhaled:* standard treatment (p. 136).
Affected eyes: standard treatment (p. 136).
Skin contact: standard treatment (p. 136). But see note on phenol p. 135.
If swallowed: standard treatment (p. 137).

Spillage   Wear face-shield or goggles, and gloves. Mix with sand and transport to a
disposal   safe, open area for burial. Site of spillage should be washed thoroughly with water and soap or detergent.

Xylidines   *mixo-* [1300-73-8]

Most of the common xylidines are red to dark-brown liquids (3,4-xylidine consists of pale brown crystals); insoluble in water.

TOXIC BY INHALATION, IN CONTACT WITH SKIN
AND IF SWALLOWED
DANGER OF CUMULATIVE EFFECTS

Avoid breathing vapour. Avoid contact with skin and eyes. RL (all isomers) (skin) 2 ppm (10 mg m^{-3}).

Toxic effects   Excessive breathing of the vapour or absorption through the skin may cause headache, drowsiness, cyanosis, mental confusion and, in severe cases, convulsions. The xylidines are dangerous to the eyes and the above effects may also be experienced if they are taken by mouth. **Chronic effects** Prolonged exposure to the vapour or slight skin exposures over a period may affect the nervous system and the blood, causing fatigue, loss of appetite, headache and dizziness.

Hazardous   Ignition on contact with fuming HNO_3 (B1106).
reaction

First aid   *Vapour inhaled:* standard treatment (p. 136).
Affected eyes: standard treatment (p. 136).
Skin contact: standard treatment (p. 136).
If swallowed: standard treatment (p. 137).

Spillage disposal   Wear face-shield or goggles, and gloves. Mix with sand and shovel mixture into glass, enamel or polythene vessel for dispersion in an excess of dilute hydrochloric acid (1 volume of concentrated acid diluted with 2 volumes of water). Allow to stand, with occasional stirring, for 24 hours and then run acid extract to waste, diluting greatly with running water and washing the sand. The sand can be disposed of as normal waste.

Yttrium   [7440-65-5]

RL 1 mg m^{-3}.

Zinc (metal)   [7440-66-6]

Hazardous reactions   A Zn dust explosion occurred during sieving of hot dry material; a mixture of As_2O_3 with an excess of Zn filings exploded on heating; Zn powder reacts with CS_2 with incandescence; a paste of Zn powder and CCl_4 will burn after ignition; zinc forms pyrophoric Grignard-type compounds with bromoethane and is expected to react explosively with chloromethane; warm Zn powder incandesces in F_2 and Zn foil ignites in cold chlorine if traces of moisture are present; the metal reacts explosively when heated with anhydrous $MnCl_2$, and violently with KO_2, TiO_2 and ZnO_2; zinc dust residues from reduction of nitrobenzene are often pyrophoric when dry (presence of $ZnCl_2$ promotes pyrophoricity); As, Se and Te all react on heating with Zn powder with incandescence; interaction of Zn powder and sulphur on heating considered too violent as a school experiment; mixtures of Zn dust with potassium chlorate or ammonium nitrate are liable to explode on impact, *etc*; a scoop contaminated with flake NaOH ignited Zn dust; in contact with air and limited amounts of water, zinc dust will generate heat and become incandescent; moisture ignites mixtures with NH_4NO_3/NH_4Cl; exothermic reaction of Zn dust/10% NaOH in air, or dust with hexachloroethane/ethanol (B1409–1412).

Zinc chloride   [7646-85-7]

White deliquescent powder or lumps; mp $\sim$290 °C; soluble in water.

CAUSES BURNS

Avoid inhalation of dust or fumes, and contact with skin. RL (fume) 1 mg m^{-3}.

Toxic effects   A moderate irritant of skin or mucous membranes; major exposure may lead to dermatitis, asthma and inflammation of the cornea.

First aid   *Fumes inhaled:* standard treatment (p. 136).
Affected eyes: standard treatment (p. 136).
Skin contact: standard treatment (p. 136).
If swallowed: standard treatment (p. 137).

Spillage   Wear face mask or goggles, and gloves. Mop up with plenty of water and run
disposal   to waste, diluting greatly with running water.

Zinc chromate [13530-65-9]

Suspected carcinogen RL (as Cr) 0.05 mg m^{-3}.

Zinc dihydrazide [25546-98-9]

Hazardous   Explodes at 70 °C (B1004).
reaction

Zinc oxide [1314-13-2]

RL (fume) 5 mg m^{-3}.

Zinc permanganate [23414-72-4]

Hazardous   Reactions of solid zinc permanganate with organic compounds are more
reactions   violent than with $KMnO_4$, but may be moderated by supporting the oxidant
on silica (B1277).

Zinc peroxide [1314-22-3]

Hazardous   Hydrated peroxide explodes at 212 °C; mixtures with Al or Zn powders burn
reactions   brilliantly (B1359).

Zirconium (metal) [7440-67-7]

Hazardous   Pyrophoric hazards of Zr powder are well documented; mixture of Zr
reactions   powder and CCl_4 exploded on heating, or spongy Zr cold; reacts vigorously
with potassium chlorate or nitrate or with CuO or lead oxides; reacts
explosively with alkali metal hydroxides or carbonates on heating; hydrated
sodium tetraborate and the alkali metal chromates, dichromates, molyb-
dates, sulphates or tungstates react violently or explosively with the metal;
Zr powder, damp with 5–10% water, may ignite; although water is used to
prevent ignition, the powder, once ignited, will burn under water more
violently than in air ; metal or dust will burn or explode in CO_2 or CO_2—N_2
mixtures (B1412–1414).

Zirconium and compounds

RL (as Zr) 5 mg m^{-3}.

Zirconium dicarbide [12070–14–3]

Hazardous reactions Ignites in cold fluorine, in chlorine at 250 °C, bromine at 300 °C and iodine at 400 °C (B322).

Zirconium dichloride [13762–26–0]

Hazardous reaction Ignites in air when warm (B987).

Zirconium tetrachloride [10026–11–6]

Hazardous reaction Ignited lithium metal strip (B1007).

Chapter 9

Precautions Against Radiations

Introduction

For convenience, radiations are divided into two broad groups – ionising and non-ionising. The first group can cause electrons to be dislodged from the atoms of which they form part as a result of the passage of radiation through matter, and include x-rays, α, β, and γ rays, neutrons, electrons and positrons. Non-ionising radiations cannot do this, but may nevertheless be hazardous, and they include ultraviolet and infrared radiations, ultrasonics, light (e.g., from lasers) and microwaves. This chapter deals, in the light of the Ionising Radiations Regulations (1985)[1], with safety during the kinds of work with these radiations likely to be encountered in chemical laboratories. Precautions against x-rays are covered briefly, but those against radioactive materials are dealt with more fully. No attempt is made to discuss the precautions necessary when large quantities of highly radioactive materials are handled as this gives rise to many special problems and also may well involve the construction of special laboratories on which expert advice should be sought. It also covers, but more briefly, the kinds of hazard which can arise from non-ionising radiations in laboratories, for which there is at present no legislation in the United Kingdom except the general provisions of the Health and Safety at Work, etc. Act (1974).[2] An indication is given as to where more detailed technical guidance can be found.

Ionising Radiations

Characteristics

The characteristics of the main types of ionising radiation are as follows:

X-Rays. These are electromagnetic rays, like light, but each photon has a much shorter wavelength, *i.e.*, a higher energy. They are produced outside atomic nuclei whenever moving electrons, or other charged particles, are stopped, or when in an atom their energy changes sufficiently. In practice, they are generated by accelerating electrons in a vacuum tube (the x-ray tube) with a high voltage, and then stopping them suddenly at the 'target', when most of their energy is lost as heat but some appears as x-rays. X-Rays are physically identical with γ-rays (*see below*) except in their mode of production, but because of the great range of voltage that can be used for accelerating the electrons, they may be softer (*i.e.*, of lower energy) or harder than any known γ-rays. x-Rays in the laboratory are usually generated at between 10 kV and 100 kV.

α-Rays. These are streams of α-particles, each the nucleus of a helium atom travelling at a high speed. An α-particle is emitted by certain radioactive nuclides, especially among the heavy elements, occasionally accompanied by a γ-ray. The energies of α-rays usually reach several MeV, and they ionise so intensely that they are easily stopped, even by a sheet of paper.

β-Rays. These are electrons emitted by many radioactive nuclides, often in conjunction with γ-radiation. A typical β-emitter is radioactive phosphorus (^{32}P) which emits β-particles with a range of energies from zero to 1.7 MeV, but no γ-radiation. These β-rays are much more penetrating than α-rays, and at 1.7 MeV will require some 8 mm of water or tissue to stop them completely.

γ-Rays. These are physically the same as x-rays, but emitted by the nuclei of certain radioactive nuclides, each yielding γ-rays of characteristic energy or energies. They may be soft (*e.g.*, ^{125}I, at 0.027 and 0.035 MeV) or hard (*e.g.*, ^{60}Co at 1.17 and 1.33 MeV). Most γ-emitters also emit β-radiation or α-radiation, but because these are relatively soft they can easily be removed by filtration, leaving only the γ-rays. The γ-radiation can be reduced in intensity rather than stopped by absorbing material; that from ^{60}Co γ-rays, for instance, is reduced by about 50 per cent on passing through a slab of lead some 15 mm thick.

Bremsstrahlung. This is a form of x-radiation generated when β-rays are stopped. It is of significance only for intense sources which emit β-rays but not γ-rays; when γ-rays are emitted, these are much more intense than the bremsstrahlung.

Neutrons. Some radioactive materials, such as plutonium, produce neutrons spontaneously, but not copiously. In the laboratory, neutron sources are usually composed of a mixture of an α-emitter, such as radium-226 or polonium-210, or a high-energy γ-emitter like antimony-124, with beryllium. Atoms of beryllium, when bombarded in this way, undergo nuclear transformation that results in the emission of a neutron. Commercial neutron generators are also available; in these, atoms of deuterium or tritium are accelerated by at least 100 kV and strike a target containing deuterium. Under these conditions, the D–D nuclear reaction generates neutrons of about 5 MeV energy and the D–T reaction neutrons of about 14 MeV. Such neutrons, when slowed down, can be used, for example, for activation analysis.

Positrons. Positrons are similar to electrons, but have a positive charge. They are emitted by certain radioactive nuclides, such as ^{11}C or ^{64}Cu, and are very short-lived. After being slowed down by ionisation, a positron and an electron combine to annihilate one another, but in the process two photons are emitted in exactly opposite directions. This is called annihilation radiation, and both photons have an energy of 0.51 MeV.

Quantities and units

Radiation quantities and units are defined by the International Commission on Radiation Units and Measurements. This Commission's recommendations cover a much wider field than concerns us here, and a fuller discussion of the units, quantities and symbols *etc* will be found in the Commission's reports.[3]

The activity of a quantity of a radioactive material is normally stated in becquerels (Bq), an amount of radioactive material in which one radioactive disintegration occurs per second being said to have an activity of 1 Bq. This is an extremely small unit, so that amounts normally handled often have activities of MBq (*i.e.*, 10^6 Bq), and sometimes GBq (10^9 Bq). Specific activities may be expressed in units such as Bq (MBq, *etc*) per gramme of material. Until recently, the unit of activity used was the curie (Ci), where 1 Ci = 37 GBq, 1 mCi = 37 MBq, and 1 μCi = 37 kBq.

The rate of disintegration of a radioactive nuclide varies over a wide range, but for any single nuclide it is constant, unaffected by temperature or any other factor. It is usually expressed in terms of the half-life, *i.e.*, the time taken for the activity of a specimen to decrease to one-half its initial value. Half-lives as short as 10^{-9} s, and as long as 10^9 years, are known.

When radiation is absorbed in a material, the *absorbed dose* is given by the radiation energy absorbed per unit mass of the material. The unit of absorbed dose is the gray (Gy), which corresponds to an energy absorption of 1 J kg^{-1} of the material, and dose-rate may be expressed in such units as Gy h^{-1}.

The biological effect of a given dose of radiation depends on a number of factors, such as the total time during which the various doses of radiation are received, the dose rate during irradiation, and the type of radiation. Heavy particles such as neutrons or protons produce greater biological damage for a given dose than do x-, γ- or β-rays. For this reason, for protection purposes the term *dose equivalent* (H) is introduced, where $H = D \times \bar{Q} \times N$, D being the dose in Gy, $\bar{Q}$ the quality factor expressing the relative biological effect of the radiation considered, and N the product of all other specified modifying factors (currently assigned the value $N = 1$). The dose equivalent is expressed in sievert (Sv) and dose limits are given in this unit. For all x-, β- and γ-ray work, however, any absorbed dose in Gy is numerically equal to the corresponding dose equivalent in Sv. For other radiations, accepted values of quality factor are given in Table 9.1

Until recently, other units in use were

Roentgen (or Röntgen) as a unit of *exposure*,

Rad, as a unit of *absorbed dose*, where 100 rad = 1 Gy and

Rem, as a unit of *dose equivalent*, where 100 rem = 1 Sv,

and these terms will still be found used. In much scientific literature the word 'dose' (applied to ionising radiations) is used interchangeably for absorbed dose and dose equivalent, unless the context indicates which is meant.

Table 9.1 Quality factors

Radiation	Value of $\bar{Q}$
X-rays, γ-rays and electrons	1
Neutrons, protons and singly-charged particles of rest mass greater than one atomic mass unit, of unknown energy	10
α-particles and multiply-charged particles (and particles of unknown charge) of unknown energy	20

Effects of radiation

Radiation effects are conveniently divided into two categories:

(*a*) Stochastic effects, which are essentially random in occurrence. The larger the dose, the more likely is a stochastic effect to take place; if it does, the severity of the effect is totally unrelated to the amount of radiation that caused it. Examples are the production of cancers, and of genetic mutations.

(*b*) Non-stochastic effects, where if the amount of radiation exceeds a certain threshold dose the effect will occur. Its severity depends on the amount of radiation delivered above the threshold level. Examples are the production of skin reddening or erythema, and epilation. Non-stochastic effects require quite large doses to cause any detectable effect whatever; skin reddening may sometimes occur after as little as 3 Gy or so delivered in a short time; some other effects required doses of at least tens of Gy.

ICRP standards
The first national attempt to coordinate action on radiation was made in 1921 with the setting up of the British X-Ray and Radium Protection Committee, whose first recommendations were issued later that year. Within months, a corresponding committee had been set up in the USA, and other countries followed. At the International Congress of Radiology in Stockholm in 1928 an International Committee was set up, and this later became the present International Commission on Radiological Protection consisting of 12 experts in the field, chosen without regard to nationality. Its publications[5] cover a wide range of radiation protection topics, and appear regularly under the title *Annals of the ICRP*. In 1977, it issued Publication 26,[5c] its latest basic document setting out radiation protection philosophy and laying down standards. All radiation protection legislation, recommendations, codes of practice *etc* anywhere in the world are based on the ICRP Recommendations, which have the highest possible standing in the field.

Today, most industrialised countries of the world have regulations or laws governing radiation protection, or at least have national guidelines; those that do not, use the ICRP Recommendations. In the absence of legislation or national guidance it is difficult to ensure compliance. Other international bodies are involved in radiation protection, and issue publications on the subject, such as the International Atomic Energy Agency,[6] the World Health Organisation,[7] the International Electrotechnical Commission[8] and the Nuclear Energy Agency of the Organisation for Economic Co-operation and Development[9] at an inter-governmental level. International Scientific bodies with an interest in the field include the International Radiation Protection Association, the International Congress of Radiology and the International Organisation for Medical Physics. Many scientific journals carry papers on radiation protection in relation to their specialist interest, and several[10] are devoted entirely to this field, often with an emphasis on nuclear energy aspects. Scientific societies with an interest in the field often publish occasional papers[11] which are helpful to others. Authoritative bodies in other countries produce helpful literature on radiation protection, and mention must be made in particular of that published by the National Council on Radiation Protection and Measurements of the USA.[12]

General principles of radiation protection
Any occupational exposure to ionising radiation simply adds to the ionising radiation we all receive, whether we like it or not, from natural background radiation, including cosmic rays. This addition can be reduced to any desired extent by appropriate measures, but the greater the reduction the greater the

cost of doing so, and this cost can sometimes increase alarmingly for quite small reductions in radiation exposure. Clearly, a balance must be found between an adequate reduction, and an acceptable cost. As a result of these and similar considerations, the ICRP[5c] has proposed a system of dose limitation based on three factors, and this system has been followed by most other bodies in the field—

(a) No practice shall be adopted unless its introduction produces a positive net benefit;

(b) all exposures shall be kept as low as reasonably achievable (ALARA), economic and social factors being taken into account; and

(c) the dose equivalent to individuals shall not exceed the limits recommended for the appropriate circumstances by the Commission.

It should be noted that where people are exposed to radation for medical purposes, (a) and (b) still apply, but the requirement of (c) is withdrawn.

Where occupational exposure may result from radioactive material entering the body, all practicable steps need to be taken to prevent this; the use of fume cupboards, special clothing, or even respirators, may be necessary. Where occupational exposure may result from the exposure of the body to external radiation, there are three main methods of reducing the doses received–

1 Distance from the source. For practical purposes, the radiation intensity may be assumed to be reduced inversely as the square of the distance from the source.

2 Time for an operation. When other factors are unchanged, the dose received is proportional to the time taken for an operation.

3 Use of shielding. The reduction of radiation by shielding depends on the material used, its thickness, and the radiation energy from the particular radioactive material considered.

For a particular procedure, it may be necessary to evaluate the optimum combination of distance, time and shielding, and as this depends on many factors, no generalisations can be made. The use of remote tools will enable handling distances to be increased for such tasks as opening ampoules or screw-cap bottles, or operating burette taps or syringes, but if badly-designed these can slow up the work so much that any advantage is lost. Similarly, the introduction of shielding can reduce radiation doses dramatically, but unless well-designed and properly used can slow up a procedure so much that greater doses are received as a result. It is not intended to imply that these measures should not be used, but rather that expert advice needs to be sought on the optimum combination appropriate to a particular situation.

Legislation

As a result of European Directives,[13] the countries of the European Community (including the UK) are required to control radiation hazards by legislation of a particular kind. In the UK, this has been done through the Ionising Radiations Regulations (1985)[1a] and the legislation in other countries of the EEC will be essentially similar. That in countries outside the EEC will follow the same general pattern, as all such legislation is based ultimately on the Recommendations of the ICRP[5]. However, the UK regulations will here be taken as typical, and be described in some detail in relation to chemical laboratories that use ionising radiations.

The regulations have been made under the provisions of the Health and Safety at Work, *etc* Act (1974)[2] (*see* Chapter 2), and are legally binding; contravention becomes a criminal act. However, they are accompanied by an Approved Code of Practice[1b], which describes in greater detail what steps are generally to be taken to enable the regulations to be observed. It is, however, open to a person concerned to use different methods of complying with the regulations in particular situations, provided he is prepared to argue in a court of law that his method was at least as effective as that described in the Approved Code of Practice. By this means, there is an in-built element of flexibility to allow problems to be solved where necessary by unconventional means, so long as they are effective. In addition, several volumes of Guidance Notes are to be published for different areas of radiation work, and these include suggestions for techniques and equipment which are useful, but not directly related to the legal requirements.

The main requirements of the Regulations are binding on the 'employer'. This may be a company, a University, a Health Authority, a dentist (or partnership of dentists), *etc*, but is the body legally responsible for undertaking the work with ionising radiation. When the 'employer' is a body corporate, it will no doubt delegate much of the detailed work required by the regulations to named officers or administrators, or to a body such as the Radiological Protection Committee of an Institution; however, the 'employer' is ultimately responsible for complying with the regulations. Only a few of the requirements are mentioned here, where the emphasis is on the impact of the regulations on the people working in a laboratory. If more detailed information is needed, the Regulations and Approved Code of Practice should be consulted. Every employer (unless his use of ionising radiations is trivial) must appoint a radiation protection adviser or RPA (full-time or part-time, employee or consultant) to advise him on compliance with the regulations, and this will include advice on radiation protection matters in the departments and laboratories where ionising radiations are used.

Among the responsibilities placed on an employer is the necesssity to 'take steps to restrict so far as reasonably practicable' the amount of radiation received by his employees, and also by any others who may be affected by his

563

radiation work, and in particular to ensure that no doses received exceed a dose limit. The dose limits are listed in Table 9.2. In addition, an employee must not expose himself or anyone else to more radiation than is necessary for his work, must use any protective equipment provided, and must notify his employer of any defects in it. He must also notify his employer if he suspects

Table 9.2 Dose limits

PART I – Dose limits for the whole body
1. The dose limit for the whole body resulting from exposure to the whole or part of the body, being the sum of the following dose quantities resulting from exposure to ionising radiation, namely the effective dose equivalent from external radiation and the committed effective dose equivalent from that year's intake of radionuclides, shall in any calendar year be—
a) for employees aged 18 years or over   50 mSv
b) for trainees aged under 18 years   15 mSv
c) for any other person,   5 mSv

PART II – Dose limits for individual organs and tissues
2. Without prejudice to Part I of this Schedule the dose limit for individual organs or tissues, being the sum of the following dose quantities resulting from exposure to ionising radiation, namely the dose equivalent from external radiation, the dose equivalent from contamination and the committed dose equivalent from that year's intake of radionuclides averaged throughout any individual organ or tissue (other than the lens of the eye) or any body extremity or over any area of skin, shall in any calendar year be—
a) for employees aged 18 years or over,   500 mSv
b) for trainees aged under 18 years,   150 mSv
c) for any other person,   50 mSv

3. In assessing the dose quantity to skin from contamination or external radiation, the area of skin over which the dose quantity shall be averaged shall be appropriate to the circumstances but in any event shall not exceed 100 cm^2.

PART III – Dose limits for the lens of the eye
4. The dose limit for the lens of the eye resulting from exposure to ionising radiation, being the average dose equivalent from external and internal radiation delivered between 2.5 mm and 3.5 mm behind the surface of the eye in any calendar year shall be—
a) for employees aged 18 years or over,   150 mSv
b) for trainees aged under 18 years,   50 mSv
c) for any other person,   30 mSv

PART IV – Dose limits for the abdomen of women of reproductive capacity
5.The dose limit for the abdomen for women of reproductive capacity who are at work, being the dose equivalent from external radiation resulting from exposure to ionising radiation averaged throughout the abdomen, shall be 13 mSv in any consecutive three month interval.

PART V – Dose limits for the abdomen of pregnant women
6. The dose limit for the abdomen for pregnant women who are at work, being the dose equivalent from external radiation resulting from exposure to ionising radiation averaged throughout the abdomen, shall be 10 mSv during the declared term of pregnancy.

Notes: 1. These dose limits do not apply to radiation received for the purposes of medical exposure, which includes approved research procedures.
2. 'Women of reproductive capacity' means any woman who—
a. cannot be said to be unlikely to receive less than 13 mSv in any three month interval, and
b. is medically certified as of reproductive capacity.

that he or anyone else has received more radiation than necessary. Students and trainees are treated as if they were employees.

The regulations require an employer to treat as 'classified' anyone who is likely to receive radiation above 3/10ths of a dose limit, since such a person could receive radiation approaching the dose limit itself – a dose which must not be exceeded – and therefore extra precautions need to be taken. A 'classified' person must have had a medical examination before being classified, and then a review of health at least annually; he must be monitored (*e.g.*, by a film badge or TLD – see below) and the records of doses must be retained for at least 50 years.

Attention must be focussed on any areas where high doses may be received; any accessible place where the radiation dose in a working year may exceed what is referred to below as 'Dose Level A', *i.e.*, 3/10ths of a dose limit for employees aged 18 or over, requires special consideration. The RPA must consider every such area, taking into account relevant factors such as intermittency of operation or of occupancy, in deciding whether the employer should designate it a 'controlled area'. Similarly, if the dose may exceed 1/3rd of Dose Level A, he will have to consider whether it should be designated a 'supervised area.' Where open radioactive materials are in use, the factors to be taken into account will include likely air concentration of the radioactive material, and any surface contamination. A controlled area must be delineated, normally by natural barriers such as walls and doors, with warning notices at points of entry. Once a controlled area is established, it remains a controlled area until specified precautions have been taken, such as the isolation of any x-ray machine from the power supply, the removal of radioactive sources to safe keeping and the removal of contamination. Thus, for example, it will be generally possible for a laboratory to be de-controlled at night to avoid any possibility of cleaners having to be classified. The only people allowed into a controlled area are—

- *a.* Classified persons
- *b.* Other people (including visitors) who conform to a written 'system of work' so designed that—
  - (*i*) employees aged 18 or over do not receive more than Dose Level A, and
  - (*ii*) other people do not receive more than the appropriate dose limit.

In any case, the employer must be able to demonstrate (by personal dose measurement or otherwise) that these conditions are being met.

Local rules must be drawn up so that their observance will ensure that the regulations are observed in the particular local circumstances. They will be similar to, but different in emphasis from, any existing local rules, and probably much longer and more comprehensive, if only because they must include a description of every controlled and supervised area. Where appropriate, each Department, or laboratory, may have its own set of local rules, but a complete collection for the establishment must be available at a specified place. In order to supervise the working of the local rules, Radiation Protection Supervisors (RPS) will be appointed corresponding roughly with the previous Departmental Radiation Supervisors or Radiological Safety Officers. These people will be directly involved in the radiation work, will have to understand the legal requirements and the local rules, will probably be in a line management position but must, in any case, be capable of supervising radiation protection and have the necessary authority to do so.

Anyone working with ionising radiations, or involved in that work, must have received adequate information, instruction and training to enable them to do so safely, and according to the regulations.

Where radioactive materials are used, there are additional requirements. There must be adequate records to account for all radioactive materials on the premises and their whereabouts, so that losses can be quickly identified. In general, the records would be expected to identify the radioactivity in each room separately. When not in use, radioactive materials must be kept in a suitable store. Similarly, radioactive materials being moved from one place to another must be in a suitable container; if being transported by vehicle, or across public areas such as car parks, the container must conform to the transport requirements of the I.A.E.A.[6d] Where large amounts of radioactive materials are handled, it may be necessary for changing facilities and showers to be installed and used. It will always be necessary to wash the hands when leaving any area controlled because of contamination, and to leave in it any clothing that has become contaminated.

Before any work with ionising radiation starts, an employer must assess the hazards likely to arise in the event of any foreseeable occurrence, including accidents, and must make plans to prevent them and also, if they should happen, to minimise the consequences. If in any such occurrence anyone would be likely to receive a dose above a dose limit, contingency plans must be prepared for inclusion in the local rules, and rehearsals held where appropriate.

Any laboratory using radioactive materials is likely already to have a certificate of authorisation from the Department of the Environment for the disposal of radioactive waste, and also (unless in an NHS hospital or other government establishment) be registered for the use of radioactive materials, and these arrangements are not affected by the Ionising Radiations Regulations. However, if as a result of any occurrence radioactive material is lost, spilt or disposed of other than as authorised by the Department of the Environment, the employer must be informed at once, as he may be required to notify the Department of the Environment, the Health and Safety Executive and/or the Police.

When x-ray equipment is used other than for medical purposes it must be in a shielded enclosure or cabinet; no-one (or any part of anyone) is allowed in the enclosure during an x-ray exposure except (under stringent conditions) to initiate or terminate the exposure. There must be an automatic warning when radiation is being generated and also (unless the enclosed volume is less than 0.2 m^3) a warning when radiation is about to be generated. Effective interlocks are required to prevent any exposure when a door is open. Where an enclosure is large enough for anyone to be shut inside, there must be an alarm system, a 'panic button' to stop the exposure immediately, and easy means for rapid exit. All control points must be outside the enclosure; fluoroscopy by direct vision of a fluorescent screen is not permitted.

Equipment used for x-ray optics (crystallography, *etc*) must conform to requirements similar to the above. If a shutter is used, its position needs to be indicated clearly. Access to the enclosure while the x-ray tube is energised is permitted (when this is essential) only under stringent conditions.

This description of the contents of the regulations and the Approved Code of Practice is necessarily brief (the original draft runs to over 160 pages of A4 typescript) and selective, and reference should be made to the full documents for more details. There will be Guidance Notes for various areas of work, including medical and dental practice which will cover laboratories in hospitals, and render the old Code of Practice[14] obsolete. At the time of writing, it is not known whether similar guidance notes for Research and Teaching will be published, as indicated in the 1983 Consultative Document.[15] If not, some help can be obtained from the Medical and Dental Guidance Notes and also from the old Research and Teaching Code of Practice,[16] where this does not conflict with the regulations or the Approved Code of Practice.

Whether the regulations are being observed will, in the first instance, be determined by the local Health and Safety Inspector. He has power to issue an 'improvement notice' requiring an employer to carry out specified measures within a stated time or face prosecution, or a 'prohibition notice' which prohibits an employer from carrying out specified procedures until stated improvements have been carried out. He can prosecute anyone who, in his view, has contravened any of the regulations, and it will then be a matter for the courts to decide whether or not a contravention has occurred. A modified procedure is laid down where government establishments (including NHS hospitals) claim Crown immunity from prosecution.

Radiation emergencies that are reasonably foreseeable must have contingency plans prepared for them. Some emergencies are not reasonably foreseeable, and the Approved Code of Practice only mentions them to include the bizarre. If such an emergency does arise, responsibility for deciding what action to take lies with those on the spot at the time, and they are required to make every effort to observe the regulations and Approved Code of Practice as far as is possible in the circumstances. It may nevertheless happen in particular situations that these suggest action clearly not for the optimum well-being of those involved. It is obvious that action should then be taken in all good faith that is sensible and appropriate to the circumstances, even if it results in some contravention of the regulations. While an Inspector is entitled to prosecute every contravention, there is no obligation on him to do so where he feels it would be inappropriate in the circumstances.

General precautions

Many of these are covered, some in considerable detail, in the regulatory package. In addition, it will be necessary for Trades Union safety representatives to be brought into discussions on radiation safety, and for them to be satisfied that the local rules are drafted in such a way as to ensure adequate protection for all concerned. Everyone needs to understand that 'adequate safety' is not merely a matter of seeing that no-one receives doses above the dose limits. Neither is it related to short hours of work, long holidays, or extra pay because of the 'dangers' of radiation. It is not always a matter of massive shielding, although this is sometimes necessary. It is primarily a matter of ensuring that all doses received are as low as is reasonably practicable.

Warning signs need to be of a kind which are immediately understood by those who need to do so. The use of the word 'danger' is to be deprecated as radiations are rarely 'dangerous' as the word is normally understood; 'caution' is less emotive and more accurate. The international trefoil sign in black and yellow[17c,d] often needs to be supplemented by, for instance, 'Do not enter when red lamp is alight' outside an x-ray room; 'Radiopharmacy; no entry without authority' where large amounts of open radioactive materials are in use; or 'Radioisotope Laboratory' for a room where small amounts are used. In addition, reference may be made to the publications of IAEA,[6] WHO,[7], NCRP[12] and the British NRPB.[18] None of these, of course, have been written to conform to the Ionising Radiations Regulations (1985),[1] and must be interpreted in the light of this fact.

Precautions in work with x-rays

X-Rays are normally produced in a special tube so shielded that no appreciable emission of radiation occurs outside the 'useful beam', but the intensity inside this beam may be very great, so that due precautions need to be taken even against the radiation scattered from the beam by any material irradiated. It is of particular importance that the useful beam of x-rays should at no time irradiate directly any part of any individual; when necessary, protective clothing should be worn.

It is also necessary to ensure that any beam of x-rays is adequately absorbed; if it is pointing vertically downwards, a floor (unless of wood) is generally an adequate absorber; if it is horizontal, brick or lead sheet is generally used, but the thickness depends on the radiation energy and the amount of use of the machine. For radiation up to 100 kV, it is unusual for more than 2 mm of lead, or 230 mm (9 in) of brick, or 150 mm (6 in) of concrete to be required, and in many situations half these thicknesses are adequate.[19]

X-Rays are reduced in intensity by material, but the greater the density, and the greater the atomic number, the greater the attentuation. Thus a given attentuation can be achieved with either iron or lead, but if iron is used a greater mass, as well as a greater thickness, will be required although the cost may in some circumstances be less.

X-Rays are frequently used for crystallographic purposes, the diffraction of the radiation in different directions giving an indication of the internal structure of a crystal. For this purpose, a small but intense beam of radiation is required, often of very soft (*i.e.*, low photon energy) radiation. It is particularly important to ensure that the x-ray tube is not energised whenever adjustments need to be made, and that hands cannot come near the x-ray beam. Automatic shutters can be used, or warning lights wired into the same circuit as the high-tension transformer but, since familiarity breeds contempt, constant vigilance is necessary.[20]

Small mobile x-ray machines are sometimes used in research laboratories. The useful beam should never be larger than is necessary for the required irradiation; restricting cones of different sizes, or adjustable diaphragms, should be available. The useful beam should be directed away from people and adjacent occupied areas.

Precautions in work with sealed sources
Radiation from sealed radioactive sources is much less intense than from an x-ray machine. They may be used, for example, to ionise gases, for the removal of electrostatic charges or in certain types of analytical apparatus; for measurements of scatter or absorption, as in thickness gauges; or for irradiating solutions or living organisms in order to produce radiation changes. It is necessary to ensure that the amount of radiation from such sources reaching working positions is minimal. Some characteristics of a few of the more commonly-used radionuclides are given in Table 9.3.

Often a source is incorporated in a piece of equipment so arranged that the radiation can be turned on and off as required, usually by moving the source by remote control into and out of a protected housing. Any equipment of this kind must be designed to 'fail safe'. If it is electrically operated, the equipment should automatically return the source to the 'safe' position in the event of a power failure, possibly by a spring, but it is usually desirable to make provision so that in any case this operation can be carried out manually—by a long tool, if necessary.

All such sources must be entered in the appropriate register: they should therefore have identifying numbers or marks and be audited at set intervals. Checks need to be made periodically to ascertain that the sealing of each

Table 9.3 Main properties of some radioactive nuclides

Nuclide	Half-life	Max. energy of main β-rays (MeV)	Main γ-ray energy (MeV)	Critical organ	ALI (Bq) Oral	ALI (Bq) Inhalation	DAC (Bq.m^{-3})	K-factor†
^{3}H	12.3 a	0.018	—	Whole body	3×10^9	3×10^9	8×10^5	—
^{14}C	5730 a	0.159	—	Fat	9×10^7	9×10^7	4×10^4	—
^{22}Na	2.6 a	0.54	0.51	Whole body	2×10^7	2×10^7	10^4	12.0
^{24}Na	15 h	1.39	1.27 1.37 2.75	GI tract	10^8	2×10^8	8×10^4	18.4
^{32}P	14.3 d	1.7	—	Bone	2×10^9	10^7	6×10^3	—
^{35}S	87 d	0.167	—	Testis	2×10^8	8×10^7	3×10^4	—
^{42}K	12.4 h	3.6	1.56	Stomach	2×10^8	2×10^8	7×10^4	1.4
^{45}Ca	163 d	0.25	—	Bone	6×10^7	3×10^7	10^4	—
^{47}Ca	4.5 d	2.0	1.3	Bone	3×10^7	3×10^7	10^4	5.7
^{51}Cr	27.7 d	—	0.32	Whole body	10^9	3×10^8	3×10^5	0.16
^{55}Fe	2.7 a	—	0.0059	Spleen	3×10^8	7×10^8	3×10^4	—
^{59}Fe	45 d	0.46	1.10 1.29	Spleen	3×10^7	10^7	5×10^3	6.4
^{58}Co	71 d	0.485	0.51 0.81	Whole body	2×10^9	2×10^9	1×10^6	5.5
^{60}Co	5.3 a	0.31	1.17 1.33	Whole body	5×10^7	3×10^7	10^4	13.2
^{64}Cu	12.7 h	0.66	0.51	Spleen	4×10^8	8×10^8	3×10^5	1.2
^{90}Sr	29 a	0.54	—	Bone	10^6	10^5	60	—
^{125}I	60 d	—	0.035	Thyroid	10^6	2×10^6	10^3	—
^{131}I	8 d	0.61	0.36	Thyroid	10^6	2×10^6	7×10^2	2.2
^{132}I	2.3 h	2.14	0.95	Thyroid	10^8	3×10^8	10^5	11.8
^{137}Cs	30 a	0.51	0.66	Whole body	4×10^6	6×10^6	2×10^3	3.3
^{226}Ra*	1600 a	3.15	2.2	Bone	7×10^4	2×10^4	10	8.2

ALI – Annual limit on intake in order to give a person, represented by reference man, the limit set by ICRP for each year of occupational exposure.

DAC – Derived air concentration, *i.e.*, the ALI divided by the volume of air inhaled by reference man in a working year (2.4×10^3 m^3)

* Together with daughter products.

† Exposure rate at 1 cm from a point source of 1 mCi in r h^{-1}.

source remains satisfactory. Procedures should ensure that sources are never handled by the bare fingers. The short distances between radioactive materials and the basal layer of the skin lead to extremely high dose-rates even with sources of moderate radioactivity. A rigid drill using long forceps or similar instruments should be standard practice. Even so, the dose to the hands can be quite high, and tongs are sometimes used that incorporate a right-angle bend so that the radioactive material can be manipulated behind a lead block. In such a situation, it may be necessary to reduce the dose to the eyes by viewing through a lead-glass window, or indirectly through a mirror.

The loss of a source is always a serious matter, especially if it contains one of the more hazardous radioactive materials, such as ^{90}Sr or ^{226}Ra. Where such sources are used, appropriate detecting equipment should be available so that if one is lost it can then more easily be detected and recovered. Radium needles have been recovered intact in this way from incinerators, drains and demolished buildings; one such needle exploded and contaminated a section of a hospital so severely that several rooms had to be demolished and rebuilt.

Precautions in work with open radioactive sources
When radioactive material is open, more care must be exercised, because of the greatly increased risk of contamination, inhalation and ingestion. The degree of care is related not only to the amount of activity and toxicity of the material, but also to the form in which it is manipulated. Because of its specific concentration in the thyroid gland, radio-iodine is particularly dangerous in this context, as also are α-ray emitters, and the radio-strontiums. Certain types of laboratory equipment are known to contain a radioactive source in this category simply to produce ionisation in gases.

Cleanliness is essential in all radioisotope laboratories, even more so than in many other types of laboratory. Care should be taken to avoid spills (for example, by conducting certain operations over a large drip-tray) and to clear up at once any spillage that does occur; if the spillage dries, active material can become airborne. When possible laboratories should be used solely for radioisotope work, but even when this is not possible specified sections of laboratories should be so reserved. To reduce the risk of internal deposition of radioactivity, no smoking, drinking, eating or application of cosmetics should be allowed. Work with radioactive materials at widely different levels of activity should be so segregated as to reduce the possibility of cross-contamination affecting experimental results. The counting of radioactive samples should not be carried out in the same room as the manipulation of even modest quantities of radioactivity.

When appreciable amounts of γ-active materials are used, shielding will be necessary. This can take many forms, depending on the work to be done, but in general it is more economical in material to put the protective shielding as close as possible to the radioactive material. On occasion a few lead pots will be almost as effective as a bench fully protected with hundreds of pounds of interlocking lead bricks, but sometimes there is no alternative to the latter. Concrete or barytes blocks are often used, being cheaper than lead but more bulky. Such protective devices need to be planned and commissioned in full knowledge of the local circumstances. Much information on the attenuation of different radiations in different materials can be found in a number of publications.[19]

Frequently, contaminated articles need to be transported for disposal. Some of these, such as animal excreta, contaminated filter papers or other absorbent paper, can be suitably sealed for the purpose in polythene bags. Polythene sheeting may also be used to line drip-trays and the like, so that if a spill does occur absorbent paper may be used to mop it up and then folded into the polythene sheet for disposal.

It should be borne in mind that polythene can be porous to halogens or organic halides, and thus radio-iodine in polythene bags may diffuse through the bags in an hour or two. Similarly, laboratories using radioactive materials should have drains of materials other than polythene.

Glassware used to hold radioactive solutions frequently becomes contaminated, and occasionally this contamination may be very difficult to remove completely. Where recovery of the articles is desired, they should be cleaned with appropriate precautions (*see* p. 77) in a chromic acid bath reserved for this purpose. It is in any event an excellent precaution against cross-contamination to mark glassware so that one article is used only for one radioisotope, different colours can be used for indicating different radioisotopes used in the laboratory. For glassware used to contain solutions of high radioactivity (such as a pipette for making up dilution of a stock solution on arrival) *two* marks can be used, so that this article is not used subsequently for very dilute tracer solutions.

In working with radioactive dusts or vapours, it is obviously necessary to use an adequate fume cupboard so vented that any radioactivity discharged will not enter the same or any other premises in appreciable amount. This may require the use of a filter, or scrubbing, in the air exhaust, together with a powerful fan and a discharge well clear of the roof of the building[21]. In certain circumstances, it may be necessary to check that even under gale conditions no blow-back of activity into the laboratory can occur, and in general the principle to be established is that air-flow is always from low-activity areas to areas of high activity. This is usually impossible for the preparation of radiopharmaceuticals, and a laminar flow cabinet will be required.[22]

Newcomers to radioisotope work need, of course, to familiarise themselves with radiation safety, as outlined here. Before they embark on work with radioactive materials, especially in the more toxic categories, they should first of all carry out the sequence of operations they intend to use with inactive material, and then possibly with a relatively non-toxic material such as ^{32}P or

^{24}Na. They should manipulate radioactive materials in much the same way as they would dangerous bacteriological materials. Indeed, the techniques used to prevent radioactive contamination are similar to those used to maintain sterility in a hospital operating theatre.

Contaminated surfaces which cannot be discarded must be cleaned, and the Codes of Practice contain official recommendations for this. The magnitude of the problem depends on the radioactive material concerned and its chemical form. Simple washing with water or a detergent may suffice. If, however, the radioactive material is of high specific activity, minute traces may be absorbed on the surface, which may be removed most easily by washing with an inactive 'carrier' solution of the same substance. Sometimes rubbing, or scrubbing, is necessary, but if the radioactive material has soaked into a surface it may be impossible to remove it; a choice must be made between sealing it in (which may be practicable with small quantities and where the half-life is short) and dismantling the laboratory. For this reason, plain wooden benches, even if waxed, are to be avoided, and surfaces of stainless steel, or a plastic tested for ease of decontamination of the radioactive materials to be used are much to be preferred.

All disposals of radioactive waste come under the control of the Radiochemical Inspectorate of the Department of the Environment, as a result of the Radioactive Substances Act of 1960 and regulations made under the Act. In general the amounts of radioactivity permitted to be disposed in dustbins and the like are very small, typically up to 40 kBq in any one article and up to 200 kBq in any one bin. The amounts permitted to be incinerated on the premises are fairly small, being limited by the anticipated emission of radioactivity from the incinerator chimney and also by the anticipated activity left in ash for dustbin disposal. The amounts permitted for disposal in drains are much larger, being limited by comparing the anticipated average radioactivity in the sewage leaving the institution with the permitted average radioactive content of drinking water. With certain exceptions (*eg* hospital laboratories) institutions using radioactive materials also need to be registered with the inspectorate for this use, and conditions may be laid down by the Inspectorate.

Measuring instruments
The use of measuring and monitoring instruments in connection with ensuring compliance with the Ionising Radiation Regulations is itself subject to the regulations and approved codes. A laboratory issuing personnel dosemeters (film badges, for example), needs to be approved for the purpose by HSE, and approval will only be given to laboratories which maintain high technical standards. Similarly, radiation and contamination monitors will have to be tested and calibrated by approved methods, such as those recommended by the British Calibration Service.[23]

Monitoring instruments fall into the three broad classes below.

Personnel monitors. The type most widely used is the film badge in which a radiographic film is used in an appropriate holder. One model is described in a British Standard[17c] but other types are also in use. The film blackening at the end of an appropriate period (usually four weeks) is a measure of the radiation dose if the type and energy of the radiation is known, and this can be deduced from the film itself by the different blackening under the various filters in the holder. Small film-badge services are usually grossly uneconomic, and at any rate a laboratory or establishment with less than 500 people wearing them should approach the National Radiological Protection Board or a local hospital, nuclear power station *etc*, which runs an approved laboratory. However, certain crystals, such as lithium fluoride, have the property that irradiation with ionising radiation causes some electrons to change to an excited state where they remain until the material is heated; they then return to the ground state with emission of a photon of light. Such material therefore can be used for radiation measurement when properly calibrated. A thermo-luminescent dosemeter (TLD) is particularly useful when a dosemeter must be small, or in an awkward position, and they are used, for instance, to measure finger-tip doses. They can also be used for personnel dosemetry generally to replace film-badges, and they have the advantage that the system can be automated to a greater extent.

Pocket dosemeters. There are various types available. One, in general shaped like a thick pen, incorporates its own measuring system; when it is held up to the light, the dose received can be read off on a scale. Another is like a small pocket calculator and incorporates a Geiger counter. When radiation above a preset level is received, an audible alarm is heard. One form of this instrument contains a digital readout to indicate the dose received. These dosemeters are extremely useful for obtaining an immediate warning of appreciable dose rates, particularly during new or emergency procedures where it is not possible to predict which part of the procedure is likely to deliver most radiation. They cannot, however, be used for the formal measurement of radiation received by classified radiation workers, for which film badges or TLDs from approved laboratories must be used.

Survey monitors. These are designed for two main purposes: the surveying of radiation levels and radioactivity. In each case, a monitor may be intended simply to indicate the presence of, or to locate, radiation or radioactivity, or on the other hand to carry out a measurement of the amount present. In the latter case, if the measurement is made to ensure that regulations are observed, the monitor will need to have been through a specified test and calibration procedure.

Radiation monitors are usually based on an ionisation chamber or proportional counter detector, especially if a measurement is required, but sometimes a Geiger counter is used where only an indication is required as this type of monitor can be made much more sensitive, and scintillation counters can be made more sensitive still. Particular attention needs to be paid to ensuring that the monitor is suitable for the radiation type and energy to be measured and that the appropriate calibration factor is used.

Contamination monitors are available for the measurement of surface or airborne contamination. Surface contamination may need to be checked on benches, floors, gloves, skin and laboratory equipment generally. The reading obtained will clearly depend on the radionuclide involved, and also on the nature of the surface. It will in some circumstances be necessary to carry out a controlled contamination of a surface similar to that contaminated, using a known amount of the relevant nuclide, in order to obtain a calibration of the instrument reading for this particular situation. A suitable monitor should be permanently mounted near a wash-hand basin used to wash hands after they may have been contaminated. It is then easy to check the hands before and after washing—and after rewashing as necessary. Monitors are available into which both hands and feet can be inserted to check the presence of contamination, although the installation of such equipment will only be justified in most laboratories where large amounts of radioactive materials are being handled or there are special circumstances.

Where surface contamination is caused by soft β-emitters, such as ^{3}H and ^{14}C, it may be difficult or impossible to detect or measure the contamination with the usual type of monitor. In such a situation, the surface may be wiped with a suitable swab soaked in a liquid likely to remove some of the contamination; the swab is then put into a scintillation counter to carry out a measurement of the radioactivity removed. Conventionally, unless other information is available, it is usual to assume that the swab has removed 10 per cent of the radioactivity on the surface swabbed.

Airborne contamination monitors are also available to check the radioactivity in the air, but clearly they only measure the air in the vicinity of the sampling point. It may not be easy to ensure that this is at the place where the worker is exposed. Such monitors are required in any laboratory where radioactive gases or vapours are in use, or where there is likelihood of radioactive material becoming airborne. In this connection, it must be remembered that a radioiodide can easily become airborne if it comes into contact with chemicals which facilitate reduction to elemental iodine.

Precautions against non-ionising radiations

Introduction

The hazards of non-ionising radiations (NIR) have received much less publicity than those of ionising radiations, which have provoked a wide emotional response among the population as a whole. As a result, no international standards comparable with the ICRP Recommendations have been established. Indeed, the ICRP itself does not deal with NIR at all. The only body to deal in a similar way with NIR at the time of writing is the International Non-Ionising Radiations Committee (INIRC) of the International Radiation Protection Association (IRPA), which is publishing a series of guidelines.[24] The World Health Organisation (WHO) has also taken the lead, producing its 'Environmental Health Criteria' in relation to NIR in collaboration with INIRC.[7h] In Great Britain, the National Radiological Protection Board includes NIR within its remit. Since the various NIRs have such different properties, hazards and precautions, they must be examined separately.

It must be realised that, as for ionising radiations, any recommended dose level or dose limit cannot represent a level at which hazards begin, or at which hazards are known to be serious. It represents a level below which responsible bodies recommend that all exposures should be kept, if only a matter of prudence, in order to restrict biological effects to a reasonable minimum. Efforts should therefore be made to see that exposures are restricted to the minimum level readily achievable which is below the recommended exposure limits.

Ultraviolet radiation
This may give rise to superficial eye damage and to burning of the skin, not unlike that produced by overexposure to the sun. A few seconds' exposure of the eye to an unscreened arc at a distance of several yards may cause an extremely painful condition generally known as 'eye flash'. The characteristic symptoms are a feeling as of sand in the eyes accompanied by intense pain, intolerance to light, watering of the eyes and possibly temporary loss of vision.

It should be noted that individuals carrying certain genetic traits (*e.g.*, porphyria) are much more sensitive to ultraviolet than others. In addition, certain chemicals (some of which are contained in cosmetics and medicaments) can also cause a substantial increase in sensitivity. Further, in a laboratory where ultraviolet radiation is produced in a confined space, there is a possible danger to persons from an excessive concentration of ozone, which can have an irritant action on the upper respiratory system.

Ultraviolet radiation occurs in sunlight, and is generated artificially for many purposes in industry, laboratories and hospitals. It is divided conventionally into three regions, as follows—

UVA 315–400 nm wavelength
UVB 280–315 nm
UVC 200–280 nm

Radiation in the UVA region (the near-ultraviolet spectral region) is widely used in industry, and causes relatively little biological hazard, but is not often produced without the presence of UVB and/or UVC, which are more hazardous. The most comprehensive standard for occupation exposure comes from the USA[25] and is supported by the WHO[7h,14] among other authorities. According to this, the intensity of UVA incident on the unprotected skin or eye should not exceed 10 W m^{-2} for exposure periods greater than 10^3 s (*i.e.*, approx. 16 min); for shorter periods, the total radiation exposure should not exceed 10^4 J m^{-2}.

Radiations in the UVB and UVC regions, on the other hand, have biological effects varying markedly with wavelength, these being greatest at about 270 nm – a wavelength very close to 254 nm, the principal emission line from quartz low pressure mercury vapour lamps. The recommended limits for 8-hour exposure are as given in Table 9.4. Strictly, where the ultraviolet radiation used contains a spectral mixture, an evaluation needs to be carried out separately for each section of the spectrum. If this is too laborious, it may be necessary to assume, for safety, that all the radiation is in the spectral band of highest biological effectiveness.

For exposures of less than 8 h, the *intensity* of exposure may be allowed to increase so long as the *total exposure* given in Table 9.4 for 8 h is not exceeded. Thus a mercury lamp emitting 254 nm radiation for 8 h may be allowed to give a total exposure of 60 J m^{-2}, corresponding to a mean intensity of $\dfrac{60}{8 \times 60 \times 60}$

Table 9.4. Maximum permissible exposures (MPE) for ultraviolet radiation (200–315 nm)

Wavelength (nm)	MPE (J m^{-2})
200	1000
210	400
220	250
230	160
240	100
250	70
254*	60
260	46
270	30
280	34
290	47
300	100
305	500
310	2000
315	10000

* Mercury lamp resonance line.

W $^{-2}$, or 2 mW m^{-2}, since 1 watt $= 1$ Joule per second. It follows that such a lamp operated for only 1 h per day could be allowed to operate at an intensity of $\dfrac{60 \text{ J m}^{-2}}{60 \times 60 \text{ s}} = 17$ mW m^{-2}, and if such a lamp was required to operate at an intensity of 6 W m^{-2}, it could only be allowed to do so in any day for $\dfrac{60 \text{ J m}^{-2}}{60 \text{ W m}^{-2}} = 10$ s.

It should be noted that these exposures and intensity limits are intended to apply in all cases to exposure of the unprotected eye, and to be a guide to skin exposure depending on an individual's skin colour, previous exposure to ultraviolet radiation, and tendency to react strongly to ultraviolet. The exposure limits given correspond roughly to 10–15 min exposure to full summer sunlight in Britain at midday.

Protective cabinets or screens should be placed around the source of emission and a screen of ultraviolet-absorbing glass should be imposed between the worker and the source of radiation. It is easier to do this if the area of ultraviolet radiation is limited to the minimum necessary. Approved goggles should be worn and, when necessary, hands and forearms should be protected by cotton material or suitable barrier creams. To guard against the danger from ozone already mentioned, the room should be well ventilated.

Infrared radiation
Since infrared radiation is readily absorbed by surface tissue it does not inflict deep injuries. In the case of the eyes, the heat absorbed by the lens of the eye from infrared radiation is not readily dispersed, and a cataract may be produced. This condition is often known as 'chain maker's eye', but it is also found in glassblowers and furnace attendants.

There are no established guidelines for safe exposure to infrared (IR) radiation, except where this radiation is emitted by lasers. However, the WHO report on lasers[7h,23] suggests that the exposure levels used for IR lasers in the USA[26] can be applied to broad-beam sources. The permitted level for IR laser beams for periods exceeding 10 s is 1 kW m^{-2}, based on the assumption that (as of course is the case with lasers) the total irradiated area is small. Where a substantial area of the body is exposed to infrared radiation, the exposure level must be lower because of the total amount of heat given to the body, and a level of 100 W m^{-2} is suggested. Even this may be uncomfortably high if the irradiation is not confined to one side of the body, and there is also a high ambient temperature. Higher levels of IR radiation can be tolerated at 0 °C in winter than, say, at 30 °C in summer.

All sources of intense infrared radiation should be shielded as near to the source as practicable by heat absorbing screens, to prevent the radiation entering the eyes of the workers. Approved goggles or eye-shields should be worn, and since this may result in less light entering the eye, the general illumination of the laboratory should be increased appropriately.

Lasers

There are many laboratory uses of lasers, and these can be dangerous as the beam energy is concentrated into a small area. Even at low intensity, damage to small areas of the retina of the eye can all too easily result from a laser beam, especially if it is viewed in an optical instrument without proper attenuation, and recovery may be slow or non-existent. At high powers, a laser beam can cause severe and rapid damage to skin. Laser beams can still be hazardous after reflection from intercepting surfaces, or even after scattering by diffuse reflection.

To indicate the degree of hazard associated with any particular instrument, lasers are classified according to the scheme given in Table 9.5[17e], and each piece of laser equipment should be labelled accordingly; similar classifications are used in other countries.[26]

The degree of protection necessary clearly depends on the wavelength(s) of light emitted, the size of beam, and the beam energy. Any laser equipment should be designed to include appropriate safety measures such as a protective housing, interlocks to restrict access (if necessary, with key control), a shutter to intercept emission and an automatic warning sign of laser operation. In addition, laser equipment should be appropriately labelled, and all relevant information given to the user to facilitate safe operation.

In addition to lasers which produce visible light (400–700 nm), some

Table 9.5 Classification of laser equipment

Class	Hazard	
1	Inherently safe	Maximum Permissible Exposure (MPE)* levels cannot be exceeded.
2	Low risk	Emission limited to 1 mW for less than 0.25 s between 400 and 700 nm; hazards are prevented by aversion reflexes including the blink response.
3A	Low risk	Limit up to 5 x that for Class 2; viewing by the unaided eye is safe, but the use of optical instruments may be hazardous
3B	Medium risk	Emission limit higher still; any direct viewing may be hazardous, but not viewing by diffuse reflection.
4	High risk	Emission limit higher still; viewing by diffuse reflection may be hazardous. Skin injuries and fire hazard are also possible.

* For MPE levels, see BS 4803.[17e]

produce ultraviolet radiation (200–400 nm) or, commonly, infrared radiation (700 nm–1 m).[24b] The use of lasers emitting ultraviolet or long infrared radiation involves a hazard to the cornea, while those emitting in the near infrared region involve a hazard to the retina; in addition, there may be skin hazards.

Anyone responsible for the safe operation of laser equipment other than that in class 1 should be familiar with the guidance in BS 4803[17e] and that given by WHO.[7h,23]

Ultrasound

Ultrasonic radiation is being increasingly used for many purposes, including cleaning in special baths. Where this radiation is used at high intensity, biological results can certainly be expected; indeed, it is used in certain medical procedures to destroy tissue. The energy intensity capable of causing damage is said to vary with frequency and with duration and frequency of exposure. In certain circumstances, it is alleged[27] that intensities as low as 4 mW cm^{-2} can cause a decrease in DNA synthesis, but whether this is the limiting effect is open to question. Certainly intensities above the cavitation threshold can be expected to produce tissue damage, and this level is said to vary from 0.08 to 6 W cm^{-2} depending on frequency.

Several proposals have been formulated for exposure limits for airborne acoustic ultrasound energy in the workplace, and these have been collated and summarised by WHO.[7h,22] The IRPA proposals[24b] may be taken as typical, and, for a total exposure time exceeding 4 h per day, these are for a value of 80 dB for frequencies up to 20 kHz, and one of 100 dB for higher frequencies, relative to a level of 20 μPa, or 10^{-12} W cm^{-2} – a level of sound just audible. For exposures of 1 to 4 h, these figures may be increased by 3.1 dB, and for exposures of less than 1 h by 9 dB. These are quite high intensities, and at lower intensities any hazards will be negligible or non-existent. Concern has been expressed at possible hazards from ultrasound to the foetus of a pregnant woman, but general scientific opinion seems to be that the evidence so far produced does not need to modify present diagnostic procedures.

Radiofrequency and microwave radiation

The microwave region of the electromagnetic spectrum is increasingly used for a variety of purposes, among them rapid heating. In particular, microwave ovens have been known to leak dangerous amounts of microwave energy when the doors are not properly shut. The health hazards of this form of radiation were discussed at an International Symposium sponsored by the World Health Organisation and at subsequent meetings.[28,7h,16]. The radiations in this group have been arbitrarily divided into microwaves (of frequencies of 300 MHz to 300 GHz and wavelengths from 1 m to 1 mm) and radiofrequency radiation (100 kHz to 300 MHz and wavelengths from 3 km to 1 m in air). Their interaction with biological materials is complex, and generally results in a rise of temperature, although this by itself does not seem to explain all the biological effects which have been described. It is not yet possible to set threshold levels, or dose–effect relationships, even in experimental animals. There have been some incidents of accidental acute over-exposure in man, but many patients have successfully received treatment using these radiations over a restricted range of frequency, intensity and duration for many years.

Thermal effects are considered to be unlikely at intensities below 1 mW cm^{-2}, and methods of measurement are discussed. Much more research is thought necessary before any conclusions are reached on cumulative or delayed effects, but in many situations the organ most at risk appears to be the eye lens because of the possibility of cataract formation. There are also said to be non-thermal effects such as headache and nausea. Different countries have set very different permissible radiation levels, which have been collated by WHO.[7h,16] At one end of the scale, in the UK and the USA radiation levels of 10 mW cm^{-2} are thought to be the acceptable limits for continuous daily exposure. At the other end of the scale, the USSR takes non-thermal effects much more seriously, and the corresponding limit is 0.01 mW cm^{-2}, although 0.1 mW cm^{-2} is permitted for up to 2 h, and 1 mW cm^{-2} for up to 20 min, provided protective glasses are worn. This disparity between East and West is now some decades old, and efforts are still being made to reach agreement on common limits. There is already wide agreement on the ways in which exposure to these radiations can be prevented or reduced.

References

1 a *Ionising Radiations Regulations (1985)*. HMSO, London.
  b *The Ionising Radiations Regulations—Approved Code of Practice* (1985). HMSO, London.
2 *Health and Safety at Work, etc. Act (1974)*. HMSO, London.
3 Reports of the International Commission on Radiation Units and Measurements. 7910 Woodmont Avenue, Washington DC, 20014, USA.
  a *Radiation Protection Instrumentation and its application* (No 20, 1971).
  b *Conceptual basis for the determination of dose equivalent* (No 25, 1976).
  c *Radiation Quantities and Units* (No 33, 1980).
  d *Determination of dose equivalents resulting from external radiation sources.* (No 39, 1985).
4 *X-ray and radium protection.* Preliminary Report of the Committee. *J. Röntgen Soc.*, 1921, **17**, 100.
5 Recommendations of the International Commission on Radiological Protection. Pergamon Press, Oxford and New York.
  a *Protection against ionizing radiation from external sources.* (Nos 15 and 21, 1976).
  b *Handling and disposal of radioactive materials in hospitals.* (No 25).
  c Recommendations of the ICRP (No 26, 1977).
  d *Limits for intakes of radionuclides by workers* (No 30, 1979).
  e *Protection against ionising radiation in the teaching of science* (No 36).
  f *A compilation of the major concepts and quantities in use by ICRP* (No 42).
6 Publications of the International Atomic Energy Agency, Vienna.
  a Safety Series No 1. *Safe handling of radioisotopes* (revised 1973).

b Safety Series No 2. *Health Physics Addendum to No. 1* (1960).

c Safety Series No 3. *Medical Addendum to No 1* (1960).

d Safety Series No 6. *Regulations for the safe transport of radioactive materials* (revised 1979).

e Safety Series No 7. *Notes on certain aspects of No. 6* (1961).

f Safety Series No 8. *Use of film badges for personnel monitoring* (1962).

g Safety Series No 9. *Basic safety standards for radiation protection* (1962).

h Safety Series No 30. *Manual on safety aspects of the design and equipment of hot laboratories* (1981).

i Technical Report Series No 133. *Handbook on calibration of radiation protection instruments* (1971).

7 Publications of the World Health Organisation, Geneva.

a *Medical supervision in radiation work.* Technical report series No. 196, 1960.

b *Radiation hazards in perspective.* Technical report series No. 248, 1962.

c *Public health responsibilities in radiation protection.* Technical report series No. 254, 1963.

d *Public health and medical use of ionizing radiations.* Technical report series No. 306, 1965.

e *Use of ionizing radiation and radionuclides on human beings for medical research, training and non-medical purposes.* Technical report series No. 611, 1977.

f *Manual on radiation protection in hospitals and general practice:*

   Vol. 1—Basic protection requirements, 1974.

   Vol. 2—Unsealed sources, 1975.

   Vol. 3—X-ray diagnosis, 1975.

   Vol. 4—Radiation protection in dentistry, 1977.

   Vol. 5—Personnel monitoring services, 1980.

g *Action procedure in the event of radiological accident.* Collection 83.02 (1983).

h Environmental Health Criteria—

   No. 14. *Ultra-violet radiation* (1979).

   No. 16. *Radiofrequency and microwaves* (1981).

   No. 22. *Ultrasound* (1982).

   No. 23. *Lasers and optical radiation* (1982).

8 Publications of the International Electrotechnical Commission, 1 rue de Varembe, Geneva.

a *Portable prospecting radiation meters with Geiger–Müller counter tube (linear scale instruments).* Publication 421 (1973).

b *Portable prospecting radiation meters with gamma-ray scintillation detectors (linear scale instruments).* Publication 460 (1974).

c *Low energy X or gamma radiation portable exposure rate meters and monitors for use in radiological protection.* Publication 463 (1974).

d *Alpha, beta and alpha–beta contamination meters and monitors.* Publication 325 (2nd ed., 1981).

e *Radiation protection equipment for the measuring and monitoring of airborne tritium.* Publication 710 (1981).

9 Nuclear Energy Agency, Organisation for Economic Co-operation and Development, 2 rue Andre-Pascal, 75775 Paris, France.

a *Assessment and recording of worker doses* (in preparation).

b *Concepts of collective dose in radiological protection* (1985).

10 For example—

a *Health Physics*, Pergamon Press, Oxford and New York.

b *Journal of the Society for Radiological Protection*, Editor, Dr. J.H. Jackson, 43 Marsham Street, London SW1P 3PY.

c *Radiation Protection Dosimetry.* Nuclear Technology Publishing, P.O. Box 7, Ashford, Kent.

11 For example—

Practical radiation protection dosimetry. Conference report series No 14, Hospital Physicists' Association, 47 Belgrave Square, London SW1X 8QX.

12 Reports of the National Council on Radiation Protection and Measurements, 7910 Woodmont Avenue, Washington DC, 20014, USA.

a *Control and removal of radioactive contamination in laboratories* (No 8, 1951).

b *A manual of radioactivity procedures.* (No 28, 1961).

c *Safe handling of radioactive materials.* (No 30, 1964).

d *Structural shielding design and evaluation for medical use of X and gamma rays of energies up to 10 MeV.* (No 49, 1976).

e *Instrumentation and monitoring methods for radiation protection.* (No 57, 1978).

f *Management of persons accidentally contaminated with radionuclides.* (No 65, 1980).

g *Radiofrequency electromagnetic fields—properties, quantities, and units, biophysical interaction and measurements.* (No 67, 1981).

13 Official publication of the European Communities, L-2985 Luxembourg.

a Council Directive 76/579 of 1 June 1976 laying down the revised basic safety standards for the health protection of the general public and workers against the dangers of ionizing radiation. OJ No L 187, 12.7.1976.

b Council Directive 80/836 of 15 July 1980 amending the Directives laying down the basic safety standards for the health protection of the general public and workers against the dangers of ionizing radiation. OJ No L 246, 17.9.1980.

c Council Directive 84/466 amending Directive 80/836. OJ No L 265, 5.10.1984.

14 Ministry of Health, *Code of practice for the protection of persons exposed to ionizing radiations arising from medical and dental use.* London, HMSO, 1972.

15 Health and Safety Executive and National Radiological Protection Board, HMSO, London. *Draft Guidance Notes for the protection of persons exposed to ionising radiations in research and teaching* (1983).

16 Department of Employment and Productivity, *Code of practice for the protection of persons exposed to ionising radiation in research and teaching.* London; HMSO, 1968, (revised 1974).

17 British Standards Institution, 2 Park Street, London W1.

a *Film Badges for personnel radiation monitoring.* BS 3664 (1963).

b *The testing, calibration and processing of radiation monitoring films.* BS 3890 (1965).

c *Colours for specific purposes.* BS 3810 (1964).

d *A basic symbol to denote the actual or potential presence of ionising radiation.* BS 3510 (1968).

e *Radiation Safety of laser products and systems.* BS 4803 (1983).

Part 1—General

Part 2—Specification of manufacturing requirements for laser products.

Part 3—Guidance for users.

18 National Radiological Protection Board, Chilton, Didcot, Oxfordshire.

For example—*Cost benefit analysis in the optimisation of protection of radiation workers—a consultative document.* (1982).

19 See, for example, references 5a. 5b. 6b. 12d. and the Handbook of Radiological Protection. Part 1—Data. HMSO, London (1971).

20 B.E. Stern, *Health Phys.* 1970, **19,** 133.

21 K. Everett and D. Hughes, *A guide to laboratory design.* London: Butterworth, 1975.

22 a *Guidelines for the preparation of radiopharmaceuticals in hospitals.* British Institute of Radiology, Special Report No. 11, 1975.

b *The hospital preparation of radiopharmaceuticals.* Hospital Physicists' Association, Scientific Report Series 16, 1976.

c *Good pharmaceutical manufacturing practice applied to the hospital preparation of radiopharmaceuticals.* Notes for guidance. DHSS, 1977.

d *Facilities for the preparation of radiopharmaceuticals.* Nuclear Medicine Communications, 1980, **1,** 54.

23 Publications of the British Calibration Service, National Physical Laboratory, Teddington, Middlesex, TW11 0LW—General Criteria for laboratory approval;

a 802 *Calibration of radiological instruments* (January 1977).

b 803 *Provision of personal dosimetry services* (October 1977).

c 811 *Calibration of radiological protection level instruments: X-, gamma- and beta-rays* (January 1977).

d 813 *Calibration of radiological protection level instruments: neutrons* (September 1977).

e 821 *Provision of personal dosimetry services using film dosemeters for beta, gamma, X- and thermal neutron radiations* (October 1977).

f 822 *Provision of personal dosimetry services using nuclear emulsion film dosemeters for neutron radiations.*

g 823 *Provision of personal dosimetry services using thermoluminescent dosemeters for beta, gamma, X- and neutron radiations* (October 1977).

h 6601 *Calibration of radiological instruments at protection and therapy levels* (January 1977).

24 Publications of the International Non-Ionising Radiations Committee (INIRC) of the International Radiation Protection Association (IRPA), BP No. 6, F-92260 Fontenay-aux-Roses, France.

a *Interim guidelines on limits of human exposure to airborne ultrasound. Health Physics;* 1984, **46**, 969.

b *Interim guidelines on limits of exposure to radiofrequency electromagnetic fields in the frequency range from 100 kHz to 300 GHz. Health Physics,* 1984, **46**, 975.

c *Guidelines on limits of exposure to ultraviolet radiation of wavelengths between 180 nm and 400 nm (incoherent optical radiation) Health Physics* (in the press).

d *Guidelines on limits of exposure to laser radiation of wavelengths between 180 nm and 1 mm. Health Physics* (in the press).

25 *Occupational exposure to ultraviolet radiation.* Washington DC: US Department of Health, Education and Welfare, 1972.

26 a *A guide for control of laser hazards.* Cincinnati: The American Conference of Governmental Industrial Hygienists, 1973.

b A.J.H. Goddard, *Phys. Bull.* 1976, **27**, 245.

27 N. Prasad, R. Prasad, S.C. Bushong, L.B. North, E. Rhea, *Lancet,* 1976, **i**, 1181.

28 *Biologic effects and health hazards of microwave radiation.* Warsaw: Polish Medical Publishers, 1974.

Chapter 10

An American View

Chemical laboratories comprise four major components: scientific knowledge or data; physical facilities and services; chemicals and ancillary equipment; humans. Each of these four basic elements has inherent strengths and weaknesses. If safer experimentation is to be achieved, a realistic and impartial appraisal and appreciation of the interplay of these four elements is recommended.[1,2,3]

Chemical knowledge or science

Chemical knowledge is never complete or absolute. Dissemination of even incomplete knowledge to those who need it most is far less than universal. Chemical Abstracts Service Registry recognizes some seven million chemical entities, and countless combinations and mixtures are widely used in practice, with new data emerging almost daily in journals, meeting papers, dissertations, and reports of specific investigations.

Chemistry is still a developing science, with no limits in sight but human ingenuity. The chemical/human interface, long ignored or simply dismissed with lip-service, now has surfaced as vital to the future of the profession and to society at large. The science itself, its risk/benefit analyses, and the popular perception by the public of chemistry as constructive and essential to human well-being is being re-examined in many quarters and at various levels of society, including the legal, the regulatory, and the sociological. Incidents where serious tragedies or off-site injuries have occurred, such as Texas City, Flixborough, Seveso, Pernis, and Bhopal, have shaken public confidence in the credibility of science and industry to a significant degree, and society now insists on more complete and fail-safe control measures and emergency co-ordination in all chemically-related operations at all times. Further regulations, which in the past have been less than completely successful in controlling technical problems, may be expected in the near future. We seriously question this approach.[4a,b,5,6,7,8]

While our knowledge of chemicals and their inherent properties which create potential hazards is extensive, new data frequently raise disquieting and unresolved questions. For example, a recent report noted that a 500 ml glass bottled sample of acryloyl chloride, inhibited with 0.05% phenothiazine, exploded with no apparent warning in a laboratory fume cupboard (hood). The ambient temperature in the laboratory approached 50 °C. The sample had been airmailed from Switzerland to Iraq during the summer and was plainly labelled as requiring refrigeration to 4 °C.[9] Another supplier, in the United States, noted that he had made and shipped the material for seventeen years with no known unusual incidents. It remains at this writing whether the basic cause of this incident was the temperature without refrigeration, contamination, or inadequate inhibition, or a combination of these factors. Uncontrolled polymerizations are major hazards in both laboratories and plants.[10] Another example in incomplete knowledge, or its application, involves the recently recognized hazards of perchlorate-doped conducting polymers, such as polyacetylene.[11] The instability of the perchlorate ion, especially at elevated temperatures and higher concentrations, in the presence of organic material, is extensively documented.[12,13]

Unusual exposures also can bring to light actions of chemicals whose properties were not well documented. The unique neurotoxicity of 1-methyl-4-phenyl-1,2,3,6-tetrahydropyridine, commonly designated as MPTP, was first recognized when young drug addicts developed severe symptoms of Parkinson's disease after injecting impure synthetic heroin made in a clandestine chemical laboratory. The action of this compound on animals is now being utilized in models to study the mechanisms of Parkinson's disease, which now afflicts more than half a million Americans.[14]

Another example of inadequate dissemination of knowledge was recently reported in the explosion of butan-2-ol during distillation. Professor Doyle has noted that it does not seem to be generally known that some alcohols may pose a danger of peroxide formation similar to that of ethers. There are apparently no warnings of this danger on the product labels or in laboratory manuals,[15a] though this had in fact been reported 9 years previously.[15b]

Hazard assessment and communication

Even when high hazards are recognized, analysed, and accepted, the risk/benefit analyses may be flawed, and can lead to belief that the risk is relatively insignificant. When making a risk/benefit analysis involving several unknowns, incomplate knowledge of all the variables, or false data, may lead the reviewer to make an error of judgement. The explosion of the Space Shuttle 'Challenger' on January 28, 1986 following difficulties in containing the emissions from the solid fuel booster, doubtless permitted liquid hydrogen to combine with liquid oxygen stored in adjacent tanks in the shuttle's main thrust system.[16] The potential hazards of liquefied flammable gases, such as liquid hydrogen, which expands 777 times between the liquid and the gaseous state, have not always been fully appreciated.[17] Three decades ago, this writer attempted, with very limited success, to convince scientific staff members in a major research center that liquid hydrogen, even in litre quantities, was a highly hazardous substance in a laboratory. Since scientific personnel are oriented to experimentation, we proposed experiments under controlled conditions, where litre amounts of liquid hydrogen would be deliberately spilled and ignited, using a motion picture camera to record the results. The tests were even more spectacular than we had hoped, and the staff members involved became very safety conscious in seconds. Footage from these tests is included in the motion picture, '*Chemical Boobytraps,*' which has become a useful visual aid in the training of laboratory personnel.[18] Written data sheets[19] or analyses on paper do not have the same impact on personnel as demonstrations[20a] and electronic presentation.[20b] With the newer techniques of computer software and video tapes, such training methods are becoming increasingly popular.

To illustrate that literature searches are not always completely dependable, a recent letter to *Chem. Eng. News.*[21a] notes that while preparing *N,N*-dimethylchloromethyleniminium chloride by the vacuum removal of sulphur dioxide from a mixture of thionyl chloride and DMF (Vilsmeier complex), a sudden exotherm and pressure rise occurred in a 400 litre reactor in a pilot plant. Spontaneous decomposition to produce sulphur, sulphur dioxide, hydrogen chloride, and dimethylcarbamoyl chloride (a potent animal carcinogen) had occurred, but fortunately the mixture was contained entirely within the reactor. A previously published report[21b] of the sudden exotherm of a thionyl chloride/DMF mixture standing at room temperature was inadvertently not abstracted by Chemical Abstracts in 1977 and thus failed to show in the literature searches conducted through CAS sources.

Physical facilities and services

Safety and adequacy of laboratory facilities have long been recognized as areas of concern.[22] Lees and Smith recently have recorded the need for and suggested mechanisms for updating physical facilities in older laboratories.[23] In addition to fume cupboards (hoods), which receive extensive treatment in this context,[24] such widely accepted traditional protection as fire doors in laboratory buildings to inhibit or prevent the spread of smoke and flames merit further practical investigations to improve their efficiency. Other details for improving the efficiency of older fume cupboards (hoods) have been reviewed by Saunders, and can be implemented with only nominal expense.[25] A fume cupboard (hood) declared as adequate for the specific task in hand should then be used properly and be checked regularly and maintained by competent professional engineers who have the equipment and facilities to make appropriate performance tests and adjustments. If potentially hazardous materials, such as concentrated nitric or perchloric acids are used routinely, the fume cupboards deserve specific precautions, preferably dedication and weekly cleaning. Hoods are not intended as storage areas, nor are they designed as evaporators for the disposal of volatile waste solvents.[26] To use them incorrectly or for such non-essential purposes, is to impair their ability to perform their intended use, namely to protect personnel from undesirable or injurious emissions. Under no circumstances should laboratory apparatus or equipment be placed on the floor in front of fume cupboards (hoods), since in any emergency situation, disaster is almost inevitable, especially during nights and off-hours.

Chemicals

Chemicals, with their containers, reaction vessels, and ancillary equipment and instrumentation, are the *sine qua non* of a chemical laboratory. Chemicals must be treated with full respect from intial ordering and delivery to ultimate disposal. Unfortunately, such respect is often lacking or is incomplete. Many chemicals, perhaps most, have more than one characteristic or hazard potential which mertis serious consideration in handling, reaction, and disposal from the safety viewpoint.[27] While fire potential (flammability or inflammability) is often recognized, known, and respected, less obvious hazards, such as the long-term effects of inhalation of vapours or gases may not be as widely appreciated. This is especially true if the effects of chronic (repeated) low-level exposures become obvious only after months or years. To illustrate, benzene (benzol) has been recognized both in England and in America as requiring adequate control or substitution in many laboratories, since it is a familiar volatile flammable solvent. Less appreciated, however, are the serious health effects of inhalation.[28,29,30]

The British Health and Safety Executive (HSE) limits the exposure to benzene vapors to an 8-hour time weighted average or 10 ppm of 30 mg m^{-3}.[31] The American Occupational Safety and Health Agency (OSHA) permissible exposure limit is also 10 ppm, even though OSHA in 1978 classified benzene as a human carcinogen and attempted to regulate it as it has with other legal carcinogens.[32a,b] On appeal, the US Supreme Court in 1980 ruled that OSHA had not adequately documented the claim of benzene as a human carcinogen, and restored the previous exposure limits. We do know for certain that benzene is a myelotoxic agent (destructive to the bone marrow), producing aplastic anaemia (generally unresponsive to specific antianaemia therapy), with leukopenia (reduction of the number of leukocytes in the blood, the count being 5000 or less) and thrombocytopenia (a fixed increase in the number of circulating blood platelets), since the bone marrow may become hypoactive (abnormally diminished activity) or hyperactive (abnormally increased activity) and may not always correlate with the peripheral blood needs. Again, our knowledge is incomplete, since even legally 'safe' exposure levels are in doubt. OSHA specifically regulates nineteen other materials as carcinogens,[33] and the benzene regulation is being reconsidered.

For 20 years the HSE reprinted annually in its entirety the list of threshold limit values recommended the previous year by the American Conference of Governmental Industrial Hygienists (ACGIH).[34] Although this is not an official governmental agency, the recommendations of that group are widely accepted both in the United States by the OSHA and its advisory group, the National Institute for Occupational Safety and Health (NIOSH) and in other countries. The attempt at listing 'safe' limits first appeared in 1945, as developed by Professor Warren Cook, who recognized that within certain (poorly defined) limits the human system can tolerate or eliminate foreign substances, and that a practical yardstick for occupational exposures was necessary. The original OSHA exposure limits in 1971 were based on the 1969 ACGIH listing. The HSE recognized the need for improvement of the ACGIH approach as applied to British industry, and has evolved, after much study and co-ordination a slightly different system which includes control limits (judged after detailed consideration of the available scientific and medical evidence to be 'reasonably practicable' for the whole spectrum of work activities in Great Britain, and which should not normally be exceeded), and recommended limits (which are considered to represent good practice and realistic criteria for the control of exposure, plant design, and engineering controls, and, if necessary, the selection and use of personal protective equipment).[31] Further exchange of data and of experiences between HSE, OSHA, ACGIH, NIOSH and comparable groups in other countries would doubtless further refine the concept of safe exposure limits, and improve the validity of the data on which they are based to mutual benefit.

Considerable difference in reaction to impending regulation appears in the US compared to the UK. An interesting document has appeared recently analysing the control of vinyl chloride monomer (VCM) after information became available in 1974 that the monomer produced liver damage and is a causal agent of angiosarcoma (tumor) of the liver. Emergency action and emergency limits followed rapidly in several countries. The US producers attacked the proposed exposure limits from a legal standpoint, and a highly acrimonious discussion ensued. On the other hand, in the UK (and three other major producing countries), more reasonable action prevailed and compromises resulted without the necessity for time-consuming and costly litigation. The exposures to VCM were reduced as required in all countries, and the industry has prospered since.[35] The tendency to institute legal action to delay or negate safety, health, and environmental actions, including tort claims against toxic exposures, real or alleged, appears more frequent in the US than in the UK. Recognizing this tendency, the American Chemical Society sponsors a professional liability (malpractice) insurance plan for member chemists and chemical engineers.[36]

The human element and training

The fourth basic element in chemical laboratories is the most complex and least understood. Even less is really known about the human aspect and its ramifications than about the other three elements we have discussed. Certainly we know that the general topic of safety is still not received with acclamation or enthusiasm in all circles even today. Part of this luke-warm attitude may arise from the concept of academic freedom, which has been long cited by academics seeking to protect themselves from those they fear will want to impose limits on their total freedom of action. Carried to extremes, such 'freedom' can be a significant barrier to intelligent reactions, especially where human life is involved. In part, this apparent hostility may arise from previous generations of academics, who had little incentive to give safety more than lip-service and students were actually told 'this is the price you pay for being a chemist'. Times have changed, and slowly more understanding of the economic, legal and moral aspects of human management, in the laboratory as well as other areas where chemicals are handled and disposed, is gaining ground. While previously few, if any, laws or pressure for legal and moral responsibility existed, the situation has changed in recent years, and many of the general concepts and techniques of safety have also matured. Safety specialists, industrial hygienists, occupational nursing and medical specialists, and environmental engineers are not 'do-gooders', but practically oriented professionals well aware of the economic, legal and emotional interplay of the laboratory or work place.

Information made available in a rational form, and explained in detail will give hazard control a positive, rather than a negative, meaning. For example, if one understands the action of the oxides of nitrogen on the respiratory system, in self interest one will respect and properly control concentrated nitric acid and its possible reactions and decomposition products.[37] For another, if we know that nitromethane reacts with calcium hypochlorite with great vigour, separation of these two substances in storage or use is clearly in order.[38] The lists of such incompatibilities are extensive and still incomplete.[39] We hope for their more extensive use.

The importance and need for more adequate instruction in chemical hazards, even at the high school and college level, has long been recognized.[40] However, it has not received the degree of interest it merits, either in academe or industry.

Regulations

Finally recognizing the importance of adequate safety information and training, OSHA has published a Hazard Communication regulation.[41] Intended primarily to apply to manufacturing categories (Standard Industrial Classes 20–39), it also has considerable significance for chemical laboratories and to chemists in general. Also known as 'Right-to-know', this regulation, which is to be fully implemented by May 25th, 1986, requires that all hazardous substances, regardless of quantity, be labelled with adequate precautionary labels presenting the basic information necessary for hazard control. Every container of hazardous materials regardless of size, must be labelled properly, using words or symbols to convey: (*a*) chemical name; (*b*) name, address, and emergency phone number of the company that makes or imports the chemical; (*c*) any important storage or handling instructions; (*d*) physical hazards; (*e*) health hazards; (*f*) basic personal protection needed for working with the chemical, and immediate first aid. The label is intended to be read completely before the user first opens the container, or moves it, and the label must not be defaced or removed.

In addition to labelling, each hazardous material delivered must be accompanied by a material safety data sheet (MSDS). These sheets are intended to elaborate on the significant hazardous properties and to include control measures or other precautions necessary to protect personnel by presenting details on identity, mixture ingredients, physical and chemical characteristics, and physical hazards, with recommendations on how to handle emergency situations and ultimate disposal methods. Such items as fire aspects and extinguishment, skin contact, reactions with water and other chemicals and mixtures, acute and chronic inhalation exposure limits (where established), proper personal protective equipment such as specific type and composition of gloves, use of barrier creams if indicated,[42] aprons, laboratory coats, conductive or acid-resistant shoes, head, eye and face protection, respiratory protection (type of respirator, mask, or self-contained breathing apparatus (SCBA) and limitations of each in use with each exposure level),[43] first aid, medical monitoring, and ultimate disposal—all of these factors are to be supplied in appropriate detail. Where knowledge gaps exist, this must be so noted. The MSDSs must be kept in the laboratory or workplace for quick and easy reference by all personnel at all times. As a supplement, computerized data bases are now becoming available which facilitate prompt retrieval of information where and when needed. This could be of great assistance to emergency response personnel, such as security, fire, and medical teams.

The Hazard Communication regulation also specifies that personnel must be given orientation to this material in training sessions, so they are fully aware of the data now available and its implications and importance to them. The laboratory or employer is to reduce the training program to writing, and to keep an updated list of hazardous chemicals on hand for emergency control purposes if required.[44]

The right of the employer to maintain control of secret information is respected in that medical personnel (occupational health physicians and nurses) who need specific information for treatment of an injured employee must be provided with it, but the employer can insist on a security pledge from the medical personnel after the immediate emergency is terminated.

The OSHA Hazard Communication regulation is the latest in a growing number of 'Right-to-know' laws which several states and major cities have placed into action. In some states, the laboratories or other organizations using or handling hazardous materials must report to proper authorities (such as the state or local health, fire, and police departments) the essential hazard information on the substances on their property. This is to assist pre-emergency planning, which is becoming an important part of public safety, especially when one considers that emergencies may occur at any time, often when informed technical personnel and management are not available on site to give the essential information to the responding emergency officers.[45,46]

Other considerations

In addition to the items highlighted above, many other aspects of safety in the chemical laboratory should be noted. For example, fires and explosions continue to be major considerations in our society, as well as in laboratories, in spite of all the studies and understanding of the fire triangle. No recent and comprehensive data of this type in America is known to the writer. However, a safety survey of practices in Canadian laboratories has been reported. Based on 38 replies (1·5% response) to a questionnaire, the survey tabulation concludes that laboratories differ widely in their approach to safety. It is interesting that during the last year (1984), 43% of the laboratories reporting noted a total of 16 chemical-related accidents, and 72% reported a total of 26 accidents non-related to chemicals.[47]

The most comprehensive study with which we are familiar regarding safety in academic institutions was reported by J.F.J. Kibblewhite for the UK Southern Universities Health and Safety Advisory Group. The ten member universities, with an annual 'population' of 20000 employees and 46000 students, combined their experiences for four years, and compared it with the HSE experiences for British industry. The conclusions are very interesting: (1) the procedures used for reporting accidents and dangerous occurrences are good and should provide reliable and consistent information; (2) accident rates for university employees and students (including notifiable incidents) are very low both in absolute terms and in relation to other sectors of UK industry; (3) the annual incident rates for accidents requiring more than local first-aid treatment for employees and students and the annual numbers of dangerous occurrences both show downward trends; (4) the numbers of fatalities (1) and major injuries (38) amongst 20000 employees and 46000 students over the four year period 1980–83 are minimal; (5) the principal areas of concern in which improvements in safety performance could be made are: (*a*) causes, such as 'falls on level', 'handling', 'handling glass/sharps', 'striking against object', and 'struck by object'; (*b*) accidents involving 'catering', 'cleaning/domestic', 'farm workers', 'grounds/gardening', 'maintenance', 'portering', 'security', and to a lesser extent, 'technical' employees; which occur in 'communal areas', 'catering areas' and 'laboratories'; *i.e.*, the majority of accidents are 'every-day' events involving mainly manual employees occurring predominantly in non-technical locations. These accidents also involve the majority of notifiable incidents which have more serious associated injuries; (6) accidents involving academic employees and postgraduate students carrying out research in laboratory areas are comparatively rare, both in terms of the numbers occurring and the seriousness of injuries. The main areas of concern relating to research and teaching laboratories are the reduction of 'handling glass/sharps' accidents amongst undergraduate students, post-graduate students and technical employees working in laboratories, and 'spillage/releases' amongst relatively inexperienced undergraduate students in teaching laboratories. In commenting on regulations, the study noted that the conclusions reached substantiate the observations by The Royal Society regarding safety in research, and confirm that the adoption of 'factory standards' in research areas is both unnecessary and inappropriate. However, the high risk potential—bearing in mind the extensive and diverse use of biological agents, chemical substances, sources of radiation, *etc*— should not be underestimated in research work where the frontiers of knowledge are being explored and the potential effects are frequently unknown. The study recommends more complete and detailed studies involving all universities in the UK. The study should be read in its entirety by any person interested in safety in the academic community.[48]

It is to be hoped that the British and American Societies will conduct comprehensive and detailed fact-finding surveys to determine where the real problems exist, and to suggest more effective control measures, since government regulation does not seem to be appropriate or effective for laboratories.

Another area of concern is that of alcohol and drug abuse. In recent times some studies have suggested that one out of ten employees has a serious substance abuse problem. This is obviously not conducive to the health or safety of personnel. It is noticeable that tobacco products are being questioned at long last as more frequently 'no smoking' becomes the rule and practice in laboratories, public buildings, and in offices.

A further area of concern is the awareness that even protective equipment may not be without hazards. It has been noted that the light weight and high charging pressure of aluminium cylinders have resulted in their widespread acceptance and use with self-contained breathing apparatus (SCBA). NIOSH estimates that more than half of the SCBAs of 30- and 60-minute duration in regular use today are equipped with aluminium cylinders, usually fibre-glass wrapped. Such cylinders, manufactured under Department of Transportation (DOT) exemptions DOT-E 7235 and DOT-E 8059 (including 2216 and 4500 psi, 105 and 306 bar, respectively) may, on ageing, develop neck cracks and may leak breathing gas from an unattended cylinder. If undetected, this loss of breathing gas could be dangerous to the user. Based on this, NIOSH recommends that where SCBA are equipped with fibre-glass wrapped aluminium cylinders, inspection for cylinder pressure should be made at least weekly for stored units. When used on a daily basis, as in fire fighting, cylinder pressure should be checked daily and immediately before use. If a leak is suspected, the cylinder and cylinder valve should be tested as prescribed in American National Standard Z88.5-1981, 'Practices for Respiratory Protection for the Fire Service,' Section 6.2.4.2.[49]

Medical surveillance

The importance of medical surveillance of laboratory and plant personnel has not always received the attention needed to promote health and safety as a preventive matter. The impending surplus of physicians in the US may encourage more physicians to discover occupational medicine as a field of interest. This is a relatively limited field in the US and currently physicians are becoming neighbourhood doctors and making house calls. Medicine in the US is, according to one authority, becoming industrialized with health medical organizations (HMOs) entering more actively in the field. One HMO provides care for 745 000 workers and their dependents. Noting that the health of man/ working environment in the central theme of occupational health, Dinman has suggested that there is a serious need for better co-ordination of the work of the medical team (physician and nurse) with the industrial hygienist. This can also assist in the delineation of new occupational hazards by detecting early warning signs and symptoms.[51] The safety engineer should be included in this team, as well as the environmental engineer.

In the UK, the Employment Medical Advisory Service is an official organization of doctors and nurses whose job is to give advice about occupational health. It is part of the Medical Division of the Health and Safety Executive, and helps to prevent ill health caused by work, advises people with health problems about the kind of work which suits them or which they should avoid, and carries out investigations which help the HSE to develop policies for improving occupational health.[52]

In conclusion, it is hoped that greater co-ordination and co-operation will occur, not only between professional groups but also between countries, such as the United Kingdom and the United States, and others. To do otherwise is to waste precious resources, to say nothing of even more valuable human lives.

Acknowledgements:
The writer is very grateful for input supplied on short notice by M. Leng, A. Schaffer and R.A. Smith of Dow Chemical Co., I.K. O'Neill of the International Agency for Research on Cancer, Joseph Clark of the National Technical Information Service, Diana Caudwell, Senior Employment Nursing Advisor of the Health and Safety Executive, William H. Norton of J.T. Baker Chemical Co., Robert Gleason, Consultant, Hadley, Mass., H. Napadensky and J.N. Keith, IIT Research Institute, G. Baumann, International Occupational Safety and Health Information Centre in Geneva, Barbara Gallagher, Nancy Gleboff and E. Klinefelter, American Chemical Society, to Barry B. Boyer, State University of New York at Buffalo, Mary Seaver, American Occupational Nurses Association, M. Renfrew of the University of Idaho, and to Don A. Rolt, Health and Safety Executive, London. F.J. Kibblewhite, Safety Officer to the University of Reading, UK and R. Sidor, General Electric Corporate Research and Development, made major inputs to this chapter.

Harry Allen Fawcett merits special appreciation for introducing the writer to the advantages of word processing, which greatly eased the preparation of this Chapter.

References

1 *Health and Safety in the Chemical Laboratory: Where do we go from here?* Special Publication No. 51. London, The Royal Society of Chemistry, 1984.
2 H.H. Fawcett, *Safer experimentation* in *Safety and accident prevention in chemical operations*, ed. H.H. Fawcett and W.S. Wood, Chichester and New York, Wiley/Interscience, 2nd edn., 1982, pp. 421–488.
3 H.H. Fawcett, *Laboratory workers* in *Encyclopaedia of occupational safety and health*. Geneva, International Labor Organization, 1983, 3rd edn., pp. 1177–1179.

4 a L. Caglioti, *The two faces of Chemistry.* Cambridge, Mass., MIT Press, 1983.
  b A.C. Nixon, *The reputation of chemists, Chem. Eng. News,* 1985, Sept. 9, pp. 2 and 3.
5 US Dept. of Labor, *Protecting people at work: A reader in occupational Safety and Health.* Washington, DC, Government Printing Office, 1980, 0-309-404. *See also* P. Early, *What's a life worth?, Washington Post Magazine,* 1985, June 9, pp. 10–13 and 36–41 (cost/benefit estimates from 1–3.5 million dollars for regulatory calculations).
6 C.L. Aydelotte, *Occup. Health Safety,* 1986, **55**(1), 18–23.
7 National Research Council, *Toxicity testing: strategies to determine needs and priorities.* Washington, DC, National Academy Press, 1984.
8 The US and UK regulatory systems have been discussed elsewhere and will only be referenced here: Royal Society of Chemistry, *Health and Safety in the Chemical Laboratory,* 1984, pp. 71–91 (US) and 92–105 (UK). *See also* C.G. Veljanovski, *Regulatory enforcement: an economic study of the British Factory Inspectorate, Law & Police Quart.,* 1983, **5**(1), 75–96; H. Saunders, *Chemical laboratory safety and the impact of OSHA, Chem. Eng. News,* 1976, May 24, 15–27; D.E. Miller, *Occupational safety, health, and fire index,* New York, Marcel Dekker, 1976; H.H. Fawcett, *The OSHA communication regulation or 'Right-to-know', J. Chem. Educ.,* 1986 (in the press). *See also* G.Z. Nothstein, *The law of occupational safety and health.* New York, The Free Press; London, Collier Macmillan, 1981; *Health and Safety at Work etc. Act 1974: Some legal aspects and how they will affect you* and Leaflet SHW28, *Protecting people at work,* available from Health and Safety Commission, 1 Chepstow Place, Westbourne Grove, London W2, HMSO, or booksellers.
[Federal Compliance officers from OSHA performed 71 303 inspections in 1985, compared to 71 731 in 1984. 55 per cent of the federal inspections were in the construction industry and 72 per cent were programmed inspections for targeted high-hazard industries and workplaces. Of the 119 706 alleged violations and penalties of $9.2 million proposed by the agency less than 3% were contested by employers. 1443 of the inspections were in response to accidents at the site; 8608 were in response to complaints; 3853 were referral inspections, and 1464 were follow-ups. Approximately 5 million workplaces are estimated to exist in the US.]
9 T.M. Pyriadi, *Chem. Eng. News,* 1985, **63**(44), 4.
10 T.E. Griffith, *Chem. Eng. News,* 1985, **63**(50), 2.
11 J.-E. Osterhold and P. Passiniemi, *Chem. Eng. News,* 1986, **64**(6), 2.
12 A.A. Schilt, *Perchloric acid and perchlorates,* Columbus, Ohio, G. Frederick Smith Chemical Co., 1979.
13 L. Bretherick, *Handbook of reactive chemical hazards.* Borough Green and Stoneham, MA, Butterworths, 1985, 3rd edn., pp. 906–917.
14 R. Cowan, *Research Resources Reporter,* 1986, **10**(2), 1–5, US Dept. of Health and Human Services, NIH, Bethesda, MD.
15 a R.R. Doyle, *Chem. Eng. News,* 1986, **64**(7), 4. b ref. 13, p. 466.
16 M. Heylin, *Chem. Eng. News,* 1986, **64**(5), 3–6.
17 T.N. Veziroglu *et al.,* in *Proceedings of the international symposium on hydrogen systems,* Beijing, China, 7–11th May, 1985. Oxford, Pergamon, 1986.
18 General Electric Co., *Chemical boobytraps,* 10-min 16 mm sound colour motion picture, General Electric Staging Area, 705 Corporation Park, Scotia, NY 12302, 1959.
19 Data Sheet I-700-82, *Hydrogen usage in the laboratory,* National Safety Council, Chicago, IL 60611, 1982.
20 a L. Summerlin and J. Ealy, *Chemical demonstrations: a sourcebook for teachers.* Washington, D.C., ACS, 1985.
   b J. Brown, *Pollut. Eng.,* 1986, **28**(1), 18–28; *see also On-line access to chemical standards information, Chem. Eng. Progr.,* 1986, **82**(1), 7–27.
21 a D.A. Livingston and M.S. Joshi (the Upjohn Co), letter to the Editor of *Chem. Eng. News,* 7 April, 1986.
   b M.J. Spitulnik, *Chem. Eng. News,* 1977, **55**(31) 31.

22 D.L. Snow and H.H. Fawcett *Occupational health and safety*, Chapter 7 in *Laboratory planning for chemistry and chemical engineering*, ed. H.F. Lewis. London, Chapman & Hall; New York, Reinhold, 1962.

23 R. Lees and A.F. Smith, *Design construction and refurbishment of laboratories*. Chichester, Ellis Horwood, for the Laboratory of the Government Chemist, 1984.

24 British Standard DD80, 1982 (in 3 parts).

25 G.T. Saunders, *Updating older fume hoods*, *J. Chem. Educ.*, 1985, **42**(6), A179, reprinted in *CHAS Notes*, 1985, Vol. III, No. 3, 2–4; *see also* Data Sheet I-687-80, *Laboratory fume hoods*, National Safety Council, Chicago, 1980.

26 *ACGIH Industrial Ventilation Manual*, American Conference of Governmental Industrial Hygienists, Cincinnati, OH, 1985.

27 H.H. Fawcett, *Hazardous and toxic materials: safe handling and disposal*. Chichester and New York, Wiley/Interscience, 1984.

28 *Occupational Diseases: a guide to their recognition*, DHEW Publication No. (NIOSH) 77-181, pp. 235–238. Cincinnati, US Dept. of Health Education and Welfare, revised edn., 1977.

29 R. Cook, *Fever*, New York, G.P. Putnam's Sons, 1982.

30 *Criteria document for benzene*. DHEW Publication No. (NIOSH) 74-137, p. 78. Cincinnati, National Institute for Occupational Safety and Health, 1974; *see also Guidelines for the laboratory use of carcinogens*, Publication No. DHH (NIH) 81-2385. Bethesda, National Institute of Health, 1981, *and*, D.B. Walters, ed., *Safe handling of chemical carcinogens, mutagens, teratogens and highly toxic chemicals*. Ann Arbor Science Press, 1980.

31 *Occupational exposure limits 1985*, Guidance Note EH 40/85. London, Health and Safety Executive, HMSO, 1985.

32 a Permissible Exposure Limits are included in the OSHA Regulations as 29 CFR 1910.1001 through 1910.1045, by US Dept. of Labor. On February 10th, 1978, in the *Federal Register*, OSHA published pages 5918–5970, Part 1910.1028 Benzene: "No employee exposed to excess of 1 ppm Time Weighted Average as 8-hour average, and no exposure to concentration in excess of 5 ppm as averaged over any 15 min. period." As noted, this was invalidated by the Supreme Court.
    b Following the Supreme Court ruling, the permissible exposure limit for benzene returned to 10 ppm, but is currently under re-evaluation by OSHA.

33 The OSHA Cancer Policy in *Federal Register* 1980, **45**, 5001–5296; *see also* H.A. Milman and E.K. Weisburger, *Handbook of Carcinogen testing*: Park Ridge, NJ, Noyes Publications, 1985, *and* M. Sitting, *Handbook of toxic and hazardous chemicals and carcinogens*: Park Ridge, NJ, Noyes Publications, 2nd edn., 1985.

34 American Conference of Governmental Industrial Hygienists, 6500 Glenway, Building D-5, Cincinnati, OH 45211, USA.

35 J.L. Babaracco, *Loading the dice, a five-country study of vinyl chloride regulation*. Boston, Harvard Business School Press, 1985. In the US, VCM was regulated to a limit of 1 ppm for 8-hour exposure, and a ceiling of 5 ppm averaged over one year as measured either continuously or by permanent sequential means, with a 7 ppm exposure limit for 8 hours or 8 ppm for one hour.

36 Details of the American Chemical Society Liability (Malpractice) Insurance are available from Earl Klinefelter, American Chemical Society, 1155 16th St., NW., Washington, D.C. 20036.

37 Ref. 13, pp. 1100–1129.

38 Ref. 13, pp. 159–162.

39 In addition to ref., 13, *see also* publications of the National Fire Protection Association, Batterymarch Square, Quincy, MA 02269 (especially *NFPA 45, 49, 491, 325 and 704M*, all of which deal with chemicals).

40 H.H. Fawcett, *J. Chem. Educ.*, 1949, **26**(2), 108.

41 *Federal Register*, 1983, **48**(228), 53280–53348. *See* B. Meier, *Use of right-to-know is increasing public's scrutiny of chemical companies, Wall Street Journal*, 1985, May 23, 10; *see also* H.H. Fawcett, *The OSHA hazard communication standard, J. Chem. Educ.*, 1986, **63**(3), A70–73.

42 Gloves and protective creams must be used with great care in selection and use. *See* H.M. Niles, *Occup. Health Safety*, 1985, **55**(12), 34–36.

43 Chapter 5 (Personal protective equipment) and 6 (Respiratory protective equipment), pp. 87–148 in ref. 27.

44 *Right-to-know training film: Employee workbooks and leader's guide for training in OSHA hazard communication*, Business and Legal Reports, Bureau of Law and Business Inc., 64 Wall Street, Madison, CT 06443-9988, 1986.

45 EPA chemical emergency preparedness program: interim guidance revision 1, 9223. 0–1A, US Environmental Protection Agency, Washington, DC 20460, 1985.

46 K.J. Tierney, *Developing a community-preparedness capability for sudden emergencies involving hazardous materials*, Chapter 34, pp. 759–788 in ref. 2.

47 M. Torchia, *Can. Chem. News*, 1985, **37**(3), 16, reprinted in *CHAS Notes* (Div. of Chemical Health and Safety of the ACS), Vol. III, No. 3, July–Sept. 1985, pp. 4 and 5. *See also* D. Bush, *Impressions of safety in universities in the USA, J. Chem. Educ.*, 1979, **56**(4), A161–166.

48 J.F.J. Kibblewhite, *Analysis of university accidents*, Southern Universities Health and Safety Advisory Group, *Safety Practitioner*, 1984, December, 12–21.

49 J.B. Moran, *Respirator users notice, inspection of certain aluminium cylinders for breathing-gas pressure*, January 17th, 1986, National Institute for Occupational Safety and Health-ALOSH, 944 Chestnut Ridge Road, Morgan Town, WV 26505, USA.

50 E. Paris, *Hippocrates meets Adam Smith*, in *Forbes*, 1986, Feb. 10, 63–66.

51 B.D. Dinman, *The physician, nurse and industrial hygienist – the occupation health protection team*, in *The industrial environment: its evaluation and control*. Washington, DC, US Government Printing Office, No. 19810357561, 1985.

52 *An introduction to the employment medical advisory service*, Booklet HSE5 C200 6/84, available from St Hugh's House, Stanley Precinct, Bootle, Merseyside L20 3QY, UK: *see also* D.J. Kilian, Chapter 9 (pp. 173–188), I.R. Tabershaw, Chapter 10 (pp. 189–194), A.J. Murphy, Chapter 11 (pp. 195–210) in ref. 2.

Index

This index is designed to help readers make quick reference to topics in the book, with the exception of the coloured alphabetically-arranged monograph section of Chapter 8.